G.M. SMITH · PHYTOPLANKTON

BIBLIOTHECA PHYCOLOGICA

Herausgegeben von
J. CRAMER

BAND 30

Phytoplankton of the Inland Lakes of Wisconsin

by

GILBERT M. SMITH

Reprint 1977

J. CRAMER
In der A.R. Gantner Verlag Kommanditgesellschaft
FL-9490 VADUZ

Printed in Germany
by Strauss & Cramer GmbH, 6945 Hirschberg II
ISBN 3-7682-1134-7

WISCONSIN GEOLOGICAL AND NATURAL HISTORY SURVEY

W. O. HOTCHKISS, Director

BULLETIN NO. 57 SCIENTIFIC SERIES NO. 12

Phytoplankton of the Inland Lakes of Wisconsin

Part I

Myxophyceae, Phaeophyceae, Heterokonteae, and Chlorophyceae exclusive of the Desmidiaceae.

BY

GILBERT MORGAN SMITH

MADISON, WIS.

PUBLISHED BY THE STATE

1920

Wisconsin Geological and Natural History Survey

Replacement

TABLE OF CONTENTS

PREFACE

The general name plankton is given to the organisms which live free-floating in the deeper parts of lakes and ponds; the plant individuals being designated the phytoplankton and the animal the zooplankton. Investigations have shown that there is a very special algal flora living exclusively in the plankton, in addition to the chance wanderers from the littoral, or marginal, region which do not multiply when in the plankton. These will be designated as the *euplanktonts* and *tychoplanktonts* respectively in this work. There is still another group of organisms found in lakes, namely those which thrive equally under littoral or pelagic conditions. This group will be referred to as the *facultative planktonts.* Attempts have been made to differentiate between organisms of these three groups by the Wests, Teiling, and others, although a different terminology has been used. Since the question of a proper habitual classification for any particular species will always be a matter of individual opinion, I have indicated at the end of the description my conception of the natural habitat of the organisms found in the lakes of this state.

It is only within the past twenty-five years that the phytoplankton has been studied, and our knowledge of the various members constituting this specialized flora has come almost exclusively from European phycologists. William West, G. S. West and Fritsch in England; Chodat and Bachmann in Switzerland; Lemmermann, Schmidle, Schröder and Volk in Germany; von Keissler, Pascher and Woloszynska in Austria-Hungary; and Wesenberg-Lund in Denmark have been the chief students of European phytoplankton. The central African lakes have been investigated by G. S. West, Schmidle, and Woloszynska. G. S. West and Playfair have studied the phytoplankton of Australia. With the exception of Miss Snow's work on Lake Erie nothing was known concerning the taxonomy of North American phytoplankton at the time the present work was undertaken, the pioneer work on the microscopy of the Massachussetts and Brooklyn water supplies being of little value to the systematist since determinations were not carried beyond the genus.

These studies of Wisconsin lakes, carried out under the auspices of the Wisconsin Geological and Natural History Survey, were commenced in the fall of 1913. Field work was continued during the summers of

1914–15–16–17 and all lake areas of the state have been visited during the progress of the investigation. Collections have been made from the lakes around Madison and Oconomowoc during the entire season that they are open. The June, August and early September flora of the northeastern lakes is represented, while the northwestern lakes were visited for two successive seasons during August. Only single collections were made from the northern lakes so that the data for any particular one of them are undoubtedly very fragmentary.

The limits of the survey are determined by the political boundaries of the state and not by ecological or regional unity. There is a continuation of the southeastern lake group in northeastern Illinois; of the northeastern lake group in the northern peninsula of Michigan; and of the northwestern lake area in Minnesota. However, variation in depth and surface area together with variation in chemical content of Wisconsin lakes gives a sufficient range of habitat to cover conditions found in almost all other parts of the world if temperature and altitude are excluded. The cosmopolitan nature of phytoplankton, therefore, makes this work of value to the student of these microörganisms in other sections of the United States or even other continents, and he will probably find a majority of the species in any region represented in the Wisconsin flora. There is a temptation to include species that have not been observed in order to round out the work for the student in other parts of the world. This practice has been followed in certain algal floras and while it may be helpful to the general student, it causes endless confusion to the phytogeographer. Only those species are included in this flora which have been collected from the state.

Descriptions of plankton algae are scattered through a large number of periodical publications and transactions of learned societies, many with only a limited circulation, so that considerable effort has been spent in giving full and correct citations for the original descriptions. When subsequent descriptions have been better or have given more characteristic figures references are also made to them. A great deal of time has also been spent making camera lucida drawings since an accurate figure is frequently more valuable than a description. Drawings of the various species of a genus are generally on the same scale, but there has been no attempt to draw the different genera on the same scale. The finer details of cytological structure, such as the nature of the chloroplast or structure of the pyrenoid, are frequently lost in preserved material; so all of the Chlorophyceae, with the exception of the Desmidiaceae, the Phaeophyceae, Heterokonteae and the Myxophyceae, have been drawn from living material.

The original plan for the study of Wisconsin lake algae included those attached to rocky shores or lying among the macroscopic vegetation in the quieter portions of lakes in order to get the relation between the benthos and the plankton, but it soon became apparent that limitations of both time and space precluded an adequate presentation of the littoral flora; attention, therefore, has been confined to phytoplankton since the second season's work. The great amount of material on the phytoplankton has necessitated dividing the work into two portions; the material included in the present volume, and the Desmidiaceae and Bacillareae which are reserved for a later one. A considerable amount of work has been done already on these two groups and it is hoped that the complete report can be published shortly.

Phycologists are generally agreed on the broader principles of algal classification but naturally differ on minor details. The classification here adopted for the Myxophyceae is used almost universally. The group of flagellates with brown chromatophores is frequently considered more animal than plant. One section, the Chrysomonadineae, where there are golden brown chromatophores, seems to be sufficiently related to the Phaeophyceae to warrant placing them in a description of plankton algae and in the arrangement of the group I have followed Pascher's classification. Since Luther's establishment of the Heterokonteae, the question of whether to recognize this group as a separate class or a division of the Chlorophyceae has arisen. The evidence seems fairly clear, however, that the origin of the group is from a primitive flagellate of the *Chloramoeba* type and since this phylogenetic starting point is so markedly different from the phylogenetic starting point of the Chlorophyceae (the Polyblepharidaceae) I feel that the two should be considered distinct classes. The classification used is that of G. S. West, although it should be noted that he considers the Heterokonteae a division of the Chlorophyceae. I have also, with a few exceptions, followed West's arrangement of the Chlorophyceae. It is hoped that the keys found in this work will prove usable. They are based solely upon characters of the species described herein; so will be of no value for those not mentioned.

During the seasons of 1916 and 1917 the field work was assisted by a grant of money from the United States Bureau of Fisheries; and the extension of these studies to the lakes of northwestern Wisconsin has thus been made possible.

The University of Wisconsin,
Department of Botany,
February, 1919.

INTRODUCTION

Geography and geology of Wisconsin. The state of Wisconsin lies between Lake Superior, Lake Michigan and the Mississippi river and has an area of 56,066 square miles, or an area somewhat larger than England. Its greatest breadth is about 295 miles and greatest length 320 miles. Within its borders are numerous lakes, the total number probably reaching well into the thousands, with an estimated surface area of about 1500 square miles. They vary in size from small lakelets and ponds that cover but a few acres, to Lake Winnebago with an area of 215 square miles. The great majority of the lakes, however, have less than a square mile of surface area. There is likewise great variation in depth, but for the most part they are shallow since the deepest (Green Lake) has a maximum depth of 237 feet and only a few have a depth of over a hundred feet. Certain phycologists have tried to differentiate between the plankton floras of ponds (heleoplankton) and lakes (limnoplankton). The lakes of this state grade so imperceptibly from the "pond" to the "lake" type that it has been impossible to distinguish between the two. To set arbitrarily a surface area of a square mile and depth of 25 feet as the maximal limits for a pond would involve many difficulties; since the state presents such examples as Beasley Lake with a surface area of 0.019 square miles and a depth of 51 feet or Lake Winnebago with a surface area of 215 square miles and a maximum depth of 21 feet. I have, therefore, listed all bodies of water as lakes but have given, in Table 2, all available data on area and depth for those desiring to recapitulate into heleoplankton and limnoplankton.

There are, roughly speaking, three lake areas in the state; the southeastern, northwestern and northeastern, all of which owe their origin to glaciation. These lake areas are not uniformly distributed over the state since the glacier invading this region was divided into lobes and lakes are found in greatest abundance along the terminal and kettle moraines of these various lobes. The southwestern portion of the state possesses no lakes since it lies in a non-glaciated region. The southeastern lake area includes several scattered moderately sized lakes and groups of smaller lakes some of which are among the deepest in the state. The majority of them lie in a region of kames and pitted plains, where the pits and lake basins owe their origin to the burial and subsequent melting of large irregular blocks of ice. The Oconomowoc, Lauderdale and Waupaca lake groups are of this type. A few of the larger and deeper lakes, as Green and Mendota, are situated in valleys eroded by preglacial streams that were possibly modified by glacial action and then closed by the building of a morainal dam which impounded the waters of the present lake.

The northeastern lake group lies chiefly in Vilas, Oneida and Iron counties. There are hundreds of lakes in this area and their closeness of position may readily be inferred from the fact that while the largest of them, Trout Lake, covers only 6.5 square miles the 346 lakes and ponds in Vilas county occupy 140 square miles or over 15 percent of the area of a county nearly as large as Rhode Island. In few parts of the world are there more lakes to the square mile. Parts of the State of Minnesota, the Province of Ontario, and Finland furnish the only parallel. These lakes are small, irregularly shaped, and connected by streams with most irregular courses. They are all glacial in origin and formed either by shallow depressions of the ground moraine, hollows in outwash plains, or damming of recessional moraines. The entire region is so deeply buried by the glacial drift that no lake is due to a damming of a preglacial valley. This entire region also contains numerous swamps (muskegs) characteristics of a poorly drained area.

The northwestern lake area lies in Barron, Polk, Burnett, Washburn, and Sawyer counties. In general it is similar to the northeastern lake area in structure and origin but the lakes do not occupy so great a portion of the land's surface. It has been suggested that some of the lakes in this region were formed before the last glacial invasion or Late Wisconsin glacial epoch. Here we find a few lakes formed by a damming of preglacial valleys; Bone, Beaverdam, and Wapagossett lakes for example.

The mean average temperature of Wisconsin varies from 6.6° C. in the southern portion to 3.3° C. in the northern. The state has a mean summer temperature similar to that of France, Germany, or southeastern England (14.5° to 18°); the average winter temperature is comparable to that of northern Sweden or central Russia (–5.4° to –8.8°). All lakes in the state are frozen over during the winter, ice forming during November and December and lasting until March or April in the southern ones, while on some of the northern lakes ice appears in October and remains until the latter part of May. The midsummer surface temperature of the northern lakes is 18°–24° C. and 20°–26° in the southern lakes.

The altitude of the lakes in the northeastern part of the state varies from 1550–1700 feet above sea level; in the northwestern area 1050–1350 feet, and from 750 to 900 feet in the southeastern. These differences in elevation are too slight to be of any biological significance.

Likewise differences in the amount of mean annual precipitation, which range from 28–44 inches in various parts of the state, need no discussion since rainfall in all regions is sufficient to maintain lakes at their normal level.

The underlying geological formations are, however, of prime importance and have a marked influence on the distribution of plankton organisms, especially the plankton desmids. Within the state are found areas of igneous, metamorphic, and sedimentary rocks each of which have a different direct influence on the chemical environment of the plankton organisms. The underlying rock formations of the southeastern lake area are the Niagara limestone of the Silurian; the Cincinnati shale, Galena limestone, Lower Magnesian limestone and St. Peter sandstone of the Ordovician. In the northwestern lake region is found the Cambrian Potsdam sandstone, the Ordovician Lower Magnesian limestone, together with areas of Keeweenawan trap rock and Precambrian granites. The northeastern area is entirely Precambrian granites.

Overlying all of these regions is a mantle of glacial drift some of which is local in origin while other constituents have been transported from distant regions. This ground moraine in southeastern Wisconsin varies in thickness from a few feet on hill tops to over 400 feet in the bottom of preglacial valleys. Mechanical analyses show that 13% consists of crystalline rocks transported from Canada while the remaining 87% is local sand- and limestone. In the northeastern lake area the glacial mantle varies from 75–100 feet in thickness with a maximum depth of 350 feet at the southern boundary of the area. This glacial accumulation is not the result of one ice invasion but of several; the last, and the one which left the most material, came from the northeast and transported the conspicuous sand deposits of Vilas and northern Oneida counties from the sandstone region of northern Michigan. Previous ice invasions came from the northwest and their deposits, in some instances, have not been disturbed by succeeding invasions. None of the drift in this region is calcareous so that all soils are acid in character. The mantle of drift in the northwestern area varies from 500 to a few feet in thickness and is composed of material from granitic and sandstone regions. There is, however, a certain amount of lime in the drift at the southern limit of the lake area.

Chemical composition of the lake waters. Birge and Juday* have made investigations on the chemical composition of the waters of Wisconsin lakes in connection with their biochemical studies. They divide the waters into three general classes; soft, medium, and hard. These

---

* Birge, E. A. and Juday, C. The inland lakes of Wisconsin. The dissolved gases of the water and their biological significance. Bull. 22 Wis. Geol. and Nat. Hist. Survey. (Sci. Ser. 7.) pp. 1–259. 1911.

varying degrees of hardness are due to the combination of the carbon dioxide with calcium and magnesium to form carbonates, the fixed ($CaCO_3$ and $MgCO_3$) and the half-bound ($CaCO_3.H_2CO_3$ and $MgCO_3.H_2CO_3$). In neutral and acid lakes the two types of carbon dioxide are present in equal amounts but in alkaline lakes there is an excess of fixed carbon dioxide. This is due to the ability of the chlorophyll bearing organisms to draw upon the half-bound carbon dioxide as a source of supply after the disappearance of the free carbon dioxide. The amounts of magnesium and calcium vary from lake to lake and there is an almost continuous series from those having the softest waters to those having the hardest. They consider those lakes whose average fixed carbon dioxide does not exceed 5 cc. per liter as soft water lakes, while medium waters have 6–22 cc. of carbon dioxide per liter and the hard waters from 23–50 cc. per liter. The carbon dioxide content varies with the depth at which the sample is taken and also with the season of the year; some of the medium class lakes, for example, have a sufficiently low content at times to qualify as soft water lakes. All of the soft and medium waters, with the exception of Devils lake in Sauk county, are located in the northeastern and northwestern lake areas and many of them have sufficient free carbon dioxide to give an acid reaction with phenolphthalein. The hardwater lakes are all in the southeastern area and the upper stratum in them is alkaline to phenolphthalein for the greater part of the year.

Aside from the question of the hardness of the water complete mineral analyses of a few lakes are also available. These results are embodied in Table 1 and represent average conditions. They are taken from the data given by Birge and Juday and their data show that there is a seasonal variation and a vertical variation in the amounts of certain elements present. These variations in vertical distribution are well shown in their Fig. 7, p. 106 (*l. c.*) The lakes of southeastern Wisconsin contain a rich flora of Chlorophyceae, Myxophyceae and Phaeophyceae, both in quantity and quality. In the northern lake areas the total volume of the plankton is, generally speaking, considerably smaller and quite different in character. These lakes have a considerable desmid flora in many instances, while with the exception of one or two species desmids are not found in the hard waters of the southeastern region. The volume of phytoplankton that can be produced in a lake is primarily dependent upon the available supply of carbon dioxide for photosynthesis. In the northern lakes this supply is chiefly the carbon dioxide dissolved in the water but in the southern lakes there is in addition considerable quantities of half-bound carbon dioxide which may be drawn upon after the free carbon dioxide is exhausted. It is not surprising, therefore, that the southern lakes with greater supplies of potential food materials have a greater production of algae. There are certain factors modifying the growth of algae in hard water lakes; in those deep ones which become thermally stratified in the summer most of the algae are restricted to the epilimnion. The thickness of the epilimnion is in part dependent upon the size and shape of the lake, since the wind is more effective in keeping water in circulation on large surfaces. Algae are continually sinking to the bottom and giving rise to carbon dioxide and nitrogenous products. In the deep

TABLE 1.—*Results of mineral analyses stated in parts per million. These show average content of various depths and at different dates.*

	SiO_2	Fe_2O_3 Al_2O_3	Ca	Mg	Na	K	CO_3	HCO_3	SO_4	Cl
Devils	2.2	0.7	3.2	1.1	....	....	1.0	8.1	8.1	8.2
Trout	13.0	1.1	6.7	2.5	0.8	0.3	0.0	28.0	6.4	3.0
Kawaguesaga	16.5	6.2	7.2	2.8	2.6	1.7	0.0	37.0	7.9	2.7
Owen	9.9	2.7	13.0	3.6	....	....	0.5	33.3	0.0	1.9
Rainbow	17.5	2.2	22.2	15.5	2.3	3.3	4.0	105.0	8.9	4.3
Mendota	4.4	1.4	23.3	22.7	3.7	2.2	2.5	105.2	13.5	3.0
Geneva	9.5	1.1	20.7	26.9	4.4	2.5	3.7	110.5	13.2	4.8
Okauchee	17.6	1.5	34.3	25.9	3.2	1.4	0.0	66.0	13.9	1.9
Beasley	16.5	1.5	38.9	21.8	2.4	2.0	1.5	111.0	2.0	3.1

lakes these products of decay are chiefly in the hypolimnion and are not immediately available for reutilization by other algae. In shallow lakes without thermal stratification these products are immediately available for the vegetative activities of algae. Shallow lakes, therefore, support proportionally larger algal floras. The restriction of the desmids to the soft water lakes was first pointed out by the Wests and their observations have been confirmed in Wisconsin (Birge and Juday, *l. c.* p. 138). An examination of the chemical analyses of such soft water lakes as Devils, Trout and Kawaguesaga (Table 1) shows that in these lakes which contain numerous desmids the calcium and magnesium content is relatively small. The influence of calcium and magnesium on the distribution of desmids may be that of a direct inhibition of growth or the absence of these elements may permit an acid condition which is favorable for the growth of desmids. Data on organic matter and distribution of nitrogen in lakes of the state are lacking. The scarcity or abundance of available nitrogen is doubtless an important factor in the volume of algae produced in a lake, and it is a well known fact that as regions become more thickly settled and drainage entering the lake becomes richer in nitrogenous material the growth of plankton algae becomes more luxuriant.

Methods of collection and study. All plankton collections have been made with nets. In general both the bolting silk net and the cotton disc net were used. The bolting silk net of No. 20 bolting cloth (new No. 25) is that described by Juday*. The cotton disc net is made by attaching a filter designed for purifying drinking water (Jones Modern Filter made by the Jones Mfg. Co., Boston, Mass.) to the bottom of an eighteen inch cone of canvas whose mouth is held open by a light metal ring ten inches in diameter. In using this filter a special cotton disc supplied by the manufacturer is placed in the filter and the whole apparatus drawn through the water. The algae collect on the disc and when there is a sufficient accumulation the disc is taken from the filter and shaken in a bottle with a small amount of water which removes the algae. This net catches many of the smaller organisms that pass through the bolting silk (the nannoplankton) and it is also very useful in obtaining samples by dipping when no boat is available. Whenever possible samples have been gathered by towing the nets at the surface and all collections have been made in the deep parts of all lakes to avoid, as far as possible, the littoral flora. No attempt has been made to study the vertical distribution of the algae in any lake.

In studying the flora of any region a field laboratory was established and with that as a center excursions were made to the surrounding lakes. By traveling from lake to lake in an automobile collections could be made in the morning and the material studied the same afternoon and the following day. In the northeastern lake area where there are no roads in many cases, but where lakes are close together, they can be visited by portaging a canoe from lake to lake. Collections have been made from row boats, where they were obtainable, but in a few instances collections had to be made by repeatedly dipping the net after wading to where the water was 3–4 feet deep or from rafts. Upon returning to the temporary laboratory equipped with compound microscope, camera lucida, and drawing materials, the living specimens were studied and numerous pencil camera lucida drawings made of all doubtful, interesting, or rare organisms, together with notes on occurrence and ocular micrometer measurements of known forms. No observations were made on desmids or diatoms at this time. The study of the Chlorophyceae, Myxophyceae and Phaeophyceae in the living condition is particularly important since many of them do not preserve well and certain features such as cilia, chloroplasts, and contractile vacuoles are lost when material is preserved. All drawings of one species are mounted on the same sheet and corresponding notes on another. These are then filed in species and genus covers in the usual

* Juday, C. Limnological apparatus. Trans. Wis. Acad. Sci. Arts & Lett. 18^2: 566–592. 1916.

manner for herbarium specimens so that all data on any particular species are immediately accessible. The figures for the plates of this work have been redrawn from camera lucida sketches made in field laboratories. The confined space of a vial is not favorable for algae and the more delicate forms, as *Uglenopsis,* frequently go to pieces after standing a few hours. It is impossible to have the vials uncorked while transporting them from station to station but they should be opened immediately on return to field headquarters. I have also found it helpful to keep them as cool as possible and have stored them in hotel refrigerators until ready to make observations. Formalin has proven a satisfactory means of preserving the desmids and diatoms until their study can be taken up during the winter months.

The following list of simple reagents is helpful in studying the various structures of living algae. Cilia of motile forms are more easily seen when the cells have been rendered immobile by mounting in a 1% cocaine solution. Mounting in a dilute aqueous acid fuchsin also helps. Pyrenoids can be recognized by the grouping of the starch grains when the latter have been stained with iodine. It is difficult to strain pyrenoids of living cells. Gelatinous envelopes are best demonstrated by Errera's method of mounting in dilute India ink which gives the colorless envelope the appearance of a halo around the cells in contrast to the dark background of water containing the ink. Nuclei can be demonstrated by Strasburger's solution of methyl green in 1% acetic acid. Simple microchemical tests are Sudan III for fats, iodine for starch, zinc chloriodide for cellulose, and Ruthineum Red for pectin.

Lakes studied. The following table gives a list of lakes in the state from which plankton samples have been taken. All known hydrographic maps, morphometric data, general descriptions of many lakes and general maps of the various lake areas have been given by Birge and Juday* so that any one wishing further data concerning a particular lake is referred to their work. They have not, however, mapped or discussed many of the smaller lakes of the northern regions. Since most of the lakes in the state were named by the settlers of the region considerable poverty of vocabulary is noticed and Round, Mud, Devils, Long, and Sand are repeated many times. In other instances Indian (Chippewa) names have been retained as the white man settled the region. In these also there is a certain amount of duplication and Sishebogema (lake with many bays) or Pokegama (side or branch lake) may be cited as examples. It is too bad that these names have not been retained to a greater extent, but certain of the English names like Razorback, Crawling Stone, Lost Canoe, Tenderfoot or Fishtrap are not

* Birge, E. A. and Juday, C. The inland lakes of Wisconsin. Bull. 27: Wis. Geol. & Nat. Hist. Survey. (Sci. Ser. 9). pp. 1–137. 1914.

without interest. This is not the place, however, for the legends on the origin of these peculiar names nor for a discussion of the occurrence of several French names in the list. In order to avoid misunderstanding the town and range in which the lake occurs is given. When more than one lake with the same name has been visited, a serial number has been added and both the lake name and serial number are used in giving stations for the various algae.

TABLE 2.—*Table giving available data on location and size of lakes visited during the course of this investigation. The first column refers to the page on which a map of the lake may be found in Birge and Juday, The inland lakes of Wisconsin, the hydrography and morphometry. Bull. 27. Wis. Geol. and Nat. Hist. Survey.*

Page mapped	Lake	County	Town	Range	Length (Miles)	Maximum breadth (miles)	Maximum depth (Feet)	Area (Acres)
......	Adelaide............	Vilas.........	43N	5E	.47	.32	68.9	45.0
......	Amicoy............	Rusk.........	34N	8W	1.06	.30	....	....
120	Balsam............	Polk..........	34N	17W	3.00	1.00	18.0	1,785.6
......	Bass (1)............	Vilas.........	40N	11E	.50	.24	....	....
......	Bass (2)............	Vilas.........	42N	6E	.93	irr.	....	....
......	Bass (3)............	Burnett......	40N	14W	....	....	....	....
......	Bear (1)............	Barron.......	34N	14W	.45	.40	....	....
......	Bear (2)............	Polk..........	33N	17W	.88	.59	....	....
120	Bear (3)............	Barron-Washburn.	36N	12W	4.50	1.00	....	....
......	Bear Trap...........	Polk.........	33N	17W	1.05	.48	....	....
104	Beasley............	Waupaca.....	21N	11E	.25	.10	51.2	13.5
......	Beaver (1)..........	Vilas....	44N	5E	.44	.25	....	....
40	Beaver (2)..........	Waukesha...	8N	18E	1.10	.44	49.0	305.4
120	Beaverdam..........	Barron.......	35N 36N	14W 13W	3.80	.75	91.9	1,267.2
......	Bettine............	Vilas.........	43N	7E	.75	.56	....	....
120	Big................	Polk..........	32N 33N	18W	.80	.65	....	....
120	Big Butternut.......	Polk..........	36N	17W	1.20	.60	21.3	384.0
120	Big McKenzie.......	Burnett......	40N	14W	2.25	1.05	....	....
116	Big St. Germain.....	Vilas.........	40N	8E	2.62	1.25	....	....
116	Birch (1)............	Vilas.........	43N	5E	1.50	.74	....	....
120	Birch (2)............	Washburn...	37N	10W	.90	.70	57.4	243.2
120	Birch Island.........	Burnett......	40N	14W	2.10	irr.	....	....
......	Blake...............	Polk..........	35N	16W	1.90	irr.	....	....
......	Bloom..............	Burnett......	37N	18W	....	....	....	....
......	Blue Gill............	Vilas.........	43N	7E	.12	.09	....	....
120	Bone...............	Polk..........	35N 36N	16W	4.80	1.00	49.2	2,054.4
116	Boulder............	Vilas.........	42N	7E	2.25	1.20	19.7	614.4
74	Browns.............	Racine.......	2N 3N	19E	1.25	.75	23.0	390.4
......	Bullhead...........	Vilas.........	40N	6E	.31	.26	....	....
75	Camp...............	Kenosha.....	1N	20E	1.50	1.00	19.7	256.0
......	Canteen............	Vilas.........	43N	7E	.31	.17	....	....
116	Carroll.............	Vilas.........	39N 40N	7E	1.30	.47	23.0	288.0
	Carson (see Flora)							

TABLE 2.—(Continued)

Page mapped	Lake	County	Town	Range	Length (Miles)	Maximum breadth (miles)	Maximum depth (Feet)	Area (Acres)
116	Catfish	Vilas	39N 40N	10E 11E	2.87	irr.	19.7	902.4
116	Catherine	Vilas	38N	6E	2.25	irr.	27.9	608.0
75	Center	Kenosha	1N	20E	.72	.35	....	....
......	Chain	Polk	35N	13W	....	....	....	....
120	Chetac	Sawyer	37N 38N	9W	6.40	1.10	....	....
120	Chetek	Barron	33N	10W	3.25	irr.	....	....
120	Clam	Burnett	38N 39N	16W	3.84	1.92	....	....
......	Clare	Polk	35N	16W	.55	.36	....	....
116	Clear	Oneida	39N	7E	2.10	irr.	82.0	921.6
116	Clear Crooked [1]	Vilas	42N	7E	1.40	1.12	32.8	672.0
......	Cochrane	Vilas	43N	8E	.69	.36	....	....
116	Crab	Vilas	43N	6E	2.12	1.35	44.3	1,318.4
116	Cranberry	Vilas	40N	11E	1.25	.70	11.5	332.8
116	Crawling Stone	Vilas	40N	5E	2.40	1.65	93.0	1,600.0
......	Crooked	Burnett	38N	16W	1.44	.44	....	....
......	Curtis	Oneida	39N	6E	.76	.25	....	....
120	Deer	Polk	34N	17W 18W	2.80	.70	39.4	806.4
70	Delavan	Walworth	2N	16E	3.75	1.10	56.7	1,813.2
......	Des Moines	Burnett	41N	14W	1.00	.30	....	....
30, 32	Devils (1)	Sauk	11N	6E	1.25	.61	43.2	360.8
120	Devils (2)	Barron	36N	11W	1.60	.40	....	....
......	Devils (3)	Vilas	43N	8E	.44	.26	....	....
......	Diamond	Polk	37N	17W	.60	.60	....	....
120	Duck	Barron	36N	13W	.50	.30	....	....
......	Dummy	Barron	36N	13W	.60	.25	....	....
......	Dunham	Burnett	38N	17W	....	....	....	....
......	Dutchmans	Burnett	38N	17W	....	....	....	....
74	Eagle	Racine	3N	20E	1.31	1.00	13.1	486.4
120	East	Polk	33N	18W	.74	.52	....	....
75	Elizabeth	Kenosha	1N	19E	2.62	.87	26.2	800.0
116	Fence	Vilas	40N	5E	3.25	2.10	93.0	3,492.0
120	Fish	Burnett	40N	14W	1.30	.55	....	....
116	Fishtrap	Vilas	42N	7E	1.50	.62	29.5	326.4
116	Flora	Vilas	43N	7E	1.31	irr.	....	....

[1] Listed as Clear Lake by Birge and Juday.

TABLE 2.—(Continued)

Page mapped	Lake	County	Town	Range	Length (Miles)	Maximum breadth (Miles)	Maximum depth (Feet)	Area (Acres)
116	Found	Vilas	40N	8E	1.06	.69	....	....
52	Fowler	Waukesha	8N	17E	.94	.44	50.0	83.5
66	Geneva	Walworth	1N 2N	16E 17E	7.50	2.00	142.0	5,451.3
......	George	Vilas	44N	5E	.56	.33	....	....
120	Granite	Barron	36N	13W	1.60	.30	....	....
......	Grassy	Vilas	42N	7E	1.31	.64	....	....
92	Green (1)	Green Lake	15N 16N	12E 13E	7.40	2.00	237.0	7,342.4
76	Green (2)	Walworth	4N	16E	1.09	.66	56.8	277.9
120	Grindstone	Sawyer	40N	8W 9W	3.40	2.10	47.6	3,462.4
120	Half Moon	Polk	35N	17W	1.77	.50	....	....
116	Harris	Vilas	44N	5E	1.51	irr.	....	....
......	Helen (1)	Vilas	40N	11E	.58	.14	....	....
......	Helen (2)	Vilas	44N	5E	.47	.20	....	....
116	High	Vilas	42N 43N	7E 8E	2.00	1.00	31.2	761.6
......	Hill	Oneida	39N	6E	.18	.14	....	....
......	Hillman	Barron	34N	14W	.60	.25	....	....
75	Hooker	Kenosha	1N	20E	.75	.37	23.0	64.0
120	Horse	Polk	33N	18W	1.22	.40	....	....
120	Horseshoe[2] (1)	Polk	34N	15W	2.50	1.00	59.0	800.0
120	Horseshoe (2)	Polk	33N	17W	.80	.62	....	....
......	Howey	Polk	34N	17W	.25	.20	....	....
120	Island	Rusk	33N	8W	1.75	.65	47.6	435.2
118	Jag	Vilas	42N	6E	1.00	irr.	....	....
......	Jennie	Vilas	43N	6E	.44	.34	....	....
......	Jones	Vilas	43N	8E	.58	.21	....	....
116	Kawaguesaga	Oneida	39N	6E 7E	4.25	irr.	55.8	2067.2
26	Kegonsa	Dane	6N	10E 11E	3.00	2.25	31.4	3,145.2
......	Kimball	Washburn	42N	13W	1.68	.80	....	....
......	Kitten	Vilas	43N	7E	.34	.19	....	....
120	Lac Court Oreilles	Sawyer	39N 40N	8W 9W	5.65	1.90	67.2	5 248.0
116	Lac du Flambeau	Vilas	40N	4E 5E	2.00	irr.	49.2	1,280.0
52	Lac la Belle	Waukesha	8N	17E	2.70	1.12	46.4	1,137.5
116	Laura	Vilas	41N	8E	1.50	.90	39.4	640.0
......	Lake of the Woods	Burnett	38N	17W	....	....	....	....

[2] Listed as Big Horseshoe by Birge and Juday.

TABLE 2—(Continued)

Page mapped	Lake	County	Town	Range	Length (Miles)	Maximum breadth (Miles)	Maximum depth (Feet)	Area (Acres)
......	Lily	Burnett	41N	14W	....	....	....	....
......	Lindy	Burnett	40N	14W	....	....	....	....
......	Little Bass (1)	Vilas	42N	6E	.28	.19	....	....
......	Little Bass (2)	Polk	34N	17W	....	....	....	....
......	Little Bear	Vilas	43N	6E	.69	.44	....	....
......	Little Butternut	Polk	36N	17W	.60	.35	....	....
......	Little Crooked	Vilas	42N	6E	1.06	irr.	....	....
......	Little Doctor	Burnett	38N	17W	.45	.30	....	....
......	Little Rice (1)	Rusk	33N	8W	.70	.40	....	....
118	Little Rice (2)	Vilas	42N	7E	.76	.38	....	....
120	Little Wood	Burnett	38N	18W	.82	.48	....	....
......	Long (1)	Burnett	41N	14W	1.85	.35	....	....
120	Long (2)	Polk	34N	17W	1.70	.55	24.6	384.0
......	Long (3)	Vilas	40N	5E	1.40	.60	55.8	448.0
104	Long (4)	Waupaca	21N	11E	.94	.36	77.7	108.4
......	Loon	Barron	34N	14W	1.00	.45	....	....
116	Lost	Vilas	40N	8E	1.69	.86	....	....
118	Lost Canoe	Vilas	42N	7E	1.25	.38	41.0	204.8
120	Loveless	Polk	34N	17W	1.00	.32	21.3	153.6
......	Lynx	Vilas	43N	7E	.75	.66	....	....
......	Marion	Vilas	44N	5E	.37	.19	....	....
104	Marl	Waupaca	21N	11E	.31	.19	60.6	21.3
75	Mary	Kenosha	1N	19E	1.20	.70	26.2	275.2
12	Mendota	Dane	7N 8N	9E	5.90	4.60	84.0	9,729.8
......	Mercer	Oneida	39N	5E	1.30	.69	....	....
......	Meronk	Burnett	40N	14W	....	....	....	....
......	Meta	Vilas	40N	11E	.64	.26	....	....
76	Middle	Walworth	4N	16E	1.60	.45	50.0	284.0
120	Middle McKenzie[3]	Burnett	40N	14W	1.10	.90	....	....
76	Mill	Walworth	4N	16E	.75	.49	50.0	296.4
20	Monona	Dane	7N	9E 10E	4.16	2.40	74.0	3,482.7
......	Mud (1)	Polk	34N	17W	.66	.28	....	....
......	Mud (2)	Vilas	40N	5E	.81	.56	....	....
114	Mud (3)	Vilas	39N	7E	.87	.36	....	....

[3] Listed as McKenzie lakes.

TABLE 2—(Continued)

Page mapped	Lake	County	Town	Range	Length (Miles)	Maximum breadth (Miles)	Maximum depth (Feet)	Area (Acres)
104	Mud (4)	Waupaca	21N	11E	.13	.06	32.0	9.7
......	Mudhen.............	Burnett......	38N	17W	1.68	1.10	....	....
116	Muskallonge.........	Vilas.........	41N	7E	1.93	irr.	....	....
120	Nancy	Washburn ...	42N	13W	2.00	1.25	....	....
54	Nagawicka	Waukesha ...	7N	18E	2.75	1.12	94.4	917.6
54	Nashotah (Upper)...	Waukesha ...	7N	17E	.81	.40	57.2	136.7
54	Nashotah (Lower) ..	Waukesha ...	7N	17E	.79	.25	46.2	100.3
......	Nell	Vilas.........	44N	5E	.46	.15	....	....
54	Nemahbin (Upper)..	Waukesha ...	7N	17E	1.05	.56	62.0	271.1
54	Nemahbin (Lower)..	Waukesha ...	7N	17E	.93	.60	35.2	265.5
......	Nicaboyne...........	Burnett......	41N	14W	....	....	....	....
116	No Mans.............	Vilas.........	44N	5E	1.00	.69	....	....
......	North Twin..........	Polk..........	33N	16W	.92	.30	....	....
......	Number One.........	Sawyer	37N	9W	.25	.18	....	....
......	Oak	Burnett......	40N	14W	.70	.45	....	....
46	Oconomowoc	Waukesha...	7N	17E	1.75	.85	62.6	631.3
46	Okauchee............	Waukesha...	8N	17E 18E	2.37	1.80	94.0	1,057.0
......	Old Taylors..........	Waupaca	22N	11E	....	....	....	....
104	Otter..................	Waupaca	22N	11E	.40	.08	40.0	14.5
120	Owen.................	Bayfield	44N 43N	7W	6.25	irr.	88.6	2,032.0
75	Paddock	Kenosha.....	1N	20E	.62	.37	31.0	9.5
116	Palmer...............	Vilas.........	43N	8E	1.75	1.25	14.8	614.4
116	Pardee...............	Vilas.........	44N	4E	.94	.87	....	....
63	Pewaukee............	Waukesha ...	7N	18E 19E	4.50	1.20	45.0	2,298.0
......	Pike..................	Polk..........	33N	16W	1.18	.30	....	....
40	Pine (1).............	Waukesha....	8N	18E	2.30	1.05	90.0	755.7
......	Pine (2)..............	Polk..........	33N	17W	.66	.28	....	....
......	Pine Tree............	Polk..........	32N	16W	.92	irr.	....	....
116	Plum.................	Vilas.........	41N	7E 8E	4.20	1.25	50.8	1,088.0
120	Pokegama (1)	Washburn....	42N	12W	2.40	.60	23.0	505.6
116	Pokegama (2)	Vilas.........	40N 41N	5E	3.25	1.90	42.6	1,216.0
120	Pokegama (3)	Barron.......	33N	10W	3.50	.70	....	....
......	Poor Farm...........	Polk..........	33N	16W	....	....	....	....
104	Pope	Waupaca.....	21N	11E	.22	.11	40.6	16.2
120	Poplar	Polk..........	33N	18W	.72	.38	....	....

TABLE 2.—(Continued)

Page mapped	Lake	County	Town	Range	Length (Miles)	Maximum breadth (Miles)	Maximum depth (Feet)	Area (Acres)
8	Poygan	Waushara	19N	13E	7.70	3.50	10.8	10,992.2
120	Prairie	Barron	33N 34N	11W	6.50	.70	....	....
104	Rainbow (1)	Waupaca	22N	11E	.62	.55	95.1	137.5
116	Rainbow (2)	Vilas	44N	5E	.63	.44	....	....
116	Razorback	Vilas	41N	8E	1.10	.90	31.1	352.0
......	Red Bass	Vilas	43N	7E	.40	.13	....	....
......	Reserve	Sawyer	39N	8W	1.00	.55	....	....
......	Rice (1)	Barron	35N	11W	2.75	.60	....	....
......	Rice (2)	Vilas	42N	6E	.45	.30	....	....
......	Rock	Vilas	44N	5E	1.12	.37	....	....
......	Rooney	Burnett	40N	14W	....	....	....	....
......	Rose	Vilas	40N	7E	1.14	.25	....	....
......	Round (1)	Vilas	43N	6E	.62	.40	....	....
......	Round (2)	Polk	33N	18W	.58	.41	....	....
......	Round (3)	Polk	32N	17W	.26	.20	....	....
104	Round (4)	Waupaca	22N	11E	.55	.35	66.6	106.2
......	Rozen	Vilas	40N	7E	.68	.50	....	....
116	Rudolph	Vilas	43N	7E	.52	.22	....	....
120	St. Croix	Douglas	45N	11W 12W	4.00	.55	19.7	1,030.4
......	Sand (1)	Burnett	38N	16W	....	....	....	....
120	Sand (2)	Burnett	40N	15W	1.70	1.45	....	....
120	Sand (3)	Polk	33N	18W	1.22	.45	....	....
116	Sand[4] (4)	Vilas	42N	7E	1.65	.87	67.2	793.6
......	Sanford	Vilas	43N	6E	.77	.40	....	....
120	Shell	Washburn	37N 38N	12W 13W	3.25	2.50	49.2	3,200.0
120	Silver (1)	Barron	36N	13W	....	....	....	....
75	Silver (2)	Kenosha	1N	20E	1.25	.87	42.6	582.0
58	Silver (3)	Waukesha	7N	17E	.97	.56	44.0	231.9
......	Sishebogema	Vilas	39N 40N	5E	2.50	irr.	....	....
58	Soft	Waukesha	7N	17E	.32	.20	....	....
......	Speese	Oneida	39N	5E	.32	.31	....	....
116	Squirrel	Oneida	39N	5E	3.75	1.00	33.8	1,568.0
......	South Crab	Vilas	43N	6E	.68	.20	....	....
120	Spirit	Burnett	37N	18W	1.70	1.00	24.6	640.0
116	Star	Vilas	41N	8E	2.12	1.60	26.2	1,152.0

[4] Listed as White Sand Lake by Birge and Juday.

TABLE 2—(Continued)

Page mapped	Lake	County	Town	Range	Length (Miles)	Maximum breadth (Miles)	Maximum depth (Feet)	Area (Acres)
......	Streitor	Vilas	42N	6E	.44	.28	....	....
......	Sunday	Oneida	39N	5E	.75	.37	....	....
......	Tamarack	Vilas	43N	5E	.56	.40	....	....
......	Tank	Vilas	0N	6E	.24	.08	....	....
104	Taylor	Waupaca	22N	11E	.45	.31	55.7	51.2
116	Tenderfoot	Vilas	43N	8E	1.60	1.50	27.9	646.4
......	Tied Canoe [5]	....	......	......	....	....	....	....
......	Travis	Polk	33N	17W	....	....	....	....
116	Trout	Vilas	41N 42N	6E 7E	4.50	2.40	115.0	4,160.0
116	Turtle (North)	Vilas	43N	5E	1.25	.80	47.6	640.0
116	Turtle (South)	Vilas	43N	5E	2.25	.63	43.0	793.6
120	Turtle (Lower)	Barron	34N	14W	1.50	.45	....	....
120	Turtle (Upper)	Barron	34N	14W	2.30	.40	23.0	480.0
120	Vermilion	Barron	35N	13W	1.70	1.20	....	....
120	Wapagossett [6]	Polk	33N	16W 17W	3.50	1.20	27.9	1,452.8
24	Waubesa	Dane	6N 7N	10E	4.20	1.40	36.6	2,034.4
......	White Ash	Polk	34N	16W	1.77	.40	....	....
120	Whitefish	Sawyer	39N	9W	2.60	.65	73.8	896.0
116	White Sand	Vilas	41N	5E	2.75	1.00	....	....
......	Whitney	Vilas	42N	6E	.56	.45	....	....
116	Wildcat	Vilas	43N	7E	.87	.60	....	....
120	Wild Goose	Polk	34N	17W	.95	.64	14.8	256.0
92	Winnebago	....	15N 20N	17E	28.00	10.40	21.0	137,708.0
116	Wolf	Vilas	42N	7E	1.21	.84	....	....
104	Youngs	Waupaca	21N	11E	.09	.07	11.6	3.0

[5] See Lost Canoe.

[6] Listed as Sucker Lake by Birge and Juday.

KEY TO THE GENERA BASED UPON THE VEGETATIVE CHARACTERS.

1\. (35) Coloring matter not restricted to definite plastids.......... ....Class Myxophyceae 2

2\. (22) Cells solitary or in colonies, never in filaments........... ....Order Coccogoneales 3

3\. (12) Cells solitary or in colonies that never contain more than a few cells.......... 4

4\. (7) Cells spherical or hemispherical 5

5\. (6) Cells without a definite arrangement..........Chroococcus p. 27

6\. (5) Cells in a flat plate..........Merismopedia p. 31

7\. (4) Cells longer than they are broad.......... 8

8\. (11) Individual cells not enclosed by a special gelatinous vesicle 9

9\. (10) Ends of cells rounded..........Rhabdoderma p. 46

10\. (9) Ends of cells pointed..........Dactylococcopsis p. 47

11\. (8) Individual cells or groups of cells enclosed by a gelatinous vesicle ....Gloeothece p. 45

12\. (3) Cells in colonies that contain many cells when mature..... 13

13\. (18) Cellular arrangement within colony definite 14

14\. (15) Cells forming a flat plate..........Merismopedia p. 31

15\. (14) Cells forming a hollow sphere one cell in thickness.......... 16

16\. (17) Center of colony with radiating gelatinous strands......... .... Gomphosphaeria p. 36

17\. (16) Center of colony without radiating gelatinous strands...... .... Coelosphaerium p. 33

18\. (21) Cells spherical 19

19\. (20) Cells densely aggregated..........Microcystis p. 38

20\. (19) Cells some distance from one another..........Aphanocapsa p. 41

21\. (18) Cells longer than they are broad..........Aphanothece p. 43

22\. (2) Cells in simple or branched filaments..Order Hormogoneales 23

23\. (30) Trichomes without heterocysts......Family Oscillatoriaceae 24

24\. (29) Trichomes solitary or in flocculent readily dissociating masses 25

25\. (26) Trichomes containing but one spirally twisted cell.......... .... Spirulina p. 49

26\. (25) Trichomes containing more than one cell.......... 27

27\. (28) Sheaths of filament firm, generally projecting beyond the trichome ....Lyngbya p. 52

28\. (27) Sheath delicate or lacking, not projecting beyond the trichome .... Oscillatoria p. 50

29\. (24) Trichomes laterally united to form free-floating bundles which do not dissociate readily..........Trichodesmium p. 54

30\. (23) Trichomes with heterocysts.......... 31

31. (34) Trichomes not attenuated..............Family Nostocaceae 32
32. (33) Filaments solitary or interwoven to form flocculent masses of small size................................Anabaena p. 55
33. (32) Filaments laterally joined to form small plate-like colonies of macroscopic size......................Aphanizomenon p. 61
34. (31) Trichomes attenuated to a hair-like point.......Gloeotrichia p. 63
35. (1) Coloring matter in definite chromatophores................ 36
36. (57) Chromoplasts golden brown in color...................... 37
37. (38) Wall heavily impregnated with silica and definitely marked with grooves or rows of dots........Bacillarieae (see part 2).
38. (37) Wall not heavily impregnated with silica.................. ....Class Phaeophyceae 39
39. (52) Cells generally motile, rarely forming palmella or rhizopodial stages.................Division Euchrysomonadinae 40
40. (45) Cells with one flagellum..............Order Chromulinales 41
41. (42) Cells naked.................................Chrysamoeba p. 67
42. (41) Cells with a differentiated periplast........................ ....Family Mallomonadaceae 43
43. (44) Cells solitary...................................Mallomonas p. 67
44. (43) Cells in colonies........................Chrysosphaerella p. 69
45. (40) Cells with two flagella.................................. 46
46. (47) Flagella of equal length..........................Synura p. 70
47. (46) Flagella of unequal length..........Order Ochromonadales 48
48. (51) Cells in a definite receptacle..........Family Dinobryaceae 49
49. (50) Receptacles homogeneous, without growth rings..Dinobryon p. 72
50. (49) Receptacles with successive growth rings......Hyalobryon p. 75
51. (48) Cells never in a definite receptacle...........Uroglenopsis p. 71
52. (39) Cells usually immobile, rarely motile.................... 53
53. (54) Cells spherical to ovoid, embedded in a gelatinous mass.... .... Phaeococcus p. 76
54. (53) Cells with pseudopodial processes........................ ....Division Rhizochrysidinae 55
55. (56) Cells solitary or in irregular colonies..........Rhizochrysis p. 77
56. (55) Cells in a linear series....................Chrysidastrum p. 78
57. (36) Chromoplast grass-green or yellowish-green in color........ 58
58. (69) Chromoplast yellowish-green, with oil not starch the assimilation product ....Class Heterokonteae 59
59. (60) Vegetative cells motile..................Chlorochromonas p. 80
60. (59) Vegetative cells immobile 61
61. (68) Cells solitary or in colonies, never in filaments............. ....Order Heterococcales 62
62. (63) Cells epiphytic ....Peroniella p. 81
63. (62) Cells free-floating 64
64. (65) Length of cells several times the diameter.....Ophiocytium p. 85
65. (64) Length of cells less than twice the diameter............... 66
66. (67) Cells irregularly distributed throughout an ovoid, gelatinous envelope .... Chlorobotrys p. 82
67. (66) Cells arranged about a common center and enclosed by an irregular, gelatinous envelope..............Botryococcus p. 33
68. (61) Cells united to form simple filaments............Tribonema p. 87
69. (58) Chloroplast grass-green, generally with pyrenoids and starch ....Class Chlorophyceae 70

70. (71) Cells solitary or in families, divided into two symmetrical halves with a chloroplast in each half. Median region generally constricted..Family Desmidiaceae (See Part 2)
71. (70) Cells not divided into two symmetrical halves and without median constrictions 72
72. (85) Cilia normally present in vegetative cells..Order Volvocales 73
73. (84) Anterior end of cells with two cilia 74
74. (75) Cells solitary ....Chlamydomonas p. 90
75. (74) Cells in colonies ....Family Volvocaceae 76
76. (77) Colony a flat plate ....Gonium p. 94
77. (76) Colony a hollow sphere 78
78. (79) Cells mutually compressed ....Pandorina p. 95
79. (78) Cells some distance from one another 80
80. (83) Colony with not more than 256 cells...................... 81
81. (82) Asexual reproduction by division of all cells to form daughter colonies ....Eudorina p. 95
82. (81) Asexual reproduction by division of certain cells to form daughter colonies ....Pleodorina p. 96
83. (80) Colony generally containing many cells............Volvox p. 97
84. (73) Anterior end of cells with four cilia...............Carteria p. 92
85. (72) Cilia not present in vegetative cells 86
86. (183) Cells solitary or in colonies, never in simple or branched filaments ....Order Protococcales 87
87. (122) Cells solitary 88
88. (121) Cells free-floating 89
89. (92) Cell shape angular 90
90. (91) Angles simple or with spines, never with long setae........ .... Tetraëdron p. 115
91. (90) Angles with tufts of long setae..............Polyedriopsis p. 124
92. (89) Cells spherical, ovoid, or elongate 93
93. (116) Length of cells not more than three times the diameter.... 94
94. (109) Cells spherical 95
95. (102) Cell wall smooth 96
96. (99) Cells embedded in a gelatinous sheath.................. 97
97. (98) Gelatinous sheath lamellated..................Gloeocystis p. 100
98. (97) Gelatinous sheath not lamellated..........Planktosphaeria p. 103
99. (96) Cells not embedded in a gelatinous sheath 100
100. (101) Wall lamellated ....Gloeotaenium p. 115
101. (100) Wall not lamellated ....Chlorella p. 108
102. (95) Wall sculptured, denticulate, or with long spines or setae.. 103
103. (108) Setae or long spines covering the wall 104
104. (107) Wall with setae 105
105. (106) Setae the same thickness throughout, or gradually tapering .... Golenkinia p. 127
106. (105) Setae visibly thickened in the lower third..Acanthosphaeria p. 128
107. (104) Walls with long spines.................Echinosphaerella p. 128
108. (103) Denticulations or ridges covering the walls......Trochiscia 108
109. (94) Cells ovoid to reniform 110
110. (111) Cells reniform ....Nephrocytium p. 114
111. (110) Cells ovoid 112
112. (113) Walls without setae or spines....................Oocystis p. 110
113. (112) Walls with spines or setae 114

114. (115) Setae somewhat tapering....................Lagerheimia p. 129
115. (114) Setae not tapering............................Franceia p. 131
116. (93) Length of cells several times the diameter................ 117
117. (118) Ends of cells produced into setae..............Schroederia p. 136
118. (117) Ends of cells not produced into setae 119
119. (120) Chloroplast with an axial row of pyrenoids....Closteriopsis p. 136
120. (119) Chloroplast without or with one pyrenoid..Ankistrodesmus p. 134
121. (88) Cells epiphytic or epizoötic....................Characium p. 175
122. (87) Cells in colonies 123
123. (142) Colonies enclosed by a conspicuous gelatinous sheath...... 124
124. (125) Cells connected by branching remains of old cell walls...... .... Dictyosphaerium p. 104
125. (124) Cells not connected by branching remains of old cell walls.. 126
126. (137) Cells spherical to ovoid or reniform 127
127. (136) Chloroplast cup to disc-shaped and parietal............... 128
128. (129) Cells with pseudocilia.........................Tetraspora p. 102
129. (128) Cells without pseudocilia................................ 130
130. (131) Colonies never containing more than eight cells..Gloeocystis p. 100
131. (130) Colonies always with at least eight cells.................. 132
132. (133) Cells ovoid to reniform.....................Gloeocystopsis p. 102
133. (132) Cells spherical 134
134. (135) Chloroplasts of old cells cup-shaped..........Sphaerocystis p. 101
135. (134) Chloroplasts of old cells disciform........Planktosphaeria p. 103
136. (127) Chloroplasts star-shaped and central..........Asterococcus p. 103
137. (126) Cells elongate 138
138. (141) Cells straight 139
139. (140) Colonies with cells in groups of two...........Elaktothrix p. 139
140. (139) Colonies with cells in groups of 2–4–8..........Quadrigula p. 137
141. (138) Cells curved ....Kirchneriella p. 140
142. (123) Cells not enclosed by a gelatinous sheath.................. 143
143. (156) Colonies always forming a flat plate 144
144. (155) Number of cells, except in rare cases, greater than two..... 145
145. (152) Cells quadrately arranged.................................. 146
146. (147) Quadrate arrangement apparent in end view of colony...... .... Tetradesmus p. 150
147. (146) Quadrate arrangement apparent in front view of colony.... 148
148. (149) Cells without setae or spines..................Crucigenia p. 144
149. (148) Cells with spines or setae 150
150. (151) Long setae or spines on outer face of cells....Micractinium p. 125
151. (150) Short spines on outer face of cells.............Tetrastrum p. 149
152. (145) Cells not quadrately arranged 153
153. (154) Colony a flat circular or oval disc..............Pediastrum p. 166
154. (153) Colony a single or double row of cells.........Scenedesmus p. 150
155. (144) Number of cells in colony always two..........Euastropsis p. 174
156. (143) Colony not forming a flat plate 157
157. (166) Colony enclosed by old cell wall 158
158. (165) Wall enclosing colony greatly expanded and homogeneous.. 159
159. (162) Wall with spines or setae 160
160. (161) Setae somewhat tapering....................Lagerheimia p. 129
161. (160) Setae not tapering............................Franceia p. 131
162. (159) Wall without spines or setae............................. 163
163. (164) Cells ovoid ....Oöcystis p. 110

164. (163) Cells ovoid to reniform....................Nephrocytium p. 114
165. (158) Wall lamellated but not expanded............Gloeotaenium p. 115
166. (157) Colony not enclosed by old cell wall...................... 167
167. (176) Length of cell not more than twice the breadth............ 168
168. (171) Remains of old cell walls at center of and holding colony together 169
169. (170) Cells spherical ....Westella p. 106
170. (169) Cells ovoid to reniform..................Dimorphococcus p. 106
171. (168) Colonies without remains of old cell wall at center........ 172
172. (173) Outer face of cells with long setae............Micractinium p. 125
173. (172) Outer face of cells with short spines or without ornamentation 174
174. (175) Cells without spines on outer face............Coelastrum p. 160
175. (174) Cells with short spines on outer face...........Sorastrum p. 162
176. (167) Length of cells several times the breadth................. 177
177. (178) Colony forming a closed hollow sac..........Hydrodictyon p. 165
178. (177) Colony not forming a closed hollow sac................... 179
179. (180) Cells radiating from a common center.........Actinastrum p. 164
180. (179) Cells not radiating from a common center................ 181
181. (182) Cells straight or arcuate.................Ankistrodesmus p. 134
182. (181) Cells markedly curved........................Selenastrum p. 132
183. (86) Cells in simple or branched filaments.................... 184
184. (195) Cells with one to eight chloroplasts, chloroplasts never reticulate 185
185. (192) Chloroplasts parietal 186
186. (191) Chloroplasts laminate to disciform...................... 187
187. (188) Cells in simple filaments...........................Ulothrix p. 179
188. (187) Cells in branching filaments............................. 189
189. (190) Setae confined to terminal cells or lacking...Stigeoclonium p. 180
190. (189) Setae occurring on any cell of filament.......Aphanochaete p. 181
191. (186) Chloroplast a spiral band.........................Spirogyra p. 185
192. (185) Chloroplast central 193
193. (194) Chloroplast an axial plate......................Mougeotia p. 184
194. (193) Chloroplasts stellate ....Zygnema p. 184
195. (184) Chloroplasts numerous or united to form a single reticulate mass .... Cladophora p. 182

Class **MYXOPHYCEAE.**

Cells solitary, in colonies of definite or indefinite shape, or in filaments but always surrounded by a gelatinous to tough envelope which is generally hyaline but may be colored. Plants living a free-floating, sessile, epiphytic or endophytic existence. Cell shape various; spherical, hemispherical, ovoid, bacilliform, disciform or angular. Cells with the coloring matter localized at the periphery but not in definite, visible chromatophores. Color of cells typically blue-green but at times grey, yellowish, reddish, brownish or nearly grass-green. The color variation being due to different amounts of the three component pigments, carotin, chlorophyll, and phycocyan; the latter being found only in this class of plants. Cells containing a single nucleus of a primitive type (the central body) that contains linin and chromatin but is without a membrane or nucleole.

Reproduction always asexual, either by vegetative division of cells; fragmentation of the filaments into smaller portions (hormogones) which may move a short distance by spontaneous movement and then come to rest; or by special non-motile reproductive bodies (gonidia and resting spores).

KEY TO THE ORDERS.

Cells solitary or in colonies, never in filaments................ Coccogoneales
Cells in simple or branched filaments......................... Hormogoneales

Order **COCCOGONEALES.**

Cells rarely solitary, generally in colonies of regular or irregular shape, and with the cells definitely or indefinitely arranged within a copious gelatinous envelope. Colonies living a free-floating, sessile, epiphytic or endophytic existence. Cells differentiated into a basal and distal portion or without such differentiation. Cell shape various; spherical, ovoid, cylindrical, acicular, or angular.

Asexual reproduction by vegetative division of the cells or, in rare cases, by a division of the cell contents into non-motile gonidia. Resting cells and sexual reproduction unknown.

There are two families in the order only one of which, the Chroococcaceae, is found in the plankton.

Family *CHROOCOCCACEAE.*

Cells not differentiated into an apical and basal portion; rarely solitary, generally in colonies of definite or indefinite shape and always enclosed by a gelatinous sheath. Cell shape various, and cellular arrangement within the colonial envelope definite or indefinite.

Asexual reproduction by vegetative cell division and the fragmentation of the colonies. Gonidia formation occasionally taking place in certain genera.

KEY TO THE GENERA.

Cells solitary or in colonies that never contain more than a few cells.
- Cells spherical or hemispherical.............................CHROOCOCCUS
- Cells longer than they are broad.
  - Individual cells or groups of cells enclosed by a gelatinous vesicle ....GLOEOTHECE
  - Individual cells or groups of cells without vesicles.
    - Ends of cells rounded...............................RHABDODERMA
    - Ends of cells pointed..........................DACTYLOCOCCOPSIS

Cells always in colonies that contain many cells when mature.
- Cellular arrangement within colony definite.
  - Cells forming a flat plate..............................MERISMOPEDIA
  - Cells forming a hollow sphere one cell in thickness.
    - Center of colony with radiating gelatinous strands.GOMPHOSPHAERIA
    - Center of colony without radiating strands......COELOSPHAERIUM
- Cellular arrangement within colony indefinite.
  - Cells spherical.
    - Cells densely aggregated ....MICROCYSTIS
    - Cells some distance from one another..............APHANOCAPSA
  - Cells longer than they are broad.......................APHANOTHECE

CHROOCOCCUS Nägeli 1849.

Cells spherical, usually hemispherical for some time after division; solitary or 2–4–8–16 or more united in colonies enclosed by a hyaline or colored, homogeneous or lamellated sheath; free-floating, sessile, or epiphytic. Individual cells of colony at times surrounded by a homogeneous or lamellated sheath. Cell contents grey, grey-green, blue-green, olive-green, yellowish, orange, reddish or violet in color; homogeneous or granular.

Reproduction by vegetative division of cells in three directions.

KEY TO THE SPECIES.

Sheaths enclosing cells homogeneous.
- Cells solitary, in twos, and in fours..........................C. MINUTUS
- Cells in colonies of 4–8–16 or more.
  - Individual cells or groups of cells some distance from one another ....C. DISPERSUS
  - Individual cells or groups of cells fairly close together..C. LIMNETICUS

Sheaths enclosing cells distinctly lamellated.
  Diameter of cell and sheath not exceeding 40 μ..............C. TURGIDUS
  Diameter of cell and sheath over 60 μ.......................C. GIGANTEUS

CHROOCOCCUS MINUTUS (Kützing) Nägeli. Pl. 1, Fig. 1.

Gatt. einz. Algen 46. 1849.
Protococcus minutus Kützing, Tabulae Phycol. 1: *pl.* 5.

Cells spherical or hemispherical, solitary or in colonies of 2–4 enclosed by a wide, spherical, homogeneous, gelatinous sheath. Cell contents pale blue-green to grey, homogeneous or granulose. (Tychoplanktont).

Diam. cells (without sheath) 5–7 μ, (with sheath) 6–12 μ; length (without sheath) 5–10 μ, (with sheath) 5–13 μ.

Wolf (rrr).*

CHROOCOCCUS DISPERSUS (v. Keissler) Lemmermann. Pl. 1, Fig. 2.

Arkiv. f. Botanik 2, No. 2: 102. 1904.
Chroococcus minor var. *dispersus* v. Keissler, Verh. d. zool.-bot. Ges. Wien 52: 311, *fig.* 6. 1902.

Cells spherical, 4–16 or more united to form spherical, ovoid, or irregular colonies. Individual cells some distance from one another, or in compact groups that lie some distance from one another within the colony. Individual cell sheaths frequently confluent with the colonial sheath. Cell contents homogeneous, pale to bright blue-green. (Euplanktont).

Diam. cells (without sheath) 3–4 μ; (with sheath) 6 μ. Distance between cells or groups of cells 15–20 μ.

Browns (rr), Pokegama (1) (rr).

A species that is closely related to *C. limneticus* but one where the number of cells in the colony is usually greater, and where the individual cells or groups of cells are some distance from one another.

var. **minor** var. nov. Pl. 1, Fig. 3.

Cells smaller than the typical form and with pale grey contents. (Euplanktont).

Diam. cells 1. 75–2. 5 μ.

Long (3) (rrr), Middle McKenzie (rr), Palmer (rr).

---

* The relative abundance of the individual species is shown by the following symbols; (aaa) very abundant, (aa) abundant, (a) fairly abundant, (ccc) very common, (cc) common, (c) fairly common, (s) fairly scarce, (ss) scarce, (sss) very scarce, (r) fairly rare, (rr) rare, (rrr) very rare.

CHROOCOCCUS LIMNETICUS Lemmermann. Pl. **1**, Fig. **4**.

Bot. Cent. **76**: 153. 1898; Forschungsbr. a. d. Biol. Stat. zu Plön **7**: 132, *pl.* 1, *figs.* 22–23. 1899.

Mature cells spherical to subspherical, 4–32 in a free-floating spherical to ovoid colony. Cells fairly close together and equidistant from one another within the colony, sometimes in groups of 2–4 but with the groups lying close together. Cell division in all three planes but a majority of the cells dividing in the same plane and giving the colony a somewhat tabular appearance. Many hemispherical cells present because of the slow maturation after division. Individual cell sheaths homogeneous and frequently confluent with colonial envelope. Cell contents homogeneous; grey, pale to bright blue-green, or olive-green. (Euplanktont).

Diam. cells (without sheath) 6–12 μ, (with sheath) 8–14 μ.

Balsam (r), Bear (1) (rr), Bear (2) (rrr), Bear Trap (sss), Beasley (rr), Beaver (1) (r), Beaver (2) (ss), Big Butternut (r), Birch (1) (rr), Birch (2) (ss), Birch Island (s), Blake (sss), Bloom (s), Boulder (rrr), Browns (s), Camp (rr), Catfish (rr), Chetac (r), Chetek (rrr), Cranberry (sss), Dunham (rr), Dutchmans (r), Eagle (rr), East (rr), Fish (rr), Found (rr), Fowler (rr), Grassy (rr), Grindstone (rr), Half Moon (rr), Horseshoe (1) (r), Island (rr), Kawaguesaga (rr), Kimball (rr), Lac Court Oreilles (rr), Lac la Belle (r), Little Bass (2) (ss), Little Butternut (rr), Little Rice (2) (rr), Little Wood (sss), Loveless (r), Mary (rr), Middle McKenzie (r), Mill (rr), Mud (1) (rr), Mud (2) (rr), Mudhen (r), Nicaboyne (r), North Twin (rr), Oconomowoc (rr), Old Taylors (rr), Pewaukee (rr), Pike (rr), Pine Tree (rr), Pokegama (1) (r), Pokegama (2) (rrr), Pokegama (3) (rr), Poor Farm (rr), Poplar (ss), Reserve (sss), Rice (1) (r), Round (3) (rr), Sand (1) (s), Sand (3) (sss), Sand (4) (rr), Shell (sss), Silver (2) (r), Spirit (r), Squirrel (r), Tenderfoot (rrr), Travis (rrr), Upper Nashotah (rr), Upper Nemahbin (rr), White Ash (ss), Whitefish (s), Wildcat (rr), Winnebago (r), Wolf (rr).

This is preëminently a plankton species. It is widely distributed throughout the state and may be the dominant member of the Myxophyceae in lakes with very soft water, but it never occurs in sufficient abundance to cause the familiar phenomenon of blooming. The cells are quite variable in both color and size.

var. SUBSALSUS Lemmermann. Pl. **1**, Fig. **5**.

Forschungsbr. a. d. Biol. Stat. zu Plön **8**: 84. 1901; Arkiv f. Botanik **2**: No. 2: 101, *pl.* 1, *fig.* 9. 1904.

Cells spherical to subspherical, 8–16–32 in a somewhat spherical or broadly ellipsoidal colony with a wide gelatinous envelope. Cell contents homogeneous; pale to bright blue-green. (Euplanktont).

Diam. cells (without sheath) 3.5–4.5 μ; (with sheath) 4.5–5.5 μ.

Big Bass (sss), Blake (rr), Lac Court Oreilles (ss), Owen (ss), Paddock (r).

The chief distinction of this and the following varieties is in the size of the cells.

var. CARNEUS (Chodat) Lemmermann. Pl. **1**, Fig. **6**.

Arkiv för Botanik 2, No. 2: 101. 1904.

Cells spherical to hemispherical, 4–8–16 irregularly distributed throughout a spherical to ovoid gelatinous matrix. Individual sheaths of cells confluent with the colonial envelope. Cell contents homogeneous; color variable, ochraceous, grey, blue-green or olive green. (Euplanktont).

Diam. cells (without sheath) 7–9 μ.

Big Bass (ss), Harris (r), Mendota (rr), Nancy (r), Oconomowoc (rr).

var. DISTANS G. M. Smith. Pl. **1**, Fig. **7**.

Bull. Torr. Bot. Cl. 43: 481, *pl.* 26, *fig.* 26. 1916.

Cells spherical to hemispherical, 8–32 in a colony and lying some distance from one another within a spherical to ovoid, hyaline, homogeneous, gelatinous, colonial envelope. Individual cell sheaths confluent with the colonial envelope. Cell contents homogeneous, grey to pale blue-green in color. (Euplanktont).

Diam. cells (without sheath) 6.5–7 μ.

Bear Trap (s), Birch Island (rr), Browns (sss), Cochrane (r), Cranberry (sss), Lindy (rr), Muskallonge (rr).

var. ELEGANS G. M. Smith. Pl. **1**, Fig. **8**.

Trans. Wis. Acad. Sci., Arts, & Lett. 19[1]: 619. 1918.

Cells circular to semicircular in front view, somewhat flattened in side view. Colonies enclosed by a spherical, homogeneous, hyaline, gelatinous envelope. Individual sheaths of cells conspicuous. Cell contents homogeneous; generally bright blue-green in color. (Euplanktont).

Diam. cells (without sheath) 18–22 μ; (with sheath) 20–26 μ.

Delavan (rr).

CHROOCOCCUS GIGANTEUS W. West. Pl. **2**, Fig. **1**.

Jour. Roy. Micr. Soc. 1892: 741, *pl.* 10, *figs.* 59–60. 1892.

Cells ordinarily hemispherical, rarely becoming spherical or ovoid after division. Solitary or in colonies of two (rarely 3–4) enclosed by a thick, hyaline, markedly lamellate, gelatinous envelope. Envelope of individual cells conspicuous, lamellated. Cell contents distinctly granulose, bright blue-green. (Tychoplanktont).

Diam. cells (without sheath) 54–58 μ; (with sheath) 67–70 μ.

Mary (rrr).

CHROOCOCCUS TURGIDUS (Kützing) Nägeli. Pl. 1, Fig. 9.

Gatt. einz. Algen 46. 1849; G. S. West, Brit. Freshw. Algae 352, *fig.* 166_B. 1904.

Cells hemispherical, spherical, ovoid, or somewhat flattened by mutual compression. Cells solitary or in colonies of 2–4 (rarely 8) enclosed by a thick, hyaline, generally lamellated, gelatinous sheath. Individual cells also with hyaline, lamellated sheaths. Cell contents homogeneous or finely granulose; pale to bright blue-green or rarely brownish in color. (Tychoplanktont).

Diam. cells (without sheath) 8–32 μ; (with sheath) 13–40 μ.

Beasley (rrr), Beaver (2) (rr), Browns (rr), Hillman (rr), Horse (rrr), Oconomowoc (rr), Otter (rr), Rudolph (rr), Sand (2) (rr), Upper Turtle (rr), Waubesa (rrr).

MERISMOPEDIA Meyen 1839.

Colonies always forming flat, curved, or irregularly twisted plates one cell in thickness with the constituent cells in rectilinear series and usually in groups of four. Cells spherical to subspherical with rounded ends, hemispherical for some time after division. Individual sheaths of cells rarely distinct, generally confluent with the copious, hyaline, homogeneous, colonial envelope. Number of cells in colony 4—4000, cell number a multiple of two in young colonies, not a multiple of two in old colonies. Cell contents generally homogeneous, rarely with pseudovacuoles. Color varying from grey through light to bright blue-green, rarely rose pink to violet.

KEY TO THE SPECIES.

Cells bright blue-green .... M. ELEGANS
Cells pale blue-green.
    Cells some distance from one another.......... ....M. PUNCTATA
    Cells close together (not mutually compressed).
        Diameter of cells 1. 5–2 μ................................M. TENUISSIMA
        Diameter of cells 3–6 μ......................................M. GLAUCA

MERISMOPEDIA ELEGANS A. Braun. Pl. 2, Fig. 5.

in Kützing, Species algarum 472. 1849; G. S. West, Brit. Freshw. Algae 348: *fig.* 162_C. 1904.

Colonies small or large, of 16–4000 cells regularly arranged in rectilinear series and groups of four; with the cell number a multiple of two in young colonies but becoming irregular in arrangement and number in old colonies. Colonies flat except when containing many cells. Cells spherical to broadly ovoid, with homogeneous, bright blue-green contents. (Facultative Planktont).

Cells 5–7 μ broad, 5–9 μ long.

Carson (rr), Center (rr), Green (1) (rr), Kegonsa (rr), Little Bear (rrr), Mendota (r), Monona (r), Mud (1) (rr), Nancy (r), Upper Nashotah (rrr), Waubesa (rr).

This alga is a good example of a typical benthitic organism that is able to thrive under limnetic conditions. It never becomes a conspicuous member of the plankton but the colonies attain a much larger size and I have observed specimens 60–70 cells broad and 60–90 cells long. There is a certain amount of irregularity in such colonies so exact determinations of cell numbers are impossible.

var. **major** var. nov. Pl. 2, Fig. 6.

Colonies usually containing many cells. Cell shape the same as in the typical form. Cell contents homogeneous, bright blue-green. (Euplanktont).

Cells 10–11 μ broad, 12–14 μ long.

Beaverdam (r), Lindy (rr).

MERISMOPEDIA GLAUCA (Ehrenberg) Nägeli. Pl. 2, Fig. 4.

Gatt. einz. Algen 55, *pl.* 1, *fig.* D^1. 1849.

Colonies generally containing less than 64 cells, compactly and regularly arranged. Cells spherical to broadly ellipsoid, hemispherical for some time after division. Cell contents homogeneous, generally pale blue-green. (Facultative Planktont).

Diam. cells 3–5 μ.

George (rr), Jones (rrr), Lac Court Oreilles (rrr), Loon (rrr), Meta (rr), Mill (rr), Waubesa (rr).

This species resembles *M. punctata* in the size of the cells but differs in the distance between the cells.

MERISMOPEDIA PUNCTATA Meyen. Pl. 2, Fig. 3.

Arch. f. Naturgesch. 5[2]: 67. 1839; Kützing, Tabulae Phycol. 5: 13, *pl.* 38, *fig.* 3. 1855.

Colonies small, with the cells some distance from one another but regularly arranged in rectilinear series. Cells spherical to broadly ovoid or hemispherical. Cell contents homogeneous, grey to pale blue-green. (Tychoplanktont).

Diam. cells 2.5–3.5 μ.

Big Bass (rrr).

MERISMOPEDIA TENUISSIMA Lemmermann. Pl. 2, Fig. 2.

Bot. Cent. **76**: 154. 1898; Forschungsbr. a. d. Biol. Stat. zu Plön **7**: 132, *pl.* 1, *fig.* 21. 1899.

Colonies rectangular, free-floating, generally with 16–100 cells lying close together. Rectilinear arrangement of cells soon breaking down and not evident when there are more than 32 cells in a colony. Colonies distorted when of any size. Cells spherical to ovoid, hemispherical for some time after division. Individual sheath of cells distinct or confluent with colonial envelope. Cell contents homogeneous; grey to pale blue-green. (Facultative Planktont).

Diam. cells (without sheath) 1.5–2 μ.

Camp (rr), Little Doctor (rr), Long (3) (r), Mary (rr).

The irregular cellular arrangement is very common in older colonies and is not due to the size of the colony, since the species is one of the smallest of the genus, but to irregularities in cell division.

COELOSPHAERIUM Nägeli 1849.

Colonies microscopic, free-floating; spherical, ellipsoid, reniform, or irregular; enclosed by a delicate, hyaline, homogeneous or radially fibrillar, gelatinous envelope. Cells spherical, ovoid, ellipsoid, sub-pyriform and arranged in a single layer to form a hollow sac just beneath the periphery of the colonial envelope. Cell contents homogeneous and grey to bright blue-green; or reddish and containing numerous pseudovacuoles.*

Reproduction by vegetative cell division in three directions.

---

*These structures, which are also called gas vacuoles, are found in certain plankton species of the Myxophyceae. They are minute bodies, which are reddish through their refractive power, that occur in large numbers in all cells of the colony. Various functions have been ascribed to them, the most probable being a means of protection against too intense light, or a giving of greater buoyancy to the colony.

KEY TO THE SPECIES.

Cells spherical.
Diameter of cells not exceeding 4 μ.................. C. KUETZINGIAUM
Diameter of cells over 5 μ.............................. C. DUBIUM
Cells ovoid to ellipsoid.................................. C. NAEGELIANUM

COELOSPHAERIUM KUETZINGIANUM Nägeli. Pl. 3, Figs. 4–5.

Gatt. einz. Algen 54, *pl.* 1, *fig.* C. 1849.

Colonies spherical to subspherical, enclosed by a very hyaline, homogeneous, gelatinous envelope. Cells spherical to subspherical, forming a single layer, with the individual cells or pairs of cells some distance from one another or close together. Cell contents homogeneous, without pseudovacuoles; usually grey to pale blue-green, rarely bright blue-green. (Facultative planktont).

Diam. cells 2.25–4 μ; diam. colonies 20–90 μ.

Balsam (rr), Bass (2) (rr), Bear (3) (r), Bear Trap (rrr), Beaverdam (ss), Big (ss), Big Butternut (sss), Birch Island (rrr), Blake (ss), Bone (rr), Browns (rr), Carson (r), Catfish (r), Clam (ss), Cochrane (sss), Crab (rr), Cranberry (ss), Crawling Stone (sss), Deer (rr), Diamond (rrr), Elizabeth (s), Eagle (ss), Fence (sss), Fish (rr), Grindstone (rr), Green (2) (rr), Half Moon (r), High (sss), Hillman (ss), Horseshoe (1) (rrr), Kegonsa (s), Kitten (rr), Island (ss), Lac la Belle (rr), Lindy (rr), Little Bear (rrr), Little Butternut (r), Little Crooked (rr), Little Rice (1) (ss), Little Wood (s), Long (2) (rr), Long (3) (rr), Loveless (sss), Mary (s), Mendota (s), Mercer (rrr), Mill (rrr), Mud (1) (cc), Mud (2) (r), Mudhen (s), Nancy (rrr), Nicaboyne (rr), North Turtle (ss), Oconomowoc (r), Owen (rrr), Paddock (ss), Palmer (ss), Pardee (cc), Pewaukee (rr), Pine (2) (rr), Pine Tree (rr), Pokegama (1) (sss), Pokegama (2) (r), Pokegama (3) (rrr), Poplar (ss), Reserve (ss), Rice (1) (rr), Rock (sss), Round (3) (sss), Sand (1) (rr), Sand (3) (rr), Sand (4) (r), Silver (1) (ccc), Silver (2) (ss), South Turtle (ss), Spirit (ss), Sunday (r), Tamarack (rr), Tenderfoot (sss), Tied Canoe (rrr), Upper Nemahbin (rrr), Upper Turtle (sss), White Ash (sss), Whitefish (s), White Sand (sss), Wildcat (rr), Wild Goose (rr), Winnebago (rr), Wolf (r).

The gelatinous sheath is generally so delicate that it is necessary to use Errera's method of mounting colonies in dilute India ink to demonstrate its presence. The specific limits of this species as established by different authors frequently includes *C. Naegelianum* as a synonym. *C. Kuetzingianum* and *C. Naegelianum* are two well defined species that differ in colony shape, cell shape, occurrence of pseudovacuoles, structure of colonial envelope, and relative abundance. *C. Kuetzingianum* is a species that is widely distributed throughout the state but one which rarely becomes a dominant member of the plankton and never occurs in sufficient abundance to cause blooming.

COELOSPHAERIUM DUBIUM Grunow. Pl. 3, Fig. 7.

in Rabenhorst, Flora Eur. Algarum 2: 55. 1865; Schmula, Beibl. z. Hedwigia 37: (47), *figs.* 1–2. 1898.

Colonies spherical to irregular, with three or four smaller colonies frequently enclosed by a common gelatinous envelope. Colonial envelope firm, hyaline, and homogeneous. Cells in a single layer and very close together at the periphery of the colonial envelope. Cell contents homogeneous or with pseudovacuoles; blue-green in color. (Facultative Planktont).

Cells 5–7 μ in diameter. Single colonies up to 150 μ in diam; compound colonies up to 300 μ.

Mud (3) (rr).

COELOSPHAERIUM NAEGELIANUM Unger. Pl. 3, Fig. 6; Pl. 4, Fig. 1.

Denkschr. d. kais. Ak. d. Wiss. Wien 7: 196. 1854; Leitgeb, Mitt. d. naturw. Ver. Steiermark 2, Heft 1: 72, *pl.* 2, *figs.* 1–6. 1869; Borge, Bot. Not. 1900: 10, *pl.* 1, *figs.* 5a–5b. 1900.

Gomphosphaeria Naegeliana (Unger) Lemmermann. Kryptogamenfl. d. Mark Brandenburg 3, Algen 1: 80, 44, *fig.* 19. 1907; Bachmann, Arch. f. Hydrobiol. u. Planktonkde. 3: 60, *figs.* III1–III5. 1908.

Colonies spherical, ovoid, reniform or irregular; with a wide, hyaline, gelatinous envelope that is at times radially fibrillar in structure. Cells ovoid to ellipsoid, radially arranged in a single layer some distance beneath the periphery of the colonial envelope, usually close together but not mutually compressed. Cell contents rarely homogeneous and pale to bright blue-green or grass-green; usually with numerous pseudovacuoles and reddish to brownish in color. (Euplanktont).

Cells 2–3 μ broad; 3.5–6 μ long.

Adelaide (ss), Amicoy (ss), Balsam (ss), Bear (1) (s), Bear (2) (rr), Bear (3) (s), Bear Trap (sss), Beaverdam (cc), Big (ccc), Big Butternut (r), Big McKenzie (ss), Big St. Germaine (rr), Birch (1) (ccc), Birch (2) (aa), Birch Island (rrr), Blake (rr), Bloom (ss), Bone (ss), Boulder (rr), Browns (rr), Camp (c), Catfish (aa), Center (aa), Chetac (aa), Clam (c), Clare (rr), Crab (sss), Cranberry (aa), Devils (1) (rr), Devils (2) (ss), Deer (rr), Duck (c), Dummy (rrr), Elizabeth (cc), Fish (sss), Found (ss), Fowler (ss), Granite (ss), Grindstone (rr), Half Moon (s), Hillman (s), Hooker (rr), Horseshoe (1) (ss), Horseshoe (2) (rrr), Island (cc), Jennie (rr), Kawaguesaga (sss), Kitten (rrr), Kegonsa (aa), Lac du Flambeau (rr), Lac la Belle (r), Laura (aa), Lindy (rr), Little Bass (2) (rr), Little Butternut (rr), Little Rice (1) (cc), Long (1) (rr), Long (2) (a), Loon (rr), Lost (aa), Lower Nemahbin (r), Marl (rr), Mendota (cc), Mercer (ss), Middle McKenzie (cc), Mill (rr), Minocqua (rr), Monona (aa), Mud (1) (cc), Mud (3) (aa), Mudhen (rr), Nancy (c), Nicaboyne (r), No Mans (aa), North Twin (cc), North Turtle (aa), Number One (ss), Oconomowoc (rr), Okauchee (r), Otter (rr), Paddock (c), Palmer (ss), Pardee (rr), Pewaukee (ss), Plum (ss), Pokegama (1) (c), Pokegama (2) (sss), Pokegama (3) (a), Poplar (r), Prairie (s), Reserve (ss), Rainbow (r), Rice (1) (cc), Rock (a), Rooney (aa), Round (1) (rrr), Round (2) (ss), Round (3) (ss), St. Croix (cc), Sanford (r), Sand (1) (ss), Sand (2) (rr), Sand (3) (rr), Sand (4) (rr), Silver (2) (c), Sishebogema (aa), South Turtle (sss), Upper Nashotah (rr), Upper Nemahbin (rr), Upper Turtle (s), Wapagossett (cc), Waubesa (cc), White Ash (rr), Whitefish (cc), White Sand (rrr), Wildcat (rrr), Winnebago (cc), Wolf (rrr).

The radial fibrillar structure at the periphery of the gelatinous envelope is also found at times in the center of the colony. These radiations have been compared by some to the gelatinous strands found in *Gomphosphaeria* and the species transferred to that genus. I have preferred to keep the alga in the genus where it was first described. This alga is one of the important members of the Myxophyceae in the formation of blooms but it is rarely found as the sole constituent of any bloom. Bachmann's figures of the alga are quite characteristic.

GOMPHOSPHAERIA Kützing 1836.

Colonies microscopic, spherical to ellipsoid, embedded in a copious, hyaline, solid, homogeneous, gelatinous envelope. Cells spherical to pyriform, solitary or in pairs, generally lying some distance from one another and forming a layer one cell in thickness towards the periphery of the colonial envelope. Central portion of colonial envelope containing a denser core from which projections radiate to the base of the cells. Individual sheaths of cells distinct or confluent with the colonial envelope. Cell contents homogeneous, rarely granulose; grey to light or bright blue-green.

Reproduction by alternate vegetative division of cells in three planes.

KEY TO THE SPECIES.

Cells spherical .... G. Lacustris
Cells pyriform to cardioid .... G. Aponina

Gomphosphaeria lacustris Chodat. Pl. 4, Fig. 5.

Bull. Herb. Boiss. 6: 180, *figs.* 1_A–1_G. 1898; W. & G. S. West, Trans. Roy. Irish Acad. **33**, Sec. **B**: 112, *pl.* 11, *figs.* 24–27. 1906.

Colonies small, generally spherical or subspherical, rarely reniform; with a hyaline, homogeneous, gelatinous envelope. Cells spherical to broadly ellipsoid, irregularly spaced to form a single layer one cell in thickness beneath the periphery of the colonial envelope. Individual cell sheaths confluent with colonial envelope. Central portion of colony containing a core of denser radiating gelatinous strands that run to the base of the cells. Cell contents homogeneous; grey to pale or bright blue-green, rose colored in rare instances. (Euplanktont).

Cells 1.5–2.5 μ broad, 1.5–4 μ long. Colonies rarely over 40 μ in diam.

Bear (1) (rr), Beaverdam (c), Big McKenzie (ss), Blake (sss), Cochrane (r), Deer (rr), Diamond (r), Dunham (c), Fish (c), Half Moon (rr), Island (rr), Kawaguesaga (r), Lac Court Oreilles (ss), Lindy (sss), Little Bass (2) (rr), Long (1) (rr), Loon (rr), Mendota (s), Mercer (ss), Middle McKenzie (ss), Mudhen (ss), Nicaboyne (r), North Twin (r), Pardee (rr), Plum (rr), Pokegama (3) (rr), Sand (1) (r), Sand (2) (rr), Sand (4) (rr), Spirit (r), Squirrel (sss), Star (sss), Wolf (r).

The colonies of this species resemble those of *Coelosphaerium Kuetzingianum* in cell shape and arrangement but differ in the radiating strands of gelatinous material running from the center of the colony to the base of the cells. The gelatinous envelope is very delicate and cells frequently break away leaving the central mass of gelatinous radiations floating by itself.

GOMPHOSPHAERIA APONINA Kützing. Pl. **4**, Figs. **2–3**.

Decades **16**: No. 151. 1836; G. S. West, Brit. Freshw. Algae 349, *fig.* 163_B. 1904.

Colonies small, spherical to subspherical, with a fairly thick, hyaline, homogeneous, gelatinous envelope. Cells pyriform to cuneate, cardioid just before division; sparsely scattered and in a single layer some distance beneath the periphery of the colonial envelope. Individual sheaths of cells generally distinct, attached at the base to the dense gelatinous strands radiating from the center of the colony. Cell contents homogeneous or finely granulose; without pseudovacuoles; generally a bright blue-green. (Tychoplanktont).

Cells 4–5 μ wide at the greatest breadth, 8–15 μ long. Diam. colonies 50–90 μ.

Beaverdam (sss), Big McKenzie (rrr), Browns (rr), Center (rr), Eagle (rr), Elizabeth (rr), Hillman (rr), Kegonsa (rrr), Lower Nemahbin (rr), Mendota (rr), Mary (rr), Monona (rrr), Mudhen (rr), North Turtle (rrr), Otter (rrr), Pewaukee (rr), Pine (1) (rrr), Plum (rrr), Silver (2) (rr), Silver (3) (rr), South Turtle (rr), Waubesa (r).

var. CORDIFORMIS Wolle. Pl. **4**, Fig. **4**.

Bull. Torr. Bot. Cl. **9**: 25, *pl.* 13, *fig.* 11. 1882.

Cells considerably larger than in the typical form and markedly heart shaped instead of pyriform. Individual sheaths of cells conspicuous. Cell contents bright blue-green. (Tychoplanktont).

Greatest breadth of cells 6–13 μ; length 9–20 μ.

Bloom (rrr).

The cells of the Wisconsin individuals are somewhat longer than Wolle found them, his measurements for the length of the cell varying from 9–16 μ.

MICROCYSTIS Kützing 1833.

Colonies free-floating; spherical, elongate, irregular, or perforate to form reticulate or clathrate masses of microscopic or macroscopic size. Frequently occurring in sufficient quantities to give the water a verdigris-green color. Envelope of colonies generally copious, delicate to firm. Cells small, spherical, distributed in a solid mass throughout the colony and densely aggregated. Individual cell sheath confluent with colonial envelope. Cell contents homogeneous or with numerous pseudovacuoles; light to dark blue-green, olive-green, or rose colored.

Reproduction by vegetative division of cells in three directions.

Certain species of the genus occur in prodigious quantities and cause the familiar blooming of lakes. These species contain pseudovacuoles, while species without pseudovacuoles rarely occur in quantity.

KEY TO THE SPECIES.

Cells with numerous pseudovacuoles.
  Several colonies enclosed by the same envelope..........M. ICHTHYOBLABE
  Colonies not differentiated into several within the same envelope.
    Colony entire, not perforate..........................M. FLOS-AQUAE
    Colony saccate and perforate..........................M. AERUGINOSA
Cells without pseudovacuoles.
  Diameter of cells less than 1.5 μ..............................M. INCERTA
  Diameter of cells 2–3 μ..M. PULVERA

MICROCYSTIS ICHTHYOBLABE Kützing.

Phycol. generalis 170. 1843.
Polycystis ichthyoblabe Kützing, Tabulae Phycol. **1**: 7, *pl.* 8. 1845.

Colonies spherical or irregular, fragmented into several distinct sub-colonies, each with its own gelatinous envelope, and with the sub-colonies enclosed by a common gelatinous envelope. Cells spherical, close together. Individual cell sheaths confluent with colonial envelope. Cell contents pale to bright blue-green. Generally containing numerous pseudovacuoles. (Facultative Planktont).

Diam. cells 2–4 μ.

Catfish (rr), Clear (rr), Kawaguesaga (ss), Minocqua (r).

MICROCYSTIS FLOS-AQUAE (Wittrock) Kirchner. Pl. **5**, Fig. **1**.

in Engler & Prantl, D. nat. Pflanzenfam. **1** (Abt. **1**[a]): 56, *fig.* 49_N. 1900.

Colonies spherical to ovoid, rarely elongate; frequently reaching a macroscopic size. Cells compactly and evenly distributed within a very delicate, hyaline, homogeneous, gelatinous envelope. Margins of colony not sharply defined. Cell contents rarely homogeneous and grey to pale blue-green; usually with pseudovacuoles and reddish in color. Plant mass yellowish to verdigris-green. (Euplanktont).

Diam. cells 3.5–6.5 μ.

Bear (2) (rr), Big St. Germain (rr), Birch (1) (c), Birch Island (sss), Browns (cc), Catfish (ss), Center (ss), Cochrane (sss), Duck (ss), Dunham (rr), Fishtrap (rr), Found (sss), Fowler (rr), High (rr), Horseshoe (1) (rr), Kegonsa (cc), Loveless (r), Lower Nashotah (rr), Lower Nemahbin (r), Mendota (aa), Mercer (rrr), Monona (aa), Mud (3) (cc), Nagawicka (rr), No Mans (r), North Turtle (ss), Oconomowoc (ss), Paddock (ss), Palmer (ss), Pardee (ss), Plum (ss), Pokegama (1) (ss), Rainbow (r), Rock (s), Rose (ss), Sand (3) (r), Shell (rr), Sishebogema (ss), South Turtle (ss), Squirrel (rr), Star (ss), Tamarack (ss), Tenderfoot (rr), Upper Nashotah (rrr), Wapagossett (ss), Waubesa (ss), Whitefish (ss), Winnebago (cc).

MICROCYSTIS AERUGINOSA Kützing. Pl. **5**, Figs. **2–3**.

Tabulae Phycol. **1**: 6, *pl.* 8. 1845.

Clathrocystis aeruginosa (Kützing) Henfrey, Trans. Micr. Soc. London N. S. **4**: 53, *pl.* 4, *figs.* 28–36. 1856 (in Quart. Jour. Micr. Sci. **4**: 1856): Bachmann, Arch. f. Hydrobiol. u. Planktonkde. **3**: 61, *figs.* IV^1–IV^6. 1908.

Colonies spherical to subspherical when young, later becoming saccate and ultimately reticulate and clathrate. Gelatinous envelope of colony homogeneous, hyaline, and indistinct. Colonies frequently occurring in sufficient quantity to color water a bright green. Cells spherical to subspherical, evenly distributed throughout the colony and close together. Cell contents rarely homogeneous and grey to pale or bright blue-green; usually with numerous pseudovacuoles and reddish. (Euplanktont).

Diam. cells 3–4 μ.

Balsam (ss), Bear (3) (rr), Big Butternut (ss), Bloom (ccc), Bone (ss), Browns (cc), Center (rr), Clam (r), Crooked (rrr), Delavan (sss), Desmoines (ss), Devils (2) (rr), Duck (rr), Dutchmans (rrr), Eagle (ccc), Elizabeth (cc), Fish (ss), Fowler (rr), Grassy (ss), Grindstone (rr), Half Moon (ss), Hooker (ss), Horseshoe (1) (rr), Kawaguesaga (rr), Kegonsa (a), Kimball (ss), Lac du Flambeau (rr), Lac la Belle (rr), Lake of the Woods (r), Little Bass (2) (sss), Little Butternut (cc), Little Crooked (ss), Little Wood (ss), Loveless (ss), Mendota (a), Meronk (ss), Mill (sss), Monona (cc), Mud (2) (r), Mudhen (ss), Nancy (ss), Nicaboyne (ss), North Turtle (ss), North Twin (cc), Oconomowoc (rrr), Old Taylors (r), Otter (rr), Owen (rrr), Pardee (ss), Pewaukee (ss), Pokegama (2) (rr), Pokegama (3) (ss), Prairie (sss), Rice (1) (rr), Rooney (r), Sand (1) (ss), Sand (4) (rr), Shell (rr), Silver (2) (ss), Travis (rr), Upper Turtle (ss), Waubesa (cc), White Ash (ss), Wildcat (ss), Wolf (rr).

var. MAJOR (Wittrock) G. M. Smith. Pl. **4**, Fig. **6**.

Trans. Wis. Acad. Sci., Arts, & Lett. **18**[2]: 535. 1916.

Mature colonies elongate; irregular, saccate, or clathrate. Colonial envelope much firmer than in typical form and with larger cells. (Euplanktont).

Diam. cells 5.5–6.5 μ.

Center (rr), Mendota (ss), Nancy (sss), St Croix (rr), Vermilion (cc).

MICROCYSTIS INCERTA Lemmermann. Pl. **5**, Fig. **4**.

Abh. Naturw. Ver. Bremen **17**: 342. 1903; Kryptogamenfl. d. Mark Brandenburg **3**; Algen **1**: 76, 44, *fig*. 12. 1907.

Colonies spherical or somewhat elongate, enclosed by a hyaline, homogeneous, and very delicate gelatinous envelope. Cells small, very close together but equidistant from one another. Cell contents homogeneous, without pseudovacuoles; grey to pale blue-green in color. (Facultative Planktont).

Diam. cells 1–2 μ.

Camp (rr), Chetek (r), Hooker (s), Prairie (r), Whitefish (rr), Winnebago (c).

The cells of this species are very similar in appearance to those found in *Aphanocapsa elachista* and *A. delicatissima* but their compact arrangement shows that the alga is a *Microcystis* and not an *Aphanocapsa.*

MICROCYSTIS PULVERA (Wood) Migula.

in Thomé, Flora v. Deutschl., Deutsch.–österr. u. d. Schw. **6**, Kryptogamen **2**, Algen **1**: 36. 1907.

Anacystis pulverus Wood. Wolle, Freshw. Algae of U. S. 329, *pl*. 210, *fig*. 25. 1887.

Colonies spherical or elongate, with a very delicate, hyaline, gelatinous envelope. Cells spherical, very closely crowded together and forming a solid mass. Cell contents homogeneous, without pseudovacuoles; grey, pale olive-green, or blue-green in color. (Facultative Planktont).

Diam. cells 2–3 μ.

Browns (rr), Mill (rrr), Pardee (ss), Sunday (r), Winnebago (sss).

Since the specific difference between this and the foregoing species is mainly one of cell size it might be better to consider *M. incerta* a variety of *M. pulvera.*

APHANOCAPSA Nägeli 1849.

Cells united to form spherical, ovoid, or amorphous colonies of microscopic or macroscopic size. Colonies, hyaline, yellow, brown, or blue-green in color. Cells spherical, equally distributed throughout the colony but always some distance from one another; generally solitary, rarely in clusters. Individual sheaths of cells confluent with the homogeneous, hyaline or colored, colonial envelope. Cell contents homogeneous; usually pale to bright blue-green or grey in color.

Reproduction by vegetative division of cells in three directions.

KEY TO THE SPECIES.

Cells growing within gelatinous envelope of other algae........A. ENDOPHYTICA
Cells forming free-floating or sessile colonies.
  Cell contents grey to pale blue-green.
    Diameter of cells 3.5–4.5 μ..............................A. PULCHRA
    Diameter of cells less than 3 μ.
      Cell diameter not exceeding 1 μ................A. DELICATISSIMA
      Cell diameter 1.5–2.5 μ...........................A. ELACHISTA
  Cell contents brightly colored, generally blue-green.
    Diameter of cells 3.5–5 μ..............................A. GREVILLEI
    Diameter of cells 5–6 μ...............................A. RIVULARIS

APHANOCAPSA DELICATISSIMA W. & G. S. West. Pl. 2, Fig. 7.

Jour. Linn. Soc. Bot. **40**: 431, *pl.* 19, *figs.* 2–3. 1912.

Colonies small, spherical, ovoid or irregular, free-floating. Colonial envelope gelatinous, homogeneous, hyaline or ochraceous. Cells spherical, very small, more or less remote from one another. Cell contents homogeneous, grey to pale blue-green. (Euplanktont).

Cells .5–.75 μ in diam.; colonies 15–30 μ broad, 15–50 μ long.

Big Butternut (r), Chain (r), Little Butternut (ss), Loveless (ss), Mendota (s), Poor Farm (r).

The cells of this species are smaller than many bacteria and during the first two year's study of the plankton flora these small colonies were considered bacteria and not recorded so that the record for distribution is far from complete. When the colonies are of any considerable size the mass of cells has a decided blue-green color so that the organism must be called a blue-green bacterium or a blue-green alga of bacterial size. The latter view seems to be the more logical. It is very probable that the bacteria have been derived from the blue-green algae and the Wests' discovery of blue-green algae of bacterial size is very suggestive.

APHANOCAPSA ELACHISTA var. CONFERTA W. & G. S. West. Pl. 2, Fig. 8.

Jour. Linn. Soc. Bot. 40: 432, *pl.* 19, *fig.* 1. 1912.

Colonies small, spherical, ovoid, or irregular; with the cells enclosed by a gelatinous, homogeneous, hyaline or ochraceous envelope. Individual sheaths of cells confluent with colonial envelope. Cells spherical, fairly close together. Cell contents homogeneous, grey to pale blue-green. (Euplanktont).

Diam. cells 1.6–2.0 μ. Colonies up to 75 μ broad and 100 μ long.

Birch Island (rr), Crawling Stone (rrr), Deer (ss), Lindy (r), Nancy (r), Pine Tree (ss), Silver (2) (rr).

var. **planctonica** var. nov. Pl. 3, Fig. 3.

Cells spherical, some distance from one another. Sheath enclosing colony always hyaline. Contents of cells pale grey. (Euplanktont).

Diam. cells 2–3 μ.

Bear (1) (s), Bear Trap (rr), Poplar (ss), Round (3) (rr), Sand (1) (r).

Differing from the typical form in the larger cell size and grey color of the cell contents.

Aphanocapsa endophytica sp. nov. Pl. 2, Fig. 10.

Colonies endophytic within the gelatinous sheath of *Microcystis* colonies and with their envelopes confluent with the envelope of the host colony. Cells spherical, equidistant from one another or in small clumps. Cell contents homogeneous, grey to bright blue-green. (Euplanktont).

Diam. cells 2 μ.

Poor Farm (ss).

The occurrence of bacteria in the gelatinous envelope of colonial Myxophyceae is a familiar phenomenon to the limnologist. The blue-green color of the cells endophytic within the envelopes of *Microcystis* colonies in Poor Farm Lake precludes their recognition as bacteria. (See G. S. West's remarks concerning a similar endophytic *Aphanothece* in Jour. Linn. Soc. Bot. 40: 432. 1912).

APHANOCAPSA PULCHRA (Kützing) Rabenhorst. Pl. 2, Fig. 9.

Flora Eur. Algarum 2: 49. 1865.

Colonies macroscopic or microscopic, free-floating or sessile, generally spherical to ovoid when found in the plankton. Sheath of colony gelatinous, firm, hyaline, homogeneous; with the cells some distance from one another. Individual cell sheaths confluent with the colonial envelope. Cell contents homogeneous, grey to pale blue-green. (Tychoplanktont).

Diam. cells 3.5–4.5 μ.

Beaver (r), Beaverdam (sss), Harris (rrr), Lac Court Oreilles (ss), Winnebago (rr).

APHANOCAPSA GREVILLEI (Hassall) Rabenhorst. Pl. **3**, Fig. **1**.

Flora Eur. Algarum **2**: 50. 1865; G. S. West, Brit. Freshw. Algae, 351, *fig.* 165_A. 1904.

Colonies free-floating or sessile; spherical to ellipsoid when free-floating, dirty green in color. Cells spherical, solitary or in pairs and somewhat crowded. Cell contents homogeneous or finely granulose, bright blue-green in color. (Facultative Planktont).

Diam. cells 3.5–5 μ.

Beaverdam (rr), Birch (ss), Camp (ss), Chetac (sss), Muskallonge (rr), Rice (1) (rr), Silver (1) (rrr), Whitefish (ss).

APHANOCAPSA RIVULARIS (Carm.) Rabenhorst. Pl. **3**, Fig. **2**.

Flora Eur. Algarum **2**: 49. 1865; Cooke, Brit. Freshw. Algae 210, *pl.* 86, *fig.* 2. 1884.

Colonies free-floating or sessile; spherical, tubular, or irregular in shape. Cells spherical, solitary or in pairs and some distance from one another within the colony. Individual sheaths of cells confluent with the hyaline, homogeneous, colonial envelope. Cell contents homogeneous or finely granulose; bright blue-green in color. (Tychoplanktont).

Diam. cells 5–6 μ.

Star (rrr).

APHANOTHECE Nägeli 1849.

Colonies free-floating, sessile, or endophytic; spherical or expanded into amorphous masses and enclosed by a firm, copious, homogeneous, hyaline or colored envelope. Cells oblong to cylindrical, with rounded ends. Cells forming a solid mass distributed throughout the colonial envelope but lying some distance from one another. Individual cell sheaths indistinct, confluent with the colonial envelope. Cell contents homogeneous or finely granulose, never with pseudovacuoles; grey, light to bright blue-green, or olive-green in color.

Reproduction by vegetative division of cells in three planes, but with cell divisions always taking place at right angles to the long axes of cells.

KEY TO THE SPECIES.

Colonies endophytic in gelatinous envelope of other algae.. ....A. NIDULANS v. ENDOPHYTICA

Colonies not endophytic.

 Cell contents grey to pale blue-green, never bright blue-green.

  Colonies ovoid to spherical, never perforate.............A. NIDULANS

  Colonies reticulate to clathrate........................A. CLATHRATA

 Cell contents bright blue-green.

  Cells less than 5 μ broad................................A. STAGNINA

  Cells more than 5 μ broad................................A. PRASINA

APHANOTHECE CLATHRATA W. & G. S. West. Pl. 6, Fig. 3.

Trans. Roy. Irish Acad. **32**, Sec. B: 111, *pl.* 10, *figs.* 9–11. 1906.

Colonies microscopic; free-floating; elongate and irregular in shape, usually perforate and reticulate. Gelatinous colonial envelope hyaline and homogeneous. Cells very small, bacilliform, straight or curved, embedded in a solid mass in the colonial envelope. Cell contents homogeneous, grey to light blue-green. (Euplanktont).

Cells 0.6–1.0 μ broad, 3.5–4.5 μ long.

Lindy (r), Little Crooked (sss), Nancy (ss), Pokegama (1) (r), Poor Farm (rr), Sand (1) (ss), Wolf (ss).

APHANOTHECE NIDULANS P. Richter. Pl. 6, Fig. 1.

Bot. Not. **1884**: 128. 1884.

Colonies spherical, ovoid, or "nest-shaped"; enclosed by a homogeneous, hyaline, gelatinous envelope. Cells broadly cylindrical, straight or curved, of small size and fairly close together. Individual sheaths of cells confluent with the colonial envelope. Cell contents homogeneous, grey to light blue-green. (Facultative Planktont).

Cells 1.0–1.5 μ broad, up to 3.5 μ long.

Beaver (2) (ss), Deer (ss), Lower McKenzie (sss), Pokegama (sss), Tenderfoot (r).

Certain species of this genus resemble bacterial colonies in much the same manner as the minute-celled species of *Aphanocapsa*. Here again the color of the cell contents is good evidence of the algal nature of the organism.

var. ENDOPHYTICA W. & G. S. West. Pl. 5, Fig. 6.

Jour. Linn. Soc. Bot. **40**: 432, *pl.* 19, *fig.* 14. 1912.

Cells embedded in gelatinous envelope of other colonial Myxophyceae. Gelatinous envelope of colony confluent with envelope of host colony. Cells narrow, at least twice as long as they are broad. Cell contents homogeneous; generally light blue-green, rarely grey. (Euplanktont).

Cells 0.8–1.5 μ broad, 2.5–4.5 μ long.

Endophytic in sheaths of *Coelosphaerium Kuetzingianum* and *Micro-Cystis aeruginosa.*

Birch Island (r), Chetac (ss), Round (3) (rr).

APHANOTHECE PRASINA A. Braun. Pl. **5**, Fig. **5**.

in Rabenhorst, Flora Eur. Algarum 2: 65. 1865; Cooke, Brit. Freshw. Algae, 218, *pl.* 88, *figs.* 3a–3b. 1884.

Colonies spherical to cylindrical, reaching a diameter of 3–4 cm. under favorable conditions; bright green in color. Cells evenly distributed throughout the colony, broadly cylindrical and with rounded ends. Individual cell sheaths confluent with the colonial envelope. Cell contents homogeneous, bright blue-green. (Tychoplanktont).

Cells 5–6.5 μ broad, 7.7–11 μ long.

Beaverdam (r), Nancy (rr), Pokegama (1) (rrr).

The determination of this and the following species is open to question since specific differences in these species are based in part upon the macroscopic appearance of the colony when growing attached to the substratum. The determination of plankton species is from immature colonies and must rest upon cell shape, size, and color.

APHANOTHECE STAGNINA (Sprengel) A. Braun. Pl. **6**, Fig. **2**.

in Rabenhorst, Algen No. 1572; Cooke, Brit. Freshw. Algae 219, *pl.* 88, *fig.* 4. 1884.

Colony spherical, ellipsoidal, or cylindrical; generally macroscopic, 0.5–20 mm in diameter, pale verdigris-green. Cells broadly cylindrical, their length never more than 1½ times their breadth. Individual sheaths of cells confluent with the hyaline, homogeneous, gelatinous, colonial envelope. Cell contents homogeneous; pale to bright blue-green. (Tychoplanktont).

Cells 3–5 μ broad; 5–8 μ long.

Lake of the Woods (rr), Whitefish (rr).

GLOEOTHECE Nägeli 1849.

Colonies free-floating or sessile, of comparatively few cells enclosed by a hyaline or variously colored, homogeneous, gelatinous envelope. Individual sheaths of cells prominent, homogeneous or slightly lamellated, sometimes 2–8 cells enclosed by a single sheath. Cells elongate to broadly cylindrical, with rounded ends and irregularly distributed throughout the colonial envelope or with their longer axes pointing in the same general direction. Cell contents homogeneous or finely granulose; grey, pale to bright blue-green, or olive green in color.

Reproduction by vegetative division of cells at right angles to their long axes.

GLOEOTHECE LINEARIS Nägeli. Pl. 6, Fig. 5.

Gatt. einz. Algen 58, *pl.* 1, *fig.* G². 1849.

Colonies small; dull yellow to flesh color or dirty olive-green; enclosed by a broad, hyaline, homogeneous, gelatinous envelope. Cells elongate-cylindrical, straight or arcuate, with rounded ends. Sheaths containing one cell only, broad, hyaline, having the same general outline as the cell. Cell contents homogeneous; pale to bright blue-green. (Tychoplanktont).

Cells (without sheath) 1.3–2.5 μ broad, 10.5–18 μ long; (with sheath) 4–6 μ broad, 13–22 μ long.

Harris (r), Muskallonge (rr).

var. **composita** var. nov. Pl. 6, Fig. 6.

Cells shorter and 2–8 enclosed by a single, broad, firm, homogeneous sheath. Individual sheaths of single cells or groups of cells not confluent with the colonial envelope. (Tychoplanktont).

Cells (without sheath) 3.0–3.5 μ broad, 4–8 μ long; sheaths 10–12 μ broad, 20–35 μ long.

Deer (rrr).

RHABDODERMA Schmidle and Lauterborn. 1900.

Colonies free-floating, small, of comparatively few cells enclosed by a hyaline, homogeneous, gelatinous envelope. Cells cylindrical, with rounded apices, straight or arcuate; with their long axes pointing in the same general direction. Individual sheaths of cells confluent with colonial envelope. Cell contents homogeneous; grey to bright blue-green in color.

Reproduction by vegetative division of cells, the divisions taking place at right angles to the long axes of the cells.

RHABDODERMA LINEARE Schmidle and Lauterborn. Pl. 6, Fig. 4.

Ber. d. D. Bot. Ges. **18**: 148, *pl.* 6, *figs.* 8–11. 1900.

Specific characters as those of the genus. (Euplanktont).

Cells 2–3 μ broad, 6–12 μ long.

Sand (2) (rrr), Whitefish (rr).

This genus is closely related to *Gloeothece,* the chief difference being a lack of individual sheaths around the cells.

DACTYLOCOCCOPSIS Hansgirg 1888.

Colonies microscopic, free-floating or lying among other algae; embedded in a copious, hyaline, homogeneous, acicular to fusiform, gelatinous envelope. Cells elongate, acicular and with pointed ends; straight arcuate, sigmoid, or spirally twisted around one another. Individual cell sheaths confluent with colonial envelope. Cell contents homogeneous or with the color bearing area localized in a definite part of the cell; grey to bright blue-green in color.

Reproduction by vegetative division of cells at right angles to their long axes.

KEY TO THE SPECIES.

Length of cells never more than sixteen times the breadth, cells frequently bent .... D. RAPHIDIOIDES

Length of cells at least twenty times the breadth, cells always straight .... D. ACICULARIS

DACTYLOCOCCOPSIS RHAPHIDIOIDES Hansgirg. Pl. 6, Fig. 7.

Prodr. d. Algenfl. v. Böhmen 139, *fig.* 49A. 1892; Lemmermann, Kryptogamenfl. d. Mark Brandenburg **3**, Algen **1**: 50. 1907.

Colonies containing 2–8 or more cells. Cells fusiform; straight, arcuate or sigmoid; frequently in pairs with one end rounded and the other pointed. Colonies enclosed by a copious, hyaline, homogeneous envelope. Cell contents homogeneous, pale blue-green. (Facultative Planktont).

Cells 1–3 μ broad, 5–25 μ long.

Deer (sss), Long (2) (rrr), Mercer (rr), Reserve (rr), Rice (1) (rr).

Hansgirg gives eight as the maximal number of cells found in the colony but I have observed colonies in which there are as many as a hundred cells. These cells are also proportionally longer than he states and may be, in extreme cases, sixteen times as long as they are broad. In this respect they follow the cell dimensions delimited by Lemmermann.

DACTYLOCOCCOPSIS ACICULARIS Lemmermann. Pl. 6, Figs. 8–10.

Ber. d. D. Bot. Ges. **18**: 309. 1900; Kryptogamenfl. d. Mark Brandenburg **3**, Algen **1**: 51, 44, *fig*. 5. 1907.

Cells solitary or in colonies enclosed by a wide, hyaline, homogeneous, gelatinous sheath. Cells elongate, acicular, with poles attenuated to fine points; straight or slightly arcuate. Cell contents homogeneous and pale blue-green; or with color bearing region restricted to one side of cell. (Euplanktont).

Cells 2–2.5 μ broad, 55–80 μ long.

Browns (cc).

The localization of the color bearing portion at one side of the cell is quite abnormal for a member of the Myxophyceae, but, as I have stated elsewhere (Trans. Wis. Acad. Sci., Arts, & Lett. **19**^{1}: 620. 1918) notwithstanding this fact the alga belongs to the Myxophyceae.

Order **HORMOGONEALES.**

Cells always in filaments that generally contain a number of cells. Filaments simple or branched; solitary or united to form definitely or indefinitely shaped aggregates. Free-floating, sessile, epiphytic, or endophytic. Cells serially arranged to form trichomes, which when enclosed by a sheath are known as filaments. Trichomes either the same thickness throughout, attenuated towards both extremities, or attenuated from base to apex; straight, arcuate, spirally or irregularly twisted. Sheaths of filaments tough and leathery; or gelatinous, homogeneous to lamellated, hyaline to variously colored, and distinct or confluent with the colonial envelope. Cell shape various. Cell contents homogeneous, or with numerous pseudovacuoles; grey, light to dark blue-green, olive-green, reddish-green, ochraceous, orange, violet, or black in color.

Reproduction, aside from vegetative cell division, by hormogones and resting spores.

KEY TO THE FAMILIES.

Trichomes without heterocysts*.......................... OSCILLATORIACEAE
Trichomes with heterocysts.
 Trichomes not attenuated .... NOSTOCACEAE
 Trichomes attenuated to a hair-like point .... RIVULARIACEAE

Family *OSCILLATORIACEAE.*

Trichomes free-floating, sessile, epiphytic, or endophytic; of one or more cells forming simple or branched filaments with one or more trichomes within a single sheath. Sheaths of various forms; copious, gelatinous, hyaline or colored, homogeneous to lamellated; firm, narrow and homogeneous or lamellated; or entirely lacking. Filaments straight or spirally twisted. Cells disciform, cylindrical, or barrel-shaped. Apical cell of trichomes frequently conical and capitate. Heterocysts never present in the trichomes. Cell contents homogeneous, granulose, or with pseudovacuoles; variously colored.

Reproduction by vegetative division of cells and the formation of hormogones.

KEY TO THE GENERA.

Trichomes solitary or in flocculent, readily dissociating masses.
 Trichomes containing but one cell............................ SPIRULINA
 Trichomes containing more than one cell.
  Sheaths of filaments firm, generally projecting beyond the trichome .... LYNGBYA
  Sheaths delicate or lacking, not projecting beyond trichome .... OSCILLATORIA
Trichomes laterally united to form free-floating bundles which do not dissociate readily .... TRICHODESMIUM

SPIRULINA Turpin 1827.

Trichome consisting of a single, elongate, cylindrical cell that is not attenuated but with rounded ends. Filaments without sheaths and twisted to form loose or compact regular spirals. Cell contents homogeneous or granulose; light to dark blue-green in color.

Reproduction by vegetative division of cells at right angles to the long axis. No hormogone formation possible.

---

*Cells commonly larger than the vegetative cells and with heavy walls that have a conspicuous nodule at each pole. Cell contents a homogeneous, pellucid, non-cytoplasmic substance.

KEY TO THE SPECIES.

Turns of spirals not over 5 μ in diameter........................... S. MAJOR
Turns of spirals over 10 μ in diameter........................... S. PRINCEPS

SPIRULINA MAJOR Kützing. Pl. 7, Fig. 1.

Phycol. generalis 183. 1843; Gomont, Ann. Sci. Nat. **7** Sér., Bot. **16**: 251, *pl.* 7, *fig.* 29. 1892.

Filaments free-floating and solitary, or forming a greenish scum. Trichomes in a more or less loose spiral that is straight or bent. Cell contents homogeneous; bright blue-green. Sheath of filament lacking. (Tychoplanktont).

Trichomes 1.2–1.7 μ in diameter. Spirals 2.5–4 μ broad, 2.7–5 μ between turns.

White Ash (rrr).

SPIRULINA PRINCEPS (W. & G. S. West) G. S. West emend. Pl. 7, Fig. 2.

Jour. Linn. Soc. Bot. **38**: 179, *pl.* 9, *fig.* 5. 1907; W. & G. S. West, Trans. Linn. Soc. Bot. **2** Ser., Bot. **6**: 205. 1902.

Trichomes solitary, in loose spirals that may be straight or curved. Ends of trichomes blunt or rounded. Cell contents homogeneous or with a few scattered granules; bright blue-green. (Tychoplanktont).

Trichomes 3–5 μ broad. Spirals 8.8–16 μ broad, 9.5–12 μ between turns.

Poor Farm (rrr).

OSCILLATORIA Vaucher 1803.

Trichomes possessing a peculiar power of spontaneous movement of an oscillating, rolling, or gliding type; free-floating; solitary or in flocculent masses; sometimes forming strata in moist places. Trichomes unbranched, cylindrical, entirely without or with barely perceptible sheaths. Ends of trichomes straight, slightly attenuated, but never attenuated to a point. Terminal cells rounded or calyptrate. Cell shape cylindrical to disciform. Cell contents usually homogeneous or granulose, rarely with pseudovacuoles. Color of cells variable.

Reproduction by hormogones.

KEY TO THE SPECIES.

Apex of trichome gradually tapering............................O. PROLIFICA
Apex of trichome not tapering ....O. TENUIS

OSCILLATORIA PROLIFICA (Greville) Gomont. Pl. 7, Figs. 3–5.

Ann. Sci. Nat. **7** Sér. Bot. **16**: 205, *pl.* 6, *fig.* 8. 1892; Olive, Trans. Wis. Acad. Sci., Arts, & Lett. **15**[1]: 124–134. 1905; Hyams and Richards, Technology Quart. **14**: 302–310, *figs.* 1–8. 1901.

Filaments rarely solitary, generally irregularly matted to form free-floating flocculent masses. Plant masses blood-red to dirty purple, becoming lilac when dried. Trichomes straight, elongate, not constricted at cross walls of cells, and slightly tapered at apex with the terminal cell of mature trichomes obtuse-capitate. Cells cylindrical, slightly longer than they are broad. Cell contents with numerous reddish pseudovacuoles, sometimes with rows of granules at the cross walls. (Euplanktont).

Cells 2.2–5 μ broad, 4–6 μ long.

Geneva (aa), Pine (1) (aa).

This species is closely related to *O. rubescens* D. C., but if Gomont's monograph of the Oscillatoriaceae must be taken as the starting point for nomenclature of the group the two are distinct. Both species are peculiar in that they produce blooms in lakes at a time when the water is very cold or covered with ice. The masses of filaments do not have the usual yellowish-green color but are blood red to dirty purple and the phenomenon is spoken of in Europe as "bloody" or "Burgundy" bloom.

Olive has described the occurrence of such a bloom in Pine lake of this state and discussed the literature of the subject. The alga also appeared in profusion in Lake Geneva this (1918) spring at the time when the ice was beginning to disappear. It was not frozen in the ice but grew luxuriantly in the water below the ice and gave a deep reddish color that was especially noticeable in the cracks between the ice. The alga had not been noted in the lake before this time and its sporadic appearance is quite like the sporadic appearance of *O. rubescens* in the Swiss lakes. The excessive growth of *O. prolifica* is not wholly correlated with cold water since it has remained present throughout the summer, but not in such quantities as early in the spring. This abundant growth of the alga in the summer has also been noted in a Massachusetts lake by Hyams and Richards and furnishes a physiological as well as a morphological distinction between *O. prolifica* and *O. rubescens*.

Oscillatoria tenuis C. A. Agardh. Pl. 7, Fig. 6.

Algarum decades **2**: 25. 1813; Gomont, Ann. Sci. Nat. **7** Sér. Bot. **16**: 220, *pl.* 7, *figs.* 2–3. 1892.

Filaments solitary or in floccose masses that are pale blue-green in color. Trichomes straight, slightly constricted at the cross walls. Terminal cells the same diameter as those in the median portion of the trichome, straight or curved; apical cells rounded but not capitate. Cell shape disciform. Cell contents homogeneous except for a row of granules along the cross walls. (Tychoplanktont).

Cells 4–10 μ broad, 2.6–5 μ long.

Monona (rr).

var. tergestina (Kützing) Rabenhorst. Pl. 7, Figs. 7–8.

Flora Eur. Algarum **2**: 102. 1865.

Trichomes narrower than the foregoing. (Tychoplanktont).

Cells 4–6 μ broad.

Geneva (ss).

LYNGBYA C. A. Agardh. 1824.

Filaments solitary or interwoven to form floccose masses; free-floating or in an expanded stratum. Sheaths of filaments narrow, firm, hyaline or colored, homogeneous or lamellated, generally projecting for some distance beyond the trichome. Filaments unbranched, cylindrical; straight, flexed, or twisted into regular spirals. Trichomes generally the same thickness throughout, rarely slightly attenuated at the apices; sometimes constricted at the cross walls. Apical cell with distal end rounded or slightly capitate. Cell contents homogeneous, granulose, or with numerous pseudovacuoles; grey, pale to bright blue-green or variously colored.

Reproduction by the formation of hormogones.

KEY TO THE SPECIES.

Filaments not over 4 μ in thickness, with pale blue-green to grey cell contents.
- Filaments straight..L. Limnetica
- Filaments in regularly coiled spirals........................L. Contorta

Filaments over 10 μ in diameter, contents deeply colored.............L. Birgei

Lyngbya limnetica Lemmermann. Pl. 7, Figs. 9–11.

Bot. Cent. **76**: 154. 1898; Kryptogamenfl. d. Mark Brandenburg **3**, Algen **1**: 135, 102, *fig.* 8. 1907.

Filaments free-floating, generally solitary; straight or slightly flexed. Trichomes not constricted at the cross walls. Apical cells the same thickness as the other cells, with distal end rounded but not capitate. Sheaths firm, hyaline, very narrow, extending for some distance beyond the end of the trichome. Cells cylindrical, at least twice as long as broad. Cell contents homogeneous; generally grey, rarely blue-green in color. (Euplanktont).

Cells 1–1.5 μ broad, 3–8 μ long.

Bear (1) (c), Geneva (ss), Soft (rr), Upper Nashotah (rrr).

The length of the individual cells of Wisconsin specimens exceeds that given by Lemmermann but it is very probable that I am dealing with the same species that he names *L. limnetica*. Several plankton "species" have been described which are closely related and it is very probable that future investigations will show that all of these straight, very narrow, colorless "species" should be included under one collective specific name.

LYNGBYA CONTORTA Lemmermann. Pl. **7**, Figs. **12–13.**

Forschungsbr. a. d. Biol. Stat. zu Plön **6**: 202, *pl.* 5, *figs.* 10–13. 1898; G. S. West Jour. of Bot. **47**: 246, *pl.* 498, *figs.* 5_A–5_B. 1909.

Filaments free-floating, solitary; occasionally straight, generally twisted to form regular, loose or compact spirals or helices. Sheaths firm, very thin, extending for some distance beyond the trichomes. Trichomes the same thickness throughout, not constricted at the cross walls, apical cells rounded but not capitate at the distal ends. Cell contents homogeneous, grey to pale blue-green. (Euplanktont).

Cells 1.5–2 μ broad, 3–6 μ long. Spirals 15–20 μ broad, 6–14 μ between the turns.

Mendota (rr), Pine Tree (ss), Poor Farm (s), White Ash (cc).

The specific distinction of this species is based in large part upon the spiral twisting of the filament. Since certain normally straight filaments of other algae become spirally twisted under limnetic conditions (vide W. & G. S. West, Proc. Roy. Soc. London Ser. **B, 81**: 169, *fig.* 1. 1909) the validity of spiral twisting as a specific character arises. *L. contorta* was found in considerable quantity at different stations and the vast majority of filaments were spirally twisted. On the other hand numerous filaments with cells of the same size that showed no signs of twisting were collected from other stations. I think that Lemmermann is correct in considering these distinct species and that the spiral twisting of filaments in plankton species of the Myxophyceae is a sufficiently constant character to warrant the establishment of species in both this genus and *Anabaena.*

LYNGBYA BIRGEI G. M. Smith. Pl. **7**, Figs. **14–15.**

Bull. Torr. Bot. Cl. **43**: 482, *pl.* 26, *fig.* 28. 1916.

Filaments free-floating, solitary, never forming floccose masses; straight or slightly flexed. Sheaths firm, at times of considerable thickness, hyaline, generally homogeneous, rarely lamellated, extending for some distance beyond the trichome. Trichomes not attenuated at apices and without constrictions at the cross walls. Terminal cells rounded but not capitate at distal ends. Cell shape disciform. Cell contents rarely homogeneous and with grey to olive green contents; usually with numerous pseudovacuoles. (Euplanktont).

Cells 18–23 μ broad, 2–5.5 long. Sheaths .5–4 μ thick.

Balsam (rr), Bear Trap (cc), Big (cc), Deer (rrr), Elizabeth (ss), Fowler (rrr), Half Moon (c), Kegonsa (aa), Lac la Belle (rr), Little Crooked (r), Little Wood (rr), Loveless (rr), Lower Nashotah (rr), Mary (ss), Mendota (cc), Mercer (rr), Monona (cc), Mudhen (rr), Nagawicka (rr), Oconomowoc (rr), Pope (rrr), Round (3) (sss), Sand (1) (r), Spirit (sss), Squirrel (sss), Upper Nashotah (cc), Upper Turtle (rr), Wapagossett (cc), Waubesa (cc), Wolf (r).

This is undoubtedly the alga that Trelease (Trans. Wis. Acad. Sci., Arts, & Lett. **7**: 124, *pl.* 10, *fig.* 6. 1888) reported from Lake Mendota as *Lyngbya Wollei* Farlow. *L. Wollei* has been shown to be a *Plectonema* (*P. Wollei* Farlow) and Trelease's determination is undoubtedly a mistake.

TRICHODESMIUM Ehrenberg 1830.

Trichomes laterally joined to form spindle- to scale-shaped, free-floating colonies that are not enclosed by a gelatinous sheath. Trichomes cylindrical, straight or spirally twisted, of the same thickness throughout or with apices slightly attenuated. Apical cell undifferentiated, or rounded and slightly capitate. Sheaths of trichomes very delicate. Cells disciform to barrel-shaped. Cell contents homogeneous, granulose, or with numerous pseudovacuoles. Heterocysts and resting spores lacking.

Reproduction by the formation of hormogones.

TRICHODESMIUM LACUSTRE Klebahn. Pl. **8**, Fig. **1.**

Forschungsbr. a. d. Biol. Stat. zu Plön **3**: 13. 1895; Flora **80**: 271, *pl.* 4, *figs.* 31–33. 1895.

Trichomes laterally joined to form plate-like, free-floating masses that are not enclosed by a gelatinous envelope. Filaments with a very delicate, homogeneous sheath. Trichomes straight, not attenuated at the apices, and with terminal cells rounded but not capitate. Cells disciform, with straight or markedly convex sides. Cell contents occasionally homogeneous and pale to bright blue-green in color; generally with numerous pseudovacuoles. Trichomes without heterocysts or spores. (Euplanktont).

Cells 5–7 μ broad, 5–7 μ long. Colonies up to 1 mm. in length.

Amicoy (rrr), Beaverdam (r), Island (rr), Lake of the Woods (rrr), Little Bass (1) (rr), Mud (1) (s), Nicaboyne (rr), Rice (1) (rr), Wild Goose (rrr).

The general appearance of the colony at once suggests an *Aphanizomenon* species with cells shorter and more rounded than the familiar *A. flos-aquae.* I have examined a sufficient number of colonies to feel confident in confirming Klebahn's statement that there are no spores or heterocysts in the alga under consideration. This being the case it is impossible to refer the species to *Aphanizomenon,* but it should be considered a freshwater species of the common *Trichodesmium* of tropical seas.

Family NOSTOCACEAE.

Filaments free-floating, sessile, epiphytic, or endophytic; solitary or in colonies of microscopic to macroscopic size that are definitely or indefinitely shaped. Trichomes always unbranched; straight, flexed, spirally twisted, or irregularly interwoven. Sheath rarely firm and narrow; generally copious, gelatinous, hyaline or colored, and distinct or confluent with the colonial envelope. Trichomes generally the same thickness throughout and with spherical to cylindrical cells. Heterocysts either terminal or intercalary, solitary or catenate. Resting spores variously shaped, adjacent to or remote from the heterocysts; formed in all genera. Cell contents homogeneous, granulose, or with pseudovacuoles; grey or variously colored.

Reproduction by fragmentation of colony, hormogones, and resting spores.

KEY TO THE GENERA.

Trichomes solitary or interwoven to form flocculent masses of small size .... ANABAENA

Trichomes laterally joined to form small plate-like colonies.... APHANIZOMENON

ANABAENA Bory 1822.

Filaments free-floating or forming a delicate mucous layer; solitary or in floccose colonies. Trichomes the same thickness throughout or slightly attenuated at the apices; straight, spiral, circinate, or aggregated without definite order. Sheaths delicate or copious, gelatinous, hyaline, homogeneous, rarely confluent with the colonial envelope; sometimes entirely lacking. Cells spherical to barrel-shaped, rarely cylindrical, never disciform. Heterocysts approximately spherical, always intercalary, generally solitary and with several in a filament. Spores variously shaped, contiguous to or remote from heterocysts, solitary or in short catenate series. Cell contents homogeneous, granulose, or with numerous pseudovacuoles; grey, blue-green, or variously colored.

Reproduction by hormogones and spores.

KEY TO THE SPECIES.

Filaments straight or slightly curved.
  Spores spherical .... A. PLANCTONICA
  Spores longer than they are broad.
    Length of spores not over twice their breadth.
      Breadth of spores less than twice that of vegetative cells.
        Spores oval to elliptic in optical cross section.
          Vegetative cells less than 8 μ broad.................... A. AFFINIS
          Vegetative cells over 10 μ broad...................... A. LIMNETICA
        Spores hexagonal in optical cross section.............. A. MACROSPORA
      Breadth of spores more than twice the vegetative cells..... A. LEVANDERI
    Length of spores at least three times their breadth.......... A. BORNETIANA
Filaments spirally, circinately, or irregularly twisted.
  Trichomes solitary, not forming floccose masses.
    Curving of trichome a definite spiral...................... A. SPIROIDES
    Curving of trichome not a definite spiral.
    Length of spore at least four times breadth..A. AUGSTUMALIS I. C. MARCHIA
      Length of spore about twice the breadth................. A. CIRCINALIS
  Trichomes forming a floccose mass.
    Spores in a mass at center of colony.................... A. LEMMERMANNI
    Spores not in a mass at the center of the colony.
      Trichomes loosely associated, individual filaments sometimes in spiral turns.............................. A. CIRCINALIS
      Trichomes greatly and irregularly twisted to form compact masses .... A. FLOS-AQUAE

ANABAENA PLANCTONICA Brunnthaler. Pl. 8, Fig. 2.

Sitzenbr. d. kais. Ak. d. Wiss. Wien **112**, Abt. **1**: 292. 1903.

Filaments free-floating, solitary, straight, or slightly bent. Trichomes enclosed by a copious, hyaline, very gelatinous sheath. Vegetative cells spherical, hemispherical, or broadly elliptic. Heterocysts spherical. Spores solitary, remote from or adjacent to heterocysts, spherical or slightly elongate, somewhat broader than the vegetative cells. Cell contents rarely homogeneous, usually with numerous pseudovacuoles. (Euplanktont).

Vegetative cells 9–15 μ broad; heterocysts 12–14 μ broad; spores 12.5–20 μ broad, 12.5–30 μ long.

Birch (1) (cc), Center (cc), Cranberry (cc), Devils (2) (ss), Rock (cc).

Many juvenile spores of other species are spherical but the spores of this species never become appreciably elongated.

ANABAENA AFFINIS Lemmermann. Pl. 8, Fig. 7.

Abh. Nat. Ver. Bremen **14**: 261, *pl.* 1, *figs.* 12–13, 16–17. 1897.

Filaments solitary, free-floating, straight or slightly flexed; with a wide, homogeneous, hyaline, faintly discernible gelatinous envelope. Mature vegetative cells generally spherical; rarely with homogeneous contents, usually with pseudovacuoles. Heterocysts spherical. Spores generally remote from heterocysts, short or elongate cylindrical, with rounded ends and covered with a thick, smooth wall. (Facultative planktont).

Cells 5–6 μ broad; heterocysts 7.5–10 μ broad; spores 9.5–12 μ broad, 17–26 μ long.

Lindy (rr), Lost (rr), North Turtle (c), Pardee (sss), Poplar (ss), Sand (1) (s).

ANABAENA LIMNETICA G. M. Smith. Pl. 8, Fig. 8.

Bull. Torr. Bot. Cl. **43**: 481, *pl.* 26, *fig.* 27. 1916.

Filaments free-floating, solitary, straight or slightly flexed; with a copious, homogeneous, hyaline sheath. Vegetative cells spherical or with poles flattened. Heterocysts spherical. Spores smooth-walled, solitary, broadly ovoid, remote from or adjacent to the heterocysts. Cell contents rarely homogeneous, generally with numerous pseudovacuoles. (Euplanktont).

Cells 12–15 μ broad; heterocysts 10–14 μ broad; spores 17–20 μ broad, 20–30 μ long.

Bear Trap (ss), Birch (1) (cc), Soft (ss), Tamarack (cc), White Ash (ss).

ANABAENA MACROSPORA var. ROBUSTA Lemmermann. Pl. 8, Fig. 6.

Bot. Cent. **76**: 154. 1898.

Filaments solitary, free-floating, straight or slightly bent, frequently attaining a considerable length; with a copious, hyaline, homogeneous, gelatinous envelope. Mature vegetative cells spherical or subspherical with flattened ends. Heterocysts spherical. Spores solitary or in pairs, adjacent to or remote from heterocysts, cylindrical with conical ends and hexagonal in optical cross section. Spore wall smooth and thick. Cells containing numerous pseudovacuoles. (Euplanktont).

Cells 12–16 μ broad, 9–12 μ long; heterocysts 12–16 μ broad; spores 17–19 μ broad, 30–34 μ long.

Camp (cc), Squirrel (ss).

ANABAENA LEVANDERI Lemmermann. Pl. 8, Figs. 4–5.

Ber. d. D. Bot. Ges. **24**: 536. 1906.

Filaments solitary, free-floating, straight or slightly flexed, not enclosed by a gelatinous sheath. Trichomes cylindrical, slightly constricted at the cross walls of the cells. Vegetative cells twice as long as they are broad, cylindrical, with rounded ends. Heterocysts hyaline, spherical or ellipsoid. Spores solitary, adjacent to or remote from the heterocysts, twice as broad as the vegetative cells, ellipsoid in shape. Cell contents rarely homogeneous and pale blue-green to grey; usually with numerous pseudovacuoles. (Euplanktont).

Cells 4–6 μ broad, 11–33 μ long; heterocysts 6.5–8 μ broad, 6.5–9.5 μ long; spores 8–15 μ broad, 19–45 μ long.

Chain (sss).

The dimensions of all specimens found in Chain lake have been near the minimal limits set by Lemmermann.

ANABAENA BORNETIANA Collins. Pl. 8, Fig. 3.

Erythrea **4**: 120. 1896.

Filaments free-floating, straight or slightly curved. Vegetative cells spherical or slightly shorter than they are broad, with homogeneous pale blue-green contents. Sheaths of trichomes very delicate. Heterocysts spherical to elongate. Spores elongate cylindrical, with rounded ends, adjacent to and on both sides of, or remote from the heterocysts. (Tychoplanktont).

Vegetative cells 12 μ broad; heterocysts 13–14 μ broad, 13–20 μ long; spores 15–20 μ broad, 50–90 μ long.

George (rr).

Only a few filaments of this species were found in George lake and these contained but few spores which were not always adjacent to the heterocysts. The species may, therefore, not be *A. Bornetiana* for Collins states that the spores are always adjacent to the heterocysts.

ANABAENA AUGSTUMALIS var. MARCHICA Lemmermann. Pl. 8, Figs. 9–10.

Forschungsbr. a. d. Biol. Stat. zu Plön **12**: 147. 1905.

Filaments solitary, free-floating, more or less bent but not circinate, without a gelatinous envelope. Vegetative cells spherical, broadly ellipsoid or cylindrical; with numerous pseudovacuoles. Heterocysts spherical or slightly elongate. Spores remote from heterocysts, elongate cylindrical, with rounded ends. Spore wall delicate, smooth, and hyaline. (Facultative planktont).

Cells 5–7 μ broad, 5–9.5 μ long; heterocysts 8–9.5 μ broad, 8–14 μ long; spores 9.5–12 μ broad, 40–70 μ long.

Bone (ss), Center (cc).

ANABAENA SPIROIDES var. CRASSA Lemmermann. Pl. **9**, Figs. **1–3.**

Bot. Cent. **76**: 155. 1898; Kryptogamenfl. d. Mark. Brandenburg **3**, Algen **1**: 188, 159, *figs.* 15–16. 1907.

Filaments solitary, free-floating; forming a regular, evenly spaced spiral enclosed by a very transparent, homogeneous, gelatinous envelope. Cells spherical or with flattened ends. Heterocysts spherical, hyaline or pale yellow. Spores solitary, remote from the heterocysts, generally broadly ovoid. Cell contents rarely homogeneous, usually with numerous pseudovacuoles. (Euplanktont).

Cells 11–15 μ broad; heterocysts 10–17 μ broad; spores 20–25 μ broad, 27–42 (generally 27–31) μ long. Spirals 50–60 μ broad and 45–55 μ between turns.

Big Butternut (cc), Big St. Germain (cc), Granite (cc), Hooker (c), Kegonsa (a), Long (2) (sss), Little Butternut (cc), Mendota (cc), Monona (aa), North Twin (sss), Round (3) (ss), South Turtle (c), Vermilion (ccc).

ANABAENA CIRCINALIS (Kützing) Rabenhorst. Pl. **9**, Figs. **4–5.**

Algen No. **209**. 1852; Phillips, Grevillea **9**: 5, *pl.* 134, *figs.* e–g. 1880; Bornet et Flahault, Ann. Sci. Nat. **7** Sér. Bot. **7**: 230. 1888.

Anabaena Hassallii (Kützing) Wittrock. Bot. Not. **1882**: 58. 1882.

Filaments free-floating, rarely solitary, curved or sigmoid; usually greatly and irregularly twisted to form floccose masses. Filaments with or without a copious, hyaline, gelatinous envelope. Cells spherical or appreciably flattened at the poles, generally with numerous pseudovacuoles. Heterocysts spherical to subspherical. Spores usually remote from the heterocysts, cylindrical, with rounded ends when mature, frequently curved and with smooth walls. (Euplanktont).

Cells 8–14 μ broad; heterocysts 8–10 μ broad; spores 16–18 μ broad, 26–30 μ long when mature.

Center (ss), Chain (sss), Lac la Belle (rr), Mercer (rr), Nagawicka (rr), Pike (r), Upper Nashotah (rr), Wapagossett (sss), Winnebago (ss).

This species differs from *A. flos-aquae* in the loose looping of the trichomes, the larger vegetative cells and the broader spores.

var. MACROSPORA (Wittrock) de Toni. Pl. **9**, Fig. **6**; Pl. **10**, Fig. **1**.

Sylloge algarum **5**: 445. 1907.

Anabaena Hassallii var. *macrospora* Wittrock, Bot. Not. **1882**: 59, *figs*. 7-9. 1882.

Vegetative cells somewhat smaller but with much narrower and proportionally longer spores.

Cells 7–8 μ broad; heterocysts 7.5–10 μ broad; spores 9–10.5 μ broad, 28–42 μ long.

Camp (sss).

ANABAENA FLOS-AQUAE (Lyngbye) de Brebisson. Pl. **10**, Figs. **2–4**.

Mém. de la soc. Acad. de Falaise **1835**: 36. 1836; Kützing, Tabulae Phycol. **1**: 51, *pl.* 94, *fig.* 4. 1848; Bornet et Flahault, Ann. Sci. Nat. **7** Sér. Bot. **7**: 228. 1888.

Filaments free-floating, densely coiled, irregularly twisted and interwoven to form a floccose colony. Cells spherical or cylindrical with convex sides; generally containing numerous pseudovacuoles. Heterocysts broader and longer than the vegetative cells. Spores cylindrical, with rounded ends, usually curved, solitary or in short catenulate series; generally adjacent to, rarely remote from the heterocysts. (Euplanktont).

Vegetative cells 4–8 (generally 5.5) μ broad; heterocysts 6–9 μ broad; spores 7–13 μ broad, 20–50 μ long.

Bear (1) (cc), Birch Island (r), Browns (cc), Camp (ss), Catfish (ss), Catherine (rr), Cochrane (sss), Crab (r), Crawling Stone (rrr), Crooked (sss), Des Moines (rr), Devils (1) (r), Eagle (sss), Fish (rr), Grassy (r), Green (2) (sss), Half Moon (rr), Hooker (r), Horse (rrr), Jag (rr), Jennie (ss), Kegonsa (rr), Lac Court Oreilles (ss), Lac la Belle (r), Little Crooked (rr), Little Wood (r), Long (1) (rr), Long (3) (rr), Lower Nashotah (rr), Lynx (sss), Mary (rr), Mendota (c), Mercer (rr), Mill (rr) Mud (2) (c), Nancy (s), Oconomowoc (r), Okauchee (rr), Palmer (rr), Pine Tree (rr), Poplar (rr), Round (1) (rr), Round (3) (r), Rudolph (rrr), Sanford (c), Shell (rrr), Sishebogema (rr), South Crab (rr), Squirrel (rr), Upper Nashotah (r), Upper Nemahbin (rr), Waubesa (c), Wildcat (sss), Wolf (r).

var. TRELEASII Bornet et Flahault. Pl. **10**, Figs. **5–7**.

Ann. Sci. Nat. **7** Sér. Bot. **7**: 230. 1888.

Anabaena mendotae Trelease, Trans. Wis. Acad. Sci., Arts, & Lett. **7**: 123, *pl.* 10, *fig.* 5. 1889.

Cells smaller and with spores somewhat narrower and proportionally longer. (Euplanktont).

Vegetative cells 3–5 μ broad, 4.5–12 μ long; heterocysts 4.5–6 μ broad, 6–8 μ long; spores 6–8 μ broad, 20–40 μ long.

Mendota (aa), Waubesa (s).

ANABAENA LEMMERMANNI P. Richter. Pl. **10**, Fig. **8**; Pl. **11**, Fig. **1**.

Forschungsbr. a. d. Biol. Stat. zu Plön **10**: 153. 1903.

Anabaena flos-aquae Klebahn non (Lyngbye) de Breb., Flora **80**: 267, *pl.* 4, *figs.* 21–22. 1895.

Filaments free-floating; irregularly or circinately interwoven to form a spherical or broadly ellipsoid mass, with filaments frequently radiating in loops from a common center. Vegetative cells spherical or slightly elongate, usually containing numerous pseudovacuoles. Heterocysts spherical, approximately the same size as the vegetative cells. Spores cylindrical, with rounded ends, with inner side straight and the outer convex; rarely remote from heterocysts, generally forming short catenate series on both sides of heterocysts and lying in a dense mass at the center of the colony. The spore masses and heterocysts frequently persisting long after the disappearance of the vegetative cells. (Euplanktont).

Vegetative cells 5.5–7 μ broad, 5–8 μ long; heterocysts 6–7.5 μ broad; spores 8–11 μ broad, 19–33 μ long.

Crawling Stone (rr), Diamond (r), Kawaguesaga (rr), Long (rr), Marl (r), Oconomowoc (r), Paddock (cc), Palmer (r), Pardee (ss), Rainbow (2) (rr), Reserve (r), Rice (1) (s), Sand (2) (rr), Taylor (ss), Trout (s).

The species differs from *A. flos-aquae* in the regular looping of the trichomes from a common center and the development of a large mass of spores at the center of the colony. The persistence of the spores after the disappearance of the vegetative cells is very common and certain of the records for the species are based upon observation of the spore masses only.

APHANIZOMENON Morren 1838.

Trichomes straight slightly flexed, or curved; in lateral contact with one another to form small, macroscopic, free-floating, feathery, plate or scale-like opaque colonies. Trichomes destitute of a gelatinous sheath, sometimes constricted at cross walls of cells, and slightly attenuated at the apices. Cells cylindrical, with flattened ends. Heterocysts intercalary, cylindrical. Spores elongate, cylindrical and with flattened ends; solitary, adjacent to or remote from the heterocysts.

Reproduction by the formation of hormogones and resting spores.

APHANIZOMENON FLOS-AQUAE (L) Ralfs. Pl. **11**, Figs. **2–4**.

Ann. & Mag. of Nat. Hist. **2** Ser. **5**: 340, *pl.* 9, *fig.* 6. 1850; Klebahn, Flora **80**: 271, *pl.* 4, *figs.* 27–30. 1895.

Trichomes straight or slightly curved, laterally joined to form a flat flake- or scale-like plate. Cells cylindrical, with flattened ends, about three times as long as they are broad, generally with numerous pseudovacuoles. Heterocysts cylindrical, with straight or slightly convex sides, without a distinct polar nodule. Spores elongate, cylindrical, with rounded or somewhat flattened ends, generally some distance from the heterocysts. (Euplanktont).

Cells 4–6 μ broad, 5–15 μ long; heterocysts 5–7 μ broad, 7–20 μ long; spores 6–8 μ broad, 35–80 μ long.

Amicoy (aa), Balsam (ss), Bear (3) (rr), Beaverdam (rr), Big (r), Big Butternut (rr), Birch (1) (c), Bone (rr), Camp (r), Catfish (sss), Center (rrr), Chetac (aaa), Cranberry (ss), Crawling Stone (rrr), Deer (rrr), Dunham (rr), Found (rr), Fowler (rr), Granite (a), Green (2) (ss), Half Moon (sss), Hillman (sss), Hooker (cc), Horse (rrr), Horseshoe (1) (rr), Island (c), Kawaguesaga (ccc), Kegonsa (cc), Lac la Belle (rr), Little Bass (2) (rr), Little Butternut (s), Little Rice (1) (s), Little Wood (rr), Long (3) (rr), Lost (ss), Loveless (ss), Lower Nashotah (rr), Mendota (s), Mud (3) (rr), Nagawicka (rr), North Twin (rrr), Oconomowoc (rr), Okauchee (rr), Pardee (cc), Pewaukee (rr), Pike (rr), Plum (r), Pokegama (2) (rrr), Pokegama (3) (aaa), Poplar (rrr), Prairie (aaa), Reserve (rr), Rice (1) (ss), Round (1) (sss), Round (2) (ss), St. Croix (ccc), Sand (1) (r), Sand (3) (r), Sand (4) (r), Sanford (sss), Shell (rrr), South Turtle (sss), Squirrel (a), Streitor (rr), Tied Canoe (sss), Upper Nashotah (rr), Upper Nemahbin (r), Upper Turtle (rr), Vermilion (r), Waubesa (s), White Ash (rr), Whitefish (rr), White Sand (ss), Winnebago (r).

When the alga occurs in any quantity the lake appears to be filled with small pieces of macerated grass and it is one of the few plankton genera that can be determined without the aid of a microscope. The colony is so opague that the heterocysts cannot be seen while the colony is intact but after a mount has been standing a few minutes the colonies begin to dissociate into the component filaments and observation of the heterocysts and spores is then an easy matter. The colonies are frequently sterile and it is generally only during the colder months of the year that there is a formation of spores.

Family *RIVULARIACEAE*.

Filaments rarely solitary, generally forming sessile or free-floating spherical, hemispherical, or amorphous colonies. Trichomes gradually attenuated from a broad basal cell to a hair-like apical cell. Sheaths copious, homogeneous or lamellated, hyaline or colored, firm or gelatinous; enclosing one or more trichomes. Cells hemispherical, spherical, or disciform with convex sides at the basal portion of the trichome, elongate cylindrical in the apical region. Cell contents variously colored; homogeneous, granulose, or with numerous pseudovacuoles. Trichomes simple or with "false branching". Heterocysts solitary; usually at base of trichomes, rarely intercalary. Spores, when present, solitary and adjacent to heterocysts.

Reproduction by formation of trichomes and spores.

GLOEOTRICHIA J. G. Agardh 1842.

Filaments always in colonies; free-floating or sessile; spherical to hemispherical, tough and leathery to gelatinous, hyaline or colored. Trichomes attenuated to fine hair-like processes at the distal end, enclosed by gelatinous sheaths that are distinct at the base and confluent at the apex. Filaments containing one trichome, but at times showing false branching. Heterocysts spherical to hemispherical, solitary; at the base of the trichomes. Spores cylindrical, solitary, always adjacent to the heterocysts. Vegetative cells convexly disciform to short cylindrical at base of trichomes, elongate cylindrical with straight sides at apex of trichomes.

Reproduction by hormogones and resting spores.

GLOEOTRICHIA ECHINULATA (J. E. Smith) P. Richter. Pl. **11**, Figs. **5–6.**

Forschungsbr. a. d. Biol. Stat. zu Plön **2**: 31, *figs.* 1–8. 1894.

Conferva echinulata J. E. Smith, Eng. Bot., *pl.* 1378. 1804.

Rivularia echinulata J. E. Smith, Eng. Bot. index; P. Richter, Forschungsbr. a. d. Biol. Stat. zu Plön **2**: 45. 1894; Lemmermann, Kryptogamenfl. d. Mark Brandenburg **3**, Algen **1**: 255, 245, *figs.* 1, 6. 1907.

Gloeotrichia pisum Bornet et Flahault (p.p.) non Thuret, Bull. Soc. Bot. France **31**: 80. 1884; Ann. Sci. Nat. **7** Sér. Bot. **4**: 366. 1886; Trelease, Trans. Wis. Acad. Sci., Arts, & Lett. **7**: 124–125 (with fig.) 1889.

Colonies always free-floating, spherical; up to 2 mm. in diameter. Filaments radially arranged about a common center but becoming irregular in arrangement on slight pressure. Sheaths hyaline, homogeneous, distinct at base, apex frequently confluent with colonial envelope. Trichomes broad at base, attenuated to long, colorless, hair-like processes. Lowest vegetative cell of trichome hemispherical; those a short distance from base disciform with markedly convex sides; those in middle of trichome cylindrical; while the cells in the apical region are elongate cylindrical. Basal cells completely filled with cytoplasm, apical cells with cytoplasm restricted to the median portion. Cell contents rarely homogeneous, usually with numerous pseudovacuoles. Heterocysts at base of trichome, spherical. Spores, when present, adjacent to heterocysts, elongate cylindrical with rounded ends.

Cells at base of trichome 8–10 μ broad; at apex 1–2 μ broad; heterocysts 7–10 μ broad; spores 8–10 μ broad, 40–54 μ long.

Bass (3) (rr), Bear (3) (rr), Bear Trap (rr), Beaverdam (rr), Birch (2) (ss), Birch Island (r), Bone (rr), Carroll (rr), Catherine (sss), Chetac (ss),

Clare (ss), Crawling Stone (rrr), Des Moines (cc), Diamond (ss), Duck (rr), Dummy (rr), Dutchmans (ss), Fish (ss), Fowler (sss), Grindstone (rr), Half Moon (ss), Horseshoe (1) (ss), Horseshoe (2) (rr), Howey (rr), Island (ss), Jennie (rr), Kegonsa (aa), Kimball (rr), Lac Court Oreilles (ss), Lake of the Woods (r), Lindy (rr), Long (1) (sss), Loon (r), Mendota (s), Monona (cc), Nancy (ss), North Twin (r), Mercer (rrr), Middle McKenzie (ss), Mud (3) (aa), Oconomowoc (rr), Owen (rrr), Razorback (aa), Reserve (rr), St. Croix (rr), Sand (1) (rr), Sand (3) (rr), Shell (rr), Soft (rr), Trout (cc), Waubesa (cc), Whitefish (sss), Wild Goose (rr), Winnebago (r).

A bloom of this alga can be recognized at once by the minute seed-like colonies surrounded by an aureole of whitish threads. It was first noted by Davis in England in 1804 and the plant figured and described by Smith in his English Botany. The various references to the alga before the recognition of a definite plankton flora have been compiled by Bornet and Flahault as well as P. Richter. Since the study of the plankton flora by phycologists the alga has been found in the lakes of most European countries. Although there seems to be but a single species of the Rivulariaceae found free-floating in lakes and ponds the nomenclatorial treatment of this species is quite varied. Bornet and Flahault affirm that it is the common *Gloeotrichia pisum,* which ordinarily grows attached to submerged plants along the shore, that has broken away and become free-floating. They hold that *Rivularia echinulata* J. E. Smith (whose dimensions are the same as those of *G. pisum*) is nothing but a free-floating *G. pisum.* In my observations on the algae of our lakes there is no correlation between the appearance of *G. pisum* and *G. echinulata* in the same lake in Wisconsin. In the region of Madison, *G. pisum* does not appear until the middle of July and yet at one time there was a heavy bloom of *G. echinulata* on the 28th of June, a time when there was no *G. pisum* along the shore. I am convinced that the sessile *Gloeotrichia* is not the one found in the plankton and that Richter is quite right in reviving the old specific name *echinulata.* Some investigators hold that *Gloeotrichia* is but a sub-genus of *Rivularia,* but I have followed Bornet and Flahault in considering the two distinct. This distinction is based upon the occurrence of spores in *Gloeotrichia* and a lack of spores in *Rivularia.*

Class **PHAEOPHYCEAE.**

Plants usually of macroscopic size and definitely shaped. Cells variously shaped and containing brown chromatophores in which a brown pigment, phaeophycin, masks the chlorophyll. Cell wall not conspicuously impregnated with silica and cells generally immobile in the vegetative phases.

Asexual reproduction by vegetative cell division, aplanospores, and zoospores.

Sexual reproduction by isogamous zoogametes; or heterogamous with a fusion of zoogametes of unequal size, or a union of a large immobile oosphere and a small motile antherozooid.

All of the complex non-motile members of this class are marine and only unicellular or simple colonial forms are found in fresh waters. These freshwater forms have the motile phase dominant, as a rule, and form the subclass Chrysomonadineae of the Phaeophyceae. Many place all of the colorless or brown chromatophored motile unicellular organisms in a class by themselves, the Flagellatae, but, as I have stated elsewhere, the group of the Chrysomonadineae seems sufficiently related to the Phaeophyceae to warrant uniting the two.

Sub-class **CHRYSOMONADINEAE.**

Cells with one or more definite, disciform, laminate, or reticulate, golden-brown chromatophores. Cells free-swimming or sessile, solitary or in colonies of definite to indefinite shape. Life cycle generally with the motile phase dominant but sometimes developing into amorphous or rhizopodial palmella stages. Cells either naked or with a gelatinous to definitely constituted periplast. Individual cells with 1–2 anterior or lateral cilia of equal or unequal length; 1–10 contractile vacuoles, generally at the anterior end; with or without an eyespot; and with one or more chromatophores. Symmetry of cells bilateral, not interrupted by a lateral or longitudinal grooving. Nutrition holophytic or saprophytic; assimilation product leucosin or fat, never starch.

Asexual reproduction by longitudinal or transverse division of cells. Resting stages with heavy walls known in many genera.

Sexual reproduction as yet unknown in the sub-class.

KEY TO THE DIVISIONS.

Motile phase dominant, rhizopodial or palmella phase rare. EUCHRYSOMONADINAE
Immobile phase dominant, motile phase known or unknown.
 Cells spherical to ovoid, embedded in a gelatinous envelope... HYDRURINAE
 Cells with pseudopodial processes.................... RHIZOCHRYSIDINAE

Division **EUCHRYSOMONADINAE.**

Cell structure and organization as noted in the description of the sub-class. Ciliate phase dominating the life history. Palmella or rhizopodial phases known but not the general rule.

KEY TO THE ORDERS.

Motile phase with one flagellum.............................. CHROMULINALES
Motile phase with two flagella.
 Flagella of the same length.............................. ISOCHRYSIDALES
 Flagella of unequal length.............................. OCHROMONADALES

Order **CHROMULINALES.**

Cells generally solitary, rarely colonial; free-floating or sessile. Motile phase with a single apical flagellum. Chromatophores reticulate, laminate, or disciform; generally two in each cell. Apical contractile vacuoles present or lacking. Cells either naked or with a definite, homogeneous or sculptured periplast.

Reproduction by vegetative division of cells. Thick-walled resting stages known in certain genera.

KEY TO THE FAMILIES.

Cells naked .... EUCHROMULINACEAE
Cells with a differentiated periplast .... MALLOMONADACEAE

Family *EUCHROMULINACEAE.*

Cells naked, without a differentiation of the periplast. Cells with sharply defined disciform to laminate chromatophores, and without contractile vacuoles. Flagellum single.

There are several genera with many species in the family but only one of them, *Chrysamoeba*, has been noted in the plankton of the state.

CHRYSAMOEBA Klebs 1893.

Cells solitary, free-floating, with a mono-flagellate motile phase and an amoeboid non-motile phase. Motile phase broadly ovoid, naked, with two elongate, laminate, golden-brown chromatophores; a large non-contractile vacuole, 2–3 small contractile vacuoles; and a single flagellum slightly longer than the cell. Eye-spot lacking.

Non-motile phase solitary or in small clumps, with several simple or branching pseudopodial processes extending in all directions. Chromatophores and vacuoles as in the motile phase.

Chrysamoeba radians Klebs. Pl. 12, Fig. 1.

Zeits. f. wiss. Zool. **55**: 406, *pl.* 18, *figs.* 1a–1b. 1893.

Specific characters as described above. (Euplanktont).

Motile phase 8–10 μ broad, 12–15 μ long; amoeboid phase (without processes) 12–17 μ broad, (with processes) 35–40 μ broad.

Round (4) (rr), Taylor (rr).

Family *MALLOMONADACEAE*.

Cells always free-swimming, solitary or colonial, never metamorphosing into a palmella or rhizopodial phase. Periplast of cells or colonial envelope with numerous small plates of silica that have long, siliceous, needle-like processes. Cells ovoid to ellipsoid, with one flagellum, a number of contractile vacuoles and two long laminate chromatophores.

KEY TO THE GENERA.

Cells solitary .... Mallomonas
Cells in spherical colonies .... Chrysosphaerella

MALLOMONAS Perty 1852.

Cells always solitary, free-swimming, with a single flagellum at the anterior end. Periplast of cells with small circular or angular siliceous discs covering the entire surface of the cell or restricted to one or both ends. Scales frequently arranged in regular transverse or diagonal rows, produced into a single, long, simple or toothed, needle-like setae. Cells containing one or more contractile vacuoles that may be apical, median, or basal in position. Chromatophores two, parietal, laminate, and golden-brown in color. Assimilation product leucosin.

Reproduction by vegetative cell division parallel to the long axis. Akinetes known in many species.

KEY TO THE SPECIES.

Setae restricted to one end of the cell.
Apical end with setae.
Cells cylindrical .... M. ALPINA
Cells ovoid .... M. TONSURATA
Basal end with setae .... M. PRODUCTA
Setae covering whole cell .... M. CAUDATA

MALLOMONAS ALPINA Pascher & Ruttner. Pl. 12, Fig. 4.

in Pascher, D. Süsswasserfl. Deutschl., österr., u. d. Schw. **2**, Flagellatae **2**: 36, *fig.* 58A. 1913.

Cells ellipsoid to elongate cylindrical, with poles rounded or pointed. Scales rhomboidal, in diagonal rows; with a single, long, straight or curved seta. Setae confined to anterior end of cells. (Euplanktont).

Cells 8–12 μ broad, 25–45 μ long.

Camp (r).

MALLOMONAS PRODUCTA (Zacharias) Iwanoff. Pl. 12, Fig. 2.

Bull. de l'Acad. Imp. d. Sc. de St. Pétersbourg **5** Sér., **11**: 250, *pl.* A, *figs.* 4–6. 1899.

Cells cylindrical, with straight or slightly convex sides, ends rounded and tapering. Scales in regular transverse to diagonal rows, diamond-shaped or laterally elongate. Setae straight or bent, confined to basal portion of cell. Chromatophores two, golden-brown, laminate, and parietal. (Euplanktont).

Cells 9–13 μ broad, 40–70 μ long.

Catfish (sss), Cranberry (ss), Crawling Stone (sss), Lac la Belle (ss), Long (1) (ss), Otter (r), South Turtle (rrr), Upper Nashotah (rr).

var. MARCHICA Lemmermann. Pl. 12, Fig. 3.

Zeits. f. Fischeri **1903**: 106. 1903.

Cells shorter and relatively broader. Setae sometimes at both ends of cells. (Euplanktont).

Cells 11–12 μ broad, 21–26 μ long.

Nancy (r), Oconomowoc (ss), Tamarack (ss).

MALLOMONAS TONSURATA Teiling. Pl. 12, Fig. 5.

Sv. Bot. Tids. **6**: 277, *fig.* 3. 1912.

Cells ellipsoid, with posterior end rounded and anterior end pointed. Setae confined to anterior third of cell, fairly short and somewhat curved. Scales not regularly arranged. (Euplanktont).

Cells 7–8 μ broad, about 18 μ long.

Waupaca (rr).

MALLOMONAS CAUDATA Iwanoff. Pl. **12**, Fig. **6**.

Bull. de l'Acad. Imp. d. Sc. de St. Pétersbourg **5** Sér., **11**: 250, *Pl.* B, *figs.* 1–8. 1899.

Cells broadly ovoid, with posterior end narrower than the anterior. Scales oval, arranged in transverse series. Setae long, covering the entire surface of the cell and frequently with minute teeth-like projections at the distal end. Cells with two long, parietal, laminate, chromatophores. (Euplanktont).

Cells 12–30 μ broad, 40–85 μ long.

Beasley (rr), Birch (1) (rrr), Browns (rr), Carson (rrr), Catherine (rr), Clare (r), Clear Crooked (rr), Crab (rr), Crawling Stone (rrr), Diamond (rrr), Hill (ss), Jennie (rr), Kitten (sss), Lac la Belle (ss), Long (3) (ss), Long (4) (rr), Lynx (rr), Mud (2) (ss), No Mans (s), North Turtle (r), Okauchee (rr), Otter (sss), Pokegama (1) (rr), Pokegama (2) (r), Red Bass (s), Rock (cc), Round (1) (sss), Rudolph (sss), St. Croix (rrr), Sand (4) (sss), Sanford (sss), South Crab (rr), Wildcat (sss), Wild Goose (rr), Wolf (rr).

CHRYSOSPHAERELLA Lauterborn 1896.

Cells ellipsoid to pyriform, in spherical colonies of microscopic size that are partially embedded in a spherical, hyaline, gelatinous envelope containing numerous, tangentially placed, minute plates of silica. Anterior ends of cells with two short, vase-shaped projections, each bearing a single, long, straight, siliceous seta of some length. Cells with one anterior flagellum; two elongate, parietal, golden-brown chromatophores; and several vacuoles.

Reproduction by vegetative cell division. Resting akinetes unknown.

CHRYSOSPHAERELLA LONGISPINA Lauterborn. Pl. **12**, Figs. **7–8**.

Zool. Anz. **19**: 16. 1896; Zeits. f. wiss. Zool. **65**: 381, *pl.* 18, *figs.* 12–16. 1899.

Specific characters as described above. (Euplanktont).

Average breadth of cells 9 μ, average length 15 μ. Colonies 50–250 μ in. diam.

Bettine (rr), Blue Gill (ss), Boulder (rrr), Canteen (r), Elizabeth (rr), Hill (aa), Lac la Belle (rr), Long (4) (rr), Lynx (r), Marl (rr), Mary (ss), No Mans (rr), Oconomowoc (ss), Okauchee (s), Otter (ss), Rainbow (ss), Round (4) (rr), Rudolph (sss), Silver (2) (ss), South Turtle (cc), Upper Nemahbin (ss), Wild Goose (ccc).

Order **ISOCHRYSIDALES**.

Cells free-floating or sessile, solitary or colonial, naked or with a definite homogeneous or sculptured periplast. Motile phase always present, palmella or rhizopodial phase lacking. Cells variously shaped, always with two apical cilia of equal length; generally with two laminate, parietal, golden-brown chromatophores and without an eyespot.

The only family represented in the flora of the state is that of the Euhymenomonadaceae.

Family *EUHYMENOMONADACEAE.*

Differing from other families of the order in having a definite sculptured periplast and an anterior vacuolar system. *Synura* the sole representative.

SYNURA Ehrenberg 1838.

Cells radially arranged in spherical to oblong-ovoid free-swimming colonies that are not enclosed by a gelatinous sheath. Individual cells broadly pyriform, with the narrower end towards the center of the colony. Outer portion of cells firm, with small spines, reticulations, or combinations of the two. Cells with two long anterior cilia of equal length, several vacuoles, two long laminate parietal chromatophores, and without an eyespot.

Reproduction by division of cells and liberation of cell contents to form amoeboid cells. Akinete formation also known.

SYNURA UVELLA Ehrenberg. Pl. **12,** Figs. **9–10.**

Infusionsthierchen 61, *pl.* 3, *figs.* 9^1–9^5. 1838.

Exterior of cells covered with fine spines that are generally limited to the anterior portion. (Facultative Planktont).

Cells 8–17 μ broad, 20–35 μ broad. Colonies up to 350 μ in diam.

Camp (ss), Canteen (cc), Crawling Stone (rrr), Fowler (cc), Jenny (rr), Kawaguesaga (rr), Little Bear (rrr), Little Crooked (rr), Long (3) (r), Lower Nemahbin (rr), Lynx (r), Mendota (ss), Mercer (rr), Mill (ccc), Mud (2) (rr), Nagawicka (cc), Nancy (cc), Nicaboyne (rr), No Mans (cc), North Turtle (rr), Oconomowoc (s), Palmer (c), Pardee (sss), Pewaukee (sss), Pope (rrr), Rainbow (1) (rr), Red Bass (ss), Round (1) (sss), Rudolph (sss), Sand (4) (rr), Sanford (ss), South Turtle (rr).

Order **OCHROMONADALES.**

Cells solitary or colonial, free-swimming or sessile, not forming palmella or rhizopodial stages. Cells naked, or with a definitely shaped receptacle; two lateral or apical cilia of unequal length, one or two golden-brown disciform to laminate chromatophores; contractile vacuoles; and an eyespot.

KEY TO THE FAMILIES.

Cells naked, at times in a gelatinous envelope but never a definite receptacle .... OCHROMONADACEAE
Cells within a definite receptacle........ DINOBRYACEAE

Family *OCHROMONADACEAE.*

Cells solitary or in colonies, always free-swimming, not enclosed by a definite receptacle but at times with a hyaline, homogeneous, gelatinous envelope. Cells with two apical cilia of unequal length, one or two golden-brown chromatophores, an eyespot and contractile vacuoles.

Uroglenopsis is the only representative in the local flora.

UROGLENOPSIS Lemmermann 1899.

Colonies free-swimming, with the cells embedded in a single layer at the periphery of a copious, hyaline, homogeneous, gelatinous matrix of spherical to broadly ellipsoid shape. Arrangement of cells radial or somewhat irregular. Cells spherical to narrowly ellipsoid; with two cilia of unequal length; two vacuoles at the anterior end; one or two laminate to disciform, parietal, golden-brown chromatophores; and with or without an eyespot.

Reproduction by vegetative cell division.

UROGLENOPSIS AMERICANA (Calkins) Lemmermann. Pl. **13**, Figs. **1–4**.

Forschungsbr. a. d. Biol. Stat. zu Plön **7**: 107. 1899; Kryptogamenfl. d. Mark Brandenburg **3**, Algen **1**: 449, 444, *figs*. 1–3, 4–6. 1908.

Uroglena americana Calkins, Ann. Rept. Mass. State. Bd. of Health **23**: 655, *pl*. 4, *figs*. 1–4. 1892; Moore, Bot. Gaz. **23**; 105, *pl*. 10, *figs*. 1–11. 1897.

Colonies free-swimming, generally spherical. Cells ellipsoid, with one basal or lateral, parietal, disciform to laminate chromatophore. Eyespot single. Cilia of unequal length, with the longer 2–3 times the length of the cell, the shorter ⅓–½ the length of the cell. (Euplanktont).

Cells 3–7 μ broad, 5–10 μ long. Colonies up to 500 μ in diam.

Beasley (cc), Bettine (ss), Browns (ss), Camp (ss), Carson (rrr), Diamond (rr), Eagle (sss), Fowler (cc), Howey (rr), Lac la Belle (ss), Long (3) (ss), Long (4) (rr), Lower Nemahbin (rr), Lynx (rr), Mendota (ss), Mill (ss), Mud (4) (cc), Nancy (rr), Oconomowoc (ss), Okauchee (rrr), Otter (ss), Paddock (s), Pewaukee (rr), Rainbow (1) (ss), Round (4) (rr), Taylor (ss), Wildcat (ss), Wolf (rr), Youngs (cc).

Family *DINOBRYONACEAE.*

Cells solitary or in colonies, free-floating or sessile. Always attached to the base of a conical to vase-like, colored or hyaline, smooth or sculptured receptacle. Cells with one or two laminate chromatophores, contractile vacuoles, an eyespot, and two cilia of unequal length.

KEY TO THE GENERA.

Receptacles homogeneous, without growth rings.................. DINOBRYON
Receptacles with successive growth rings...................... HYALOBRYON

DINOBRYON Ehrenberg 1835.

Cells free-floating, sessile, or epiphytic; solitary or in colonies; attached to the bottom of conical, campanulate, or cylindrical receptacles with open apices and pointed bases. Colonial species arbuscular, each receptacle with the pointed ends of 1–4 other receptacles in its mouth. Development of branching indefinite. Protoplast spindle-shaped, conical, or ovoid; attached by a stalk to the base of the receptacle. Anterior end of protoplast with two cilia of unequal length that project beyond the open end of the receptacle. Cells with one or two laminate, elongate, parietal, golden-brown chromatophores; several apical, lateral, or basal contractile vacuoles; and one apical eyespot. Assimilation product generally leucosin.

Reproduction by longitudinal division of cell contents to form ovoid daughter cells that migrate to the mouth of the old receptacle or leave it entirely before developing new receptacles. Akinetes known.

The specific limits of colonial plankton species of this genus are very poorly defined and it is almost impossible to draw a sharp line of demarcation between the various species. I have indicated as species the centers around which the variations seem to occur.

KEY TO THE SPECIES.

Cells solitary.
  Cell shape conical, elongate.............................. D. CALICIFORME
  Cells broadly spindle-shaped.............................. D. TABELLARIAE
Cells in colonies.
  Branching of cells markedly divergent..................... D. DIVERGENS
  Branching of cells not sharply divergent.
    Cells elongate conical.
      Sides of receptacles straight......................... D. STIPITATUM
      Sides of receptacles undulate......................... D. BAVARICUM
    Cells campanulate....................................... D. SETULARIA

DINOBRYON CALICIFORME Bachmann. Pl. **13**, Figs. **5–6.**

Arch. f. Hydrob. u. Planktonkde. 3: 82, *fig.* XI[18]. 1908.

Cells solitary; epiphytic upon the gelatinous envelope of various colonial Myxophyceae. Receptacles elongate conical, with a sharply pointed base. Protoplast the same shape as the lower two-thirds of receptacle. Chromatophores two, laminate, parietal, and golden-brown in color. Eyespot single. Cilia not as long as the cell. (Euplanktont).

Receptacles 5–6 μ broad at apex, 30–40 μ long.

Embedded in or on gelatinous envelope of *Coelosphaerium Naegelianum* and *Microcystis aeruginosa.*

Browns (sss).

DINOBRYON TABELLARIAE (Lemmermann) Pascher. Pl. **13**, Figs. **7–9.**

Süsswasserfl. Deutschl., österr. u. d. Schw. **2,** Flagellatae **2:** 66, *figs.* 105a–105b. 1913.

Dinobryon utriculus var. *tabellariae* Lemmermann, Arkiv för Botanik **2,** No. **2:** 119, *pl.* 1, *fig.* 19. 1904.

Cells solitary; epiphytic upon filaments of *Tabellaria.* Receptacle broadly spindle-shaped, with the base produced into a short stipe. Protoplast attached to side of lower portion of receptacle, ovoid in shape and with the usual contents of the genus. (Euplanktont).

Receptacles 7–10 μ broad at center, 4–5 at apex; 18–22 μ long.

Crawling Stone (r), Long (3) (rr), Sand (4) (s).

DINOBRYON BAVARICUM Imhof. Pl. **13**, Fig. **10.**

Zool. Anz. **13:** 484. 1890; Lemmermann, Bot. Not. **1903:** 78, *pl.* 3, *figs.* 3–4. 1903.

Cells in free-floating compact colonies. Receptacles of individual cells elongate-conical. Base of receptacle very much produced with sides almost parallel; expanded portion of receptacle with undulate sides. Seasonal variation pronounced, the receptacles being much longer during the summer. (Euplanktont).

Receptacles 6.5–8.5 μ broad; 45–100 μ long.

George (ss), Harris (rr), Marl (ss), Oconomowoc (ss), Speese (ss), Tied Canoe (ss), White Sand (s).

DINOBRYON STIPITATUM Stein. Pl. 13, Fig. 11.

Der Organismus der Infusionstiere **3**: *pl.* 12, *fig.* 5. 1878; Pascher, D. Süsswasserfl. Deutschl. österr. u. d. Schw. **2**, Flagellatae **2**: 73, *fig.* 118. 1913.

Cells always in narrow, fairly dense, free-floating colonies. Receptacles of individual cells elongate-conical, with bases produced into stipes with nearly parallel sides. Sides of receptacle not undulate. Individual receptacles of colony not divergent from one another. Seasonal variation chiefly in elongation of apical cells of colony. (Facultative Planktont).

Receptacles 6–8 μ broad; 56–96 μ long.

Marl (ss), Oconomowoc (cc), Round (4) (sss), South Turtle (rr).

DINOBRYON SOCIALE Ehrenberg. Pl. 13, Fig. 12.

Abh. d. k. Ak. d. Wiss. Berlin **1833**: 279. 1835; Die Infusionsthierchen 125, *pl.* 8, *fig.* 9. 1839; Pascher, Süsswasserfl. Deutschl. österr. u. d. Schw. **2**, Flagellatae **2**: 73, *figs.* 116–117. 1913.

Cells in free-floating colonies. Individual receptacles somewhat divergent from one another and producing a fairly loose colony. Receptacles conical, sometimes slightly constricted below the open mouth, with bases gradually tapering to a sharp point that is not extended to form a process. (Facultative planktont).

Receptacles 7–8 μ broad, 30–70 μ long.

Beasley (ss), Green (1) (ss), Green (2) (cc), Mill (c), North Turtle (rr), Oconomowoc (cc), Silver (2) (s), Trout (rr).

DINOBRYON SETULARIA Ehrenberg. Pl. 13, Fig. 13; Pl. 14, Fig. 1.

Abh. d. k. Ak. d. Wiss. Berlin **1833**: 280. 1835; Die Infusionsthierchen 124, *pl.* 8, *fig.* 8. 1838.

Cells in free-floating, densely branching colonies. Receptacles cylindrical-campanulate, with the convex bases tapering abruptly to a sharp point. Cells with shorter and broader receptacles than the other species. Seasonal variation not so pronounced as in other species. (Facultative Planktont).

Receptacles 10–14 μ broad, 30–44 μ long.

Devils (1) (ccc), Kawaguesaga (ss), Long (4) (ss), Lost (ss), Lower Nemahbin (ss), Rock (ss), Round (3) (ss), Winnebago (rr).

Dinobryon divergens Imhof. Pl. 14, Fig. 2.

Jahresbr. d. Naturf. Ges. Graubündens N. F. **30**: 134. 1887.

Dinobryon cylindricum var. *divergens* (Imhof) Lemmermann, Ber. d. D. Bot. Ges. **18**: 517, *pl.* 19, *figs.* 15–20. 1900.

Cells in free-floating densely or openly branched colonies. Receptacles with conical bases and the upper portion cylindrical or with slightly flaring mouths. The axis of the upper portion of receptacle forming an angle of 45–90 degrees with that of the basal portion. Sides of receptacle sometimes undulate. Seasonal variation confined to denseness of branching in colony rather than changes in receptacle shape. (Facultative Planktont).

Receptacles 7–8 μ broad, 35–50 μ long (20–27 μ in upper portion and 15–20 in basal portion).

Beasley (s), Boulder (cc), Browns (cc), Center (ss), Devils (1) (ss), Elizabeth (cc), George (r), Green (2) (rr), Hill (cc), Lac la Belle (ss), Long (2) (rr), Marl (ss), Mary (cc), Meta (r), Mill (cc), Mud (1) (ss), Nagawicka (cc), North Turtle (ss), Oconomowoc (cc), Otter (ss), Rock (ss), Round (4) (rrr), Silver (2) (cc), Speese (cc), Tamarack (ss), Taylor (ss), Tied Canoe (ss).

HYALOBRYON Lauterborn 1896.

Cells solitary or in branching colonies, free-floating or epiphytic. Receptacle shape and cell contents as in *Dinobryon* but with the wall of the receptacle containing a number of successive growth rings instead of homogeneous. Wall of receptacle more delicate than *Dinobryon*.

Hyalobryon mucicola (Lemmermann) Pascher. Pl. 14, Figs. 3–4.

D. Süsswasserfl. Deutschl., österr. u. d. Schw. **2**, Flagellatae **2**: 81, *fig.* 131. 1913.

Hyalobryon Lauterbornei var. *mucicola* Lemmermann, Ber. d. D. bot. Ges. **19**: 340. 1901; Bachmann, Arch. f. Hydrob. u. Planktonkde. **3**: 81, figs. XI^{16}–XI^{17}. 1908.

Cells solitary, epiphytic upon gelatinous envelope of colonial Myxophyceae, Chlorophyceae and Phaeophyceae. Receptacle delicate, upper portion cylindrical and with a flaring mouth, lower portion conical and with a short stipe. Growth rings of receptacle numerous. Protoplast ovoid, attached to base of receptacle, with two golden-brown chromatophores, two cilia of unequal length and an eyespot. (Euplanktont).

Receptacles 4–6 μ broad, 25–45 μ long; stipe 4–12 μ long.

Crawling Stone (r), Lac du Flambeau (sss), Long (3) (c), Mud (2) (rr), Round (1) (rr).

Division **HYDRURINEAE.**

Cells free-floating or sessile, usually in palmella-like colonies and motile only at the time of reproduction; rarely solitary, generally in gelatinous colonies of microscopic to macroscopic size that have a definite to indefinite shape. Colony capable of growth in all parts or with growing region restricted to apex. Gelatinous sheath of cells hyaline, homogeneous or lamellated. Cells ovoid to spherical, with one or two golden-brown parietal chromatophores and with or without a definite wall.

Reproduction by pyramidate, spherical, or ovoid zoospores with one or two cilia.

The members of this division are the connection between the normally motile and the highly differentiated sessile Phaeophyceae.

Order **PHAEOCAPSALES.**

Cells generally organized into free-floating, spherical, ellipsoid, or elongate colonies. Growth of colony taking place in all parts and not restricted to the apical region. Cells variously shaped and with one or two golden-brown chromatophores.

Reproduction, aside from vegetative cell division, by the formation of spherical to ovoid zoospores.

The order contains the single family of the Phaeocapsaceae and only one member of the family, *Phaeococcus,* is found in the local flora.

PHAEOCOCCUS Borzi 1892.

Cells in spherical, ovoid, or irregularly shaped free-floating colonies enclosed by a copious, hyaline, homogeneous or slightly lamellated, gelatinous envelope. Cells ellipsoid to sub-spherical or spherical. Cell number a multiple of two in young colonies, indefinite in older colonies. Cells containing one or more golden-brown chromatophores, sometimes with a red pigment spot.

Reproduction by vegetative cell division or direct metamorphosis of cell contents into biciliate ovoid to pyriform zoospores.

PHAEOCOCCUS PLANCTONICUS W. & G. S. West. Pl. **14,** Fig. **5.**

Trans. Roy. Soc. Edinb. 41: 496, *pl.* 6, *figs.* 15–16. 1905.

Cells spherical to sub-spherical, generally distributed in small clumps within a copious, hyaline, homogeneous or radially fibrillar, gelatinous envelope. Number of cells in colony generally 16–64. Cells containing 1–2 parietal, disciform to bell-shaped, golden-brown chromatophores with smooth or irregular margins. (Euplanktont).

Cells 7.2–9.6 μ in diam.; colonies 50–250 μ in diam.

Carson (ss), Clear Crooked (c), Crawling Stone (rr), Devils (1) (ss), Hillman (rr), Jennie (r), Mary (s), Mendota (sss), Pokegama (1) (rr), Sand (4) (rr), Shell (ss), Silver (2) (s), White Sand (rr).

Division **RHIZOCHRYSIDINAE.**

Algae that are known in the rhizopodial stage only, although the power to form a motile phase is presumed to be present. Cells solitary or in small colonies, always free-floating. Cells with rhizopodial processes and with or without a differentiated periplast. Cells with 1–2 golden-brown chromatophores, and with or without numerous vacuoles.

Reproduction by vegetative cell division.

This is an entirely artificial group where rhizopodial forms are placed pending further elucidation of their life histories. Since the group contains such doubtful forms no attempt has been made to differentiate the orders and families.

KEY TO THE GENERA.

Cells solitary or in irregular colonies.......................... RHIZOCHRYSIS
Cells joined in a linear series to form colonies............... CHRYSIDASTRUM

RHIZOCHRYSIS Pascher 1913.

Cells free-floating, solitary or in colonies that are naked or enclosed by a gelatinous sheath. Cells without a differentiation of the periplast but with few or numerous, delicate, pseudopodial processes. Cells with one or two golden-brown chromatophores and numerous vacuoles, of small size when united in colonies.

Rhizochrysis limnetica sp. nov. Pl. **14**, Fig. **6**.

Cells solitary, free-floating, without a differentiation of the periplast but with numerous spicule-like processes extending in all directions. Cells approximately spherical, with one golden-brown chromatophore and numerous vacuoles. (Euplanktont).

Diam. cells (without processes) 35–45 μ.

Beasley (sss), Birch Island (rr), Crawling Stone (rr), Deer (rr), Diamond (rrr), Half Moon (rr), Horse (rr), Kitten (sss), Lac du Flambeau (r), Long (3) (sss), Marl (rrr), Nemahbin (ss), Oconomowoc (rr), Okauchee (rrr),

Otter (rrr), Pokegama (2) (r), Rainbow (1) (r), Round (3) (rr), Sand (4) (sss), Taylor (r), Upper Nashotah (r), Wildcat (sss).

This species is quite unlike other members of the genus since the processes are much more delicate and the cells larger.

CHRYSIDASTRUM Lauterborn 1913.

Cells joined to one another by pseudopodial processes and arranged to form free-floating linear colonies of 2–16 cells that are not enclosed by a gelatinous envelope. Cells spherical, with the periplast undifferentiated but extending into several long, delicate, pseudopodia-like processes. Chromatophore single, central, and laminate to band-shaped.

Reproduction unknown.

Chrysidastrum catenatum Lauterborn. Pl. **14**, Figs. **7–8.**

in Pascher, D. Süsswasserfl. Deutschl., Österr. u. d. Schw. **2**, Flagellatae **2**: 91, *fig.* 144. 1913.

Specific characters the same as those of the genus. (Euplanktont).
Diam. cells (without processes) 12–15 μ; (with processes) 45–60 μ.

Kitten (rr), Mud (2) (rr), Wildcat (rrr).

The cell shape is quite like that of *Chrysamoeba* but the definite organization into a linear colony and the single laminate chromatophore are points of difference. The alga should also be compared with *Rhizochrysis planctonica* Pascher.

Class HETEROKONTEAE.

Cells microscopic or macroscopic; free-floating, sessile, or epiphytic. Cell shape spherical, ovoid, pyriform, cylindrical or irregular. Cells solitary, in colonies of regular to irregular shape, or united to form simple filaments. Colonies or individual cells with or without a gelatinous envelope. Chromatophores one to many, parietal, generally disciform, rarely laminate, yellowish-green in color and without a pyrenoid. Assimilation product oil, never starch. Cells uninucleate or multinucleate.

Asexual reproduction by vegetative cell division and a formation of zoospores. Zoospores ovoid to pyriform, with two cilia of unequal length and generally more than one yellowish-green chromatophore.

Sexual reproduction, insofar as known, isogamous, by zoogametes that are similar in structure to the zoospores.

The various genera of the Heterokontae were formerly scattered throughout the different families of the Chlorophyceae. The distinguishing characteristics of this class are the yellowish-green chromatophores without pyrenoids, with oil not starch the assimilation product; and zoospores with two cilia of unequal length. There are three orders in the class and representatives of all three are found in Wisconsin lakes.

KEY TO THE ORDERS.

Vegetative cells motile.................................. Heterochloridales
Vegetative cells immobile.
  Cells solitary or in colonies, never in filaments........... Heterococcales
  Cells united to form simple filaments.................... Heterotrichales

Order HETEROCHLORIDALES.

Cells solitary, free-swimming, ovoid to pyriform in shape. Motile phase dominant, rarely metamorphosing over into an amoeboid or palmella phase. Cells with two cilia of unequal length, one or more chromatophores of a yellowish-green color and without an eyespot. Assimilation product leucosin or fat, never starch.

Asexual reproduction by vegetative division of cells.

Sexual reproduction unknown.

The order has but one family, the Chloramoebaceae, whose characters are those of the order.

CHLOROCHROMONAS Lewis 1913.

Cells solitary, generally free-swimming, very rarely amoeboid, sometimes attached to substratum by a posterior pseudopodia-like structure. Cell shape pyriform to sub-ovoid, with an oblique depression at the anterior end. Cilia two, at the anterior end of the cell and of unequal length. Chromatophores two, yellowish-green in color, oblong with rounded ends in front view, somewhat flattened in side view. Cells with a single anterior vacuole, one nucleus, and without an eyespot. Nutrition holophytic or saprophytic.

Reproduction by longitudinal fission when in the motile phase.

CHLOROCHROMONAS MINUTA Lewis. Pl. 15, Figs. 1–3.

Arch. f. Protistenkunde **32**: 254, *pl.* 12, *figs.* 1–24. 1913.

Characters as described above. (Euplanktont).

Cells 4.5–9.5 (av. 6.3) μ long, with an average breadth of 4.7 μ. Average length of longer cilium 12.5 μ, of the shorter 2.9 μ.

Mendota.

This organism was first found by Lewis in a dish containing sections of basswood (*Tilia americana*) that had been standing for some time in a dish of tap water, which is drawn directly from Lake Mendota. I have never noted the alga in plankton collections but have no doubt but that a continued search will demonstrate its presence in Lake Mendota. During the past winter (1917–18) the alga appeared in an infusion of fern leaves which had been standing for some time in tap water. This shows that the original source of the alga was the water with which the sections were moistened and not the *Tilia.*

Order **HETEROCOCCALES.**

Cells spherical, ovoid, saccate, or elongate cylindrical; solitary or in colonies of definite or indefinite shape; free-floating or sessile. Cells of colonies embedded in a hyaline, gelatinous envelope or a tough elastic membrane. Chromatophores one to many, yellowish-green to nearly grass-green in color and with oil not starch the assimilation product.

Asexual reproduction by vegetative cell division and a formation of zoospores with cilia of unequal length.

Sexual reproduction by a fusion of zoogametes that are similar in structure to the zoospores is known in certain genera.

KEY TO THE FAMILIES.

Cells epiphytic . . . . CHLOROSACCACEAE
Cells free-floating.
  Length of cells several times the diameter . . . . OPHIOCYTIACEAE
  Length of cells less than twice the diameter.
    Cells irregularly distributed throughout an ovoid
      gelatinous envelope . . . . CHLOROBOTRYDACEAE
    Cells arranged about a common center within an
      irregular gelatinous envelope . . . . BOTRYOCOCCACEAE

Family *CHLOROSACCACEAE.*

Cells epiphytic, solitary, gregarious, or embedded in an amorphous gelatinous matrix. Chromatophores one to several, parietal, of a yellowish-green color, with oil not starch the assimilation product.

Asexual reproduction by a formation of zoospores that presumably have two cilia of unequal length.

Sexual reproduction by the fusion of isogamous zoogametes.

PERONIELLA Gobi 1887.

Cells epiphytic upon other algae, solitary or gregarious. Shape of mature cells globular to ovoid, of young cells pyriform to ovoid. Cells attached to substratum by a delicate hyaline stipe that is usually somewhat longer than the cell. Chromatophores one to two, pale yellowish-green, without pyrenoids or starch and with oil as the assimilation product.

Asexual reproduction by the division of cell contents into 2–4–8 uniciliate (?) zoospores that are liberated by a lateral rupture of mother cell wall. Zoospores swarming a short time and then coming to rest upon substratum and developing into vegetative cells.

PERONIELLA PLANCTONICA G. M. Smith. Pl. **15**, Fig. **4**.

Bull. Torr. Bot. Cl. **43**: 476, *pl.* 25, *fig.* 15. 1916.

Cells broadly ovoid to pyriform, with a long, delicate, hyaline stipe that is without a disc or other enlargement at the base. Chromatophores usually single, rarely two, pale yellowish-green, without pyrenoids or starch. (Euplanktont).

Cells (without stipe) 6–9.5 μ long; (with stipe) 15–18 μ. Stipe 8–10 μ long, 1.2 μ broad.

In gelatinous envelope of *Sphaerozosma* filaments.

Devils (1) (rr).

Family *CHLOROBOTRYDACEAE.*

Cells solitary or in free-floating colonies enclosed by a copious, homogeneous, gelatinous envelope. Colony generally with a small number of cells and with cell number a multiple of two. Cells spherical, ovoid, angular, or cylindrical; with a firm cell wall that is sometimes impregnated with a slight trace of silica; and with one to many parietal, disciform, yellowish-green chromatophores without pyrenoids.

Asexual reproduction by vegetative cell division or a formation of zoospores with two cilia of unequal length.

CHLOROBOTRYS Bohlin 1901.

Cells rarely solitary, generally in colonies of 2–4–8–16 or more embedded in a copious, homogeneous, hyaline, gelatinous envelope. Cells spherical to broadly ovoid, with a heavy wall that sometimes contains a small amount of silica. A conspicuous red pigment spot present in cells at times.

Reproduction by vegetative division of cells in three directions.

KEY TO THE SPECIES.

Cells ovoid, with few chromatophores..........................C. LIMNETICUS
Cells spherical, with numerous chromatophores..................C. REGULARIS

CHLOROBOTRYS REGULARIS (W. West) Bohlin. Pl. **15**, Fig. **10**.

Bih. kgl. Sv. Vet.-Ak. Handl. **27**, Afd. **3**, No. **4**: 34. 1901; W. & G. S. West, Jour. of Bot. **41**: 78, *pl.* 447, *figs.* 7–10. 1903.

Chlorococcum regulare W. West, Jour. Roy. Micr. Soc. **1892**: 737, *pl.* 10, *fig.* 55. 1892.

Cells spherical to subspherical; rarely solitary, generally in colonies of 2–4–8–16 enclosed by a copious, hyaline, homogeneous, gelatinous envelope. Cell wall thick, sometimes containing a trace of silica. Chromatophores yellowish-green, disciform, parietal, 6–30 in each cell. Cells at times containing a single conspicuous red pigment spot. (Tychoplanktont).

Diam. cells 10–27 μ. Diam. colonies 35–90 μ.

North Turtle (rrr).

CHLOROBOTRYS LIMNETICUS G. M. Smith. Pl. **15**, Fig. **9**.

Trans. Wis. Acad. Sci., Arts, & Lett. **19**[1]: 653, *pl.* 14, *fig.* 10. 1918.

Cells ovoid, irregularly distributed within a homogeneous, hyaline, gelatinous envelope. 10–30 cells in a colony. Chromatophores 3–4, pale yellowish-green, disciform, parietal, and without pyrenoids. (Euplankton†).

Cells 5–6 μ broad, 6–8 μ long. Diam. colonies 40–200 μ.

Camp (rr).

Family *BOTRYOCOCCACEAE*.

Cells always in free-floating colonies of definite or indefinite shape. Number of cells in colony generally indefinite, few or many. Colony enclosed by an envelope that may be hyaline or colored, gelatinous or leathery. Cells usually ovoid, rarely spherical; with 1–2 parietal, laminate to disciform, yellowish-green to nearly grass-green chromatophores without pyrenoids and with oil as the assimilation product.

Asexual reproduction by vegetative cell division. Zoospore formation or sexual reproduction unknown.

BOTRYOCOCCUS Kützing 1849.

Cells in free-floating colonies of indefinite shape. Cells closely apposed to one another and enclosed by a gelatinous or tough membrane that is sometimes wrinkled and folded. Colonies generally with cells in several aggregates that are connected by broad or delicate strands of the colonial integument. Cell shape spherical, ovoid, or cuneate. Groups of cells arranged in a single layer around a common center, or irregular in arrangement. Chromatophore single, parietal, laminate to disciform, yellowish-green to nearly grass-green in color and without a pyrenoid.

Reproduction by vegetative division of cells.

KEY TO THE SPECIES.

Cells held together by a thin gelatinous envelope................B. Sudeticus
Cells held together by an irregular tough membrane.
  Membrane completely enclosing cells........................B. Braunii
  Membrane covering base only of cells..........B. Protuberans var. Minor

Botryococcus sudeticus Lemmermann. Pl. 15, Fig. 8.

Forschungsbr. a. d. Biol. Stat. zu Plön. 4: 111, *figs.* 6–7. 1896.

Botryococcus Braunii G. S. West non Kützing, Brit. Freshw. Algae 237, *figs.* 106A–106B. 1904; G. M. Smith, Trans. Wis. Acad. Sci., Arts, & Lett. 18[2]: 545 1916.

Colonies free-floating; spherical or irregular; with cells held in botryoidal clusters by a hyaline gelatinous envelope and frequently joined to one another by broad or narrow strands of the gelatinous envelope. Cells spherical to subspherical, 16–32 or more in each botryoidal group, and arranged around a common center. Chromatophore single, cup-shaped to disciform, without a pyrenoid. (Euplanktont).

Diam. cells 6–13 μ.

Clear (rr), Waubesa (rrr).

This is the *B. Braunii* Ktz. of the preliminary list, the identification being based upon G. S. West's description of *B. Braunii* in his British Freshwater Algae. Dr. West has since informed me that the *Ineffigiata neglecta* of that work is *B. Braunii* Ktz. and the *B. Braunii* described therein is *B. sudeticus* Lemm.

BOTRYOCOCCUS BRAUNII Kützing. Pl. 15, Fig. 5.

Species algarum 892. 1849; Fresenius, Abh. Senckenb. Naturf. Ges. **2**: 239, *pl.* 11, *figs.* 27–33. 1858.

Ineffigiata neglecta W. & G. S. West, Jour. Roy. Micr. Soc. **1897**: 503. 1897; Jour. of Bot. **41**: 80, *pl.* 447, *figs.* 1–6. 1903.

Cells joined in small free-floating colonies of irregular shape that are without a conspicuous gelatinous envelope but completely enclosed by a tough, elastic, hyaline to orange colored membrane that is produced into irregular wrinkles, folds, and spines. Colonies frequently united to form compound colonies by long delicate projections of the colonial envelope. Cells ovoid or ellipsoid in shape and forming a single layer around a central cavity. Chromatophore single, cup-shaped to laminate, yellowish-green to nearly grass-green in color, without a pyrenoid. Assimilation product a reddish oil that may color the whole colony a brick red. (Facultative planktont).

Cells 3.5–5.5 μ broad, 5.7–10.5 μ long. Simple colonies up to 100 μ in diam.; compound colonies up to 1.5 mm.

Balsam (rr), Bear (3) (rr), Bear Trap (rrr), Beaver (2) (rr), Beaverdam (ss), Birch Island (rrr), Birch (2) (rr), Boulder (rrr), Browns (sss), Camp (rrr), Carson (r), Center (r), Chain (rrr), Chetac (rr), Clam (rrr), Clare (r), Cochrane (rrr), Crawling Stone (rr), Crooked (sss), Deer (rr), Delavan (rrr), Des Moines (rr), Devils (1) (r), Devils (2) (r), Devils (3) (rrr), East (rr), Elizabeth (rr), Green (1) (rr), Green (2) (rrr), Grindstone (r), High (rr), Horse (rrr), Horseshoe (1) (rr), Island (rr), Jennie (rr), Jones (rrr), Kawaguesaga (rrr), Kegonsa (r), Kimball (rrr), Kitten (rrr), Lac Court Oreilles (sss), Lac du Flambeau (rrr), Little Bass (1) (rrr), Little Butternut (rrr), Little Crooked (rrr), Little Doctor (rr), Little Rice (1) (rr), Long (1) (rrr), Lower Nashotah (r), Lynx (rr), Mary (sss), Mendota (ss), Mercer (rrr), Mill (rrr), Monona (sss), Mud (3) (rrr), Mud Hen (sss), Nancy (r), Nicaboyne (rrr), No Mans (rr), Paddock (r), Palmer (rrr), Pardee (sss), Pike (rrr), Pokegama (1) (rrr), Pokegama (2) (rrr), Poor Farm (rrr), Reserve (rr), Rice (1) (rr), Round (2) (rrr), St. Croix (rrr), Sand (1) (r), Sand (2) (rr), Sand (3) (rr), Sand (4) (rrr), Shell (rr), Silver (2) (r), Sishebogema (rr), South Turtle (rr), Spirit (rrr), Tied Canoe (rrr), Travis (rrr), Upper Nashotah (sss), Wapagossett (rr), Waubesa (r), White Ash (rrr), Whitefish (rrr), Wildcat (rrr), Winnebago (rrr), Wolf (rrr).

BOTRYOCOCCUS PROTRUBERANS var. MINOR G. M. Smith. Pl. **15**, Figs. **6–7.**

Trans. Wis. Acad. Sci., Arts, & Lett. **19**[1]: 652, *pl.* 14, *figs.* 6–7. 1918.

Cells in colonies of 4–8–16 or more held together by a tough gelatinous membrane. Colonies frequently joined to one another by long hyaline strands of the colonial envelope. Cells ovoid to cuneate-ovoid, with their bases embedded in the colonial envelope and their apices not enclosed by the envelope. Cells arranged in a single layer around a central cavity. Chromatophores single, yellowish-green, laminate, occupying the median portion of the cell. (Euplanktont).

Cells 8–9.5 μ long, 5–6.5 μ broad at the widest part.

Diamond (rrr), George (aa), Hillman (rr), Lindy (rr), Oak (ss), Pardee (rr), Sanford (rr), South Crab (s).

This species is distinguished from the foregoing by the smaller number of cells in the colony and the projection of the apex of the cells through the colonial integument.

Family *OPHIOCYTIACEAE.*

Cells sessile or free-floating, solitary or colonial. Length of cells generally several times the breadth. Cells straight, curved, sigmoid, or spiral; with or without terminal spines. Chromatophores yellowish-green; annular, parietal and few in number; or disciform, parietal and numerous; without pyrenoids and with oil as the assimilation product.

Asexual reproduction by division of cell contents to form 4–16 aplanospores or 2–8 biciliate zoospores with cilia of unequal length.

Sexual reproduction unknown.

OPHIOCYTIUM Nägeli 1849.

Cells free-floating or epiphytic, solitary or colonial; straight, arcuate, sigmoid, or spiral elongate cylinders with rounded or capitate ends that may be apiculate or attenuated into a single spine. Chromatophores pale yellowish-green, H-shaped in optical cross section, without pyrenoids and with oil as the assimilation product. Cells coenocytic. Cell wall consisting of a homogeneous cap at one end and with the remaining portion of the wall composed of a number of laminate layers.

Asexual reproduction by division of cell contents to form 4–16 aplanospores or 2–8 biciliate zoospores with cilia of unequal length.

KEY TO THE SPECIES.

Poles of cells without spines.................................O. PARVULUM
Poles of cells with spines.....................................O. CAPITATUM

OPHIOCYTIUM PARVULUM (Perty) A. Braun. Pl. **15**, Fig. **11.**

Alg. unicell. 55. 1855; Lemmermann, Hedwigia **38**: 33, *pl.* 4, *figs.* 30–33. 1899.

Cells cylindrical, several times as long as broad; straight, arcuate, or spiral; poles rounded and without spines. Chromatophores several, H-shaped in cross section, pale yellowish-green. Oil droplets rarely present. (Tychoplanktont).

Cells 3–9 μ broad.

Fishtrap (rrr).

OPHIOCYTIUM CAPITATUM Wolle. Pl. **15**, Figs. **12–13.**

Freshw. Algae of U. S. 176, *pl.* 158, *figs.* 3–7. 1887.

Cells solitary; arcuate, sigmoid, or spiral cylinders with poles rounded and the same size as or slightly larger than the median portion of the cell. Length of cell 5–10 times the breadth. Poles of cells with a single short, straight, or slightly curved spine. (Tychoplanktont).

Cells 5–10 μ broad, up to 85 μ long. Spines 5–7 μ long.

Tied Canoe (rrr).

var. LONGISPINUM (Möbius) Lemmermann. Pl. **15**, Figs. **14–16.**

Hedwigia **38**: 32, *pl.* 4, *figs.* 21–25. 1899.

Cells cylindrical; straight, arcuate, or spirally twisted, with a single long spine at each pole that is nearly as long as the cell. (Euplanktont).

Diam. cells 4.5–6 μ. Length of spines 16–50 μ.

George (rr), Loon (rr).

Order **HETEROTRICHALES.**

Cells in simple, unbranched filaments; with or without a gelatinous envelope. Cell wall firm, fairly thick, lamellated in structure and breaking down into H-shaped pieces at the time of reproduction. Chromatophores few to several, yellowish-green, disciform, and parietal. Cells usually uninucleate. Assimilation product oil, not starch.

Asexual reproduction by the formation of aplanospores or by zoospores with two cilia of unequal length.

Sexual reproduction by the fusion of isogamous zoogametes.

There is but one family in the order, the Tribonemaceae, whose characters are the same as those of the order.

TRIBONEMA Derbes and Solier 1856.

Cells in simple unbranched filaments destitute of an enveloping gelatinous sheath. Cells cylindrical or barrel-shaped, with flattened end walls. Cell wall frequently of considerable thickness and laminate in structure. Chromatophores two to several, yellowish-green, disciform, parietal, without starch and with oil as the assimilation product.

Asexual reproduction by aplanospores and biciliate zoospores that have cilia of unequal length. The cell wall pulling apart and breaking down into H-shaped pieces at the time of reproduction.

Sexual reproduction by the fusion of isogamous zoogametes.

TRIBONEMA MINUS (Wille) Hazen. Pl. **15**, Figs. **17–18**.

Mem. Torr. Bot. Cl. **11**: 185, *pl.* 25, *figs.* 7–8. 1902.

Filaments generally forming a floccose yellowish to greenish mass. Cells cylindrical or slightly swollen in the median portion, generally 2–4, rarely 6, times as long as broad. Chromatophores 2–4, yellowish-green, disciform, parietal, generally in pairs. (Facultative planktont).

Cells 5–6 μ broad, 10–35 μ long.

Green (1) (rrr), Mendota (s).

Members of this genus are frequently found in ditches but seldom in lakes. During the past season *T. minus* has appeared in considerable quantity in the plankton of Lake Mendota, although previous to this time the alga has not been found in the plankton. A similar sudden appearance of *T. bombycina* forma *depauperata* Wille, a form closely related to *T. minus,* has been noted in the plankton of Danish lakes by Wesenberg-Lund (Studier over de danske soers Plankton p. 100. Kjobenhavn, 1904).

Class **CHLOROPHYCEAE.**

Cells with the coloring matter restricted to definite bodies that are grass-green in color (chloroplasts). Cells free-floating, sessile, epiphytic, epizootic, or endophytic; solitary, in microscopic to macroscopic colonies of definite or indefinite shape that are naked or enclosed by a hyaline to colored, homogeneous or lamellated, firm or gelatinous envelope. Colonies with all of the cells morphologically alike or differentiated. Colonies growing in all directions; forming simple, laminated, or parenchymatous plates; or developing into simple or branched filaments. Cells motile throughout the whole life history, for a portion of it, or immobile at all times. Cells variously shaped; spherical, hemispherical, ovoid, ellipsoid, spindle-shaped, acicular, reniform, angular, irregular, cylindrical, conical, simple to branching tubular, or constricted to form two symmetrical halves of various shapes. Cells containing one or more disciform, laminate, cup-shaped, astral, ribbon-like, or reticulate chloroplasts that are parietal, axial, or central in position and have one or more centers for the synthesizing of starch (pyrenoids). Assimilation product oil in a few cases. Cells uninucleate or multinucleate (coenocytic); containing a central vacuole, numerous small vacuoles, or without vacuoles. Contractile vacuoles chiefly confined to the motile genera.

Asexual reproduction by vegetative cell division, aplanospores, zoospores, or akinetes.

Sexual reproduction by isogamous or heterogamous zoogametes, isogamous aplanogametes, or heterogamous motile antherozoids and immobile oospheres. Zygotes germinating immediately, or after a period of rest; with a direct or an indirect germination into the plant body.

KEY TO THE DIVISIONS

Sexual or asexual reproductive bodies with 2–4 cilia of equal length. . Isokontae
Sexual or asexual reproduction by non-ciliate reproductive bodies. . . . Akontae

Division **ISOKONTAE.**

Cells solitary or in colonies and showing the same great range of variation in shape, organization of the colony, interior structure, and method of reproduction noted in the description of the class.

Included in this division is the important plankton family of the Autosporaceae where no motile phase of reproduction is known. The

close phylogenetic connection between the Autosporaceae and the other members of the Protococcales is, however, perfectly evident.

KEY TO THE ORDERS.

Cilia normally present throughout the whole life cycle............VOLVOCALES
Cilia present at time of reproduction only.
  Cells in simple or branched filaments.
    Cells uninucleate, with a single laminate chloroplast....ULOTRICHALES
    Cells multinucleate, with one or more chloroplasts..SIPHONOCLADIALES
  Cells solitary or in simple colonies, generally of microscopic size ....PROTOCOCCALES

Order **VOLVOCALES.**

Cells normally ciliate and motile during the vegetative phases of the life cycle. Solitary or organized into colonies of definite form that generally contain a definite number of cells. Protoplasts rarely naked, usually with a differentiated outer cellulose covering that may be partially gelatinized. Shape of protoplast spherical, ovoid, disciform, pyramidate, or irregularly radiate. Cells with a single cup-shaped to irregular chloroplast, with or without pyrenoids; with 2–4 cilia at the anterior end; generally with a disciform to bacilliform eyespot, two to several contractile vacuoles, and one nucleus. Palmella stages frequently arising in the unicellular members.

Asexual reproduction by division of cell contents into 2–16 zoospores; or a development of special, or all cells of the colony into autocolonies.

Sexual reproduction by isogamous zoogametes; heterogamous zoogametes; or small motile antherozoids and large immobile oospheres.

Many hold with G. S. West (Algae 151, 1916) "that it is very doubtful if the Volvocineae have sufficient claims to be removed from the Protococcales and elevated to a group of equal rank (the 'Volvocales')". There is a very close relationship between the "palmella" stage of various unicellular Volvocales and the family Palmellaceae of the Protococcales. This represents to me, however, a close phylogenetic relationship between the two and not a justification for uniting in the same order forms with mobile vegetative cells and those which are immobile.

KEY TO THE FAMILIES.

2 cilia at the anterior end of cells.
  Cells solitary ................................CHLAMYDOMONADACEAE
  Cells in definite coenobia ....VOLVOCACEAE
4 cilia at the anterior end of cells ..........................CARTERIACEAE

Family *CHLAMYDOMONADACEAE.*

Cells normally solitary and motile during the vegetative phases of the life history. Spherical, ovoid, cylindrical, sub-acicular, compressed, or quadrately projected. Chloroplast cup-shaped to laminate, lateral or posterior in position, with or without pyrenoids. Cells uninucleate, with or without an eyespot, with two cilia of equal length that generally have two contractile vacuoles at their base. Cells at times becoming immobile and developing into amorphous colonies enclosed by a wide gelatinous envelope.

Asexual reproduction by the division of cell contents to form 2–16 biciliate zoogametes.

Sexual reproduction by the fusion of isogamous zoogametes.

CHLAMYDOMONAS Ehrenberg 1832.

Cells normally solitary and motile; though capable of coming to rest, losing the cilia, developing a gelatinous envelope and dividing vegetatively to form colonies containing an indefinite number of cells. Cells spherical, fusiform, ovoid, pyriform, or ellipsoid; with two cilia of equal length at the anterior end. Chloroplast typically cup-shaped, posterior, and with one pyrenoid. Cells uninucleate, generally with two contractile vacuoles at the base of the cilia; with or without an eyespot which, when present, lies in the anterior portion of the cell.

Asexual reproduction by division of cell contents to form 2–16 biciliate zoospores.

Sexual reproduction isogamous, by the division of cell contents into 8–64 zoogametes that fuse in pairs to form smooth-walled or sculptured zygotes.

KEY TO THE SPECIES.

Cells spherical.
- Anterior end without a beak.................................C. GLOBOSA
- Anterior end with a pronounced beak......................C. EPIPHYTICA

Cells ovoid.
- Chloroplast cup-shaped and basal............................C. SNOWII
- Chloroplast laminate, lateral, or basal.......................C. DINOBRYONI

CHLAMYDOMONAS GLOBOSA Snow. Pl. **16**, Fig. **8**.

Bull. U. S. Fish Comm. **22**: 389, *pl.* 1, *figs.* III[1]–III[5]. 1903.

Cells spherical or slightly ellipsoid, without an anterior beak. Chloroplast cup-shaped, occupying the greater portion of the cell, with one pyrenoid at the posterior end. Eyespot present, small, generally in the middle third of the cell. Cilia slightly longer than the cell and with a single contractile vacuole at their base. Palmella stages frequent and usually containing 4–8 cells. (Euplanktont).

Cells 5–10 μ broad.

Crawling Stone (rr).

Chlamydomonas epiphytica sp. nov. Pl. **16,** Figs. **11–13.**

Cells spherical to pyriform, with a very pronounced anterior beak. Chloroplast cup-shaped, forming a thin layer next to the cell wall, with a single pyrenoid at the posterior end. Eyespot conspicuous, disciform, at the anterior end of the cell. Cilia more than twice as long as the cell. Cells frequently coming to rest in the gelatinous envelope of *Microcystis* colonies but not losing the eyespot and usually retaining the cilia. (Euplanktont).

Cells 7–8 μ broad, 8–9 μ long; cilia 20–25 μ long.

Elizabeth (s), Mary (sss).

The alga is frequently found in the gelatinous envelope of *Microcystis* colonies and suggests an epiphytic *Characium* but the presence of cilia shows that it is a *Chlamydomonas.*

CHLAMYDOMONAS SNOWII Printz. Pl. **16,** Figs. **9–10.**

Skr. Vidensk.-Selsk. i Kristiana, Mat.-Naturv. Kl. **1913**[1]: No. **6**: 18. 1914.

Chlamydomonas communis Snow, Bull. U. S. Fish Comm. **22**: 388, *pl.* 1, *figs.* II[1]–II[3]. 1903.

Cells motile but frequently forming palmella stages containing a considerable number of cells. Cell shape ovoid to ellipsoid, with a very inconspicuous anterior beak. Chloroplast cup-shaped, massive in the posterior portion, with one pyrenoid at the center of the cell. Cilia slightly longer than the cell; eyespot inconspicuous, variable in position. (Facultative planktont).

Cells 6.5–8 μ broad, 10–15 μ long.

Nancy (rr).

Chlamydomonas dinobryoni sp. nov. Pl. **16,** Figs. **5–7.**

Cells rarely free-swimming, generally in large numbers and filling the empty receptacles of *Dinobryon* colonies. Cells small, ovoid to pyriform in shape, without an anterior beak; with two short cilia at the anterior end. Chloroplast disciform to hemispherical, basal or lateral in position, containing a pyrenoid. Eyespot and contractile vacuoles not observed. (Euplanktont).

Cells 2–3 μ broad, 3.5–5 μ long; cilia 6–8 μ long.

Lac du Flambeau (ss), Long (3) (s), Pokegama (2) (r).

Many of the empty *Dinobryon* receptacles in these three connecting lakes were filled with a minute green organism. These had cilia, even when in the receptacle, and on the slightest crushing of the receptacle began swimming around outside of it. Since cilia are constantly present the alga is a *Chlamydomonas* and not one of the Palmellaceae collected at the time of reproduction.

Family *CARTERIACEAE*.

Cells normally motile and with four cilia of equal length at the anterior end. Solitary, or in coenobia containing a definite number of cells. Cell shape spherical, ovoid, ellipsoid, cardioid, pyriform, or lenticular. Cells enclosed by a definite wall, and with or without a cup-shaped to lenticular chloroplast that is with or without pyrenoids. Eyespot generally present, conspicuous, and at the anterior end of the cell. Cells frequently with contractile vacuoles at the base of the cilia. Palmella stages known in the solitary members of the family.

Asexual reproduction by the division of cell contents into 2–16 tetraciliate zoospores or into autocolonies.

Sexual reproduction isogamous, by a fusion of tetraciliate zoogametes.

CARTERIA Diesing 1866.

Cells solitary, generally motile throughout the vegetative phases of the life cycle. Spherical, ovoid, ellipsoid, or cardioid in shape; with four cilia of equal length at the anterior end. Cell contents consisting of a single cup-shaped chloroplast that contains a pyrenoid; one eyespot; and two contractile vacuoles at the base of the cilia. Palmella stages containing an indefinite number of cells are found at times.

Asexual reproduction by division of cell contents to form 2–8 tetraciliate zoospores.

Sexual reproduction by the division of cell contents to form 4–64 tetraciliate zoogametes that fuse in pairs to form zygotes with smooth or sculptured walls.

KEY TO THE SPECIES.

Cells ellipsoid..C. KLEBSII
Cells cardioid..C. CORDIFORMIS

CARTERIA CORDIFORMIS (Carter) Diesing. Pl. 16, Fig. 1.

Sitzbr. d. k. Ak. d. Wiss. Wien, Mat.-Naturw. Kl. **52**[1]: 356. 1866; Dill, Jahrb. f. wiss. Bot. **28**: 353, *pl.* 5, *figs.* 42–44. 1895.

Cells cardioid in shape with four cilia at the base of the apical depression. Chloroplast cup-shaped, fairly massive, with a single pyrenoid in the posterior region. Cells with two contractile vacuoles at the base of the cilia and with a disciform eyespot in the upper third of the cell. (Tychoplanktont).

Cells 9–16 μ broad, 12–20 μ long.

Crawling Stone (rrr).

CARTERIA KLEBSII (Dangeard) Dill. Pl. **16**, Figs. **2–4**.

Jahrb. f. wiss. Bot. **28**: 353. 1895; Printz, Skr. Vidensk.-Selsk. Kristiana, Mat.-Naturv. Kl. **1913**[1]: No. **6**: 13. 1914.

Pithiscus Klebsii Dangeard, Ann. Sci. Nat. **7** Sér. Bot. **7**: 137, *pl.* 12, *figs.* 1–6. 1888.

Cells solitary, ellipsoid to ellipsoid-cylindrical, with a more or less prominent beak at the anterior end. Chloroplast cup-shaped, massive, with a large pyrenoid at the posterior end. Cells without an eye-spot and with two contractile vacuoles at the base of the cilia. Cilia about twice as long as the cells. Palmella stages containing many cells at times. (Tychoplanktont).

Cells 5–10 μ broad, 8–16 μ long.

Pokegama (1) (r).

Family *VOLVOCACEAE*.

Cells always motile and in coenobia of definite shape that contain a definite or an indefinite number of cells. Colonies enclosed in a homogeneous, hyaline, gelatinous envelope. Cells of colony all alike or differentiated into vegetative and reproductive cells. Vegetative cells spherical, ellipsoid, pyriform or disciform; with or without connecting cytoplasmic processes; always biciliate. Chloroplasts generally cup-shaped and containing one pyrenoid. Vegetative cells usually with an eyespot and two contractile vacuoles.

Asexual reproduction by the division of all or certain of the cells in the colony to form autocolonies.

Sexual reproduction by a division of all cells in the colony to form isogamous zoogametes or heterogamous zoogametes of slightly different size; or the formation of motile antherozoids from certain cells and large immobile oospheres from others. Zygotes with smooth or sculptured walls.

KEY TO THE GENERA.

Colony with cells in a flat plate.......................................GONIUM
Colony with cells forming a hollow sphere.
  Cells mutually compressed.......................................PANDORINA
  Cells some distance from one another.
    Asexual reproduction by division of all cells in colony......EUDORINA
    Asexual reproduction by division of special cells.
      At least half of cells capable of forming autocolonies...PLEODORINA
      Only a few cells capable of forming autocolonies..........VOLVOX

GONIUM Mueller 1773.

Colonies of 4–8–16 cells arranged in a flat quadrangular plate and embedded in a common gelatinous matrix or connected by broad gelatinous strands. Cells ovoid to pyriform, with a single cup-shaped chloroplast containing one pyrenoid. Each cell with two cilia of equal length, contractile vacuoles at the base of the cilia, and an eyespot. Four- and eight-celled colonies with the cilia on the same side; sixteen-celled colonies with the four central cells having their cilia on the same side and the twelve marginal cells with radially arranged cilia.

Asexual reproduction by simultaneous division of all cells in the colony to form autocolonies, or by a formation of 2–4 zoospores in each cell.

Sexual reproduction isogamous, by a fusion of biciliate zoogametes.

GONIUM PECTORALE Mueller. Pl. 16, Figs. 14–15.

Vermium terrestrium et fluviatilium etc. **1**: 60. 1773; Migula, Bot. Cent. **44**: 72, *pl.* 2, *figs.* 1–8. 1890; Harper, Trans. Am. Micr. Soc. **31**: 65, *pl.* 5, *figs.* 1–19, 23. 1912.

Colonies containing 4–8–16 cells. Usually with 16 cells arranged in two concentric series in a flat plate with truncate corners. Cells connected with one another by broad, hyaline, gelatinous processes and with open spaces between the processes; or with processes indistinct and the whole colony embedded in a homogeneous, gelatinous matrix. Outer ring with twelve cells whose long axes lie parallel to the plane of the coenobe. Inner ring of four cells with a large open space at the center and their long axes perpendicular to plane of the coenobe. Cells ovoid to pyriform, with two cilia at the anterior end; a posterior massive cup-shaped chloroplast; an anterior eyespot; and two contractile vacuoles at the base of the cilia. (Tychoplanktont).

Cells 7–11 μ broad, 10–25 μ long.

Beaverdam (rrr), Clam (rrr), Duck (rr), Fowler (rrr), Island (r), Kegonsa (rrr), Long (1) (rrr), Mendota (rrr), Number One (rr), Pike (rrr), Poplar (rrr).

PANDORINA Bory 1824.

Colonies always motile, spherical to subspherical, containing 4–8–16–32 cells mutually compressed to form a hollow sphere with a small open space at the center. Colony enclosed by a fairly copious, firm, hyaline, homogeneous, gelatinous envelope. Cells pyriform to angular by mutual compression, with pointed ends toward the center of the colony; biciliate at the flattened distal end, the two cilia lying close together while passing through the colonial envelope and then becoming markedly divergent. Chloroplast single, massive, cup-shaped, with one pyrenoid at the base. Eyespot single, generally on distal face of the cell. Cells with a definite wall and two contractile vacuoles at the base of the cilia.

Asexual reproduction by simultaneous division of all cells of the colony to form autocolonies that are liberated by a gelatinization of the colonial envelope.

Sexual reproduction by a division of each cell of the colony into 16–32 zoogametes. Zoogametes showing indications of heterogamy in the slight difference in size and motility of the pairs that fuse to form the smooth walled zygote.

PANDORINA MORUM Bory. Pl. 16, Figs. 16–17.

Encyc. meth. hist. nat. de Zoophytes **2**: 600. 1824; G. S. West, Brit. Freshw. Algae 193, *figs.* 76A–76H. 1904.

Colonies generally 16–celled and as described above. (Facultative Planktont).

Cells 8–16 μ broad; colonies 20–45 μ broad.

Bear Trap (rrr), Blake (rrr), Camp (rr), Center (cc), Chetac (rr), Chetek (ss), Deer (rrr), Duck (ss), East (rrr), Fowler (rr), Grindstone (s), Hillman (rr), Hooker (s), Kegonsa (sss), Lac Court Oreilles (ss), Lindy (rr), Little Bass (2) (rr), Little Butternut (rr), Little Rice (1) (r), Little Wood (rrr), Long (1) (rrr), Loveless (rr), Lower Nashotah (rr), Lower Nemahbin (rr), Lower Turtle (sss), Mendota (ss), Monona (s), Mud (1) (rrr), Nagawicka (ss), Nicaboyne (r), No Mans (rr), Number One (rr), Oconomowoc (rrr), Okauchee (rrr), Pokegama (3) (r), Prairie (r), Rice (1) (r), St. Croix (rr), Sishebogema (rr), South Turtle (sss), Upper Nashotah (rr), Upper Nemahbin (rr), Upper Turtle (rr), Waubesa (ss), Whitefish (rr), Winnebago (rr).

EUDORINA Ehrenberg 1832.

Colonies always motile, spherical or slightly elongate, of 16–32–64 cells lying some distance from one another and arranged to form a hollow sphere near the periphery of the homogeneous, hyaline, gelatinous envelope. Cells spherical, with or without a beak at the point of origin of the two cilia. Cilia parallel while passing through the colonial envelope and then widely divergent. Chloroplast single, cup-shaped, filling practically the whole cell and generally with several pyrenoids. Cells with a firm wall, one or two anterior contractile vacuoles, and a single conspicuous eyespot near the base of the cilia.

Asexual reproduction by a simultaneous division of all cells to form autocolonies which are liberated by a rupture of the colonial envelope.

Sexual reproduction heterogamous, dioecious, with all cells of a colony developing into large immobile oospheres or plate-like masses of 32–64 fusiform antherozoids; or monoecious with four cells forming antherozoids and the remainder oospheres. Zygote smooth-walled.

EUDORINA ELEGANS Ehrenberg. Pl. 19, Fig. 1.

Abh. d. kgl. Ak. d. Wiss. zu Berlin **1831**: 78, *pl.* 2, *figs.* 10A–10D. 1832; G. S. West Brit. Freshw. Algae 194, *figs.* 77A–77F. 1904.

Colonies generally 32–celled and as described above. (Facultative planktont).

Cells 12–24 μ in diam.; colonies 50–200 μ in diam.

Balsam (rrr), Bear (3) (ss), Bear Trap (r), Beaverdam (sss), Big Butternut (rr), Birch (2) (ss), Birch Island (rr), Boulder (cc), Bryer (rr), Camp (cc), Catherine (ss), Center (sss), Chetac (ss), Chetek (s), Clam (rr), Crawling Stone (rrr), Delavan (sss), Devils (1) (aa), Diamond (sss), Duck (ss), Dunham (ss), Dutchmans (rrr), Elizabeth (rr), Fish (sss), Fowler (rr), Granite (ss), Green (2) (rr), Grindstone (rrr), Hillman (rr), Hill (rrr), Hooker (rr), Horseshoe (1) (rr), Horseshoe (2) (rrr), Howey (rrr), Island (ss), Jennie (r), Kawaguesaga (rr), Kegonsa (ss), Lac Court Oreilles (rr), Lac la Belle (rr), Lake of the Woods (s), Lindy (rr), Little Butternut (rr), Little Wood (rrr), Long (3) (rrr), Loon (rr), Lower Nemahbin (rrr), Lower Turtle (r), Lynx (rr), Meronk (rrr), Middle McKenzie (rr), Nancy (rr), Nicaboyne (ss), No Mans (rr), North Turtle (ss), North Twin (rr), Number One (rr), Okauchee (rr), Paddock (cc), Pine (2) (rrr), Pike (r), Pokegama (2) (rr), Pokegama (3) (ss), Poplar (rr), Prairie (ss), Reserve (rr), Rainbow (rrr), Rice (1) (ss), Rooney (rr), Round (1) (r), Round (2) (rr), Round (3) (sss), Sand (1) (sss), Sand (4) (rr), Silver (2) (c), Sishebogema (ss), South Crab (rr), South Turtle (sss), Spirit (sss), Squirrel (ss), Tenderfoot (rrr), Travis (rrr), Vermilion (rr), Waubesa (sss), White Sand (rrr).

PLEODORINA Shaw 1894.

Colonies always motile; spherical to sub-spherical, with 32–128 cells lying some distance from one another just within the periphery of the homogeneous, hyaline, gelatinous, colonial envelope and not connected by cytoplasmic strands. Cells differentiated into those that are purely vegetative in character and those capable of dividing to form daughter colonies. All but four cells of the colony reproductive or about half reproductive and half vegetative. Cells spherical to ovoid in shape. Vegetative cells with a cup-shaped chloroplast containing one pyrenoid;

a large anterior eyespot; two cilia of equal length with two contractile vacuoles at their base. Reproductive cells at first like the vegetative cells, later with a more massive chloroplast that eventually contains several pyrenoids. The eyespot and cilia of reproductive cells disappearing when they are mature.

Asexual reproduction by a simultaneous division of all reproductive cells to form autocolonies.

Sexual reproduction (which has been observed in one species only) heterogamous, dioecious, by a division of all reproductive cells to form plates of 64–128 biciliate, fusiform antherozoids; or a development of reproductive cells into oospheres. Zygotes smooth-walled.

The validity of this genus has been questioned by many and the view put forward that the alga is but a stage in the life history of *Eudorina*. Grove (New Phytol. **14**: 169–182. 1915) gives a full discussion of the literature of *Pleodorina*. Although *Eudorina elegans* is found in a large number of Wisconsin lakes, *Pleodorina* has been found in but three of them. If *Pleodorina* is a developmental stage of *Eudorina*, it is strange that the "Pleodorina" stage has not been observed with greater frequency. The young colonies look very much like those of *Eudorina* and it is probably on this account that the two have been confused.

Pleodorina californica Shaw. Pl. **17**, Figs. **1–3**.

Bot. Gaz. **19**: 279, *pl.* 27, *figs.* 1–9. 1894.

Colonies containing 64–128 cells. Cells usually spherical and about half vegetative and half reproductive. Reproductive cells becoming 4–5 times broader than the vegetative cells. (Facultative planktont).

Vegetative cells 6–14 μ in diam.; reproductive cells 6–34 μ in diam. Colonies 40–400 μ in diam.

Clam (ss), Mud (1) (cc), Sand (1) (r).

VOLVOX L. 1758.

Colonies always motile, spherical to ovoid, containing a large number of cells arranged in a single layer just within the periphery of the homogeneous, hyaline, gelatinous, colonial envelope. Cells differentiated into those for vegetative purposes, asexual, and sexual reproduction. Vegetative cells close together or some distance from one another, with or without cytoplasmic connections of varying thickness. Cell shape spherical, ovoid, or disciform. Individual cells or groups of cells sometimes enclosed by their own gelatinous envelope. Cells with two cilia of equal length, 2–6 contractile vacuoles, one nucleus, one eyespot, and a cup-shaped to disciform chloroplast that contains one pyrenoid. Number of cells in a colony varying from 200–22,000.

Asexual reproductive cells few in number (rarely more than 20) and forming autocolonies by repeated division, the young colonies migrating to the center of the coenobe at a certain stage in their development.

Sexual reproduction heterogamous, monoecious or dioecious, with a development of 6–400 cells into spherical oospheres and few or many of the cells into antheridia that contain 16–256 biciliate fusiform antherozoids. Zygote with a smooth or stellate wall.

KEY TO THE SPECIES.

Cells with cytoplasmic connections.
 Cytoplasmic connections heavy............................V. GLOBATOR
 Cytoplasmic connections delicate..........................V. AUREUS
Cells without cytoplasmic connections.........................V. MONONAE

VOLVOX GLOBATOR L. Pl. 18, Figs. 3–5.

Systema natura, Ed. 10: 820. 1758; Cohn, Beitr. z. Biol. d. Pflanzen **1**, Heft 3: 93, *pl.* 2, *figs.* 1–6. 1875.

Colonies large, with many (1,000–22,000) cells that are flattened and pentagonal to octagonal in surface view. Cells connected to one another by stout cytoplasmic strands. Chloroplast of cells single, flattened and angular, with one pyrenoid. Cells with 2–6 contractile vacuoles at the base of the two cilia and with an eyespot. Gelatinous sheaths of individual cells sharply defined and (in surface view) pentagonal to octagonal by mutual compression. Generally with 8 asexual reproductive cells giving rise to autocolonies. Colonies commonly monoecious, with 12–40 (generally 20–30) oospheres enclosed by large gelatinous vesicles. Antherids a flat plate of 64–256 fusiform antherozoids with laterally attached cilia. Zygote verrucose, with conical warts. (Tychoplanktont).

Cells 2–7.5 (generally 3–5) μ in diam. Colonies 400–800 μ in diam.

Center (s), Hooker (rr), Kegonsa (rr), Monona (r), Soft (rrr), Waubesa (rrr).

VOLVOX AUREUS Ehrenberg. Pl. 18, Fig. 2.

Die Infusionsthierchen 71, *pl.* 4, *fig.* 2. 1838; Klein, Jarhb. f. wiss. Bot. **20**: 138, *pl.* 12, *figs.* 26–33. 1889.

Colonies spherical, containing a comparatively small (200–4,000) number of cells. Cells circular in surface view, connected with one another by very fine cytoplasmic strands. Individual gelatinous sheaths of cells confluent with the colonial envelope. Cells containing a spherical chloroplast, one eyespot, and a pair of contractile vacuoles at the base of the two cilia. 4–14 asexual reproductive cells giving rise to autocolonies. Colonies generally dioecious, with 3–9 oospheres and about half of the cells developing into antherids that contain 16–32 fusiform antherozoids with two terminal cilia. Zygote smooth-walled. (Facultative Planktont).

Cells 5–9 μ in diam.; colonies 200–680 μ in diam.

Beaverdam (rrr), Camp (sss), Center (ss), Delavan (ss), Eagle (rr), Elizabeth (rrr), Fowler (rrr), Jennie (rrr), Kawaguesaga (rrr), Lac du Flambeau (rr), Lac la Belle (rr), Mill (rrr), Oconomowoc (rrr), Paddock (r), Pokegama (1) (rr), St. Croix (rrr), Silver (2) (rr).

Volvox mononae sp. nov. Pl. **18**, Fig. **1**.

Colonies spherical, containing a comparatively small (400–2500) number of cells. Cells spherical, not connected with one another by cytoplasmic processes. Individual sheaths of cells confluent with colonial envelope. Cells containing a single bell-shaped chloroplast with one pyrenoid, an eye spot, and two cilia whose length is 2–3 times the diameter of the cell. 3–10 asexual colonies formed inside the coenobium and these daughter colonies in turn frequently containing young colonies. Asexual reproductive cells becoming much larger than vegetative cells before division commences. Colonies dioecious, with oogonia and antheridia in young daughter colonies only. 20–60 cells of a young colony formng typical *Volvox* antheridia that generally contain 64 antherozoids. Female colonies with 6–10 oospheres. Zygote smooth walled.

Vegetative cells 5–8 μ in diam. Colonies 50–350 μ in diam.

Monona (aaa).

Order **PROTOCOCCALES.**

Vegetative cells immobile; free-floating, sessile, or epiphytic; solitary or in colonies. Colonies amorphous and frequently embedded in a gelatinous envelope; or of definite shape and with or without a gelatienvelope. Colonies capable or incapable of increasing the number of their cells after they are formed. Cells variously shaped, generally with one chloroplast and pyrenoid.

Asexual reproduction by zoospores, fragmentation of the colony, autospores, or autocolonies. Resting akinetes known in some species.

Sexual reproduction by isogamous zoogametes known but not found in a majority of the genera.

KEY TO THE FAMILIES.

Asexual reproduction by zoospores.
 Cells epiphytic or endophytic.............................PLANOSPORACEAE
 Cells or colonies free-floating.
  Colonies capable of growth by cell division.................PALMELLACEAE
  Colonies incapable of growth by cell division...........HYDRODICTYACEAE
Asexual reproduction by autospores or division of cells.
 Cells capable of vegetative division.
  Solitary, in regular masses, or in strata.................PROTOCOCCACEAE
  In colonies held together by remains of old cell walls..DICTYOSPHAERIACEAE
 Cells incapable of division except to form autospores........AUTOSPORACEAE

Family *PALMELLACEAE.*

Cells spherical, ovoid, or reniform; rarely solitary, generally united to form colonies which are at times of macroscopic size. Colonies always embedded in a gelatinous matrix. Colonies capable of an increase in size by the vegetative division of the constituent cells. Chloroplasts generally single, cup-to disc-shaped and parietal; or star-shaped and central; with or without pyrenoids.

Reproduction by the fragmentation of the colony or by zoospores.

Sexual reproduction known in certain genera.

KEY TO THE GENERA.

Chloroplast star-shaped and central.............................ASTEROCOCCUS
Chloroplast cup- to disc-shaped and parietal.
 Cells with pseudocilia...TETRASPORA
 Cells without pseudocilia.
  Gelatinous envelope of colony lamellated.....................GLOEOCYSTIS
  Gelatinous envelope of colony homogeneous.
   Cells spherical.
    Chloroplast of old cells cup-shaped...................SPHAEROCYSTIS
    Chloroplast of old cells disciform.................PLANKTOSPHAERIA
   Cells ovoid to reniform.............................GLOEOCYSTOPSIS

GLOEOCYSTIS Nägeli 1849.

Cells spherical; solitary or in small colonies of eight cells or less. Each cell and the whole colony surrounded by a hyaline, lamellated, gelatinous sheath. Chloroplast of young cells cup-shaped, parietal, and with one pyrenoid; generally diffuse and rich in starch in old cells.

Asexual reproduction by a fragmentation of the colony through the softening of the gelatinous envelope, or by zoospores. Akinetes known.

GLOEOCYSTIS GIGAS (Kützing) Lagerheim. Pl. **19**, Fig. **2**.

Öfvers. kgl. Vet.-Ak. Förh. **40**: No. **2**: 63. 1883; G. S. West, Brit. Freshw. Algae 245: *figs.* 113F–113H. 1904.

Cells spherical; solitary or in families of 2–8; enclosed by a thick, distinctly lamellated, gelatinous envelope. Chloroplast of young cells cup-shaped, of older cells diffuse. (Tychoplanktont).

Cells 9–17 μ in diam.; colonies 45–95 μ in diam.

Mendota (rr), Oconomowoc (rr).

SPHAEROCYSTIS Chodat 1897.

Colonies spherical with the cells in groups of 4–8–16–32 and disposed towards the periphery of a hyaline, homogeneous, gelatinous envelope and with the envelope surrounding each group of cells distinct or confluent with the colonial envelope. Colonies sometimes forming palmella stages where the cells are irregularly disposed. Cells spherical; chloroplast cup-shaped or filling the entire cell, with one pyrenoid.

Reproduction by the liberation of a group of cells through the softening of the colonial envelope. Zoospores of two sizes have been observed.

SPHAEROCYSTIS SCHROETERI Chodat. Pl. **19**, Figs. **3–4**.

Bull. Herb. Boiss. **5**: 119, *pl.* 9, *figs.* 1–12. 1897; G. S. West, Jour. Linn. Soc. Bot. **39**: 76, *figs.* 9A–9F. 1909.

Characters the same as those of the genus. (Euplanktont).

Diam. cells 6–22 μ; diam. colonies 50–500 μ.

Balsam (rr), Bear (2) (ss), Bear Trap (sss), Beasley (rr), Big (sss), Big Butternut (rr), Birch (1) (rr), Birch (2) (ss), Birch Island (rr), Boulder (ss), Browns (rr), Camp (rr), Carson (rr), Catfish (rr), Center (rr), Chetac (ss), Chetek (cc), Clam (rrr), Clear Crooked (s), Crab (r), Cranberry (rr), Deer (rr), Devils (1) (sss), Eagle (rr), East (rr), Elizabeth (sss), Found (r) Fowler (rr), Green (1) (rr), Half Moon (rr), Hillman (rr), Hooker (rr), Kimball (rr), Kitten (rr), Lac Court Oreilles (rr), Lac du Flambeau (r), Lac la Belle (ss), Little Bass (1) (ss), Little Butternut (r), Little Doctor (rr), Little Wood (rr), Long (1) (ss), Long (2) (rr), Long (3) (rrr), Loveless (rr), Lower Nashotah (rr), Lower Nemahbin (cc), Lower Turtle (r), Mary (ss), Mendota (r), Meta (ccc), Mill (sss), Monona (r), Mud (1) (rr), Mudhen (sss), Nicaboyne (r), No Mans (rr), North Turtle (rr), North Twin (rr), Number One (cc), Oconomowoc (ss), Okauchee (ss), Old Taylors (rr), Pike (rr), Pine (2) (rr), Pine Tree (rr), Pokegama (1) (ss), Pokegama (3) (sss), Poplar (ss), Prairie (ss), Rainbow (rr), Rice (1) (ss), Rock (r), Round (1) (rrr), Round (2) (rr), Rudolph (rr), Sand (3) (rr), Sand (4) (rrr), Shell (ccc), Silver (2) (r), South Turtle (r), Tamarack (rr), Upper Nemahbin (rr), Upper Turtle (rr), Waubesa (rr), Winnebago (r).

Lemmermann follows Wille in placing the alga as a synonym of Braun's old genus *Gloeococcus* but I prefer to follow G. S. West's practice and consider the two genera distinct. The ordinary vegetative condition is easily recognized but the rather rare palmella stage may be mistaken for one of the limnetic *Tetraspora* species.

TETRASPORA Link 1809.

Colonies macroscopic or microscopic; attached or free-floating; gelatinous; spherical, cylindrical, expanded, or variously lobed. Cells spherical, generally in groups of four towards the periphery of the homogeneous, hyaline, gelatinous colonial envelope. Sometimes irregularly distributed within the gelatinous colonial envelope. Chloroplast cup-shaped, parietal; sometimes diffuse; with one pyrenoid. Each cell with two or four long, hyaline, gelatinous bristles (pseudocilia).

Asexual reproduction by the direct metamorphosis of vegetative cells into biciliate zoospores.

Sexual reproduction by a division of vegetative cells to form 4–8 bilciliate zoogametes that fuse in pairs.

TETRASPORA LACUSTRIS Lemmermann. Pl. **19**, Figs. **5–6.**

Bot. Cent. **76**: 152. 1898; Forschungsbr. a. d. Biol. Stat. zu Plön **7**: 118, *pl.* 1, *fig.* 13. 1899.

Colonies microscopic; spherical, elongate, or irregular in shape; always free-floating. Cells spherical, generally in groups of four and with pseudocilia 6–8 times as long as the diameter of the cell. Chloroplast single, cup-shaped, and parietal; with or without pyrenoids. (Euplanktont).

Diam. cells 7–10 μ; diam. colonies up to 300 μ.

Crawling Stone (sss), Fence (rrr), Long (3) (sss), Mud (2) (sss), Pike (rrr).

Small colonies frequently resemble palmella stages of *Sphaerocystis Schroeteri* but *Tetraspora* may be recognized by the pseudocilia. In case of doubt it is always advisable to stain with safranin to bring out the pseudocilia.

GLOEOCYSTOPSIS G. M. Smith 1916.

Cells elongate-cylindrical, with rounded ends and more or less curved. Cells aggregated in fours or eights within a homogeneous gelatinous sheath. Colonies spherical to ovoid and composed of an indefinite number of these aggregates. Colony not enclosed by a gelatinous sheath.

Chloroplast of young cells single, laminate, parietal, with one pyrenoid; diffuse in old cells.

Reproduction, aside from vegetative cell division, unknown.

GLOEOCYSTOPSIS LIMNETICUS G. M. Smith. Pl. **20**, Figs. **1–2**.

Bull. Torr. Bot. Cl. **43**: 475, *pl.* 24, *fig.* 12. 1916.

Cells 4–6 μ broad, 10–15 μ long. Cell aggregates 25–30 μ in diam.; colonies up to 125 μ in diam. (Euplanktont).

Bear Trap (rr), Beaverdam (rrr), Catfish (rr), Harris (rrr), Little Bass (2) (rrr).

A species that combines the general external morphological characteristics of *Gloeocystis* and the cell shape of *Nephrocytium*.

PLANKTOSPHAERIA G. M. Smith 1918.

Cells spherical, embedded in a copious, homogeneous, hyaline, gelatinous envelope. At first solitary, later irregularly distributed within the gelatinous envelope. Mature cells with several parietal, polygonal, flattened chloroplasts each of which contains a single pyrenoid.

Reproduction by liberation of daughter colonies through the breaking down of the gelatinous colonial envelope.

PLANKTOSPHAERIA GELATINOSA G. M. Smith. Pl. **20**, Figs. **3–6**.

Trans. Wis. Acad. Sci., Arts, & Lett. **19**[1]: 627, *pl.* 10, *figs.* 8–11. 1918.

Specific characters as described above. (Euplanktont).

Diam. cells 4.5–25 μ; colonies up to 150 μ in diam.

Browns (rr), Mendota (r).

The general appearance of the colony suggests *Sphaerocystis Schroeteri* but *P. gelatinosa* differs in the numerous parietal chloroplasts of each cell.

ASTEROCOCCUS Scherffel 1908.

Cells spherical or sub-spherical, solitary or in colonies, with a wide, homogeneous or lamellated, gelatinous envelope. Chloroplast single, central, star-shaped, the astral radiations ending in a disc at the cell wall. Pyrenoid single, at the center of the chloroplast. Certain species with two peripheral vacuoles and an eyespot.

Asexual reproduction by vegetative cell division and biciliate subspherical zoospores.

Sexual reproduction unknown.

Asterococcus limneticus G. M. Smith. Pl. 20, Figs. 7–10.

Trans. Wis. Acad. Sci., Arts, & Lett. **19**[1]: 627, *pl.* 10, *figs.* 3–6. 1918.

Colonies spherical, rarely ovoid, of 4–8–16 spherical cells widely separated from one another and lying within a hyaline, homogeneous, gelatinous matrix. Chloroplast single, star-shaped, with 4 or 8 (rarely 16) rays coming from a central mass and ending in a disc at the cell wall. Pyrenoid single, at the center of the chloroplast. (Euplanktont).

Diam. cells 7.5–35 μ; colonies up to 125 μ in diam.

Chain (r), Clam (sss), Crooked (rrr), Deer (rrr), Des Moines (rr), Dummy (ss), Fish (sss), Half Moon (rr), Hillman (rrr), Horseshoe (1) (ss), Horseshoe (2) (rr), Howey (rr), Lac Court Oreilles (ss), Lake of the Woods (rr), Long (1) (r), Nancy (r), North Twin (rrr), Owen (cc), Reserve (r), Rooney (rr), Sand (1) (r), Sand (3) (r), Shell (cc), Whitefish (sss), Wild Goose (rrr).

The shape and cellular arrangement in colonies of this species is quite similar to that of *Sphaerocystis,* but a sharp distinction exists in its star-shaped chloroplast, a type of chloroplast that is very uncommon. The alga is widely distributed in the lakes of northwestern Wisconsin.

Family *DICTYOSPHAERIACEAE.*

Cells in spherical or ovoid colonies, the number in young colonies a multiple of two, in older colonies indefinite. Colonies with or without a gelatinous envelope. Cells held together in fours or eights by the persistence of the old mother cell wall in repeatedly branching four-armed thongs or irregular threads. Chloroplast single; cup-shaped, or filling the entire cell; with one pyrenoid.

Reproduction by the division of cells into autospores which remain attached to mother colony until accidentally broken away.

KEY TO THE GENERA.

Colony enclosed in a gelatinous sheath .... Dictyosphaerium
Colony not enclosed in a gelatinous sheath.
  Cells spherical .... Westella
  Cells ovoid to reniform .... Dimorphococcus

DICTYOSPHAERIUM Nägeli 1849.

Cells spherical, ovoid, or reniform, connected with one another by cruciately or dichotomously branching threads. Colony enclosed by a hyaline, homogeneous, gelatinous envelope that is spherical or ovoid in shape. Chloroplast single, parietal, cup-shaped, with one pyrenoid.

Reproduction by the division of the cell into 2 or 4 daughter cells which remain connected to the colony by the old mother cell wall that divides into 2 or 4 parts to form the branching thread system connecting the cells. Zoospores reported but their existence is very doubtful. Resting akinetes known.

KEY TO THE SPECIES.

Mature cells ovoid...D. EHRENBERGIANUM
Mature cells spherical...**D. PULCHELLUM**

DICTYOSPHAERIUM EHRENBERGIANUM Nägeli. Pl. **20**, Figs. **11–12.**

Gatt. einz. Algen, 73, *pl.* 2, *figs.* Ea–E1. 1849.

Colonies spherical to ovoid, rarely containing more than 30 cells. Cells ovoid to ellipsoid when mature. Chloroplasts 1–2, parietal, laminate, generally with a single pyrenoid. (Facultative Planktont).

Cells 4–7 μ broad; 6–10 μ long.

Clear (rr), Kegonsa (rrr), Mendota (rr).

DICTYOSPHAERIUM PULCHELLUM Wood. Pl. **20,** Fig. **13**; Pl. **21,** Fig. **1.**

Smithsonian cont. to knowl. **19,** No. **241**: 84, *pl.* 10, *fig.* 4. 1872.

Colonies spherical to broadly ovoid, of 4–8–16–32 or many cells lying near the periphery of the hyaline, homogeneous, gelatinous envelope. Cells spherical when mature, spherical to ovoid when young; generally in fours at the ends of a repeatedly branching system of threads. Chloroplast single, parietal, cup-shaped, with one pyrenoid. (Facultative planktont).

Diam. cells 3–10 μ.

Amicoy (ss), Balsam (rrr), Bear (1) (ss), Bear (2) (r), Bettine (s), Big Butternut (r), Birch (1) (r), Birch (2) (rrr), Blake (ss), Bone (r), Boulder (rrr), Canteen (rr), Carson (sss), Catfish (r), Catherine (rrr), Center (r), Chain (rr), Chetac (ss), Chetek (ss), Crab (s), Cranberry (rr), Crawling Stone (rrr), Deer (rr), Diamond (rr), Duck (cc), Dummy (sss), Eagle (rr), Elizabeth (rrr), Fishtrap (rr), Granite (ss), Grassy (rr), Grindstone (rr), Half Moon (rr), High (rr), Hillman (ss), Horse (rr), Horseshoe (1) (rrr), Horseshoe (2) (ss), Howey (rr), Jag (rr), Kawaguesaga (rr), Kegonsa (r), Kitten (rrr), Lac Court Oreilles (ss), Lac du Flambeau (rr), Lindy (rr), Little Bass (1) (rr), Little Bass (2) (rr), Little Butternut (rr), Little Crooked (rrr), Little Rice (2) (sss), Little Wood (sss), Long (1) (rrr), Long (3) (rrr), Loon (r), Loveless (rr), Lower Turtle (r), Lynx (rrr), Mendota (ss), Mercer (rrr), Middle McKenzie (rrr), Monona (ss), Mud (2) (rr), Mudhen (rr), Nagawicka (rrr), Nancy (r), Nicaboyne (rr), North Turtle (r), Oconomowoc (r), Okauchee (rr), Old Taylors (rr), Owen (rrr), Palmer (rr), Pewaukee (rr), Pine Tree (sss), Pokegama (1) (r), Pokegama (3) (c), Poor

Farm (rrr), Prairie (c), Reserve (rr), Rice (ss), Rock (rr), Rooney (rr), Rose (sss), Rozen (rr), St. Croix (rrr), Sand (3) (rrr), Shell (cc), Sishebogema (rr), South Turtle (rr), Spirit (r), Tamarack (sss), Tenderfoot (rr), Tied Canoe (rrr), Travis (rr), Upper Nashotah (sss), Upper Nemahbin (rr), Vermilion (r), Wapagossett (rrr), Waubesa (ss), White Ash (sss), Whitefish (rr), White Sand (rr), Wild Goose (s), Winnebago (sss).

DIMORPHOCOCCUS A. Braun 1855.

Cells in groups of four and held in an irregular free-floating colony by the branching remains of the old mother cell wall. Colony not enclosed by a gelatinous envelope. Each aggregate of four cells lying in one plane with two cells ovoid to cylindrical with rounded ends, and two cells reniform to cardioid. Chloroplast of young cells single, parietal, laminate, with one pyrenoid; chloroplast of old cells occupying the entire cell.

Reproduction by the division of any cell into four daughter cells which remain attached to the colony by the threadlike remains of old cell wall until accidentally broken away.

DIMORPHOCOCCUS LUNATUS A. Braun. Pl. 21, Fig. 5.

Alg. Unicell. 44. 1855; Rabenhorst, Fl. Eur. Algarum **3**: 36, *figs.* a–c (p. 6). 1868; G. S. West, Brit. Freshw. Algae 221, *figs.* 93A–93B. 1904.

Cells of each group of four alternately arranged (in vertical view) the central cells cylindrical with rounded ends, the terminal cells reniform to cardioid. (Facultative planktont).

Cells 4–15 μ broad, 9–25 μ long. Colonies up to 150 μ in diam.

Bear (2) (rr), Bear (3) (rr), Beaverdam (rrr), Blue Gill (rrr), Catfish (rr), Des Moines (rrr), Devils (2) (rr), Horseshoe (1) (rr), Horseshoe (2) (rrr), Kegonsa (rrr), Lindy (rr), Minocqua (rrr), Mud (1) (rr), Pike (rr), Pine (1) (rr), Pine Tree (rrr), St. Croix (rrr), Shell (rrr).

WESTELLA de Wildeman 1897.

Cells small, spherical to sub-spherical, grouped in fours (rarely eights) in irregular, free-floating colonies containing 30–100 cells; the colony being held together by the non-gelatinizing remains of the old mother cell walls. Colonies without a gelatinous envelope. Chloroplast cup-shaped and parietal, or entirely filling the cell, with or without a pyrenoid.

Reproduction by the division of any cell to form 4–8 autospores that remain connected with the mother colony until accidentally broken away.

KEY TO THE SPECIES.

Groups with cells quadrate or pyramidate........................W. BOTRYOIDES
Groups with cells in a linear series................................W. LINEARIS

WESTELLA BOTRYOIDES (W. West) de Wildeman. Pl. **21**, Fig. **4**.

Bull. Herb. Boiss. **5**: 532. 1897.

Tetracoccus botryoides W. West, Jour. Roy. Micr. Soc. **1892**: 735, *pl.* 10, *figs.* 43–48. 1892.

Colonies irregularly shaped, without a gelatinous envelope, generally containing 40–80 cells. Cells spherical, pyramidately or quadrately arranged in groups of four (rarely eight). Chloroplast single, cup-shaped or entirely filling the cell; with or without a pyrenoid. (Facultative Planktont).

Diam. cells 3–9 μ.

Blake (rrr), Fishtrap (rr), Hillman (rrr), Horse (rrr), Horseshoe (1) (rr), Jag (r), Little Bass (2) (rr), Little Butternut (rrr), Little Crooked (rr), Mendota (rrr), Mill (rrr), Nancy (rr), Pine Tree (rr), Round (3) (rr), Sishebogema (rr).

var. MAJOR G. M. Smith.

Trans. Wis. Acad. Sci., Arts, & Lett. **19**[1]: 628. 1918.

Cells larger than the typical form, otherwise the same.

Diam. cells 8–13 μ.

Whitefish (rr).

Westella linearis sp. nov. Pl. **21**, Figs. **2–3**.

Colonies of irregular shape, rarely containing more than 40 cells. Cells spherical, always in groups of four and arranged in a linear series. Remains of old mother cell walls that hold the colony together very inconspicuous. Colony without a gelatinous envelope. Chloroplast single, parietal, without a pyrenoid. (Euplanktont ?).

Diam. cells 3–6 μ.

Pine Tree (rr).

In the arrangement of the cells this species is quite unlike the other members of the genus but since the cells are in groups of four that are held together by the remains of old mother cell walls it seems better to place it in this genus than in *Crucigenia*.

Family *PROTOCOCCACEAE.*

Cells solitary and spherical; or in colonies containing an indefinite number of cells that are spherical and mutually compressed or in very short irregular filaments. Cell wall delicate to heavy, smooth or variously sculptured. Chloroplast single, parietal and disciform, with or without pyrenoids. Cells capable of vegetative division.

Reproduction, aside from vegetative division of cells, by zoospores or aplanospores.

KEY TO THE GENERA.

Cell wall thin, smooth...Chlorella
Cell wall thick, sculptured..Trochiscia

CHLORELLA Beyerinck 1890.

Cells small, spherical; solitary or in small colonies of indefinite shape. Chloroplast disciform to cup-shaped, with or without a pyrenoid.

Reproduction by the division of cell contents into 2–4–8 daughter cells. Resting akinetes known.

CHLORELLA VULGARIS Beyerinck. Pl. **22,** Fig. **1.**

Bot. Ztg. **48:** 758, *pl.* 7, *figs.* 2a–2b. 1890.

Cells spherical, generally solitary. Chloroplast parietal, cup-shaped, with or without a pyrenoid. Cell wall smooth and delicate. (Facultative Planktont).

Diam. cells 5–10 μ.

Mendota (rr).

C. vulgaris is a minute alga and one that is easily overlooked. The record of its occurrence in Lake Mendota is based upon its isolation by pure culture methods from a plankton catch. It is very probable that it is present in other lakes but has been overlooked.

TROCHISCIA. Kütznig 1845.

Cells spherical to sub-spherical, solitary or in small clumps. Cell wall fairly heavy, areolate or ornamented with spines, denticulations, ridges, or other projections. Cells generally containing several parietal disciform chloroplasts each with a pyrenoid.

Reproduction by vegetative cell division or the formation of 4–16 aplanospores that are liberated by the dissolution of the old mother cell wall.

TROCHISCIA RETICULARIS (Reinsch) Hansgirg. Pl. **22**, Fig. **2**.

Hedwigia **27**: 129. 1888.

Acanthococcus reticularis Reinsch, Ber. d. D. bot. Ges. **4**: 242, *pl.* 12, *figs.* 24a–24b. 1886.

Cells generally solitary, spherical. Cell wall fairly thick and covered with linear ridges that intersect to form 50–70 polygons on the surface of the cell. (Tychoplanktont).

Diam. cells 24–32 μ.

Wildcat (rrr).

Family *AUTOSPORACEAE.*

Cells solitary or in colonies that are generally of a definite shape and without a gelatinous envelope. Cells variously shaped; with a single chloroplast and pyrenoid as a rule. Cells incapable of vegetative division to form two similar daughter cells.

Reproduction by division of cell contents into 2–4–8–16–32–64 cells which assume the shape of the mother cell before their liberation (autospores). In the colonial species the autospores from any one cell are organized to form the new colony (autocolony) before their liberation. Motile asexual reproductive bodies or sexual reproduction unknown.

KEY TO THE SUBFAMILIES.

Cells solitary.
  Cells angular .... TETRAEDREAE
  Cells not angular.
    Ornamented with long spines or setae .... MICRACTINIEAE (p. p.)
    Not ornamented with spines or setae.
      Length of cells several times the diameter .... SELENASTREAE (p. p.)
      Length of cells not more than twice the diameter .... OOCYSTEAE (p. p.)
Cells in colonies.
  Ornamented with long setae or spines.
    Coenobe capable of growth in three directions ...... MICRACTINIEAE (p. p.)
    Coenobe a flat plate .... CRUCIGENIEAE (p. p.)
  Not ornamented with setae or spines.
    Number of cells in coenobe always a multiple of two.
      Coenobe flat .... CRUCIGENIEAE (p. p.)
      Coenobe spherical or cells radiating from a common
        center .... COELASTREAE
    Number of cells not a multiple of two in every case.
      Cells elongate, straight or curved .... SELENASTREAE (p. p.)
      Cells spherical, ovoid, or reniform .... OOCYSTEAE (p. p.)

Subfamily *OOCYSTEAE.*

Cells spherical, ovoid, or reniform; solitary or in temporary colonies. Daughter cells retained within old mother cell wall for some time after their formation. Walls generally thick and cells with one to many chloroplasts.

KEY TO THE GENERA.

Cells within old mother cell wall separated by a dark gelatinous mass .... GLOEOTAENIUM
Cells within old mother cell wall not separated by a dark gelatinous mass.
  Cells ovoid to elliptic, never curved .... OOCYSTIS
  Cells ovoid to reniform, frequently curved .... NEPHROCYTIUM

OOCYSTIS Nägeli 1845.

Cells ovoid, ellipsoid, or cylindrical with rounded to somewhat pointed ends, generally symmetrical but never curved. Cell wall smooth, without spines, frequently with nodular thickenings at the poles. Chloroplasts one to many; parietal, disciform, stellate, or reticulate; with or without pyrenoids. Cells solitary or in temporary colonies of 2–4–8–16 or more enclosed by the partially gelatinized and greatly swollen old mother cell wall.

Reproduction by a formation of 2–4–8–16 autospores that remain within the old cell wall for some time. Three to four cell generations sometimes enclosed by the same wall. *Tetraedron*-like resting spores with granulate walls have been observed.

The genus differs from others in the expanded cell wall that encloses the cells in temporary colonies. Printz (Nyt. Mag. f. Naturvidensk. 1913, p. 165) has recently monographed the genus and cleared up many points by collecting the published descriptions of species; but there is great need of a series of cultural studies to determine the range of variation within the so-called "species" that are recognized at the present time.

KEY TO THE SPECIES.

Chloroplast stellate .... O. NATANS
Chloroplast disciform to laminate.
  Cells with polar nodules.
    Sides of cells concave .... O. PANDURIFORMIS
    Sides of cells convex.
      4 or more chloroplasts in each cell.
        Cells with 40–60 chloroplasts .... O. EREMOSPHAERIA

Cells with 6–25 chloroplasts.
Polar nodules conspicuous, 12–25 chloroplasts..........O. SOLITARIA
Polar nodules not prominent, 4–10 chloroplasts..........O. CRASSA
1–2 (rarely 4) chloroplasts in each cell.
Cells elliptic, length 1½ times breadth.................O. LACUSTRIS
Cells oblong-cylindrical, length 2–3 times breadth......O. SUBMARINA
Cells without polar nodules.
Poles of cells rounded.
Length of cells less than 12 μ........................O. PUSILLA
Length of cells more than 12 μ.
1–4 chloroplasts in each cell.........................O. BORGEI
8–10 chloroplasts in each cell.....................O. ELLIPTICA
Poles of cells pointed.
Old cell wall enclosing colonies rounded................O. PARVA
Old cell wall enclosing colony sharply pointed..O. GLOEOCYSTIFORMIS

OOCYSTIS PUSILLA Hansgirg. Pl. 22, Fig. 3.

Sitzbr. d. k. böhm. Ges. d. Wiss. in Prag, Jahrgang **1890**: 9. 1890; Printz, Nyt. Mag. f. Naturvidensk. **51**: 181, *pl.* 4, figs. 31–32. 1913.

Cells cylindrical, poles rounded and without nodules. Chloroplasts 2–3, parietal, laminate, without pyrenoids. Temporary colonies containing 2–4–8 cells. (Tychoplanktont).

Diam. cells 4.5–6.5 μ; length 8–12 μ.

Tied Canoe (rr).

OOCYSTIS BORGEI Snow. Pl. 22, Fig. 4.

Bull. U. S. Fish Comm. **22**: 379, *pl.* 2, *figs.* VII[1]–VII[5]. 1903.

Cells broadly elliptic, length 1½ times the breadth; poles rounded but without nodules. Chloroplasts 1–4, parietal, laminate, each containing a single pyrenoid. Cells solitary or in families of 2–4–8 within an old mother cell wall. (Euplanktont).

Cells 9–13 μ broad, 9–17 μ long.

Big (rr), Birch (1) (sss), Camp (r), Catfish (r), Devils (1) (rr), Duck (r), Found (rr), Green (2) (rr), Kawaguesaga (rr), Kitten (rrr), Long (2) (rr), Mendota (r), North Turtle (r), Pardee (rr), Rock (r), Winnebago (rr).

OOCYSTIS ELLIPTICA W. West. Pl. 22, Fig. 5.

Jour. Roy. Micr. Soc. **1892**: 736, *pl.* 10, *fig.* 56. 1892.

Cells oblong-elliptic, with poles rounded and without nodules. Chloroplasts 10–20, disciform, parietal, without (?) pyrenoids. Cells rarely solitary, generally in families of 4–8 within an old mother cell wall. (Tychoplanktont).

Cells 11–12.5 μ broad, 20–25 μ long.

Amicoy (rr).

OOCYSTIS PARVA W. & G. S. West. Pl. 22, Fig. 6.

Jour. of Bot. 36: 335. 1898; G. S. West, ibid. 37: 223, *pl.* 394, *figs.* 14–17. 1899.

Cells broadly fusiform to elliptic, ends pointed but without polar nodules. Each cell with 1–3 parietal, laminate to disciform chloroplasts, with or without pyrenoids. Cells solitary or in families of 2–4–8. (Facultative Planktont).

Cells 4–7 μ broad; 6–12 μ long.

Beaver (1) (rr), Bettine (rr), Clear Crooked (rr), Cochrane (rrr), Crab (rrr), Deer (rr), George (rr), Half Moon (rr), Hooker (rr), Little Bass (2) (rr), Lynx (rr), Mendota (rr), Muskallonge (rr), Palmer (rr), Pine (1) (rr), Rudolph (rr), Sand (1) (rrr), Stone (rr), Wolf (rr).

OOCYSTIS GLOEOCYSTIFORMIS Borge. Pl. 22, Fig. 7.

Botaniska Studier till. F. R. Kjellman 23, *pl.* 1, *figs.* 1a–1b. 1906.

Cells narrowly elliptic with ends pointed. Chloroplast single, parietal, without a pyrenoid. Cells in families of 2–4 within the old cell wall, several cell generations frequently forming a single colony. Wall enclosing colony with pointed ends but no polar nodules. (Tychoplanktont).

Breadth of cell 3–5.5 μ; length 8–15 μ.

Lac Court Oreilles (sss).

The old mother cell wall does not gelatinize as readily in this species as in others so that several cell generations are generally found in the same colony. The species is best recognized, however, by the sharply pointed ends of the colony wall.

OOCYSTIS LACUSTRIS Chodat. Pl. 22, Figs. 8–9.

Bull. Herb. Boiss. 5: 119, *pl.* 10, *figs.* 1–7. 1897.

Cells elliptic to broadly fusiform, somewhat pointed and with a nodule at the poles. Chloroplasts 1–3, laminate, parietal, with or without pyrenoids. Margins of chloroplasts sometimes irregular. Cells rarely solitary, generally in families of 2–4–8 within the old mother cell wall. (Euplanktont).

Cells 8–22 μ broad, 14–32 μ long.

Big Butternut (rr), Bone (rr), Center (r), Chain (rr), Chetek (r), Devils (2) (ss), Dummy (rr), Fowler (rr), Hillman (rr), Little Bass (2) (rr), Little Butternut (rrr), Long (2) (rrr), Long (3) (rrr), Mary (r), Mercer (rr), Palmer (rr), Pokegama (3) (rr), Prairie (rr), Reserve (rr), Sand (4) (rr), Sishebogema (rrr), Upper Nashotah (rrr), Upper Turtle (sss), White Sand (rr).

OOCYSTIS SUBMARINA Lagerheim. Pl. **22**, Fig. **10**.

Bot. Not. **1886**: 45, *fig.* 1. 1886.

Cells oblong-cylindrical, 2–3 times as long as broad, somewhat pointed and with a nodule at each pole. Chloroplasts 1–2, laminate, parietal, usually with a pyrenoid. Cells rarely solitary, generally in families of 2–4–8–16 within the old mother cell wall. (Tychoplanktont).

Cells 3–9 μ broad, 7–20 μ long.

Camp (sss), Upper Nemahbin (rr).

OOCYSTIS SOLITARIA Wittrock. Pl. **22**, Fig. **11**.

Bot. Not. **1879**: 24, *figs.* 1–5. 1879.

Cells ovoid to ellipsoid, solitary or in families of 2–4–8. Cell wall fairly thick and with conspicuous polar nodules. Chloroplasts numerous (generally 12–25), parietal, disciform to polygonal, with or without pyrenoids. (Facultative Planktont).

Cells 7–20 μ broad, 14–35 μ long.

Hooker (r), Sand (1) (rr).

OOCYSTIS CRASSA Wittrock. Pl. **22**, Figs. **12–13**.

Bot. Not. **1880**: 117. 1880.

Cells ovoid, length 1½ times the breadth, with a nodule at each pole. Chloroplasts 4–10 in number, fairly large, parietal, disciform to angular, generally with a pyrenoid. Cells solitary or in families of 2–4–8. Old cell wall enclosing the colony greatly distended. (Tychoplanktont).

Cells 10–20 μ broad, 14–26 μ long.

St. Croix (rr).

OOCYSTIS EREMOSPHAERIA G. M. Smith. Pl. **23**, Figs. **1–2**.

Trans. Wis. Acad. Sci., Arts, & Lett. **19**[1]: 630, *pl.* 14, *figs.* 8–9. 1918.

Cells ovoid, 1½ to 2 times as long as broad. Solitary or in families of 2–4. Chloroplasts numerous (40–60), parietal, lenticular, each containing a single pyrenoid. Cell wall thick, with a conspicuous polar nodule. (Facultative Planktont ?).

Cells 23–31 μ broad, 35–45 μ long.

Beaverdam (rr), Shell (sss).

OOCYSTIS PANDURIFORMIS var. MINOR G. M. Smith. Pl. **23**, Fig. **3.**

Bull. Torr. Bot. Cl. **43**: 471, *pl.* 24, *fig.* 2. 1916.

Cells oblong-ovoid, 2 to 2½ times as long as broad, sides slightly concave and poles with a conspicuous nodule. Chloroplasts numerous (15–25), parietal, disciform, and each containing a pyrenoid. Cells solitary or in families of 2–4–8. (Tychoplanktont).

Greatest diameter of cells 12–15 μ, median diameter 11–14; length 30–35 μ.

Star (rr).

This is the only species of the genus where the sides of the cells are concave.

Oocystis natans var. **major** var. nov. Pl. **23,** Figs. **4–5.**

Cells ovoid, without a conspicuous polar nodule. Chloroplasts 4–8, parietal, star-shaped or irregular, each chloroplast with a pyrenoid. More than two cell generations rarely found within the same mother cell wall. (Facultative Planktont).

Cells 16–25 μ broad, 31–38 μ long. Families up to 90x120 μ.

Birch (2) (sss), Chetac (s), Rice (1) (rr).

Lemmermann founded the genus *Oocystella* upon this stellate nature of the chloroplast (Zeits. f. Fischeri u. d. Hilfsw. **11**: 103. 1903) but the genus has not been recognized and the single species is placed in *Oocystis.*

NEPHROCYTIUM Nägeli 1849.

Cells ovoid, reniform, or oblong-elliptic; generally in colonies of 2–4–8–16 cells within the partially gelatinized remains of the old mother cell wall. Arrangement of cells in young colonies spiral, in old colonies irregular. Chloroplast single and expanded at first, later fragmenting and becoming diffuse; pyrenoid single.

Reproduction by the formation of 2–4–8–16 autospores in each cell.

NEPHROCYTIUM AGARDHIANUM Nägeli. Pl. **23,** Figs. **6–7.**

Gatt. einz. Algen 79, *pl.* 3, *figs.* Ca–Cp. 1849.

Cells reniform, generally spirally arranged in young colonies. Chloroplast single, parietal, laminate, and with one pyrenoid in young cells; diffuse in old cells. (Tychoplanktont).

Cells 2–12 μ broad, 6–28 μ long; colonies up to 60 μ long.

Bear (3) (rrr), Bear Trap (rrr), Camp (rr), Hillman (rr), Little Bass (rrr), Mendota (rr), Mill (rrr).

GLOEOTAENIUM Hansgirg 1890.

Cells spherical, ovoid, or variously flattened; solitary or united in colonies of 2–4–8 cells enclosed by a non-expanded mother cell wall and separated from one another by a dark colored gelatinous mass that lies between the cells. Colonies rectangular, with rounded corners in front view; oblong with sharp corners in side view. Cell wall thick, distinctly lamellated; old cell wall enclosing the colony frequently with folds and gelatinous discs opposite the cells. Chloroplast entirely filling cells; with or without a pyrenoid.

Reproduction by autospores and aplanospores.

GLOEOTAENIUM LOITLESBERGERIANUM Hansgirg. Pl. **23,** Figs. **8–9.**

Sitzbr. d. k. böhm. Ges. d. Wiss. in Prag, Jahrgang **1890**: 10. 1890; Transeau, Bot. Gaz. **55**: 66, *pl.* 3, *figs.* 1–24. 1913.

Cells spherical to ovoid, completely filling the space inside of old mother cell wall. (Tychoplanktont).

Diam. cells 20–30 μ; length 2-celled colonies 40–70 μ breadth 22–40 μ; thickness 20–30 μ. 4-celled colonies 40–70 μ broad, 40–80 μ long.

Big Butternut (rrr).

Sub-family *TETRAEDREAE.*

Cells always solitary, without a gelatinous envelope, flattened or isodiametric, triangular, quadrangular, or polyangular; angles simple or produced into spines or long hyaline setae.

KEY TO THE GENERA.

Cells simple or with fairly stout spines at angles.................TETRAEDRON
Cells with long hyaline setae at angles.........................POLYEDRIOPSIS

TETRAEDRON Kützing 1845.

Cells solitary, free-floating, flattened or isodiametric, triangular, quadrangular, or polyangular. Angles simple or produced into simple or furcate spines. Cell wall smooth or verrucose. Chloroplasts one to many, parietal, disciform to angular; or completely filling the cell; with or without pyrenoids.

Reproduction by division of cell contents into 4–8 autospores that are liberated by the rupture of the old mother cell wall.

KEY TO THE SPECIES.

Angles of cells never produced into processes.
  Corners of cells without spines.
    Cells triangular..T. MUTICUM
    Cells quadrangular..T. MINIMUM
  Corners of cells with spines.
    All cell corners in the same plane.
      Cells triangular..T. TRIGONUM
      Cells with more than three angles.
        Cells with four angles..................T. REGULARE v. INCUS (p. p.)
        Cells with five angles................................T. CAUDATUM
    All cell corners not in the same plane.
      Cells quadrangular.
        Pyramidal cells.......................................T. REGULARE
        Halves of cells twisted over one another.
          Wall smooth.
            Angles delicately tapering..............T. VICTORIEAE var. MAJOR
            Angles strongly convex..........T. REGULARE var. TORSUM (p. p.)
          Wall verrucose......................................T. VERRUCOSUM
      Cells five-angled.......................................T. PENTAEDRICUM
Angles at corners produced into processes.
  Processes always unbranched.................................T. HASTATUM
  Processes always branched.
    Cells flattened and processes in the same plane.
      General outline of cell H-shaped........................T. CONSTRICUM
      General outline of cell not H-shaped.
        Processes long, conspicuously branched..................T. GRACILE
        Processes short, not conspicuously branched....T. LOBULATUM (p. p.)
    Cells isodiametric and processes not in the same plane.
      Processes short...T. ENORME
      Processes long.
        With 3–5 series of branchings......T. LOBULATUM var. POLYFURCATUM
        Never more than three series of branchings.
          Diameter of cells always greater than the length
              of the processes..................................T. PLANCTONICUM
          Diameter of cell always less than the length of
              the processes.....................................T. LIMNETICUM

TETRAEDRON MUTICUM var. PUNCTULATUM (Reinsch) de Toni. Pl. **23**, Figs. **10–13**.

Sylloge algarum **1**: 600. 1889.

Polyedrium muticum var. *punctulatum* Reinsch, Notarisia **3**: 498, *pl.* 4, *fig.* 6. 1888.

Cells small, triangular, somewhat flattened. Sides of cells convex in front view. Corners sharp but without spines. Cell wall covered with small granules. (Tychoplanktont).

Diam. cells 15–20 μ; thickness 8–10 μ.

Beaverdam (rr), Deer (rrr), Sand (2) (rr).

TETRAEDRON TRIGONUM (Nägeli) Hansgirg. Pl. **23,** Figs. **14–16.**

Hedwigia **27**: 130. 1888.

Polyedrium trigonum Nägeli, Gatt. einz. Algen 84, *pl.* 4, *figs.* B1a–B1b. 1849.

Cells medium sized, triangular, flattened. Sides of cells generally concave, rarely straight. Corners with a single, stout, straight or curved spine. (Facultative Planktont).

Diam. cells (without spines) 20–30 μ; length spines 8–10 μ.

Crawling Stone (rrr), High (r), Pokegama (1) (rr), Rice (1) (rrr), Tenderfoot (rr), Tied Canoe (rrr).

var. SETIGERUM (Archer) Lemmermann. Pl. **24,** Figs. **1–4.**

Arkiv. för Botanik **2**, No. **2**: 110. 1904.

Tetrapedia setigera Archer, Quart. Jour. Micr. Sci. N. S. **12**: 365, *pl.* 21, *figs.* 14–17. 1872; Grevillea **1**: 46, *pl.* 3, *figs.* 14–17. 1872; G. S. West, Brit. Freshw. Algae 349. 1904; de Toni, Sylloge algarum **5**: 113. 1907.

Polyedrium trigonum var. *setigerum* (Archer) Schröder, Forschungsbr. a. d. Biol. Stat. zu Plön **6**: 23, *pl.* 1, *fig.* 6. 1898.

Cells small, triangular, flattened, with a conspicuous emargination on each side in front view. Corners of cells rounded and produced into a long delicately tapering spine. Chloroplast single, filling the entire cell; with one pyrenoid. (Facultative Planktont).

Diam. cells (without spine) 7–9 μ; length of spines 12–15 μ.

White Ash (r).

G. S. West and de Toni follow Archer in considering this organism a member of the Myxophyceae West stating; "*T. setigera* Archer is a beautiful little species known from N. W. Scotland and W. Ireland, which has been erroneously referred by some authors to the chlorophyceous genus Tetraëdron. The cell contents of *Tetrapedia setigera* are homogeneous and of a pale blue-green color." The cell contents of the individuals that I have observed are distinctly grass green in color and contain a single pyrenoid. This pyrenoid has not been noted by either Archer or Schröder. Since the shape and dimensions of the cells that I have found are exactly the same as those given by Archer I believe that I am dealing with the same organism that he described and one which is a member of the Chlorophyceae and not the Myxophyceae.

var. GRACILE (Reinsch) de Toni. Pl. **24,** Figs. **5–9.**

Sylloge algarum **1**: 598, 1889; Printz, Kgl. Norske Vidensk. Selsk. Skr. **1915,** No. **4**: 29, *pl.* 3, *figs.* 167–169. 1916.

Polyedrium trigonum var. *gracile* Reinsch, Die Algenfl. d. mitt. Th. v. Franken 75: *pl.* 3, *figs.* 1a–1b. 1867.

Tetraëdron proteiforme G. M. Smith non (Turner) Brunnthaler, Trans. Wis. Acad. Sci., Arts, & Lett. **19**[1]: 632, *pl.* 15, *figs.* 4–5. 1918.

Cells triangular, flattened, concavity of sides very marked. Angles of cells ending in short spines. Several parietal, disciform chloroplasts without pyrenoids. (Facultative Planktont).

Cells 23–50 μ broad; 6–10 μ thick.

Blake (rr), Chain (r), Cochrane (r), Fishtrap (rr), Horse (r), Lac Court Oreilles (rr), Little Bear (rr), Little Crooked (rrr), Loon (rr), Paddock (rrr), Palmer (rrr), Rice (1) (rr), Tenderfoot (r), Tied Canoe (rrr), White Ash (rr), Wolf (rr).

In the second list of Wisconsin algae this alga was given as *T. proteiforme* but I am now convinced that it is *T. trigonum* var. *gracile.* The cells reach a much larger size than that given by Reinsch and the excavation of the sides may be very much greater; in fact the cells are at times so delicate that they might be placed as *Cerasterias* species had not other individuals in the same collection the ordinary cell shape of *T. trigonum* var. *gracile.*

TETRAEDRON MINIMUM (A. Braun) Hansgirg. Pl. **24,** Figs. **10–13.**

Hedwigia **27**: 131. 1888; G. S. West, Brit. Freshw. Algae 231, *fig.* 101A. 1904.

Cells small, flattened, front view square or oblong with slightly rounded corners and concave sides. Side view fusiform-elliptic. Wall smooth or finely granulate. Chloroplast single, parietal, with one pyrenoid. (Facultative Planktont).

Cells 6–20 μ long, 5–8 μ thick.

Beaver (2) (rrr), Camp (rr), Deer (rrr), Hillman (rrr), Horse (rr), Little Bass (2) (rrr), Little Doctor (rr), Mendota (rrr), Okauchee (rrr), Paddock (rrr), Round (3) (rrr).

TETRAEDRON REGULARE Kützing. Pl. **24,** Fig. **14.**

Phycol. germanica 129. 1845.

Polyedrium tetraedricum Nägeli, Gatt. einz. Algen 84, *pl.* 4, *figs.* B3a–B3b. 1849.

Cells medium sized, tetragonal, generally pyramidal, sides concave, straight, or slightly convex. Angles with a single, heavy, short, blunt, spine. (Tychoplanktont).

Diam. cells (with spines) 14–45 μ.

Blake (rr).

var. INCUS Teiling. Pl. **24,** Figs. **15–16.**

Sv. Bot. Tids. **6**: 277, 274, *fig.* 12. 1912.

Polyedrium regulare Chodat non Kützing, Matér. pour la flore crypt. Suisse 1[3]: 220, *fig.* 146. 1902.

Cells small, tetragonal, flattened or pyramidal, sides slightly concave. Angles produced into a single short delicate straight or curved spine. (Tychoplanktont).

Diam. cells (without spine) 13–18 μ; length of spines 7–8 μ.

Poor Farm (rr).

This is probably what Chodat has figured as *T. regulare.* It differs from the typical form in the smaller size, constantly concave sides of the cell, and the delicate spines at the cell angles.

var. TORSUM (Turner) Brunnthaler. Pl. **24**, Figs. **17–18.**

in Pascher, Süsswasserfl. Deutschl., österr. u. d. Schw. 5[2]: 150, *fig.* 169. 1915.

Cells medium sized, tetragonal, with the two halves of the cells cruciately arranged. Sides of cells straight or convex. Angles of cells with a single short stout spine (Facultative Planktont).
Diam. cells 25–40 μ.

Horse (sss).

Over a hundred individuals were observed in the plankton catch from this lake and the twisting of the cell halves was a very constant character.

Tetraëdron victorieae var. **major** var. nov. Pl **24,** Figs. **19–22;** Pl. **25,** Fig. **1.**

Cells medium sized, four-sided, with two deeply emarginate sides dividing the cell into two cruciately arranged halves. Sides of cells, in vertical view, broadly fusiform with acute apices. All angles of cells with a single heavy straight spine (Euplanktont).
Cells (with spines) 30–60 μ long; 15–20 μ broad.

Chain (r).

The cell shape and the peculiar twisting of the two halves is identical with that of *T. victorieae* Woloszynska (Hedwigia 55: 203, *pl.* 6, *figs.* 1–4. 1914) which is known from the plankton of Lake Victoria Nyanza, Africa. Measurement of about 25 Wisconsin individuals give the following variation in size, length 30–60 μ, breadth 15–20; whereas the length of the African individuals varies from 20 to 30 μ and the breadth 10–15 μ.

TETRAEDRON VERRUCOSUM G. M. Smith. Pl. **25,** Figs. **2–3.**

Trans. Wis. Acad. Sci., Arts, & Lett. **19**[1]: 632, *pl.* 15, *figs.* 1–2. 1918.

Cells large, tetragonal, pyramidate, or with the halves cruciate. Sides convex. Angles of cell terminating in a very heavy, blunt spine. Cell wall thick, verrucose. (Euplanktont).
Diam. cells (with spines) 65–80 μ; length of spines 13–18 μ.

Bear (1) (rr).

TETRAEDRON CAUDATUM (Corda) Hansgirg. Pl. 25, Figs. 4–7.

Hedwigia **27**: 131. 1888; G. S. West, Brit. Freshw. Algae 231, *fig.* 101B. 1904.

Cells small, flattened, five-sided; four of the sides concave and the fifth deeply incised. Angles rounded and produced into a single straight spine that lies parallel with the flat side of the cell. (Tychoplanktont).

Diam. cells 6–22 μ; length of spines 1–3.5 μ.

Chain (rrr), East (rrr), Dutchmans (rrr), Horse (rr), Lake of the Woods (rrr), Mendota (rrr), Palmer (rrr), Reserve (rrr), Wild Goose (rrr).

The notch at one side of the cell varies considerably in depth but is always deeper than the incurving of the other four sides.

var. LONGISPINUM Lemmermann. Pl. 25, Figs. 8–12.

Bot. Cent. **76**: 151. 1898; Forschungsbr. a. d. Biol. Stat. zu Plön **7**: 117, *pl.* 1, *figs.* 8–9. 1899.

Spines on corners at right angles to the flat side of the cell, generally two pointing in one direction and three in the other. Spines much longer than in the typical form. (Euplanktont).

Diam. cells (without spines) 8–18 μ; length of spines 3–8 μ.

Pine Tree (r), Travis (sss)

This is a sharply defined variety, differing from the typical form in the bending of the spines so that they lie at right angles to the plane of the cell. Lemmermann's figure 9 shows this very well but for some reason the cell in his figure 8 has no spines and so is not at all characteristic.

TETRAEDRON PENTAEDRICUM W. & G. S. West. Pl. 25, Figs. 13–17.

Trans. Linn. Soc. 2nd Ser. Bot. **5**: 84, *pl.* 5, *figs.* 15–16. 1895.

Cells small, five-cornered, with four corners in one plane and the fifth lying at an angle of 30–90 degrees. Corners subacute, each with a single, short, curved spine. Chloroplast single, parietal, with one pyrenoid. (Facultative Planktont).

Diam. cells (without spines) 5.5–14 μ, (with spines) 9–20 μ.

Soft (rr).

T. pentaëdricum has not been observed since it was described from Madagascar by the Wests. Chodat considers it a synonym of *T. caudatum* (Corda) Hansg. erroneously citing it as *Tetrapedia pentaëdricum* (Matér. pour la fl. crypt. Suisse 1^3: 219. 1902). Isolation of this alga in pure culture with the retention of its distinguishing character shows that the Wests are correct in considering it a distinct species. It resembles *T. caudatum* in the size, number of corners, character of spines at the corners, and the concave sides; but differs in having four corners in approximately the same plane and the fifth at an angle with the plane of the other four. There is considerable variation in the angle at which this fifth corner lies but it is never in the same plane as the other corners. The Wests (1. c. fig. 17) also describe a variety *minima* which, as the name implies, is smaller. I find that the variations in size of cells of the typical form in a pure culture include the dimensions set for the variety *minima* so that this variety cannot be recognized.

TETRAEDRON HASTATUM (Reinsch) Hansgirg. Pl. 25, Fig. 18.

Hedwigia **27**: 132. 1888.

Polyedrium tetraedricum var. *hastatum* Reinsch, Die Algenfl. d. mitt. Th. v. Franken 77, *pl.* 5, *figs.* 3a–3b. 1867.

Cells medium sized, pyramidate, with the angles produced into long tapering and somewhat concave processes. Processes unbranched but having 2 or 3 short spines at the tip. (Tychoplanktont).

Diam. cells (with processes) 28–36 μ; diam. base of process 8 μ.

Horse (rrr).

The general shape of the cell is similar to that of *T. limneticum* but there is no branching of the processes.

var. PALATINUM (Schmidle) Lemmermann. Pl. 25, Figs. 19–21.

Ber. d. D. bot. Ges. **20**: (247). 1902.

Polyedrium hastatum var. *palatinum* Schmidle, Ber. d. D. bot. Ges. **18**: 148, 149, *pl.* 6, *figs.* 4–5. 1900.

Cells pyramidal, with convex sides, angles produced into fairly delicate non-tapering processes about as long as the diameter of the cells. Processes unbranched but with three short spines at the apex. (Tychoplanktont).

Diam. cells (without processes) 4–14 μ; length processes about 10 μ.

Horse (rr).

The processes at the angles of the cell do not seem as much a part of the cell as in the typical form. Although both the typical form and the variety *palatinum* were found at the same station the two seem distinct and without transition forms.

Tetraëdron constrictum nom. nov. Pl. **25,** Figs. **22–24.**

Tetraëdron arthrodesmiforme var. *lobulatum* Woloszynska, Hedwigia **55**: 203, *pl.* 6, *fig.* 11. 1914.

Cells medium sized, tetragonal, flattened, two sides straight and two opposite sides concave. Angles produced into narrow slightly tapering processes which give the cell a decided H-shape. Ends of processes generally dichotomously branched the branchlets ending in two short spines. (Euplanktont).

Diam. cells (without processes) 5–8 μ; length 18–25 μ; thickness 8 μ. Length of processes 10–15 μ.

Bone (rr).

Woloszynska has found these peculiar H-shaped cells in the plankton of Lake Victoria Nyanza, Africa and placed them among other varieties of *T. arthrodesmiforme* described at the same time. The variety *lobulatum* is, however, sufficiently different from the other varieties to warrant raising it to specific rank. Bone lake is the only station where it has been found but several individuals were seen in the catch from this lake.

TETRAEDRON GRACILE (Reinsch) Hansgirg. Pl. **26,** Figs. **1–5.**

Hedwigia **28**: 19. 1889.

Polyedrium gracile Reinsch, Notarisia **3**: 502, *pl.* 7, *figs.* 1a–1c. 1888.

Cells medium size, conspicuously flattened, rectangular, with flattened sides. Angles of cells produced into narrow processes that generally branch twice and end in minute spines. The primary branchings of the processes are at an angle of 90 degrees with one another and always parallel with a side of the cell. Each cell containing several, parietal disciform to laminate chloroplasts without pyrenoids. (Euplanktont).

Diam. cells (without processes) 15–30 μ; (with processes) 35–80 μ; thickness of cells 6–12 μ.

Bear (1) (rr), Bear Trap (rrr), Blake (rr), Chain (rr), Elizabeth (rrr), Hillman (sss), Loon (rrr), Pine (2) (rr), Poor Farm (rr), Round (rr), Travis (r), Wapagossett (rr).

TETRAEDRON LOBULATUM (Nägeli) Hansgirg. Pl. **26,** Figs. **6–7.**

Hedwigia **27**: 132. 1888.

Polyedrium lobulatum Nägeli, Gatt. einz. Algen 84, *pl.* 4, *figs.* B4a–B4b. 1849.

Cells medium sized, tetragonal, flattened, or pyramidal. Processes of angles with one or two short dichotomous branchings that generally do not end in spines. Sides of cells distinctly concave. Cells containing several parietal disciform chloroplasts without pyrenoids. (Tycho-planktont).

Diam. cells (with processes) 30–40 μ.

Birch Island (rrr), Long (3) (rrr), Sand (2) (rr).

The specimens from Sand lake were of particular interest since three showed the production of autospores (fig. 6). The chloroplast of old cells are frequently indistinct but they can be seen quite readily in young cells.

var. POLYFURCATUM G. M. Smith. Pl. **26,** Fig. **8.**

Bull. Torr. Bot. Cl. **43**: 480, *pl.* 26, *figs.* 21–22. 1916.

Cells tetragonal, flattened or pyramidal; with concave, straight, or slightly convex sides. Processes at angles somewhat tapering, 3–5 times dichotomously branching, the ultimate branchlets with 2–3 minute spines. (Facultative Planktont).

Diam. cells (without processes) 15–25 μ, (with processes) 35–70 μ.

Beaverdam (r), Blake (rr), Bone (rr), Clare (rr), Fish (rr), Loon (rr), Muskallonge (rrr), Palmer (r), Pardee (r), Pine Tree (rr), Rice (1) (rr), Shell (rr), Wild Goose (rrr).

Differing from the typical form in the greater branching of the processes and from *T. enorme* (Ralfs) Hansg. in the much longer processes.

TETRAEDRON LIMNETICUM Borge. Pl. **27,** Figs. **1–3.**

Bot. Not. **1900**: 5, *pl.* 1, *fig.* 2. 1900; W. & G. S. West, Trans. Roy. Soc. Edinburgh **41**: 507, *pl.* 6, *figs.* 10–11. 1905.

Cells of medium size, tetragonal, pyramidal, the four angles being produced into one or two dichotomous branchings. Terminal branchlets with 2–3 short spines. (Euplanktont).

Diam. cells (with processes) 45–85 μ; diam. base of processes 8–12 μ.

Chain (rr), Chetek (r), Clear (r), Crawling Stone (rrr), Deer (rr), Diamond (rrr), Island (rr), Kitten (rr), Little Butternut (rr), Little Crooked (rrr), Mud (2) (r), Nancy (rr), Palmer (rr), Pine Tree (rr), Poplar (rr), Spirit (rrr), Squirrel (rr), Wildcat (rr), Wild Goose (rrr), Wolf (rr).

TETRAEDRON PLANCTONICUM G. M. Smith. Pl. **27,** Figs. **4–5.**

Bull. Torr. Bot. Cl. **43**: 479, *pl.* 26, *figs.* 19–20. 1916.

Cells medium-sized, four to five sided, sides concave to convex. Angles of cells produced into narrow, once to twice dichotomously branched processes. Ultimate branchlets with 2–3 short spines. Processes broad in front view, narrow in side or vertical view. (Euplanktont).

Diam. cells (without processes) 18–30 μ; (with processes) 45–60 μ.

Cranberry (rr), Cochrane (rr), Devils (2) (sss), Granite (rr), Horseshoe (1) (rr), Lindy (rrr), Lost (rr), No Mans (rr), Prairie (rr), Rock (rr), Vermilion (r).

The chief difference between this species and *T. limneticum* is the greater size of the cells in proportion to the length of the processes.

Tetraedron enorme (Ralfs) Hansgirg. Pl. 27, Fig. 6.

Hedwigia **27**: 132. 1888; G. S. West, Brit. Freshw. Algae 231, *fig.* 101D. 1904.

Cells irregularly tetrahedric, angles produced into short processes and sides emarginate. Processes generally bilobed and ending in short spines. (Tychoplanktont).

Diam. cells 25–45 μ.

Carson (rr), Cochrane (rr), Devils (2) (r), Horse (rrr), Horseshoe (2) (rrr), Lake of the Woods (rrr), Okauchee (rrr), Pine (2) (rr).

POLYEDRIOPSIS Schmidle 1899.

Cells solitary, generally tetragonal, flattened; rarely pyramidal or with sides cruciate. Corners truncate and with 3–10 long, hyaline, slightly tapering setae. Chloroplast parietal or entirely filling the cell, pyrenoid single.

Reproduction by the formation of 2–4–8 autospores through the division of the cell contents.

Polyedriopsis spinulosa Schmidle. Pl. 27, Figs. 7–8.

Allg. Bot. Zeitschr. **5**: 17. 1899.

Tetraëdron spinulosum Schmidle, ibid. **2**: 193, *fig.* 2. 1896.

Characters the same as those of the genus. (Euplanktont).

Greatest diameter of cells 12–25 μ; length of setae about 40 μ.

Horse (rr), Prairie (rrr).

The general tendency among phycologists is to follow Wille's practice and place the alga in a special sub-genus of *Tetraëdron*. The setae on the cell angles are of the same type as those found on *Lagerheimia* and *Micractinium* so that *Polyedriopsis* forms, in my opinion, a genus parallel with *Tetraëdron* just as *Lagerheimia* does with *Oocystis*.

Sub-family *MICRACTINIEAE.*

Cells solitary or in colonies; spherical or ovoid; with long hyaline setae or spines completely covering or restricted to certain parts of the cell wall. Chloroplast single and cup-shaped; or laminate to disciform and from one to several in each cell. Chloroplasts generally with pyrenoids.

KEY TO THE GENERA.

Cells spherical.
- **Cells organized into definite colonies.................Micractinium (p. p.)**
- **Cells solitary.**
  - **Wall covered with fine hair-like setae.**
    - **Setae thickened in lower third........................Acanthospheaera**
    - **Setae not thickened in lower third.......................Golenkinia**
  - **Wall covered with delicately tapering spines..........Echinosphaerella**

Cells ovoid.
- **Cells organized into definite colonies.................Micractinium (p. p.)**
- **Cells solitary.**
  - **Setae somewhat tapering ....Lagerheimia**
  - **Setae not tapering ....Franceia**

MICRACTINIUM Fresenius 1858.

Cells spherical or ovoid, in colonies of 4–32 or more, generally quadrately arranged in one plane but sometimes tetrahedrically disposed. Cell wall delicate, with 1–7 long hyaline setae that are not thickened at the base lying on the free surface of the cells. Chloroplast single, parietal, cup-shaped; with one pyrenoid.

Reproduction by the division of contents of any cell to form a 4- or 8-celled autocolony, the young colony being liberated by a rupture of the mother cell wall into four parts. Young coenobia frequently remaining attached to the parent coenobe for some time.

KEY TO THE SPECIES.

Cells spherical ....M. Pusillum
Cells ovoid ....M. Quadrispinum

Micractinium pusillum Fresenius. Pl. **28**, Figs. **1–3**.

Abh. Senckenb. Naturf. Ges. **2**: 236, *pl.* 11, *figs.* 46–49. 1858; G. M. Smith, Bull. Torr. Bot. Cl. **43**: 477, *pl.* 25, *fig.* 18. 1916.

Richteriella botryoides (Schmidle) Lemmermann, Hedwigia **37**: 307, *pl.* 10, *figs.* 1–6. 1898.

Richteriella botryoides forma *fenestrata* Chodat, Matér. pour la flore crypt. Suisse **1**^{3}: 194, *fig.* 110. 1902.

Richteriella botryoides forma *tetraedrica* Lemmermann, Arch. f. Hydrobiol. u. Planktonkde. **5**: 307. 1910.

Colonies of 4–8–16 or more cells in groups of four and disposed in a quadrate or pyramidate manner with the groups of four in turn pyramidately or quadrately arranged. Cells spherical, with 1–5 long, hyaline, delicately tapering setae on the face not adjacent to other cells. Chloroplast single, parietal, cup-shaped; with one pyrenoid. (Euplanktont).

Diam. cells (without setae) 3–7 μ; length of setae 20–35 μ.

Blake (rrr), Bone (sss), Chain (sss), Chetek (rr), East (rr), Green (1) (rr), Hooker (rr), Horse (rr), Horseshoe (2) (rrr), Island (rrr), Little Bass (2) (r), Little Crooked (rrr), Loveless (r), Mendota (rr), Monona (cc), North Turtle (rr), Pokegama (3) (rr), Poor Farm (rr), Pope (rr), Prairie (rr), Round (1) (rr), Tenderfoot (rrr), Tied Canoe (rr), Wolf (rrr).

Flat sixteen-celled colonies with a central open space are the most common. Such colonies have been given the form name *fenestrata* but both it and the form *tetraedrica* are unworthy of recognition. On rare occasions large colonies are found; colonies containing over 750 cells having been collected from the plankton of Lake Monona.

var. ELEGANS G. M. Smith. Pl. **29**, Fig. **1**.

Trans. Wis. Acad. Sci., Arts, & Lett. **19**[1]: 631, *pl.* 12, *fig.* 4. 1918.

Cells with 5–7 setae on the exposed face. Dimensions as in the typical form. (Euplanktont).

Rice (1) (rr).

This seems to be a distinct variety since the cells of colonies from this lake all bore the large number of setae.

MICRACTINIUM QUADRISETUM (Lemmermann) G. M. Smith. Pl. **28**, Fig. **4**.

Bull. Torr. Bot. Cl. **43**: 479, *pl.* 25, *fig.* 17. 1916.

Cells ovoid, with 1–4 long, delicately tapering setae on the exposed faces. Chloroplast single, parietal, cup-shaped; with one pyrenoid. (Euplanktont).

Diam. cells (without setae) 4–7 μ, length 8–10 μ; length of setae 23–40 μ.

Bear Trap (rrr), Camp (rr), Cranberry (rrr), Eagle (rrr).

The cells of this species are somewhat larger than those of *M. pusillum* and are distinctly ovoid. The character upon which the specific name is based is quite misleading since the number of setae may vary from one to four. Colonies containing more than sixteen cells have not been observed and the cell arrangement has been quadrate in every case.

GOLENKINIA Chodat 1894.

Cells spherical, normally solitary; occasionally forming temporary colonies. Cell wall thin, sometimes enclosed in a thin gelatinous envelope. Surface of cell covered with hyaline setae that are slightly tapering but not conspicuously thickened in the lower third. Chloroplast single, parietal and cup-shaped or entirely filling the cell; pyrenoid single.

Reproduction by the formation of autospores. Zoospores have been reported but their formation is doubtful.

KEY TO THE SPECIES.

Setae 25–45 μ long..G. RADIATA
Setae 12–18 μ long..G. PAUCISPINA

GOLENKINIA RADIATA Chodat. Pl. 29, Figs. 2–3.

Jour. de Bot. **8**: 305, *pl.* 3, *figs.* 1–24. 1894.

Cells spherical, solitary, sometimes in temporary colonies of four cells. Entire cell wall covered with long, delicate, hyaline setae. Chloroplast single, cup-shaped, parietal; with one pyrenoid. (Euplanktont).

Diam. cells (without setae) 7–15 μ; length setae 25–45 μ.

Green (2) (rrr), White Ash (rr).

This and the following *Golenkinia* species have been made a subgenus of *Micractinium* by Wille and that terminology followed in the second list but the colonial and solitary habit seem to me sufficiently diverse to warrant generic distinction.

GOLENKINIA PAUCISPINA W. & G. S. West. Pl. 29, Figs. 4–5.

Trans. Roy. Irish Acad. **32**, Sec. B: 68, *pl.* 1, *fig.* 18. 1902.

Cells solitary, spherical, with few to fairly numerous, short, delicately tapering, hyaline or dark setae. Chloroplast single, completely filling the cell, with one pyrenoid. (Euplanktont).

Diam. cells (without setae) 15–18 μ; length setae 12–18 μ.

Prairie (rr).

The setae of the cells in Prairie lake were more numerous than the Wests found them in individuals from the plankton of Lough Neagh,

Ireland. They were also dark colored and are more conspicuous than in other members of the *Micractineae*.

ACANTHOSPHAERA Lemmermann 1899.

Cells spherical, solitary, with a single cup-shaped chloroplast containing one pyrenoid. Cell wall thin, without a gelatinous sheath; covered with many long hyaline setae that are thicker in the lower third and then abruptly diminish to a very fine hair.

Method of reproduction unknown.

ACANTHOSPHAERA ZACHARIASI Lemmermann. Pl. 29, Figs. 6–8.

Forschungsbr. a. d. Biol. Stat. zu Plön 7: 118, *pl.* 1, *figs.* 10–11. 1899.

Characters as those of the genus. (Euplanktont).

Diam. cells (without setae) 9–14.5 μ; length setae 30–35 μ.

Horse (rr).

The chief distinction between this genus and *Golenkinia* is the thickening of the lower part of the setae.

Echinosphaerella gen. nov.

Cells spherical, solitary. Cell wall thin, completely covered with heavy, long, hyaline, delicately tapering and slightly concave, homogeneous spines. Chloroplast single, parietal, cup-shaped; with one pyrenoid.

Reproduction unknown.

Echinosphaerella limnetica sp. nov. Pl. 29, Figs. 9–11.

Characters as described above. (Euplanktont).

Diam. cells (without spines) 9–12 μ. Spines 2.5–3 μ broad at base, 20–25 μ long.

Horse (rr).

The alga at once suggests a desmid zygospore, especially those of certain *Staurastrum* species. In all desmid zygospores the chloroplast is not definite and since the alga collected from Horse Lake has a single, well defined, cup-shaped chloroplast with one pyrenoid the presumption is that it is a vegetative cell and not a zygospore. The nearest genus is *Echinosphaeridium* Lemmermann where the setae covering the wall have a broad tapering gelatinous sheath at the base. In *Echinosphaerella* there is not a central hair-like seta with a gelatinous covering, but a homogeneous spine of considerable thickness at the base. Although the genus is provisionally placed in the Micractinieae its spines are quite different from those of other genera of the subfamily.

LAGERHEIMIA Chodat 1895.

Cells solitary; ovoid, ellipsoid, or citriform. Cell wall firm, without a gelatinous sheath; with long tapering setae that lie a short distance back from the poles or cover the entire wall. Chloroplasts 1–4, parietal, laminate; with or without pyrenoids.

Reproduction by autospores that are generally liberated at once but which may remain inside the old mother wall a short time.

KEY TO THE SPECIES.

Setae covering entire cell wall................................L. Droescheri
Setae confined to poles of the cell.
  Cells lemon-shaped..L. Citriformis
  Cells ovoid.
    Length of setae more than double that of cell..............L. Longiseta
    Length of setae less than double that of cell.
      2–4 setae at each pole....................................L. Subsalsa
      3–8 (generally 6) setae at each pole......................L. Ciliata

Lagerheimia ciliata (Lagerheim) Chodat. Pl. **31**, Figs. **1–2**.

Nuova Notarisia **10**: 90. 1895.

Oocystis ciliata Lagerheim, Öfvsers Kgl. Vet.-Ak. Förh. **39**, No. **2**: 76, *pl.* 3, *figs.* 33–37. 1882.

Chodatella ciliata (Lag.) Lemmermann, Hedwigia **37**: 310. 1898.

Cells oblong-ovoid, with 3–7, hyaline, delicately tapering setae at each pole. Chloroplasts 1–4, parietal, laminate; each with a single pyrenoid. (Euplanktont).

Breadth of cells (without setae) 6–18 μ; length (without setae) 10–21 μ. Setae 15–20 μ long.

Kegonsa (rr), Mendota (sss).

The setae of this species are not in definite whorls but distributed over the entire polar region.

var. **minor** (G. M. Smith) comb. nov. Pl. **31**, Fig. **3**.

Chodatella ciliata var. *minor* G. M. Smith. Bull. Torr. Bot. Cl. **43**: 477, *pl.* 25, *fig.* 16. 1916.

Cells ovoid, with 6–8 setae at each pole. (Euplanktont).

Diam. cells (without setae) 6–7.5 μ; length (without setae) 8–10 μ. Setae about 20 μ long.

Muskallonge (rr).

LAGERHEIMIA SUBSALSA Lemmermann. Pl. **30**, Figs. **3–4.**

Forschungsbr. a. d. Biol. Stat. zu Plön **6**: 193, *pl.* 5, *figs.* 2–6. 1898.
Chodatella subsalsa Lemmermann, Hedwigia **37**: 310. 1898.

Cells ovoid, with 2–4 delicate setae at each pole. Chloroplast single, laminate, parietal; with one pyrenoid. (Euplanktont).

Diam. cells (without setae) 2.5–8 μ; length (without setae) 5–12 μ. Setae 7.5–26 μ long.

Deer (rrr), Geneva (rr).

LAGERHEIMIA LONGISETA (Lemmermann) Printz. Pl. **30**, Figs. **8–9.**

Skr. Vidensk. i Kristiania, Mat.-Naturv. Kl. **1913**[1], No. **6**: 60. 1914.
Chodatella longiseta Lemmermann, Hedwigia **37**: 310, *pl.* 10, *figs.* 11–18. 1898.

Cells ellipsoid, with a whorl of 4–10 long, hyaline, delicately tapering setae a short distance from each pole. Chloroplast single (?), parietal; without a pyrenoid (?). (Euplanktont).

Breadth of cells (without setae) 5–8 μ; length (without setae) 9–13 μ. Setae 40–55 μ long.

White Sand (sss).

The exceedingly long setae at once distinguish the species from all others. The arrangement of the setae in a whorl is best seen in an end view of a cell.

var. **major** var. nov. Pl. **30**, Figs. **10–12.**

Cells ovoid, with somewhat pointed ends. With a whorl of 4–7 setae a short distance from each pole. Cells containing 1–2 laminate parietal chloroplasts each with a single pyrenoid. (Euplanktont).

Breadth of cells (without setae) 12–15 μ; length (without setae) 15–22 μ. Setae 45–60 μ long.

Blake (rr), Little Bass (2) (rrr).

The variety is distinguished from the typical form by the more pointed and larger cells.

Lagerheimia citriformis (Snow) comb. nov. Pl. **30**, Figs. **1–2.**

Chodatella citriformis Snow, Bull. U. S. Fish Comm. **22**: 389, *pl.* 2, *figs.* 8^1–8^3. 1903.

Cells ovoid to ellipsoid, with a small obtuse projection at each pole. Poles with a whorl of 4–8 long, hyaline, delicately tapering setae at the base of the polar projections. Chloroplast single, parietal; with one pyrenoid. (Euplanktont).

Breadth cells (without setae) 8–20 μ; length (without setae) 13–23 μ. Setae 25–35 μ long.

Horse (rrr), Kegonsa (rrr).

This is the only species with polar projections that resemble the polar nodules of *Oocystis*.

LAGERHEIMIA DROESCHERI (Lemmermann) Printz. Pl. **30**, Figs. **5–7.**

Skr. Vidensk. i Kristiania Mat.-Naturv. Kl. **1913**[1], No. **6**: 60. 1914.

Chodatella Droescheri Lemmermann, Ber. d. D. Bot. Ges. **18**: 98, *pl.* 3, *fig.* 12. 1900.

Cells ovoid to ellipsoid; with numerous hyaline delicately tapering setae covering the entire wall. Cells with 2–4 parietal, laminate chloroplasts; each with a single pyrenoid. (Euplanktont).

Breadth of cells (without setae) 5–12 μ; length (without setae) 9–16 μ. Setae 15–22 μ long.

Horse (rrr), Travis (r), Upper Nashotah (rr).

FRANCEIA Lemmermann 1898.

Cells ovoid, solitary or in colonies of 2–4; free-floating. Cells enclosed in a delicate gelatinous sheath (?). Entire cell wall covered with delicate hyaline setae that do not taper. Chloroplasts 1–3, parietal, laminate, with or without pyrenoids.

Reproduction by autospores.

FRANCEIA OVALIS (Francé) Lemmermann. Pl. **31**, Fig. **4**.

Hedwigia **37**: 308. 1898.

Phythelios ovalis Francé, Notarisia **1894**: 2, *figs*. 1–5. 1894.

Cells ovoid to elliptic; with two chloroplasts. (Euplanktont).

Breadth of cells (without setae) 7–10 μ; length (without setae) 13–17 μ. Setae 15–23 μ long.

Pine Tree (rr).

The minimal dimensions given above are somewhat smaller than those given by Francé but I feel that the Wisconsin organism should not be considered a distinct variety until the variation in size of the typical form has been determined more fully.

The recognition of the genus *Franceia* is open to question since the characters upon which Lemmermann bases the genus (number of chloroplasts, presence of a vacuole, and cell division in one plane) are ones which might be found in any *Lagerheimia* species. The setae, however, are very delicate and do not taper so that the genus is provisionally considered distinct from *Lagerheimia.*

Sub-family *SELENASTREAE.*

Cells elongate, straight or curved; with finely pointed ends or with ends rounded. Solitary or associated in colonies containing an indefinite number of cells. Colonies with or without a gelatinous envelope. Cells generally with one parietal chloroplast that is with or without pyrenoids.

KEY TO THE GENERA.

Cells solitary.
- Ends of cells produced into setae.............................SCHROEDERIA
- Ends of cells not produced into setae.
  - Chloroplast with an axial row of pyrenoids.................CLOSTERIOPSIS
  - Chloroplast never with more than a single pyrenoid. ANKISTRODESMUS (p. p.)

Cells in colonies.
- Colonies not enclosed in a gelatinous sheath.
  - Cells straight, in loose fasiculate bundles or spirally twisted around one another.................ANKISTRODESMUS (p. p.)
  - Cells curved, with convex faces towards one another........SELENASTRUM
- Colonies enclosed in a gelatinous sheath.
  - Cells curved...KIRCHNERIELLA
  - Cells straight.
    - Colonies with cells in groups of two.....................ELAKTOTHRIX
    - Colonies with cells in groups of 2–4–8....................QUADRIGULA

SELENASTRUM Reinsch 1867.

Cells broadly to narrowly acicular, semilunar; joined by their convex faces into colonies of 4–8–16 or more cells without a gelatinous envelope. Chloroplast single, parietal, laminate, on convex side of young cells; entirely filling older cells; with one pyrenoid.

Reproduction by any cell forming 2–4–8 autospores which, when liberated, have the arrangement of the cells in an adult colony. Autospores remaining attached to parent colony for some time.

The statement is generally made that there are not more than sixteen cells in a colony but I have found that in the plankton of Wisconsin lakes the colonies sometimes contain over a hundred cells.

KEY TO THE SPECIES.

Cells arcuate, never sickle-shaped.................................S. Westii
Cells arcuate to sickle-shaped.
  Diameter of cells not exceeding 5 μ.............................S. Gracile
  Diameter of cells 5–8 μ.......................................S. Bibraianum

Selenastrum gracile Reinsch. Pl. 31, Fig. 5.

D. Algenfl. d. mitt. Th. v. Franken 65, *pl.* 4, *figs.* 3a–3b. 1867.

Cells lunate to sickle-shaped, quite narrow in proportion to their length. Apices of cells acute. Chloroplast single, laminate, parietal, on convex side of cell. Without a pyrenoid. (Facultative Planktont).

Diam. cells 3–5 μ; length 15–30 μ. Distance between apices 8–20 μ.

Bear (3) (rr), Chain (rr), Hillman (rrr), Horse (rr), Monona (r), Pokegama (1) (rr), Prairie (rr), Upper Turtle (rrr), Whitefish (rr).

Differs from the next species in the more delicate cells. Possibly it should be considered a variety of *S. Bibraianum* and not a distinct species.

Selenastrum Bibraianum Reinsch. Pl. 31, Figs. 6–7.

D. Algenfl. d. mitt. Th. v. Franken 64, *pl.* 4, *figs.* 2a–2b. 1867.

Cells in spherical to ovoid colonies of 4–8–16 or more with their convex sides facing the center of the colony. Cells crescent-shaped to sickle-shaped, gradually tapering to a fairly sharp apex. Chloroplast parietal, on convex side of and almost filling the cell; pyrenoid single. (Facultative Planktont).

Diam. cells 5–8 μ; length 20–38 μ. Distance between apices 5–25 μ.

Beaverdam (rr), Blake (rrr), Bone (rr), Deer (rrr), Little Bass (2) (rrr), Loveless (rr), Mendota (rrr), Meta (r), Nancy (rr), Pewaukee (rr), Pine Tree (rr), Poor Farm (rr), Rice (1) (rr), Spirit (rrr), Squirrel (rrr), Tenderfoot (rr), Travis (rr), White Ash (sss), Wolf (rr).

All descriptions of *S. Bibraianum* state that there are but four or eight cells in the colony but I have found that in Wisconsin lakes there may be over a hundred cells in the colony, the number of cells usually ceasing to be a multiple of two after there are more than sixteen cells in the colony.

Selenastrum Westii sp. nov. Pl. 31, Figs. 8–10.

Selenastrum acuminatum G. S. West non Lagerheim, Jour. of Bot. **50**: 88, *figs.* 5a–5g. 1912.

Colonies of 2–4–8 cells irregularly arranged with their convex sides in apposition. Frequently dissociating into the individual cells when mature. Cells lunate to arcuate, never sickle-shaped; gradually tapering to a sharp point. Chloroplast single, parietal on convex side of cell; without a pyrenoid. (Facultative Planktont).

Diam. cells 1.5–2.5 μ. Distance between apices 15–18 μ.

Devils (1) (rr).

The cells of this species are quite like those of *Scenedesmus acuminatus* (Lagerheim) Chodat but their organization in the colony is quite different. West has recently investigated an organism which he considers identical with the *Selenastrum acuminatum* described by Lagerheim. I have obtained West's organism in pure culture and find that the cell shape and colony organization is as he describes. The fortunate rediscovery of Lagerheim's species from the plankton of Horse Lake shows that in the organization of the colony it is a *Scenedesmus* and that Chodat is entirely correct in removing the species from *Selenastrum.* The organism that West found is, on the other hand, a true *Selenastrum* but is not the species which Lagerheim erroneously referred to *Selenastrum.*

ANKISTRODESMUS Corda 1838.

Cells acicular to fusiform; straight, lunate, or sigmoid; ends of cells gradually tapering to a point; solitary or loosely aggregated without order in temporary colonies that are not embedded in a gelatinous envelope. Chloroplast single, parietal, sometimes fragmented into small pieces; with or without a pyrenoid.

Reproduction by the formation of several autospores in any cell.

KEY TO THE SPECIES.

Cells twisted around one another.................................A. SPIRALIS
Cells not twisted around one another...........................A. FALCATUS

ANKISTRODESMUS FALCATUS (Corda) Ralfs. Pl. **32,** Fig. **1.**

Brit. Desmidieae 180, *pl.* 34, *figs.* 3a–3b. 1848; G. S. West, Brit. Freshw. Algae 223, *fig.* 94A. 1904.

Cells acicular to narrowly fusiform, tapering very gradually to acute apices, lunate or slightly arcuate. Length 12–24 times the diameter. Seldom solitary, generally loosely associated in fasiculate bundles. Chloroplast single, generally without pyrenoids. (Facultative Planktont).

Diam. cells 1.5–3.5 μ; length 20–80 μ.

Fence (rrr), Half Moon (rrr), Horse (rr), Kawaguesaga (rrr), Lac du Flambeau (rrr), Little Crooked (rr), Lost (rr), Oconomowoc (rrr), Pine

Tree (rr), Poor Farm (rr), Poplar (rr), Rock (1) (rr), Tenderfoot (rr), White Ash (rr).

var. ACICULARIS (A. Braun) G. S. West.

Brit. Freshw. Algae 223, *figs.* 94B–94C. 1904.

Cells usually solitary, straight or slightly curved, apices acutely attenuated. Chloroplast single, parietal, with or without a pyrenoid.

Cells 2.5–4 μ broad, 36–65 μ long.

Beaverdam (rr), Little Bass (1) (rrr), Mendota (rr), Rice (1) (rr), Waubesa (rrr).

var. TUMIDUS (W. & G. S. West) G. S. West. Pl. **32,** Fig. **2.**

Brit. Freshw. Algae 224, *fig.* 94D. 1904.

Cells solitary or in small aggregates, straight or curved, inflated at the middle and with acute apices. Chloroplast single, parietal, with or without pyrenoids. (Tychoplanktont).

Cells 4.5–6.5 μ broad; 61–73 μ long.

Devils (1) (rrr), Mendota (rr).

var. MIRABILIS (W. & G. S. West) G. S. West. Pl. **32,** Figs. **3–5.**

Brit. Freshw. Algae 224, *fig.* 94E. 1904.

Cells solitary, straight, lunate, or sigmoid; with acute apices; much longer and narrower than the typical form. Chloroplast single, commonly fragmented by large vacuoles in the cytoplasm; without pyrenoids. (Euplanktont).

Cells 2–3.5 μ broad; up to 150 μ long.

Bear (3) (rr), Bryer (rr), Chain (sss), Crab (rrr), Devils (1) (sss), Devils (2) (rr), Dummy (rr), High (rr), Horseshoe (1) (rr), Reserve (rr), Squirrel (rr), Travis (rrr), Whitefish (rr).

ANKISTRODESMUS SPIRALIS (Turner) Lemmermann. Pl. **32,** Figs. **6–7.**

Arch. f. Hydrobiol. u. Planktonkde. **4:** 176. 1908.

Raphidium spirale Turner, Kgl. Sv. Vet.-Ak. Handl. **25,** No. **5:** 156, *pl.* 20, *fig.* 26. 1892.

Ankistrodesmus falcatus var. *spiralis* (Turner) G. S. West, Brit. Freshw. Algae 224. 1904.

Cells acicular, attenuated to acute apices; in colonies of four or eight and spirally twisted around one another in the middle of the cell but with the extremities free. Chloroplast single, completely filling the cell; without a pyrenoid (?). (Tychoplanktont).

Cells 1–2.5 μ broad; 20–40 μ long.

Grassy (rr), Silver (2) (rrr).

G. S. West places the alga as a variety of *A. falcatus* but the twisting of the cell seems to be such a constant character that I have followed Lemmermann's practice in considering it a distinct species.

CLOSTERIOPSIS Lemmermann 1898.

Cells solitary, acicular, with ends produced into very acute apices, without a gelatinous envelope. Chloroplast single, not interrupted at the center of the cell, occupying the greater portion of the cell and containing a single axial row of about a dozen pyrenoids.

The genus is very closely related to *Ankistrodesmus,* differing only in the greater length of the cells and the axial row of pyrenoids. It also resembles certain *Closterium* species but the chloroplast of *Closterium* is always interrupted at the center of the cell. Many feel that the single species is not sufficiently differentiated to warrant its separation from *Ankistrodesmus.*

CLOSTERIOPSIS LONGISSIMA var. TROPICA W. & G. S. West. Pl. **32,** Fig. 8.

Trans. & Proc. Bot. Soc. Edinb. **23**: 31, *pl.* 1, *fig.* 1. 1905.

Rhaphidium longissimum var. *tropicum* W. & G. S. West, Trans. Linn. Soc. 2nd. Ser. Bot. **6**: 198. 1902.

Cells acicular, poles acute but not ending in a point of hair-like fineness. (Facultative Planktont).

Cells 6–7.5 μ broad; 225–370 μ long.

North Turtle (rr).

The dimensions of specimens that I have found agree with those given for individuals from Ceylon but not with those from the Scotch plankton.

SCHROEDERIA Lemmermann 1898.

Cells solitary, straight or curved, acicular; with one apex produced into a fine seta tapering to a point, the other with a slightly broader projection terminating in a small disc. Cells without a gelatinous envelope. Chloroplast single, parietal, laminate; with one pyrenoid.

Method of reproduction unknown.

SCHROEDERIA SETIGERA (Schröder) Lemmermann. Pl. **32**, Fig. **12**.

Hedwigia 37: 311. 1898; G. M. Smith, Bull. Torr. Bot. Cl. 43: 473, *pl.* 24, *fig.* 8. 1916.

Length of cells 15–20 times breadth. (Euplanktont).

Cells 3–6 μ broad, 60–85 μ long (with spines). Length of spines 13–27 μ.

Chain (sss), Devils (1) (c), Long (2) (rr), Sand (1) (rr).

This peculiar alga has the general appearance of a *Characium*. Although hundreds of individuals have been found, it has never been found attached to other algae or to crustaceans so that it seems safe to assume that it leads a free-floating and not an epiphytic existence. The disc at one end of the setae described above has not been observed by other investigators, possibly because the alga is very minute.

Reproductive stages have never been found so that the systematic position of the genus is doubtful. Judging by the cell shape it is a close relative of *Ankistrodesmus,* a genus to which it has been referred by some.

SCHROEDERIA JUDAYI G. M. Smith. Pl. **32**, Figs. **9–11**.

Bull. Torr. Bot. Cl. 43: 474, *pl.* 24, *figs.* 9–11. 1916.

Cells smaller than the foregoing. (Euplanktont).

Cells 2.5–6 μ broad, 45–63 μ long (with spines). Length of spines 10–16 μ.

Amicoy (rr), Birch (rrr), Camp (rr), Delavan (rr), Kawaguesaga (rr), Mendota (sss), Prairie (rr), Round (1) (rrr), St. Croix (rr).

QUADRIGULA Printz 1915.

Cells generally in groups of 2–4–8 with their long axes parallel to the long axis of the colony and enclosed by a hyaline, homogeneous, gelatinous sheath. Colonies ellipsoid, 2–5 times as long as broad. Cells cylindrical to fusiform, 4–7 times as long as broad, straight or slightly curved. Chloroplast single, filling most of the cell; with or without pyrenoids.

Reproduction by 2–4–8 autospores from each cell, the fasiculate group of autospores frequently remaining within the envelope of the mother colony for some time.

KEY TO THE SPECIES.

Ends of cells rounded..Q. PFITZERI
Ends of cells pointed.
  Chloroplast notched at the center..Q. CHODATI
  Chloroplast not notched at the center......................................Q. LACUSTRIS

Quadrigula Pfitzeri (Schröder) comb. nov. Pl. **33,** Figs. **1–2.**

Rhaphidium Pfitzeri Schröder, Verh. d. Nat.-Med. Ver. zu Heidelberg N. F. **7**: 152, *pl.* 6, *fig.* 6. 1902.

Ankistrodesmus Pfitzeri (Schröder) G. S. West, Brit. Freshw. Algae 224, *figs.* 94G–94H. 1904.

Cells about 7 times as long as broad, central portion cylindrical with ends slightly tapering and rounded. Chloroplast generally filling the entire cell, with or without a pyrenoid. (Euplanktont).

Breadth of cells 3–6 μ, length 22–45 (generally about 35) μ.

Bear (1) (rr), Beaverdam (rr), Blake (rr), Chain (rr), Clear (rr), Cochrane (sss), Devils (1) (s), Fishtrap (rrr), Found (r), Fowler (rr), Harris (rr), Hillman (ss), Jennie (rr), Lac Court Oreilles (rr), Lindy (rr), Long (2) (rrr), Little Wood (rr), Mud (1) (rr), Mudhen (rr), Oconomowoc (rr), Okauchee (rr), Palmer (sss), Pardee (rr), Pine Tree (r), Poplar (r), Reserve (rr), Rice (1) (rr), Sand (1) (rr), Shell (rr), Sishebogema (rr), Travis (rr), White Ash (r), Whitefish (rr), Wild Goose (rrr).

The chief characteristic of the genus is the arrangement of the cells in groups of 2–4–8 with their long axes parallel to the long axis of the colony, together with the gelatinous envelope that surrounds the colony. Admitting the generic validity of *Quadrigula, Ankistrodesmus Chodati* and *A. lacustris* are in a much more natural position when placed with *Quadrigula* species.

Quadrigula Chodati (Tanner-Fullman) comb. nov. Pl. **33,** Fig. **3.**

Raphidium Chodati Tanner-Fullman, Bull. Herb. Boiss. **2** Ser. **6**: 156, *figs.* 1–11. 1906.

Ankistrodesmus Chodati (Tanner-Fullman) Brunnthaler, in Pascher, Süsswasserfl. Deutschl., österr. u. d. Schw. **5**^{2}: 193, *fig.* 306. 1915.

Cells linear, arcuate, or lunate; fusiform and tapering to sharp ends. Longitudinally arranged in a gelatinous colony of 4–8–16 or more cells. Cells with a single, parietal, laminate chloroplast that is notched at the center and contains two pyrenoids. (Euplanktont).

Breadth of cells 3.5–7 μ, length 30–80 μ. Colonies up to 50x250 μ.

Boulder (r), Little Bass (2) (rrr), Lac du Flambeau (rr), Sand (3) (ss).

Tanner-Fullman states that the cells are always lunate but those that I have observed were always straight or slightly curved. The notching of the chloroplast together with the two pyrenoids is such a distinct feature that there can be no doubt of the identity of the Wisconsin alga with his species.

Quadrigula lacustris (Chodat) comb. nov. Pl. **33**, Figs. **4–6.**

Raphidium Braunii var. *lacustre* Chodat, Bull. Herb. Boiss. **5**: *pl.* 11, *figs.* 9–10. 1897; Matér. pour la fl. crypt. Suisse **1**[3]: 200, *fig.* 117. 1902.
Ankistrodesmus lacustre (Chodat) Ostenfeld, Hedwigia **46**: 384. 1907.

Cells spindle-shaped, straight, with pointed ends. Cells rarely solitary, generally irregularly distributed throughout, or in groups with their long axes parallel to the long axis of the fusiform gelatinous colonial envelope. Chloroplast single, parietal, laminate, at one side of the cell or spirally twisted; generally with a single pyrenoid. (Euplanktont).

Cells 3–5 μ broad, 20–25 μ long.

Blue Gill (rr), Browns (rr), Clear Crooked (cc), Green (2) (rr), Horseshoe (1) (rr), Lynx (rr), Pokegama (2) (rr), Round (1) (rrr).

ELAKTOTHRIX Wille 1898.

Cells fusiform when mature, longitudinally arranged in pairs within a fusiform homogeneous gelatinous matrix. Chloroplast single, laminate, parietal, with one or two pyrenoids. Colonies epiphytic when young and free-floating when mature; or free-floating throughout the life of the plant.

Reproduction by a transverse division into two daughter cells which remain unchanged for some time. Akinetes known.

Wille first placed the alga as a close relative of *Actinastrum Hantzschi* Lagerheim, but later thought it a relative of *Coccomyxa* Schmidle (Wille, in Engler & Prantl, Nat. Pflanzenfam. 1^2 (Nachtr.): 36, *figs.* 18A–18E. 1909); Pascher (Susswasserfl. Deutschl., Osterr. u. d. Schw. 5^2: 220, *figs.* 25–26. 1915) places it among the algae of doubtful systematic position; while G. S. West (Algae **1**: 202. 1916) thinks it closely allied to *Ankistrodesmus*, a view which seems to me most logical since *Ankistrodesmus*, *Quadrigula* and *Elaktothrix* form a very natural evolutionary series.

KEY TO THE SPECIES.

Both poles of cells acute.. E. VIRIDIS
One pole of cell acute, the other generally rounded............. E. GELATINOSA

ELAKTOTHRIX GELATINOSA Wille. Pl. **34**, Figs. **1–3.**

Biol. Cent. **18**: 302. 1898; in Engler & Prantl, Nat. Pflanzenfam. **1**[2] (Nachtr.): 36, *figs.* 18A–18E. 1909; W. & G. S. West, The Naturalist, **1909**: 291, *figs.* 6A–6D. 1909.

Colonies free-floating, elongate fusiform to somewhat irregular; with the long axes of the cells parallel to the long axis of the colony. Cells lying in pairs, spindle-shaped; generally with one pole rounded and the other pointed. The rounded ends of a pair of cells in apposition. Chloroplast single, parietal, laminate; with one pyrenoid. Colonies rarely containing over fifty cells. (Euplanktont).

Cells 3–6 μ broad; 15–25 μ long. Colonies 10–30 μ broad; 70–160 μ long.

Crooked (rrr), Deer (rrr), Hillman (rr), Jag (rrr), Jennie (rr), Long (2) (rrr), Long (3) (rr), Loon (rrr), Mudhen (rr), Pine (2) (rrr), Sand (2) (rrr), Streitor (sss).

The dimensions of the cells and the general appearance of the colony in individuals that I have found is more like the Wests' description of the alga as found in England. Wisconsin colonies rarely contain more than sixteen cells.

ELAKTOTHRIX VIRIDIS (Snow) Printz. Pl. 33, Fig. 7.

Skr. Vidensk. i Kristiania Mat.-Naturv. Kl. **1913**[1]: No. **6**: 31, *pl.* 1, *figs.* 9–12. 1914.

Fusola viridis Snow, Bull. U. S. Fish Comm. **22**: 389, *pl.* 2, *figs.* VI[1]–VI[4]. 1903.

Colony free-floating or epiphytic, fusiform or irregular; gelatinous sheath of colony sharply defined, frequently showing remains of old cell wall at apex. Cells broadly fusiform, straight or sigmoid, with acute apices. Chloroplast single, completely filling the cell; with one or two pyrenoids. (Tychoplanktont).

Cells 6–15 μ broad, 12–35 μ long.

Soft (rr).

KIRCHNERIELLA Schmidle 1893.

Cells lunate to sickle-shaped with their apices almost touching; or elongate, vermiform, arcuate, or spiral cylinders with rounded ends. Associated without definite arrangement in colonies enclosed by a hyaline, homogeneous, gelatinous envelope. Number of cells in young colonies a multiple of two, in older colonies indefinite but rarely more than seventy-five. Chloroplast single, parietal, on convex side of or entirely filling the cell; generally with a single pyrenoid.

Reproduction by the division of any cell to form four or eight autospores that are liberated by a rupture of the old mother cell wall but remain within the gelatinous envelope of the colony.

KEY TO THE SPECIES.

Cells lunate to sickle-shaped.
  Ends of cells pointed....................................K. LUNARIS
  Ends of cells rounded....................................K. OBESA
Cells arcuate, spiral, or vermiform cylinders.
  Breadth of cells 2 μ or less.............................K. CONTORTA
  Breadth of cells over 2 μ................................K. ELONGATA

KIRCHNERIELLA LUNARIS (Kirchner) Möbius. Pl. **34**, Fig. **4**.

Abh. d. Senckenb. Naturf. Ges. **18**: 331. 1894; Chodat, Matér. pour la fl. crypt. Suisse **1**^{3}: 202, *figs.* 121^1–121^6. 1902.

Cells flattened, crescent-shaped, with apices some distance apart, about twice as long as broad, ends tapering to a blunt point; irregularly arranged in groups of four to eight within the gelatinous colonial envelope. Chloroplast single, filling the entire cell with the exception of a small part on the concave side; pyrenoid single. (Facultative Planktont).

Cells 3–8 μ broad, 6–15 μ long; colonies up to 250 μ in diam.

Bear (3) (rr), Beaverdam (rr), Birch Island (rrr), Blake (rr), Chain (ss), Chetek (rr), Cochrane (rr), Crab (rrr), Deer (rr), Devils (2) (rr), Duck (rr), Dummy (sss), Elizabeth (rr), Green (2) (rr), High (rr), Hillman (r), Horseshoe (1) (rr), Jag (rr), Lindy (rrr), Little Crooked (rrr), Little Doctor (rr), Long (3) (rrr), Loveless (rr), Lower Turtle (rrr), Lynx (rrr), Mendota (rr), Meta (rr), Mud (2) (rr), Nancy (sss), North Twin (rrr), Paddock (rr), Pardee (rr), Pine Tree (rr), Pokegama (1) (r), Pokegama (3) (sss), Poplar (rr), Prairie (sss), Round (3) (rrr), St. Croix (r), Shell (r), Spirit (rrr), Squirrel (rr), Upper Turtle (rrr), White Ash (rr).

The species differ from *K. obesa* in the pointed ends of the cells. The two are widely distributed plankton organisms but have never been found in quantity in any lake of the state.

var. DIANAE Bohlin. Pl. **34**, Fig. **5**.

Bih. Kgl. Sv. Vet.-Ak. Handl. **23**, Afd. **3**, No. **7**: 20, *pl.* 1, *figs.* 28–30. 1897.

Cells curved until the apices are almost touching, apices more pointed and frequently slightly twisted. (Facultative Planktont).

Cells 3–5 μ broad, 10–21 μ long.

Hillman (rr), Pine (2) (rr), White Ash (r).

The cell shape of this variety resembles that of *Selenastrum* but the colony differs in being enclosed by a gelatinous envelope. The apices are much sharper than in the typical form.

var. **irregularis** var. nov. Pl. **35**, Fig. **1**.

Apices of cells twisted and frequently not pointing in the same direction. (Facultative Planktont).

Cells 4–6 μ broad, 6–13 μ long.

Cochrane (rr), Pokegama (1) (r), Tenderfoot (rr).

Although the cell apices of *K. lunaris* var. *dianae* are sometimes not in the same plane, they never exhibit the twisting found in this variety. The twisting of the cells of this variety is similar to that of *K. contorta* but they have the shape and pointed ends that characterize *K. lunaris*.

KIRCHNERIELLA OBESA (W. West) Schmidle. Pl. **35**, Figs. **2–3**.

Ber. d. Naturf. Ges. zu Freiburg i. B. **7**: 16 (83). 1893; G. S. West, Brit. Freshw. Algae 226, *figs.* 96A–96E. 1904.

Cells flattened with outer side markedly convex and inner sides parallel, ends rounded and tapering slightly. Chloroplast single, filling the entire cell with the exception of a small portion on the concave side; with one pyrenoid. (Facultative Planktont).

Cells 3–8 μ broad, 6–16 μ long.

Bear (2) (rr), Bear Trap (rr), Camp (rr), Catherine (rr), Center (r), Chain (rr), Chetac (r), Crawling Stone (rrr), Devils (1) (rrr), Hillman (rr), Kitten (rr), Lac Court Oreilles (ss), Lac du Flambeau (rr), Mendota (r), Palmer (rrr), Pine Tree (rr), Pokegama (2) (rr), Poor Farm (rr), Reserve (rr), Rice (1) (rr), Rock (rrr), Sand (4) (rr), South Turtle (rr), Whitefish (rrr), Wildcat (rr), Winnebago (rr).

The cells of this species might be described as thick discs incised a third to a half the width of the cell did not other members of the genus show that this effect is produced by a curving of the cell.

var. MAJOR (Bernard) G. M. Smith. Pl. **35**, Fig. **4**.

Trans. Wis. Acad. Sci., Arts, & Lett. **19**[1]: 636, *pl.* 10, *fig.* 7. 1918.

Inner face between cell apices curved; apices not close together, rounded and slightly tapering. (Facultative Planktont).

Cells 3–5 μ broad, 8–21 μ long.

Carson (rr), Lac Court Oreilles (r), Wolf (rr).

This variety has a large open space between the apices whereas in the typical form this space is narrow and with parallel sides. The cells are also much narrower in proportion to their length. Brunnthaler (in Pascher, Süsswasserfl. Deutschl. Osterr. u. d. Schw. 5[2]: 180. 1915) has placed the alga as a form of *K. lunaris* var. *dianae* but the cell apices are too rounded to consider it a variety of *K. lunaris*.

var. APERTA (Teiling) Brunnthaler. Pl. **35**, Figs. **5–6.**

In Pascher, Süsswasserfl. Deutschl. österr. u. d. Schw. **5**²: 182, *fig.* 268. 1915.
Kirchneriella aperta Teiling, Sv. Bot. Tids. **6**: 276, *fig.* 9. 1912.

Cells with blunt, rounded apices, concave side between apices broadly V-shaped. (Facultative Planktont).

Diam. cells 6–12 μ.

Diamond (rr), Palmer (ss).

KIRCHNERIELLA CONTORTA (Schmidle) Bohlin. Pl. **35**, Fig. **7.**

Bih. Kgl. Sv. Vet.-Ak. Handl. **23**, Afd. **3**, No. **7**: 20. 1897.
Kirchneriella obesa var. *contorta* Schmidle, Flora **78**: 44, *pl.* 7, *fig.* 2. 1894.

Cells cylindrical, arcuate or in spirals that never make more than 1½ turns; ends rounded, not tapering. Cells generally lying some distance from one another within the homogeneous colonial envelope. Chloroplast single, filling the entire cell, without a pyrenoid. Cells sometimes spirally arranged in the colony; rarely more than sixteen cells in a colony. (Facultative Planktont).

Cells 0.7–2.0 μ broad, 8–14 μ long; colonies up to 60 μ in diam.

Chain (rrr), Clear (rr), Cochrane (rr), Fowler (rrr), Mendota (rrr), Parlee (rr), Pine Tree (rrr), Poor Farm (rrr).

A species which at first glance does not appear related to *K. lunaris* and *K. obesa,* but which at times shows a certain resemblance to K. *obesa* var. *major.* It and the following species are rare algae in Wisconsin. They are closely related, the chief difference being the smaller size and fewer number of cells in the colony of *K. contorta.*

KIRCHNERIELLA ELONGATA G. M. Smith. Pl. **36**, Figs. **1–2.**

Bull. Torr. Bot. Cl. **43**: 473, *pl.* 24, *fig.* 7. 1916.

Cells elongate-cylindrical with rounded ends; spirally or irregularly twisted. Colonies of 4–8–16 or more cells arranged in a snarl at the center of a copious homogeneous gelatinous sheath. Chloroplast single, parietal; without a pyrenoid. (Facultative Planktont).

Cells 2–3 μ broad, 15–25 μ long; colonies up to 100 μ in diam.

Devils (2) (rrr), Horse (rrr), Lac Court Oreilles (rr), No Mans (rr), Palmer (rr), White Ash (sss).

Sub-family *CRUCIGENIEAE.*

Cells acicular, fusiform, ovoid, triangular, or trapezoidal; in flat coenobia with the number of cells always a multiple of two. Cells quadrately arranged in coenobe or in a linear series. Coenobe generally without a gelatinous envelope. Chloroplasts 1–4, parietal, disciform or laminate; generally with one pyrenoid.

KEY TO THE GENERA.

Cells quadrately arranged.
  Quadrate arrangement apparent in front view of cells.
    Outer face of cells without spines........................CRUCIGENIA
    Outer face of cells with spines...........................TETRASTRUM
  Quadrate arrangement apparent in end view of cells........TETRADESMUS
Cells arranged in a linear series...............................SCENEDESMUS

CRUCIGENIA Morren 1830.

Coenobe free-swimming, of four cells quadrately arranged with a quadrangular open space at the center and frequently connected to other coenobia by a gelatinous envelope or the remains of the old mother cell walls. Multiple coenobia forming a flat plate one cell in thickness and containing 4, 16, 32, 64 or more cells. Cells flattened; ovoid, triangular, trapezoidal, or semicircular in front view; with a smooth cell wall. Chloroplasts 1–4, parietal, disciform to laminate, with or without pyrenoids.

Reproduction by autocolony formation, the four cells of the coenobe generally producing autocolonies at the same time.

KEY TO THE SPECIES.

Cells apiculate...C. APICULATUM
Cells not apiculate.
  Coenobe with a large rectangular open space at the center.
    Front view of cells trapezoidal...............................C. FENESTRATA
    Front view of cells semicircular.............................C. LAUTERBORNEI
  Coenobe with a small open space at the center.
    Length of coenobe greater than the breadth.
      Wall of cell the same thickness throughout.
        Outer face of cells concave...............................C. CRUCIATA
        Outer face of cells convex.
          Cells of coenobe regularly arranged.............C. RECTANGULARIS
          Cells of coenobe irregularly arranged...............C. IRREGULARIS
      Wall of cell thicker at the poles............................C. TRUNCATA
    Length of coenobe the same as the breadth.
      Cells triangular, with fairly acute corners................C. TETRAPEDIA
      Cells spherical to triangular, with rounded corners........C. QUADRATA

CRUCIGENIA RECTANGULARIS (Nägeli) Gay. Pl. 36, Fig. 3.

Recherches sur le. dév. et la class. de quelques algues vertes 100, *pl.* 15, *fig.* 151 1891.

Coenobe four-celled, cells in mutual contact at the poles and sides and with a small rectangular open space at the center. Generally united in compound coenobia containing an indefinite number of coenobes. Cells ovoid to elongate-ovoid, with 1–4 parietal, disciform chloroplasts, each with a single pyrenoid. Old mother cell wall frequently enclosing young coenobia for some time. (Facultative Planktont).

Cells 4–7 μ broad, 5–10 μ long.

Bear (2) (rr), Beaverdam (rr), Big (rrr), Birch Island (rrr), Devils (1) (r), Fowler (rrr), Half Moon (rr), Hillman (rr), Kegonsa (rr), Lac Court Oreilles (rr), Lindy (rr), Little Bass (2) (rr), Little Doctor (rr), Mill (rrr), Nancy (rr), Nicaboyne (rr), Okauchee (rrr), Old Taylors (rr), Pine Tree (rr), Sand (1) (rr), Upper Nashotah (rrr), Winnebago (rr).

CRUCIGENIA IRREGULARIS Wille. Pl. **36**, Figs. **4–5.**

Biol. Cent. **18**: 302. 1898; Nyt. Mag. f. Naturvidensk. **38**: 10, *pl.* 1, *fig.* 15, 1900; W. & G. S. West Trans. & Proc. Bot. Soc. Edinburgh **23**: 29, *pl.* 1, *figs.* 6–7. 1905.

Coenobia four-celled, with cells in contact at sides and poles, at other times without a definite arrangement. Coenobia irregularly combined into multiple coenobia. Cells ovoid; with 1–4 parietal, disciform chloroplasts; with or without pyrenoids. (Facultative Planktont).

Cells 4–9 μ broad, 6–14 μ long. Coenobia 8–14 μ broad, 15–30 μ long.

Browns (rr), Camp (rr), Fish (rr), Horseshoe (2) (rr), Mudhen (rr), Reserve (rr), Sand (3) (sss).

The cell shape is quite similar to that of *C. rectangularis* but the arrangement of cells within the coenobe is very irregular in about half the colonies one meets with. This irregular cellular arrangement coupled with the larger cell size warrants the separation of the alga from *C. rectangularis.* The alga is generally reported to be without pyrenoids but I have frequently found them in the chloroplasts of *C. irregularis.*

CRUCIGENIA CRUCIFERA (Wolle) Collins. Pl. **36**, Fig. **6.**

Tufts. coll. studies **2**: 170. 1909.

Staurogenia crucifera Wolle, Bull. Torr. Bot. Cl. **6**: 140. 1877.

Staurogenia cruciatum Wolle, Freshw. Algae of U. S. 171, *pl.* 157, *figs.* 9–11. 1887.

Coenobia four-celled, rhomboidal, with rounded ends, slightly concave sides, and a small rectangular open space at the center. Cells elongate, outer face concave, inner and those in contact with other cells straight. Chloroplast single, parietal, laminate; without a pyrenoid. (Facultative Planktont).

Cells 3.5–5 μ broad, 5–7 μ long. Coenobia 9–11 μ broad, 14–16 μ long.

Hillman (rr), Tenderfoot (rr).

The arrangement of the individual coenobia into compound coenobia is generally that which Wolle gives in his figure 9 of *Staurogenia cruci-*

atum. Collin's surmise that the reported cruciform marking of the cell wall is probably due to a division of the cell contents is correct and what Wolle calls the cell is in reality the coenobe. The remains of the old cell walls play an important part in binding together the various coenobia of the multiple coenobe.

Crucigenia truncata sp. nov. Pl. **36,** Figs. **7–9.**

Coenobe four-celled, with a small rectangular space at the center and held in compound coenobia of indefinite size by the remains of old mother cell walls. Cells quadrately arranged with bases and middle third of inner sides in mutual contact but with apices and outer faces free. Cells elongate, ovoid, with free ends truncate. Cell wall at free apex thicker than in other parts. Chloroplasts 1–4, parietal, disciform, without pyrenoids. (Euplanktont).

Cells 4–6 μ broad, 6–10 μ long. Coenobia 6–12 μ broad, 12–20 μ long. Compound coenobia up to 50x75 μ.

Chain (rr), Hillman (rr), Pine Tree (rr), Travis (rr).

The species is differentiated by its flattened cell apices and the thicker walls at the apices. The cells are always regularly arranged within the coenobe, but when in compound coenobia the arrangement is frequently irregular.

CRUCIGENIA APICULATA (Lemmermann) Schmidle. Pl. **37,** Fig. **1.**
Allg. Bot. Zeitschr. **6:** 234. 1900.

Staurogenia apiculata Lemmermann, Bot. Cent. **76:** 151. 1898; Forschungsbr. a. d. Biol. Stat. zu Plön **7:** 119, *pl.* 1, *fig.* 14. 1899; Snow, Bull. U. S. Fish Comm. **22:** 376, *pl.* 1, *figs.* V^1–V^8. 1903.

Tetrastrum apiculatum (Lemmermann) Brunnthaler, in Pascher, Süsswasserfl. Deutschl., österr. u. d. Schw. **5²:** 177, *fig.* 258. 1915; G. M. Smith, Bull. Torr. Bot. Cl. **43:** 472, *pl.* 24, *fig.* 3. 1916.

Coenobia four-celled, embedded in a delicate gelatinous sheath. Cells quadrately arranged with their long axes parallel and a small rectangular space at the center. Coenobia generally in multiple coenobia which, when containing large numbers of coenobia, are frequently irregular. Cells ovoid to triangular, with a short conical projection at the free apex and frequently another one at the side of the base away from the center of the coenobe. Each cell containing 1–4 parietal disciform chloroplasts with one pyrenoid. (Euplanktont).

Cells 3–7 μ broad, 5–10 μ long. Coenobia 6–12.5 μ broad, 9–18 μ long. Compound coenobia up to 150 μ broad.

Mendota (rrr), Palmer (rrr), South Crab (rr).

The cells of 16-celled multiple coenobia are regularly arranged but when multiple coenobia contain 150–200 cells the arrangement is frequently irregular. The gelatinous sheath enclosing the colony is very delicate and can only be seen when properly stained. This species was placed in the genus *Tetrastrum* in a preliminary list, but it is more closely related to *Crucigenia* than to *Tetrastrum.*

CRUCIGENIA QUADRATA Morren. Pl. **36,** Figs. **10–14.**

Ann. Sci. Nat. **20:** 415, *pl.* 15, *figs.* 1–5. 1830.

Staurogenia triangularis Chodat, Mém. Herb. Boiss. **1,** No. **17:** 7, *figs.* 14–19. 1900.

Crucigenia triangularis (Chodat) Schmidle, Allg. Bot. Zeitschr. **6:** 234. 1900.

Staurogenia multiseta var. *punctata* Schmidle, Ber. d. D. bot. Ges. **18:** 157, *pl.* 6, *figs.* 13–14. 1900.

Coenobia four-celled, circular in outline. Cells in mutual contact at or with very small rectangular openings at center of the coenobe. Four coenobia compressed to form rectangular multiple coenobia which in turn are joined to other sixteen-celled multiple coenobia. Cells spherical to triangular; with a smooth wall or with 1–6 minute knob-like projections on the wall. Chloroplasts 1–4, parietal, disciform, with or without pyrenoids. (Facultative Planktont).

Cells 2.5–6 μ broad, 3–7 μ long. Coenobia 5.5–17 μ in diam. 16-celled compound coenobia 12–30 μ in diam.

Carson (rrr), Chain (rr), Hillman (sss), Kitten (rrr), Lac du Flambeau (rr), Palmer (sss), Pardee (rr), Tenderfoot (rrr), White Sand (rrr).

Pyrenoids are very prominent in the chloroplasts of certain colonies while other colonies collected from the same station show no trace of them. In some of the colonies from Hillman lake small, globular, button-like structures were observed on the cell walls. Such structures have been noted by Schmidle and form the distinctive character of his variety *punctata* of *Staurogenia multiseta* [*Tetrastrum multisetum* (Schmidle) Chodat]. If the variety is to be recognized it should be called *C. quadrata* var. *punctata* (Schmidle) but I am of the opinion that it is not a sufficiently constant character to warrant recognition. There is likewise no justification for considering *C. triangularis* (Chodat) Schmidle a distinct species since it does not differ in any respect from *C. quadrata.*

CRUCIGENIA TETRAPEDIA (Kirchner) W. & G. S. West. Pl. **37,** Fig. **2.**

Trans. Roy. Irish Acad. **32,** Sec. **B:** 62, *pl.* 1, *figs.* 11–12. 1902.

Staurogenia tetrapedia Kirchner, Jahreshefte d. Ver. f. Vaterl. Naturk. i. Württemberg **1880,** *pl.* 2, *fig.* 1.

Tetrapedia emarginata Schröder, Ber. d. D. bot. Ges. **15:** 492, *pl.* 25, *figs.* 6a–6c. 1897.

Lemmermannia emarginata (Schröder) Chodat, Mém. Herb. Boiss. **1,** No. **17;** 5, *figs.* 1–8. 1900.

Lemmermannia tetrapedia (Kirchner) Lemmermann, Arkiv. för Botanik **2**, No. **2**: 157. 1904.

Coenobia four-celled, with a very small rectangular open space at the center. Frequently joined in multiple coenobia of sixteen cells that may be connected in turn to other 16-celled multiple coenobia. Cells flattened, triangular, with rounded corners. Cell contents homogeneous, or with a single parietal chloroplast, with or without pyrenoids. (Facultative Planktont).

Diam. cells 4.8–9.5 μ; diam. coenobia 8–15 μ; diam. 16-celled multiple coenobia up to 40 μ.

Palmer (rr).

Certain investigators have noted that the cell contents are at times absolutely homogeneous. This has caused Schröder to consider the alga a species of *Tetrapedia,* one of the Myxophyceae, but the Wests have given cogent reasons for considering it a *Crucigenia.* I have found a distinct parietal chloroplast and pyrenoid in the cells whereas Chodat states that there is a chloroplast but no pyrenoid. His founding of the genus *Lemmermannia* upon this single character (lack of a pyrenoid) has rightly been called by G. S. West (Brit. Freshw. Algae 217. 1904) "the climax of absurdity."

Crucigenia fenestrata Schmidle. Pl. **37**, Fig. **3**.

Allg. Bot. Zeitschr. **6**: 234. 1900.
Staurogenia fenestrata Schmidle, Allg. Bot. Zeitschr. **3**: 107, *fig.* 1. 1897.

Coenobia four-celled, rectangular, with a medium-sized opening at the center; with an inconspicuous gelatinous sheath. Frequently in multiple coenobia with the component coenobia lying close together. Cells trapezoidal, with the longest face at the outside of the coenobe. Chloroplast single, except just before division, parietal, without a pyrenoid (?). (Euplanktont).

Cells 3–5 μ broad, 5–12 μ long; coenobia 8–14 μ broad.

Bettine (rr), White Sand (rr).

The specimens that I have found do not agree in all respects with Schmidle's description since there was no gelatinous sheath surrounding the coenobe and filling the central space.

Crucigenia Lauterbornei Schmidle. Pl. **37**, Fig. **4**.

Allg. Bot. Zeitschr. **6**: 234. 1900.
Staurogenia Lauterbornei Schmidle, Allg. Bot. Zeitschr. **2**: 192, *fig.* 1. 1896.

Coenobia four-celled, cells in contact with one another at their apices only and with a large square open space at the center of the coenobe.

Coenobia joined in multiple coenobia of at least sixteen cells by the remains of old cell walls and a gelatinous envelope. Cells flattened, subhemispherical in front view with the flattened side towards the center of the coenobe. Chloroplast single, parietal, on the outer face of the cell; with one pyrenoid. (Euplanktont).

Cells 4.5–9 μ broad, 8–15 μ long. Coenobia 15–25 μ in diam; multiple coenobia up to 150 μ in diam.

Kegonsa (r), Pokegama (rr), Rice (1) (r), Upper Turtle (r).

A very distinct species and one that is readily recognized by the large opening at the center of the colony. *C. fenestrata* also has a large, rectangular open space at the center of the coenobe but the two species have quite differently shaped cells; those of *C. Lauterbornei* being semicircular in front view, and those of *C. fenestrata* trapezoidal.

TETRASTRUM Chodat 1895.

Coenobia four-celled, with cells cruciately arranged in a flat plate; at times embedded in a thin gelatinous envelope. Cells triangular to ovoid, with several spines or setae on the free face. Chloroplasts 1–4, parietal, disciform, with or without pyrenoids.

Reproduction by autocolony formation in any cell.

The genus is closely related to *Crucigenia*, differing only in the spines on the margins of the cells and an absence of an association in multiple coenobia.

Tetrastrum staurogeniaeforme (Schröder) Lemmermann. Pl. 37, Figs. 5–6.

Ber. d. D. bot. Ges. **18**: 95. 1900.

Cohniella Staurogeniaeformis Schröder, Ber. d. D. bot. Ges. **15**: 373, *pl.* 17, *fig.* 5. 1897.

Coenobia four-celled, with a very minute open space at the center of the colony. Coenobe embedded in a very delicate gelatinous envelope. Cells cruciately arranged, triangular, with 5–6 very delicate setae on the free surface. Chloroplasts 1–4, parietal, disciform, with or without pyrenoids. (Euplanktont).

Diam. cells (without setae) 3–6 μ; diam. coenobia (without setae) 7–15 μ. Setae 4–8 μ long.

Horse (rr), Mendota (rrr).

The setae on the outer face of the cells are very delicate and easily overlooked. If the setae are not taken into consideration the alga looks exactly like *Crucigenia quadrata*.

TETRADESMUS G. M. Smith 1913.

Colonies free-floating, four-celled, rarely two- or three-celled. Cells in two planes with two cells in each plane and joined along the long axes when the colony is viewed from the side. Viewed from the top cells quadrately arranged. Cells elliptic-fusiform to broadly elliptic, with outwardly divergent or straight apices. Chloroplast single, parietal and laminate or completely filling the cell; pyrenoid single, central or eccentric.

Reproduction by autocolony formation in any cell of the colony, the young coenobia being liberated by a rupture of the mother cell wall.

Tetradesmus wisconsinensis G. M. Smith. Pl. **37,** Figs. **7–11.**

Bull. Torr. Bot. Cl. **40:** 76, *pl.* 1, *figs.* 1–2, 3–20. 1913.

Cells elliptic-fusiform, apices attenuated and outwardly divergent. Pyrenoids eccentric. (Facultative Planktont).

Cells 4–6 μ broad, 12–14.5 μ long.

Blake (rrr), Mendota (rr), Nancy (rrr).

SCENEDESMUS Meyen 1829.

Coenobe a flat (rarely curved) plate of ellipsoidal, ovoid, or acicular cells with cell number always a multiple of two. Cells in lateral contact and in one or two rows. Cell wall smooth, corrugated, granulate, or spicate; with or without marginal or lateral teeth or spines. Chloroplast single, parietal, and laminate in young cells; frequently filling entire cell in older colonies; pyrenoids single, central or eccentric.

Reproduction by autocolonies from any or all cells of the coenobe, the number of cells in the colony not necessarily the same as that of the mother colony.

KEY TO THE SPECIES.

Walls smooth, without granulations, lateral ridges, terminal teeth or spines.
- Cells acicular to broadly elliptic, with pointed ends.
  - Colony a flat plate.
    - All cells in colony erect.................................S. Obliquus
    - Median cells erect, terminal cells lunate..................S. Dimorphus
  - Colony a curved or twisted plate.
    - Cells forming a curved surface........................S. Acuminatus
    - Cells alternately arranged with terminal cells not in the same plane as the others............................S. Bernardii
- Cells ovoid to cylindrical, with rounded ends.
  - Arrangement of cells a single linear or subalternating but never a double series.................................S. Bijuga
  - Arrangement of cells a double linear series.................S. Arcuatus

Wall variously sculptured or ridged, at times with terminal teeth or spines.
Longitudinal ridges on sides of the cells.
Without terminal teeth or spines........................S. ACUTIFORMIS
With terminal teeth or spines.
Terminal cells with long spines at poles..................S. ARMATUS
All cells with short teeth at poles.....................S. BRASILIENSIS
Longitudinal ridges not present on sides of the cells.
Teeth-like spines covering entire wall......................S. HYSTRIX
Teeth or spines confined to poles, outer face of terminal cells, or middle of all cells.
Cells with short teeth at poles......................S. DENTICULATUS
Cells with long spines.
Spines on both median and terminal cells.
Poles of cells only region bearing spines.
Cells ovoid to ovoid-cylindrical.........................S. LONGUS
Cells naviculoid ....S. OPOLIENSIS
Poles and median part of cells bearing spines..........S. ABUNDANS
Spines confined to terminal cells.
Cells ovoid to ovoid-cylindrical...................S. QUADRICAUDA
Cells naviculoid ....S. OPOLIENSIS

SCENEDESMUS OBLIQUUS (Turpin) Kützing. Pl. **37**, Figs. **12–14.**

Linnaea **8**: 609. 1833; G. S. West, Brit. Freshw. Algae 220, *fig.* 92A. 1904.

Cells of 2–4–8 cells arranged in a single linear or sublinear series. Cells fusiform, with acute apices; interior cells with straight sides, outer face of terminal cells slightly concave, straight, or convex. Cell wall smooth, without terminal teeth or spines. (Tychoplanktont).

Cells 3–9 μ broad, 10–21 μ long. 4-celled coenobia 10–21 μ broad, 12.5–24 μ long.

Balsam (rrr), Bear (1) (rrr), Blake (rr), Bone (rr), Grassy (rr), Hillman (rrr), Horseshoe (2) (rrr), Lindy (rr), Little Bear (rrr), Little Doctor (rr), Little Rice (2) (rrr), Loon (rr), Mendota (rrr), Pewaukee (rrr), Poor Farm (rrr), Pope (rrr), Rice (2) (rr), Sand (3) (rrr), Sand (4) (rrr), Upper Nemahbin (rrr), White Ash (rrr).

SCENEDESMUS DIMORPHUS (Turpin) Kützing. Pl. **37**, Figs. **15–17.**

Linnaea **8**: 608. 1833; G. M. Smith, Trans. Wis. Acad. Sci., Arts, & Lett. **18**[2]: 434, *pl.* 32, *figs.* 185–189; *pl.* 33, *figs.* 190–195. 1916.

Coenobia of four or eight cells arranged in a linear or alternating series. Cells fusiform with apices attenuated to delicate points. Interior cells erect, outer erect or markedly lunate. Cell wall smooth, without terminal teeth or spines. (Facultative Planktont).

Cells 2–5 μ broad, 18–23 μ long. 4-celled coenobia 18–23 μ broad, 12–25 μ long.

Dunham (rrr), Horse (r), Pine Tree (sss), Poor Farm (ss), Travis (sss).

The outwardly turned lunate terminal cells, together with the pronounced alternate arrangement distinguishes this species from the foregoing. I have given elsewhere (l. c.) the reasons for considering this a distinct species and not a variety of *S. obliquus.*

SCENEDESMUS ACUMINATUS (Lagerheim) Chodat. Pl. **38,** Figs. **1–4.**

Matér. pour la fl. crypt. Suisse **1**^{3}: 211. 1902.

Selenastrum acuminatum Lagerheim Öfvers. Kgl. Vet.-Ak. Förh. **39**: No. **2**: 71, *pl.* 3, *figs.* 27–30. 1882.

Coenobia curved, of 4 or 8 (generally 4) cells. All of the cells of a colony arcuate or lunate, fusiform, with the ends gradually tapering to a fine point. Cell wall smooth, without teeth, granulations or spines. (Facultative Planktont).

Cells 3–7 μ broad, 30–40 μ between apices.

Horse (sss).

This species should not be confused with *Selenastrum acuminatus* G. S. West non Lagerheim, a species which I have already discussed in connection with *Selenastrum Westii* G. M. Smith.

SCENEDESMUS BERNARDII G. M. Smith. Pl. **38,** Figs. **5–9.**

Trans. Wis. Acad. Sci., Arts, & Lett. **18**^{2}: 436, *pl.* 33, *figs.* 196–208. 1916.

Colonies of 2–4–8 cells. Interior cells fusiform, lunate or sigmoid, with acute apices; alternately arranged with apices in contact with apices or median portion of adjacent cells. Terminal cells fusiform to lunate, frequently at an angle to the plane of the coenobe, in contact with the interior cells by the apex only. Cell wall smooth, without terminal spines or teeth. (Tychoplanktont ?).

Cells 3–6 μ broad, 8–17 μ long.

Monona (rrr).

SCENEDESMUS BIJUGA (Turpin) Lagerheim. Pl. **37,** Figs. **18–20.**

Nuova Notarisia **1893**: 158. 1893.

Scenedesmus bijugatus (Turpin) Kützing. G. S. West, Brit. Freshw. Algae 220, *fig.* 92c. 1904.

Coenobes flat, of 2–4–8 cells in a single (rarely subalternate) series. Cells oblong-elliptic to ovoid, with broadly rounded ends. Cell wall smooth, without teeth or spines. (Tychoplanktont).

Cells 4–7 μ broad, 7–18 μ long. 4-celled colonies 7–18 μ broad, 16–25 μ long.

Fowler (rrr), Lindy (rr), Little Rice (2) (rr), Mill (rr), Poplar (rrr), Silver (2) (rr), Streitor (rr), Upper Turtle (rrr), Wolf (rr).

var. ALTERNANS (Reinsch) Borge. Pl. **38**, Figs. **10–11.**

Arkiv. för Botanik **6**, No. **1**: 57. 1906.

Scenedesmus alternans Reinsch, Abh. Senckenb. Naturf. Ges. **6**: 135, *pl.* 20, *fig.* D[5]. 1866.

Colonies flat, generally 8-celled, with a very marked alternate arrangement of the component cells. Cells ovoid to elliptic. (Tychoplanktont).

Cells 4–8 μ broad, 6–15 μ long. 8-celled coenobia 12–30 μ broad, 22–45 μ long.

Blake (rr), Devils (1) (rrr), Muskallonge (rr).

Since cultural studies have shown that the alternate arrangement of cells is a constant character it is proper to consider this a distinct variety. The variety has also been found in lakes where the typical form is absent.

SCENEDESMUS ARCUATUS Lemmermann. Pl. **38**, Figs. **12–14.**

Forschungsbr. a. d. Biol. Stat. zu Plön **7**: 112, *pl.* 1, *figs.* 2–4. 1899.

Colonies always curved, of 4–8–16 (generally 8) cells arranged in a double row. Cells ovoid, sometimes angular by mutual compression at the base. Coenobia with small interstices between the cells. Cell wall smooth, without spines or teeth. (Euplanktont).

Cells 3–9 μ broad, 9–17 μ long. 8-celled coenobia 15–25 μ broad, 15–40 μ long.

Beaverdam (rr), Blake (rrr), Camp (rrr), Center (ss), Chain (rr), Cochrane (rr), Cranberry (rr), Hillman (r), Lac Court Oreilles (rr), Lac du Flambeau (rr), Little Bass (2) (rr), Little Crooked (rr), Long (1) (rrr), Long (3) (rr), Lower Turtle (rr), Mendota (sss), Meta (rrr), North Turtle (rr), Owen (rrr), Paddock (rrr), Palmer (rrr), Pardee (rr), Pokegama (2) (rr), Reserve (rr), Round (1) (rrr), Sand (1) (rrr), Sand (4) (rrr), Shell (rrr), Sishebogema (rr), Tied Canoe (rr), Upper Turtle (rr), Wolf (r).

One of the few species of the genus that is exclusively confined to the plankton. The curving of the coenobe and the double series of ovoid cells at once separates it from all other species.

var. CAPITATUS G. M. Smith. Pl. **38**, Figs. **15–16.**

Trans. Wis. Acad. Sci., Arts, & Lett. **19**[1]: 637, *pl.* 11, *figs.* 4–5. 1918.

Cells with nodular thickenings at the poles. (Euplanktont).

Cells 5–11 μ broad, 11–23 μ long. 8-celled colonies 22–27 μ broad, 20–35 μ long.

Center (rr).

var. PLATYDISCA G. M. Smith. Pl. 39, Figs. 1–3.

Trans. Wis. Acad. Sci., Arts, & Lett. 18[2]: 451, *pl.* 30, *figs.* 101–105. 1916.

Colonies always flat, generally 8-celled. Cells oblong-elliptic and arranged in a double series. Coenobia with small interstices between the cells. (Facultative Planktont).

Cells 4.5–7.5 μ broad, 8–17 μ long. 8-celled coenobia 11–30 μ broad, 18–30 μ long.

Camp (rrr), Eagle (rr), Fowler (rr), George (rr), Horse (rrr), Kegonsa (ss), Paddock (rr), Pine (1) (rr).

An alga that has been considered a normal variation of *S. bijuga* but one which, as stated elsewhere (l. c.), seems more closely related to *S. arcuatus.*

SCENEDESMUS ACUTIFORMIS Schröder. Pl. 39, Figs. 4–6.

Forschungsbr. a. d. Biol. Stat. zu Plön 5: 45, *pl.* 2, *figs.* 4a–4b. 1897.

Coenobia flat, of 2–4–8 (generally 4) cells arranged in a linear series. Cells fusiform, with acute apices. Cell wall of interior cells with a single ridge that extends from pole to pole on each side; wall of terminal cell with 2–4 ridges. Poles of cells without spines or teeth. (Facultative Planktont).

Cells 5–8 μ broad, 16–22 μ long. 4-celled colonies 16–22 μ broad, 20–25 μ long.

Mendota (rr), Rice (1) (rrr).

Only a single longitudinal ridge can be seen in a front view of the cell but the number of ridges is readily ascertained in an end view of cells. The presence of these ridges without accompanying spines or teeth separates this from other species.

SCENEDESMUS ARMATUS (Chodat) G. M. Smith. Pl. 39, Figs. 7–10.

Trans. Wis. Acad. Sci., Arts, & Lett. 18[2]: 460, *pl.* 29, *figs.* 90–93. 1916.

Coenobia of 2–4–8 cells arranged in a linear or subalternating series. Cells ovoid to oblong-ellipsoid; with rounded apices. Terminal cells with a single long spine at each pole. Both sides of cells with a longitudinal ridge that may extend from pole to pole but is generally indistinct in the median portion of the cell. (Facultative Planktont).

Cells 4–7 μ broad, 7–16 μ long. 4-celled coenobia (without spines) 7–16 μ broad, 16–25 μ long. Spines 5–7 μ long.

Bear (2) (rr), Bear Trap (rrr), Bone (rr), Browns (rr), Carson (rr), Fishtrap (rrr), Horseshoe (2) (rr), Lac Court Oreilles (rr), Little Butternut (rrr), Little Crooked (rr), Loveless (rr), Lynx (rrr), Mendota (rrr), Pardee (rrr), Pine Tree (rr), St. Croix (rrr), Silver (rr), Upper Turtle (rrr), Wolf (rr).

var. **major** var. nov. Pl. **39**, fig. **11.**

Cells up to 9 μ broad and 25 μ long. Spines 15 μ long. (Facultative Planktont ?).

Palmer (rr).

SCENEDESMUS BRASILIENSIS Bohlin. Pl. **39**, Figs. **12–14.**

Bih. Kgl. Sv. Vet.-Ak. Handl. **23**, Afd. **3**, No. **7**: 22, *pl.* 1, *figs.* 36–37. 1897.

Coenobia flat, of 2–4–8 (generally 4) cells arranged in a single linear series. Cells cylindrical to ovoid-ellipsoid, with a single longitudinal ridge extending from pole to pole on each side of the cell. Poles with 1–4 small teeth. (Facultative Planktont).

Cells 3–6 μ broad, 11–24 μ long. 4-celled coenobia 11–24 μ broad, 12–25 μ long.

Bear (2) (r), Beaverdam (rrr), Chain (rrr), Clare (rrr), Deer (rrr), Jag (rr), Lac du Flambeau (rrr), Lake of the Woods (rrr), Lindy (rr), Little Doctor (rr), Old Taylors (rr), Pine (2) (rr), Rice (1) (rr), White Sand (rrr).

The species may be distinguished from *S. denticulatus* by the linear ridges on the sides of the cells.

SCENEDESMUS HYSTRIX Lagerheim. Pl. **39**, Figs. **15–16.**

Öfvers. Kgl. Vet.-Ak. Förh. **39**, No. **2**: 62, *pl.* 2, *fig.* 18. 1882.

Coenobia flat, of 2–4–8 cells joined in a single linear series. Cells oblong-cylindrical, with blunt conical apices. Entire surface of cells covered with minute spines. (Tychoplanktont).

Cells 3–5 μ broad, 8–18 μ long. 4-celled coenobia 8–18 μ broad, 12–25 μ long.

Rudolph (rrr).

SCENEDESMUS DENTICULATUS Lagerheim. Pl. **39**, Figs. **17–19.**

Öfvers. Kgl. Vet.-Ak. Förh. **39**, No. **2**: 61, *pl.* 2, *figs.* 13–17. 1882.

Coenobia flat, generally four-celled. Cells ovoid to ovoid-ellipsoid, arranged in a linear or subalternating series. Poles of cells without ridges but with 1–4 short spines. (Facultative Planktont).

Cells 5–11 μ broad, 7–15 μ long. 4-celled coenobia 7–15 μ broad, 15–22 μ long.

Amicoy (rr), Bear (2) (rr), Browns (rr), Chain (rr), Deer (rrr), High (r), Hillman (rr), Horse (rrr), Horseshoe (2) (rr), Lac du Flambeau (rrr), Little Doctor (rrr), Silver (3) (rr), White Ash (rr), Wolf (rr).

SCENEDESMUS LONGUS Meyen. Pl. 39, Figs. 20–22.

Nova Acta Phys.-Med. Acad. Caes. Leop. Nat. Cur. **14**: 774, *pl.* 43, *fig.* 28. 1829. G. M. Smith, Trans. Wis. Acad. Sci., Arts, & Lett. **18**^{2}: 469, *pl.* 31, *figs.* 156–158. 1916.

Scenedesmus quadricauda var. *setosus* (Kirchner) Lagerheim, Öfvers. Kgl. Sv. Vet.-Ak. Förh. **39**, No. **2**: 63. 1882.

Coenobe flat, of 2–4–8 (generally 4) cells arranged in a single linear series. Cells ovoid to oblong-cylindrical, with rounded poles. Poles of all cells with 1–2 spines. Cell wall destitute of spines except at the poles. (Tychoplanktont).

Cells 4–5 μ broad, 8–11 μ long. 4-celled coenobia 8–11 μ broad, 16–20 μ long.

Rock (rr).

var. **Naegelii** (Brebisson) comb. nov. Pl. 40, Figs. 1–2.

Scenedesmus Naegelii de Brebisson, Mem. de la soc. Imp. de Sc. Nat. de Cherbourg **4**: 158. 1856.

Scenedesmus caudatus Nägeli non Corda, Gatt. einz. Algen 91, *pl.* 5, *figs.* 2c–2d. 1849.

Colonies flat, always eight-celled. Cells cylindrical, with rounded poles. Terminal cells with spines at both poles, interior cells with spines at one pole, rarely at both poles. (Facultative Planktont).

Cells 6.6–11 μ broad, 18–33 μ long. 8-celled colonies 18–33 μ broad, 50–85 μ long.

Blake (rr), Camp (rrr), Poor Farm (rrr), White Ash (rr).

This alga was described by de Brebisson from the drawings of *S. quadricauda* by Nägeli. In my monograph of *Scenedesmus* I included it as a synonym of *S. longus* (Trans. Wis. Acad. Sci., Arts, & Lett. **18**^{2}: 469. 1916) but am now convinced that it should be considered a distinct variety. The cells are much larger than the typical form but unfortunately Nägeli gives neither measurements for the cells nor an indication of the scale of magnification of his drawings. The measurements stated above are those of cells from Wisconsin lakes. The peculiar arrangement of the spines that Nägeli figures, where three adjacent interior cells have spines on the same end and none on the other with the reverse condition in the remaining interior cells, was found to be a common but not a constant character since specimens were found with spines at both poles of interior cells.

SCENEDESMUS ABUNDANS (Kirchner) Chodat. Pl. **39**, Figs. **23–25.**

Matér. p. la fl. crypt. Suisse. **4**[2]: 77. 1913; G. M. Smith, Trans. Wis. Acad. Sci., Arts, & Lett. **18**[2]: 465, *pl.* 31, *figs.* 137–140. 1916.

Scenedesmus quadricauda var. *abundans* (Kirchner) Lagerheim, Öfvers. Kgl. Vet.-Ak. Förh. **39**, No. **2**: 64, *pl.* 2, *fig.* 20. 1882.

Coenobe flat, generally 4-celled. Cells ovoid to oblong-ellipsoid. Terminal cells with 1–2 spines at poles and additional spines on outer face between the poles. Interior cells with 1–2 spines at each pole. (Facultative Planktont).

Cells 4–7 μ broad, 7–12 μ long. 4-celled coenobia 7–12 μ broad, 15–30 μ long. Spines 4–7.5 μ long.

Beaver (2) (rr), Cochrane (rr), Diamond (rr), Grassy (rr), Horse (rr), Long (1) (rrr), Nancy (rrr), Palmer (rr), Poor Farm (ss).

var. LONGICAUDA G. M. Smith. Pl. **39**, Figs. **26–27.**

Trans. Wis. Acad. Sci., Arts, & Lett. **18**[2]: 467, *pl.* 30, *figs.* 121–125. 1916.

Cells smaller, ovoid. Spines much longer in proportion to the size of the cells. (Facultative Planktont).

Cells 3–6 μ broad, 7–9 μ long. 4-celled coenobia 7–9 μ broad, 10–15 μ long. Spines 6–10 μ long.

Camp (rr), Chain (rrr), Horse (rr), White Ash (rr).

var. BREVICAUDA G. M. Smith. Pl. **40**, Figs. **3–5.**

Trans. Wis. Acad. Sci., Arts, & Lett. **18**[2]: 468, *pl.* 30, *figs.* 126–132. 1916.

Cells ovoid, smaller. Spines much shorter and never more than three between the polar spines of terminal cells. (Facultative Planktont).

Cells 2.5–5 μ broad, 5–8 μ long. 4-celled coenobia 5–8 μ broad, 10–13 μ long. Spines 1.5–3.5 μ long.

Mendota (rr).

var. ASYMMETRICA (Schröder) G. M. Smith. Pl. **40**, Figs. **6–8.**

Trans. Wis. Acad. Sci., Arts, & Lett. **18**[2]: 468, *pl.* 27, *figs.* 45–46. 1916.

Scenedesmus quadricauda var. *asymmetrica* Schröder, Forschungsbr. a. d. Biol. Stat. zu Plön **5**: 45, *pl.* 2, *figs.* 5a–5c. 1897.

Coenobia generally 4-celled, with oblong-ellipsoid cells. Poles of terminal cells with a single spine. All cells of coenobe with a single median spine perpendicular to the plane of the coenobe and at one side of cell only. (Facultative Planktont).

Cells 2.5–4.5 μ broad, 12–15 μ long. 4-celled coenobia 12–15 μ broad, 13–18 μ long. Spines 7–10 μ long.

Blue Gill (rr), Wild Goose (rrr).

SCENEDESMUS QUADRICAUDA (Turpin) de Brebisson. Pl. **40**, Figs. **9–11.**

Mém. de la Soc. Ac. de Falaise 1835: 66. 1835; **G. S. West, Brit. Freshw. Algae** 220, *figs.* 92D–92F. 1904.

Coenobia flat, of 2–4–8–16 (generally 4) cells in a simple linear or subalternating series. Cells cylindrical-ovoid, with rounded poles. Poles of terminal cells with a single long straight or curved spine. Poles of interior cells without spines. (Facultative Planktont).

Cells 3.5–6 μ broad, 11–16 μ long. 4-celled coenobia 11–16 μ broad, 12–23 μ long. Spines 10–12 μ long.

Beaverdam (rr), Big Butternut (rrr), Blake (rr), Browns (rr), Camp (rr), Chetek (rr), Clam (rr), Clear (rr), Eagle (rr), East (rrr), Fishtrap (rr), Green (1) (rr), Hillman (rr), Horseshoe (2) (rr), Lac du Flambeau (rr), Lindy (rr), Little Rice (2) (r), Lost (rr), Loveless (rr), Lower Nashotah (rr), Lower Nemahbin (rr), Mary (rr), Meta (rr), Mud (1) (rr), Mud (2) (rr), Muskallonge (r), Nancy (rr), Okauchee (rrr), Old Taylors (rr) Pokegama (1) (rr), Pope (rr), Poplar (rr), Poor Farm (sss), Rice (1) (rr), Round (3) (rr), Sand (1) (rrr), Sand (4) (rr), Travis (ccc), White Ash (rr), Wild Goose (rrr), Winnebago (rr).

The most ubiquitous species of the genus. The presence of spines on the poles of the terminal cells only is the distinguishing feature of the species. Cultural work has shown that it is possible to recognize several distinct varieties of this species.

var. PARVUS G. M. Smith. Pl. **40**, Fig. **17.**

Trans. Wis. Acad. Sci., Arts, & Lett. 18[2]: 480, *pl.* 31, *figs.* 162–166. 1916.

Cells ovoid-cylindrical. Length of cells 2–2.5 times the breadth and equal to the length of the spines. (Tychoplanktont).

Cells 3–4 μ broad, 5.5–8 μ long. 4-celled coenobia 5.5–8 μ broad, 12–15 μ long. Spines 4.5–8 μ long.

Deer (rrr).

var. QUADRISPINA (Chodat) G. M. Smith. Pl. **40**, Figs. **15–16.**

Trans. Wis. Acad. Sci., Arts, & Lett. 18[2]: 479, *pl.* 31, *figs.* 167–171. 1916.

Cells broadly ovoid, about twice as long as broad. Spines short, about as long as the diameter of the cell. (Tychoplanktont).

Cells 3.5–8 μ broad, 8.5–15 μ long. 4-celled coenobia 8.5–15 μ broad, 16–32 μ long. Spines 2.5–5.5 μ long.

Mill (rrr), Pine (rr), Pokegama (1) (rrr), Round (1) (rr), Tied Canoe (rrr).

var. LONGISPINA (Chodat) G. M. Smith. Pl. **40,** Figs. **12–14.**

Trans. Wis. Acad. Sci., Arts, & Lett. **18**[2]: 480, *pl.* 31, *figs.* 159–161. 1916.

Cells ovoid-cylindrical. Length of cells 2.5 times the breadth. (Facultative Planktont).

Cells 3.5–5 μ broad, 8–11 μ long. 4-celled coenobia 8–11 μ broad, 12–20 μ long. Spines 7.5–10 μ long.

Horse (rr), Lost (rr), Mendota (sss), Palmer (r).

var. WESTII G. M. Smith. Pl. **41,** Figs. **3–7.**

Trans. Wis. Acad. Sci., Arts, & Lett. **18**[2]: 480, *pl.* 32, *figs.* 177–180. 1916.

Cells 4.5–8 μ broad, 16–22 μ long. 4-celled coenobia 16–22 μ broad, 25–35 μ long. Spines 12–16 μ long. (Facultative Planktont).

Bear Trap (rrr), Blake (rr), Camp (rr), Cochrane (rr), Crawling Stone (rr), Grassy (rr), Horse (rrr), Lac Court Oreilles (rr), Loon (rrr), Meta (rr), Silver (2) (rr).

var. MAXIMUS W. & G. S. West. Pl. **41,** Figs. **1–2.**

Trans. Linn. Soc. 2 Ser. Bot. **5**: 83, *pl.* 5, *figs.* 9–10. 1895.

Cells much larger and spines proportionally longer. (Facultative Planktont).

Cells 9–11.5 μ broad, 27–36 μ long. 4-celled coenobia 27–36 μ broad, 35–40 μ long. Spines 20–30 μ long.

Eagle (rr), Nicaboyne (rr), Pine Tree (rr), Poor Farm (rr), Tenderfoot (rrr), Travis (r).

SCENEDESMUS OPOLIENSIS P. Richter. Pl. **41,** Figs. **8–11.**

Zeits. f. angw. Mikr. **1**: 7, *figs.* A–E. 1896.

Coenobia flat, of 2–4–8 (generally 4) cells in a single linear series. Interior cells naviculoid, outer margin of terminal cells straight or slightly convex. Cells in lateral contact in median third only. Terminal cells with a long spine at each pole; interior cells with or without long or short spines. Polar spines greatly reduced at times. (Facultative Planktont).

Cells 5–8 μ broad, 12–28 μ long. 4-celled coenobia 12–28 μ broad, 15–30 μ long. Spines 5–28 μ long.

Palmer (rr).

The navicular shape of the cells, their lateral contact in the median third only, together with the beak-like instead of rounded apex separates this species from *S. quadricauda.*

Sub-family *COELASTREAE.*

Coenobia forming a sphere with the cells radially arranged in a single layer around a common center. The number of cells in a coenoba always a multiple of two. Cells spherical, polygonal, or elongate cylindrical.

KEY TO THE GENERA.

Cells spherical, polygonal, or flattened; length never more than twice the breadth.
  Cells without spines on the outer face........................COELASTRUM
  Cells with spines on outer face...............................SORASTRUM
Cells 3–7 times as long as broad...................................ACTINASTRUM

COELASTRUM Nägeli 1849.

Coenobe a hollow sphere of 2–4–8–16–32–64–128 cells. Cells spherical, ovoid, or pyramidal; compressed or with large intercellular spaces; enclosed in a very delicate gelatinous sheath. Daughter coenobia sometimes remaining joined in multiple coenobia by the remains of the old mother cell wall.

Reproduction by autocolony formation in any cell of the coenobe. Liberation of young coenobia by a rupture of old mother cell wall.

KEY TO THE SPECIES.

Cells spherical.
  Processes uniting the cells long..........................C. RETICULATUM
  Processes uniting the cells short.
    External face of cells without projections..............C. MICROPORUM
    External face of cells with short truncate projections...C. CAMBRICUM
Cells truncate cones...C. PROBOSCIDEUM

COELASTRUM MICROPORUM Nägeli. Pl. **41**, Figs. **12–13**; Pl. **42**, Fig. **1**.

in Braun, Alg. unicell. 70. 1855; Senn, Bot. Ztg. **57**: 53, *pl.* 2, *figs.* 11–17. 1899

Coenobia spherical to ovoid; with 8–16–32–64 cells. Cells spherical enclosed by a delicate gelatinous sheath and connected to one another by short, stout, gelatinous processes. Intercellular interstices smaller than the diameter of the cells. (Facultative Planktont).

Diam. cells (without sheath) 3–20 μ; (with sheath) 4–23 μ. Diam coenobia 20–90 μ.

Balsam (rrr), Bear (1) (rr), Bear (2) (rr), Bear Trap (rrr), Big Butternut (rrr), Blake (rrr), Bone (rrr), Boulder (rrr), Browns (rr), Center (rr), Chetac (rr), Cochrane (rr), Deer (rrr), East (rrr), Elizabeth (rr) Fishtrap (rrr), Granite (rr), Hill (rr), Hillman (rrr), Horseshoe (2) (rrr)

Island (rr), Lac du Flambeau (sss), Little Bass (2) (rr), Little Butternut (rr), Little Crooked (rrr), Little Doctor (rr), Loveless (rrr), Lower Nashotah (rrr), Lower Turtle (rr), Mary (rr), Mendota (sss), Mill (rrr), Mud (2) (rr), Nancy (rrr), Oconomowoc (rr), Otter (rrr), Pewaukee (rrr), Pine Tree (rrr), Pokegama (1) (rr), Pokegama (2) (rr), Poor Farm (rrr), Pope (rrr), Prairie (rrr), Round (1) (rrr), Sand (1) (rr), Sand (4) (rr), Shell (rrr), Silver (1) (rr), South Turtle (rr), Squirrel (rr), Tenderfoot (rr), Tied Canoe (rrr), Travis (rrr), Vermilion (rr), White Ash (rrr), White Sand (rrr), Wolf (rrr).

A species that is found in all types of lakes.

COELASTRUM CAMBRICUM Archer. Pl. **42,** Figs. **2–3.**

Quart. Jour. Micr. Sci. N. S. **8**: 65. 1868; W. West, Jour. of Bot. **31**: 99, *pl.* 333, *fig.* 14. 1893.

Coenobia spherical, of 8–16–32–64–128 (generally 32) cells. Cells spherical, with a single truncate projection on the outer face. Each cell connected to adjacent ones by short, somewhat flattened, gelatinous projections. Small circular or triangular interstices between the cells. (Facultative Planktont).

Diam. cells (with sheath) 6–21 μ, (without sheath) 5–18 μ. Diam. coenobia 20–100 μ.

Balsam (rrr), Bear (2) (rr), Beaverdam (rr), Birch Island (rrr), Blue Gill (rrr), Catfish (rr), Cochrane (rrr), Cranberry (rr), East (rr), Found (rr), Hillman (rrr), Jag (rr), Little Bass (2) (r), Mercer (rr), Middle McKenzie (rrr), Minocqua (rr), Nancy (rrr), Nicaboyne (rr), Old Taylors (rrr), Owen (rrr), Pine (1) (rr), Pine Tree (rr), Pokegama (1) (rr), Prairie (rr), Rudolph (rrr), St. Croix (rr), Silver (1) (rr), Streitor (rr), Travis (r), Wild Goose (rrr).

COELASTRUM RECTICULATUM (Dangeard) Senn. Pl. **42,** Figs. **4–6.**

Bot. Ztg. **57**: 66, *pl.* 2, *figs.* 1–10. 1899.

Coenobia spherical, of 8–16–32 cells; generally held in multiple coenobia by the remains of old mother cell walls. Cells spherical, enclosed by a narrow gelatinous sheath and connected to adjacent cells by 6–9 long, slender, tough, gelatinous processes. (Facultative Planktont).

Diam. cells (with sheath) 6–24 μ. Diam. coenobia 10–65 μ.

Bear Trap (rrr), Bear (3) (rr), Beaverdam (rrr), Big McKenzie (rrr), Birch (2) (rr), Blake (rrr), Browns (ss), Camp (rr), Fish (rr), High (rrr), Island (rr), Lac Court Oreilles (rr), Little Bass (2) (rr), Mary (rr), Mendota (rr), Mud (1) (rr), Mudhen (rr), Nancy (rr), North Twin (rrr), Palmer (rr),

The long processes connecting the cells distinguish this species from all others. Solitary coenobia are rarely met with since the long processes hinder the escape of young coenobia.

COELASTRUM PROBOSCIDEUM Bohlin. Pl. **42**, Figs. **7–8**; Pl. **43**, Figs. **1–2.**

Bih. t. Kgl. Sv. Vet.-Ak. Handl. **23**, Afd. **3**, No. **7**: 33, *pl.* 2, *figs.* 19–22. 1897.

Coenobia pyramidal or cubical, rarely polygonal; of 4–8–16–32 (generally 4 or 8) cells; frequently joined in compound coenobia of irregular shape. Cells truncate cones, at times with concave sides; with base of cone towards the center of coenobe. Interstices between cells large, polygonal. Chloroplast single, parietal, with one pyrenoid. (Facultative Planktont).

Diam. cells 6.5–17 μ; diam. 4-celled coenobia 12–35 μ; compound coenobia up to 350 μ in diam.

Bear Trap (rrr), Cochrane (rr), Durchmans (rrr), Pine Tree (aa), Round (rrr).

The appearance of this alga as the dominant organism of the plankton of Pine Tree lake is of particular interest since *Coelastrum* species have not been reported as occurring in abundance by any student of the phytoplankton. Compound coenobia were very common and it was not at all unusual to find 50–80 four- or eight-celled coenobia joined in one irregular mass. The cells of four-celled coenobia are tetrahedrically arranged in the manner that G. S. West has described them for *Coelastrum compositum* (Jour. Linn. Soc. Bot. **38**: 136, *pl.* 5, *figs.* 8–9. 1907) and I believe that *C. compositum* should be considered a synonym of *C. proboscideum.* The cellular arrangment of eight-celled coenobia is exactly as Bohlin and Senn describe. In the sixteen-celled coenobia, which are of very rare occurrence, the cells form a hollow sphere so that the general appearance of the colony resembles the other species of the genus. This difference in cell number is of minor significance, the number of cells in a colony being dependent upon the vegetative vigor of the mother cell at the time colony formation takes place.

SORASTRUM Kützing. 1845.

Cells pyriform, semilunar, or reniform; united to form spherical colonies of 8–16–32–64–128 cells. Center of colony a polyhedral body from which gelatinous strands radiate to each cell. Distal side of cells with 1, 2, or 4 spines. Chloroplast diffuse, with one pyrenoid.

Reproduction unknown; presumably by autocolonies as in *Coelastrum,* possibly by zoospores as in *Pediastrum.*

KEY TO THE SPECIES.

Spines on outer face of cells short (4–8 μ) .... S. SPINULOSUM
Spines on outer face of cells long (10–15 μ) .... S. AMERICANUM

SORASTRUM SPINULOSUM Nägeli. Pl. **44**, Figs. **4–5.**

Gatt. einz. Algen 99, *pl.* 5, *figs.* Da–Dd. 1849; G. S. West, Brit. Freshw. Algae 215, *figs.* 89A–89B. 1904.

Coenobe of 4–8–16–32 cells. Cells broadly reniform to cuneate, with four short horns at the corners of the outer face. Stipe leading to sphere at center of colony short. Chloroplast single, generally filling the entire cell; pyrenoid single. (Tychoplanktont).

Cells 8–20 μ broad, 6–18 μ long, 5–8 μ thick. Horns of adult cells 4–8 μ long.

Hillman (rrr), Horse (rrr).

SORASTRUM AMERICANUM (Bohlin) Schmidle. Pl. **44**, Fig. **1.**

Engl. Jahrb. **27**: 230. 1900: Lemmermann, Arch. f. Hydrob. u. Planktonkde. **5**: 310, *fig.* 5. 1910.

Selenosphaerium americanum Bohlin, Bih. t. Kgl. Sv. Vet.-Ak. Handl. **23**, Afd. **3**, No. **7**: 40, *pl.* 2, *figs.* 38–41. 1897.

Coenobia spherical, solitary, of 16–32–64–128 (generally 16 or 32) cells. Cells cardiate to pyramidate, somewhat flattened, with the apex towards the center of the coenobe and ending in a broad gelatinous stalk as long as the cell. Stalk cylindrical, 5–6 faceted at the base; with the facets in contact with those of other stalks and forming a hollow sphere. Outer face of cells emarginate, with a single long spine at each of the four corners. (Facultative Planktont).

Cells 7–20 μ broad, 5–20 μ long (without stalk), 4–8 μ thick. Spines of adult cells 10–15 μ long.

Devils (1) (rr), Devils (2) (rrr), Nancy (rrr), Travis (rr).

The longer spines and stalk, together with the more prominent central sphere, separates this species from the foregoing.

var. UNDULATUM G. M. Smith. Pl. **44**, Figs. **2–3.**

Trans. Wis. Acad. Sci., Arts, & Lett. $\mathbf{19}^{1}$: 640, *pl.* 13, *fig.* 7. 1918.

Sides of facets at base of stipe undulate. Dimensions as in the typical form. (Facultative Planktont).

Beaverdam (rrr).

ACTINASTRUM Lagerheim 1882.

Coenobia free-floating, of 4–8–16 (generally 8), ovoid, cylindrical, club-shaped, or drum stick shaped cells radiating from a common center. Coenobia sometimes united to form multiple coenobia of irregular shape. Each cell with a single, laminate, parietal chloroplast; with or without pyrenoids.

Reproduction by longitudinal and transverse division of the contents of any cell, the young autocolony being liberated by a rupture of the mother cell wall.

KEY TO THE SPECIES.

Center of cell twice as broad as poles.............................A. Hantzschi
Center of cells about the same width as the poles.............A. Gracillimum

Actinastrum gracillimum G. M. Smith. Pl. **43,** Figs. **3–5.**

Bull. Torr. Bot. Cl. 43: 480, *pl.* 26, *fig.* 23. 1916.

Coenobia of 4 or 8 cells radiating from a common center. Cells elongate-cylindrical, tapering but slightly to a truncate end, 7–10 times as long as broad. Chloroplast single, parietal, laminate, with or without a pyrenoid. (Euplanktont).

Cells 1.75–3 μ broad, 14–21 μ long; colonies 30–45 μ in diam.

Horse (rr), Kegonsa (rrr), Mendota (rr).

The species differs from the following in the very slight dilation of the median portion of the cell and the greater proportional length.

Actinastrum Hantzschi Lagerheim. Pl. **43,** Figs. **6–7.**

Öfvers. Kgl. Vet.-Ak. Förh. 39, No. 2: 70, *pl.* 3, *figs.* 25–26. 1882.

Coenobia of 4 or 8 cells radiating from a common center. Coenobia sometimes united to form multiple coenobia. Cells 3–6 times as long as broad, with truncate poles, ovoid-cylindrical. Diameter of middle of cells twice that of poles. Chloroplast single, parietal, laminate; with one pyrenoid. (Euplanktont).

Cells 3–6 μ broad, 10–26 μ long; coenobia up to 50 μ in diam; compound coenobia up to 200 μ in diam.

Chetek (rr), Kegonsa (r), Lower Turtle (rrr), Mendota (rr), Prairie (rr), Upper Turtle (rr), Waubesa (rrr).

Multiple coenobia are of rare occurrence and have only been found in the plankton of Chetek lake. This may be due to the fact that such colonies readily fragment into the individual coenobia and the handling incident to the collection of samples is sufficient to break up multiple coenobia.

var. ELONGATUM G. M. Smith. Pl. **43**, Fig. **8**.

Trans. Wis. Acad. Sci., Arts, & Lett. 19[1]: 636, *pl.* 12, *fig.* 3. 1918.

Colonies of 4 or 8 cells. Cells cylindrical, tapering but little towards the poles. (Euplanktont).

Cells 4–5 μ broad, 30–35 μ long.

Bear (3) (rr).

The cells of this variety are much longer than those of the typical form.

Family *HYDRODICTYACEAE.*

Cells cylindrical, or flattened and hexagonal to trapezoidal with one, two, or four projections. Coenobia of definite shape; cylindrical with a large number of cells; or disciform with 2–4–8–16–32–64–128 cells. Cells incapable of division after their formation. Chloroplast single, parietal, laminate to reticulate, frequently filling the entire mature cell; with one to many pyrenoids.

Asexual reproduction by the division of contents of any cell to form zoospores that swarm within old cell wall or within a gelatinous vesicle extruded from the cell. Zoospores becoming apposed in the shape of the adult cell when they come to rest.

Sexual reproduction by isogamous zoogametes which fuse in pairs to form angular resting cells which form colonies on their germination.

KEY TO THE GENERA.

Coenobe a hollow cylindrical tube with rounded ends..........HYDRODICTYON
Coenobe a flattened disc one cell in thickness.
  Number of cells in coenobe always 2......................EUASTROPSIS
  Number of cells in coenobe 4–8–16–32–64–128, very
    rarely 2...PEDIASTRUM

HYDRODICTYON Roth 1800.

Cells macroscopic, oblong-cylindrical with rounded ends; united to form a free-floating, reticulate, saccate coenobium. Meshes of nets 3–12 (generally 5–6) sided. Cells coenocytic; chloroplast parietal, reticulate at first, later diffuse; pyrenoids single in very young cells, several hundred in mature cells.

Asexual reproduction by the division of cell contents into very many biciliate zoospores which swarm about in and become arranged as in the mature coenobe within the old mother cell wall. Liberation of the young coenobe by a gradual gelatinization of the mother cell wall.

Sexual reproduction by zoogametes which are similar in structure to zoospores but smaller, the zoogametes fusing in pairs to form a spherical zygote after their liberation through a pore in the cell wall. Zygote germinating, after a short period of rest, into 2–5 large uni- or biciliate zoospores that form irregular polyhedral cells on coming to rest. Polyeders ultimately germinating into 200–300 zoospores that are liberated in a vesicle and form a net as in asexual reproduction.

HYDRODICTYON RETICULATUM (L) Lagerheim. Pl. **44**, Fig. **6**; Pl. **45**, Fig. **1**.

Öfvers. Kgl. Vet.-Ak. Förh. 40, No. **2**: 71. 1883; G. S. West, **Brit. Freshw. Algae** 208, *figs.* 84A–84F. 1904

Cells elongate cylindrical, forming a definite closed net. (Tychoplanktont).

Cells 5–250 μ broad, up to 1.5 cm. long. Coenobia up to 30 cm. long.

Little Bass (2) (rr), Mendota (rr), Monona (sss).

An alga that is in no sense a plankton organism but one which sometimes grows in abundance in shallow, well protected bays and so may, at times, be found free-floating in large patches a considerable distance from the shore.

PEDIASTRUM Meyen 1829.

Coenobia disciform to stellate, free-floating, of 2–4–8–16–32–64–128 cells arranged in a layer one cell in thickness. Coenobe entire, perforate, or clathrate. Marginal cells polygonal, with one, two, or four processes that sometimes terminate in a tuft of long hyaline setae. Interior cells polygonal, without processes. Cell wall smooth, granulate, or covered with a meshwork of fine ridges; without a gelatinous sheath. Chloroplasts parietal and disciform at first, later filling the entire cell; with 1–4 pyrenoids. Cells coenocytic.

Asexual reproduction by the division of contents of any cell into 2–4–8–16–32–64–128 biciliate zoospores that are extruded in a gelatinous vesicle and, after a short period of swarming, on becoming quiescent appose themselves in the same position as the mature cells of the coenobe.

A fusion of biciliate zoogametes into polyhedral resting cells has been observed. On germination these zygospores form 8–16–32 zoospores that behave as in asexual reproduction.

KEY TO THE SPECIES.

Outer face of marginal cells without projections.
  Wall smooth...P. INTEGRUM
  Wall reticulate...P. ARANEOSUM
Outer face of marginal cells with 1, 2, or 4 projections.
  Marginal cells with more than one projection.
    Cells on margin with four projections and a deep linear incision..P. TETRAS
    Cells on margin with two projections.
      Coenobe entire.
        Cell wall smooth or punctate.
          Projections side by side in front view of cell...........P. BORYANUM
          Projections above one another in front view..........P. KAWRAISKYI
        Cell wall reticulate.
          Wall covered with coarse reticulations...............P. ARANEOSUM
          Wall covered with fine reticulations.................P. SCULPTATUM
      Coenobe perforate.
        Ends of projections truncate............................P. DUPLEX
        Ends of projections incised............................P. BIRADIATUM
  Marginal cells with one projection...............................P. SIMPLEX

?EDIASTRUM SIMPLEX var. DUODENARIUM (Bailey) Rabenhorst. Pl. 45, Figs. 2–6.

'l. Eur. Algarum **3**: 72. 1868.

Monactinus duodenarius Bailey, Smithsonian cont. to knowl. **7**, Art. **3**: 14, *l.* 1, *fig.* 37. 1855.

Monactinus octonarius Bailey, *ibid.* p. 14, *pl.* 1, *fig.* 36. 1855.

Pediastrum simplex var. *clathratum* (Schröter) G. S. West, Jour. Linn. Soc. ؛ot. **38**: 134. 1907.

Coenobia circular to oval, of 4–8–16–32–64–128 (generally 8–16–32) ells. Inner face of marginal cells concave, outer face produced into single, delicately tapering, concave process frequently furnished with tuft of delicate hyaline setae. Interior cells as those of the periphery ut with shorter processes. Interstices between cells large. Cell wall mooth or finely punctate. (Facultative Planktont).

Cells 10–24 μ broad, 20–45 μ long.

Balsam (rr), Bear (1) (r), Bear Trap (rrr), Blake (r), Camp (rr), Green 1) (rrr), High (rr), Hillman (rrr), Kegonsa (rr), Little Bass (rr), Little utternut (r), Little Crooked (rrr), Lost (rr), Loveless (rrr), Mary (rr), [onona (r), Mud (2) (rr), Palmer (rrr), Pokegama (1) (sss), Rice (1) (r), t. Croix (rr), Silver (1) (rr), Spirit (rr), Upper Turtle (r), Vermilion (cc), 'apagossett (r), White Ash (rr), Winnebago (rr), Wolf (rrr).

This alga is one that has been given a number of names by various in- ɘstigators, but for the reasons stated elsewhere (Trans. Wis. Acad. Sci., .rts, & Lett. 19[1]: 642. 1918) I consider the one given above the proper ame.

Pediastrum integrum Nägeli. Pl. **45**, Fig. **7**.

Gatt. einz. Algen 97, *pl.* 5, *figs.* B4a–B4r. 1849.

Coenobia 8–16–32–64–celled, circular or oval, without perforations. Interior cells polygonal; marginal cells 5–6 sided, at times with rudiments of two projections that are sometimes reduced to a single wart, but generally without projections. Cell wall smooth or punctate. (Tychoplanktont).

Diam. cells 15–30 μ.

Soft (rrr).

var. priva Printz. Pl. **45**, Figs. **8–10**.

Skr. Vidensk. i Kristiana Mat.-Naturv. Kl. **1913**[1]: No. **6**: 73, *pl.* 5, *fig.* 147 1914.

Coenobia generally 8-celled. Marginal cells trapezoidal, without incisions or processes. Interior cells polygonal, without incisions. Cell wall smooth. (Facultative Planktont).

Diam. cells 8–10 μ. Diam. 8-celled coenobia 15–25 μ.

Palmer (rr).

The general shape of the cells is like that of *P. tetras* but since there are no incisions in the cells they cannot be placed among the varieties of *P. tetras.*

Pediastrum araneosum Raciborski. Pl. **45**, Fig. **11**.

Mem. Acad. Imp. Sci. Cracow **20**: 84. 1890.

Coenobia circular to oval, without perforations, of 8–16–32–64–128 cells. Interior cells in contact with 4–6 other cells, straight sided. Inner half of marginal cells similar to interior cells, outer half without processes but with a deep emargination. Surface of cell covered with a coarse reticulum of ridges. (Facultative Planktont).

Diam. cells 15–32 μ.

Birch Island (rr), Blue Gill (rrr), Boulder (rrr), Carson (rrr), Deer (rrr) Devils (1) (rrr), Diamond (rrr), Dutchmans (rr), Horseshoe (1) (rr), Lake of the Woods (rrr), Lindy (rr), Loon (rrr), Nicaboyne (rr), Nancy (r), Pine Tree (rr), Pokegama (1) (rrr), Poor Farm (rrr), Sand (2) (rr), Wild Goose (rr).

var. rugulosum (G. S. West) G. M. Smith. Pl. **45**, Figs. **12–13**.

Bull. Torr. Bot. Cl. **43**: 476, *pl.* 25, *fig.* 14. 1916.

Sides of cells undulate where they are in contact with one another (Facultative Planktont).

Meta (r).

PEDIASTRUM SCULPTATUM G. M. Smith. **Pl. 46, Fig. 1.**

Bull. Torr. Bot. Cl. **43**: 475, *pl.* 25, *fig.* 13. 1916.

Coenobia oval to circular, of 8–16–32 cells. Interior cells with 4–6 straight sides, with a few interstices between the cells. Marginal cells bilobed, with the lobes produced into two short divergent or parallel processes. Cell wall covered with a network of very fine ridges. (Facultative Planktont).

Diam. cells 10–15 μ. Diam. coenobia 80–150 μ.

South Turtle (rr).

PEDIASTRUM BORYANUM (Turpin) Meneghini. **Pl. 46, Figs. 2–7.**

Linnaea **14**: 210. 1840; G. S. West, Brit. Freshw. Algae 211, *figs.* 85F–85H. 1904.

Coenobia circular to ovoid, of 4–8–16–32–64–128–256 (generally 8–16–32) cells that are normally compactly arranged in concentric rings. Interior cells 4–6 sided, with straight sides. Inner half of marginal cells similar to interior cells; outer half produced into two short obtuse processes ending in short spines, slightly to deeply emarginate between the processes. Projections on outer face and their spines sometimes greatly reduced. Wall smooth or granulate. (Facultative Planktont).

Diam. cells 7.5–30 μ. Diam. 16-celled coenobia 40–100 μ.

Balsam (rrr), Bear (1) (rrr), Bear (2) (rrr), Bear (3) (ss), Bear Trap (rr), Beasley (rr), Beaver (rr), Beaverdam (ss), Big Butternut (rr), Birch Island (rr), Blake (rr), Bloom (r), Boulder (rrr), Browns (rr), Camp (r), Carroll (sss), Carson (rrr), Catherine (rr), Center (ss), Chetac (rr), Chetek (rrr), Clear Crooked (rr), Cochrane (rrr), Crawling Stone (rr), Deer (r), Des Moines (rr), Devils (1) (rr), Eagle (sss), East (rrr), Elizabeth (rrr), Fence (rrr), Fishtrap (rr), Found (rr), Fowler (rrr), Geneva (rr), Granite (rr), Grassy (rr), Green (2) (rrr), Grindstone (rr), Half Moon (rrr), Harris (rr), High (rr), Hillman (rr), Hooker (rr), Horse (rrr), Horseshoe (1) (ss), Kawaguesaga (r), Kegonsa (sss), Kitten (rr), Lac la Belle (rr), Lac Court Oreilles (rr), Lac du Flambeau (r), Lindy (rr), Little Bass (2) (r), Little Bear (sss), Little Butternut (rrr), Little Crooked (rr), Little Rice (2) (rr), Little Wood (rrr), Long (2) (rr), Long (3) (rrr), Loveless (r), Lower Nashotah (rrr), Lower Nemahbin (rr), Lower Turtle (rr), Marl (rr), Mary (ss), Mendota (ss), Monona (ss), Mud (1) (r), Mud (2) (r), Mudhen (rr), Muskallonge (rr), Nagawicka (rr), Nancy (sss), Nicaboyne (rrr), Oconomowoc (rr), Old Taylors (rr), Owen (r), Paddock (rrr), Pewaukee (rr), Pine (2) (rrr), Pokegama (1) (sss), Pokegama (2) (r), Pokegama (3) (rr), Poor Farm (rr), Poplar (rr), Prairie (rrr), Rainbow (rr), Reserve (rrr), Rice (1) (rr), Round (1) (sss), Round (2) (rr), St. Croix (rr), Sand (1) (rrr), Sand (3) (rrr), Sand (4) (sss), Silver (2) (r), Spirit (rr), Star (rr), Streitor (rr), Taylor (r), Tenderfoot (rrr), Tied Canoe (rrr), Turtle [North and South] (rr), Upper Turtle (r), Vermilion (rr), Waubesa (r), White Ash (sss), Whitefish (rr), White Sand (rr), Wildcat (rrr), Winnebago (r), Wolf (s).

A species that is almost universally present and one which varies greatly in the character of the marginal cells.

var. UNDULATUM Wille. Pl. 46, Fig. 8.

Öfvers. Kgl. Vet.-Ak. Förh. **36**, No. **5**: 28. 1879.

Sides of cells in contact with other cells undulate instead of straight. (Tychoplanktont).

Winnebago (rr).

var. LONGICORNE Raciborski. Pl. 46, Fig. 9.

Akad. Umiej. w. Krakowie **20**: 13. 1889; Printz, Skr. Vidensk. i Kristiana Mat.-Naturv. Kl. **1913**, No. **6**: 74, *pl.* 5, *fig.* 149. 1914.

Horns at the ends of processes much longer than typical form, sometimes swollen at apex. Cell wall smooth or granulate.

Diam. cells up to 40 μ. Horns 12–30 μ long.

Bone (rrr).

Raciborski recognizes the forms *glabrum* and *granulatum* but, judging by the variation in the nature of the wall in the typical form, these forms should not be recognized.

PEDIASTRUM KAWRAISKYI Schmidle. Pl. 46, Figs. 10–13.

Arb. d. Botan. Gartens zu Tiflis **2**: 269. 1897; Nitardy, Beih. z. Bot. Cent. **32**[2]: 164, *pl.* 5, *fig.* 5. 1914.

Coenobia of 16–32–64 cells that are compactly arranged. Interior cells 4–6 sided, sides straight. Inner half of marginal cells as interior cells, outer half with two short projections that lie one above the other when the coenobe is seen in surface view. Cell wall smooth. (Euplanktont).

Diam. cells 9–22 μ. Diam. 32-celled coenobia 60–100 μ.

Big Bass (r), Palmer (rr), Tenderfoot (rrr), Winnebago (r).

The cell shape is quite similar to *P. Boryanum* but the peculiar position of the processes on the marginal cells at once distinguishes this species from all others. Some investigators are inclined to the view that *P. Kawraiskyi* is but a passing condition of *P. Boryanum* that is found for one generation only. I have grown *P. Boryanum* in pure cultures for several years and have never seen a single coenobe with the projections of marginal cells lying one above the other, and so feel confident that *P. Kawraiskyi* is not a temporary growth condition of *P. Boryanum.*

PEDIASTRUM DUPLEX Meyen. Pl. **46,** Figs. **14–16.**

Nova Acta Phys. Med. Ac. Caes. Leop.-Carol. Nat. Cur. **14**[2]: 772, *pl.* 43, *figs.* 6–20. 1829.

Coenobia of 8–16–32–64–128 (generally 16–32) cells, with small lens-shaped perforations between the cells. Interior cells more or less rectangular, not in contact with one another in the central portion of the side walls. Inner half of marginal cells similar to interior cells, outer half prolonged into two short truncate processes. Cell wall smooth. (Facultative Planktont).

Diam. cells 11–21 μ.

Balsam (rrr), Bear (2) (r), Birch Island (rr), Browns (rr), Catherine (rr), Center (r), Clam (rrr), Cochrane (rr), Crawling Stone (r), Deer (rr), Des Moines (rrr), Diamond (rrr), Eagle (r), Fishtrap (rr), Found (rr), Fowler (rr), Helen (1) (rr), High (rr), Kawaguesaga (rr), Kegonsa (r), Kitten (r), Lac la Belle (r), Lindy (rr), Little Bass (2) (rr), Little Bear (r), Little Crooked (r), Little Rice (2) (rr), Long (3) (rrr), Lost (rr), Mendota (sss), Mercer (rr), Meronk (rrr), Monona (c), Nagawicka (rr), Nancy (sss), Nicaboyne (rrr), North Twin (rr), Oconomowoc (rr), Okauchee (rr), Pardee (rr), Pokegama (1) (ss), Pokegama (2) (rr), Poor Farm (r), Poplar (rrr), Round (1) (r), Round (2) (rr), Sand (4) (sss), Sishebogema (rr), Soft (r), South Turtle (sss), Taylor (rr), Tenderfoot (r), Tied Canoe (rrr), Upper Nemahbin (rr), Waubesa (ccc), White Sand (rrr), Wildcat (sss), Wild Goose (rrr), Winnebago (ss), Wolf (sss).

In many cases the marginal cells of this species and the varieties *clathratum, reticulatum* and *gracillimum* have a tuft of long hyaline setae at the apex of the projections. Their function is to give buoyancy to the coenobe when living under pelagic conditions. Petersen (Bot. Tids. **31**: 161–176. 1911) has been able to demonstrate their presence in other species by means of stains.

var. CLATHRATUM (A. Braun) Lagerheim. Pl. **47,** Figs. **1–3.**

Öfvers. Kgl. Vet.-Ak. Förh. **39,** No. **2:** 56. 1882.

Cells more deeply emarginate and perforations of disc larger. (Facultative Planktont).

Diam. cells 12–22 μ.

Balsam (rr), Bear (3) (rr), Bear Trap (rr), Beaverdam (sss), Big Butternut (rr), Blake (r), Bloom (rr), Camp (ss), Catfish (rr), Catherine (rr), Chetek (s), Cranberry (rr), East (rrr), Fish (s), Horseshoe (1) (rr), Horseshoe (2) (rr), Lac du Flambeau (rrr), Lac la Belle (r), Little Bass (2) (ss), Little Butternut (r), Little Wood (rrr), Long (1) (rrr), Loveless (sss), Lower Turtle (r), Mary (r), Mendota (r), Middle McKenzie (s), Mud (2) (rr), Mud (3) (sss), Nancy (r), Okauchee (rrr), Old Taylors (rrr), Pokegama (1) (r), Rice (1) (ss), Round (2) (rr), Sand (3) (rr), Sand (4) (r), Spirit (sss), Upper Turtle (sss), Wapagossett (rr), White Ash (r).

var. RETICULATUM Lagerheim. Pl. 47, Figs. 4–7.

Öfvers. Kgl. Vet.-Ak. Förh. 39, No. 2: 56, *pl.* 2, *fig.* 1. 1882.

Perforations of coenobe very large, oval. Interior cells sub-cruciate to H-shaped. Sides of projections of marginal cells parallel, projections generally terminating in a tuft of delicate long hyaline setae. (Euplanktont ?).

Cells 10–20 μ broad, 15–40 μ long.

Big (rr), Camp (s), Center (s), Chetek (s), Crab (rr), Hill (rr), Hooker (rr), Horse (sss), Island (sss), Kawaguesaga (rr), Little Bear (rr), Little Rice (1) (rr), Lost Bass (r), Loveless (r), Nancy (r), Paddock (rr), Palmer (r), Pardee (r), Pine (2) (rrr), Pokegama (1) (r), Pokegama (3) (ss), Prairie (ss), Reserve (rr), Rice (r), St. Croix (sss), Sishebogema (r), South Turtle (rr), Tenderfoot (rr), Waubesa (rr), Whitefish (rr), Winnebago (rr).

var. GRACILLIMUM W. & G. S. West. Pl. 47, Figs. 8–11; Pl. 48, Figs. 1–2.

Jour. of Bot. 33: 52. 1895.

Cells very narrow, never wider than the processes. Body of marginal cells curved outwards, with two long, somewhat converging processes that have parallel or slightly convex sides. Interior cells like those of margin but with somewhat shorter processes. Perforations of coenobe very large. (Euplanktont).

Cells 10–22 μ broad, 12–32 μ long. Diam. 16-celled coenobia 65–140 μ.

Clam (rr), Pine Tree (sss), Travis (r).

The perforations of the coenobe and the extension of the processes are the greatest in this variety. The varieties *clathratum, reticulatum,* and *gracillimum* are not sharply defined and it is possible, taking the collections as a whole, to arrange a series that grades almost imperceptibly from the typical form to the variety *gracillimum.* The individuals from any one lake, however, generally vary within fairly narrow limits so I feel that these varieties should be recognized instead of putting them in one collective species without varieties.

var. ROTUNDATUM Lucks. Pl. 46, Fig. 17.

Jahrb. d. Westpr. Lehrervereins f. Naturk. 2–3: 31. 1907; Nitardy, Beih. z. Bot. Cent. 32²: *pl.* 9, *fig.* 11. 1914.

Projections of marginal cells markedly convex and capitate instead of parallel. (Tychoplanktont).

Oconomowoc (rrr), Horseshoe (1) (rr).

var. COHAERENS Bohlin. Pl. **48**, Figs. **3–4.**

Bih. t. Kgl. Sv. Vet.-Ak. Handl. **23**, Afd. **3**, No. **7**: 31, *pl.* 2, *fig.* 1. 1897.
Pediastrum duplex var. *reticulatum* forma *cohaerens* (Bohlin) Brunnthaler, in Pascher, D. Süsswasserfl. Deutschl., Österr. u. d. Schw. **5**^{2}: 95, *fig.* 57e. 1915.

Coenobia circular to oval, of 16–32–64 cells, with median-sized intercellular spaces. Wall of cells covered with short, linear, concentrically arranged granules. (Facultative Planktont).

Cells 12–22 μ broad; 13–24 μ long.

Pine Tree (r).

PEDIASTRUM BIRADIATUM Meyen. Pl. **48**, Figs. **5–8.**

Nova Acta Phys.-Med. Ac. Caes.-Leop. Nat. Cur. **14**^{2}: 773, *pl.* 43, *figs.* 21–22. 1829.
Pediastrum rotula (Ehrenberg) A. Braun emend. Alg. unicell. 101, *pl.* 6, *figs.* 2–3, 5–7, 10, 12–14. 1855.

Coenobia of 4–8–16–32–64 cells, with medium sized openings between cells. Marginal cells in contact with one another at base only, with two long projections that are dilate and incised at the apex. Interior cells as those of the margin but apices of projections neither dilate nor incised, in contact with other cells at base and ends of projections only. Cell wall smooth. (Facultative Planktont).

Cells 10–22 μ broad, 15–30 μ long. Diam. 32-celled coenobia 80–150 μ.

Balsam (rrr), Bear (3) (rr), Big McKenzie (rr), Lac du Flambeau (rr), Mud (2) (rr), Nancy (rrr), Pokegama (1) (rrr), Wolf (rrr).

This species resembles *P. duplex* var. *reticulatum* in the perforation of the disc but differs in the deep incision of the marginal cell projections. Perfect colonies are seldom seen since apices of projections of certain interior cells are free and have the same dilation and incision that is found in the marginal cells. The number of cells in the colony is generally stated as 8–16–32 but the coenobia in the plankton of Wisconsin lakes are 32- or 64-celled except in very rare cases.

PEDIASTRUM TETRAS (Ehrenberg) Ralfs. Pl. **48**, Figs. **9–12.**

Ann. & Mag. of Nat. Hist. **14**: 469, *pl.* 12, *fig.* 4. 1844.

Coenobia rectangular, oval or circular; without perforations; of 4–8–16–32 (generally 4–8) cells. Outer face of marginal cells with a single deep linear to cuneate incision, the halves of the outer face slightly to deeply emarginate. Interior cells with a single deep linear incision and 4–6 straight sides. Cell wall smooth. (Tychoplanktont).

Diam. cells 8–15 μ.

Blake (rrr), Browns (rrr), Chain (rrr), Chetek (rr), Cochrane (rrr), Hillman (rrr), Horse (rrr), Horseshoe (2) (rrr), Kimball (rrr), Lindy (rrr),

Little Doctor (rr), Mendota (rrr), Palmer (rrr), Pewaukee (rrr), Pine (2) (rr), Pokegama (1) (rr), Poor Farm (rr), Tied Canoe (rrr).

The four projections in the outer face of marginal cells are due to a deep emargination of the two halves. This, coupled with the deep incision of all cells, at once separates the species from others of the genus. The alga does not thrive under plankton conditions although it is common in ditches and ponds of the state.

var. OBTUSATA Raciborski.

Akad. Umiej. w. Krakowie **20**: 32. 1889; Printz, Skr. Vidensk. i Kristiana Mat.-Naturv. Kl. **1913**[1], No. **6**: 78, *pl.* 6, *fig.* 153. 1914.

Cells larger than typical form, sometimes expanded at base. Halves of outer face of marginal cells entire or very slightly emarginate. (Tychoplanktont).

Diam. cells 14–18 μ; length 18–22 μ.

Beaverdam (rrr).

var. TETRAODON (Corda) Hansgirg. Pl. **48**, Figs. **13–14**; Pl. **49**, Figs. **1–2**.

Prodr. d. Algenfl. v. Böhmen **1**: 112. 1888.

Euastrum tetraödon Corda, Almanach de Carlsbad **9**: 238, *pl.* 2, *fig.* 9. 1839.

Coenobia generally 8-celled. Incision of cells deep, and with projections adjacent to incision very pronounced on marginal cells. (Facultative Planktont).

Diam. marginal cells 8–15 μ; length 12–18 μ.

Pine Tree (rr), Travis (r).

There is no question that my figures of the organism are identical with the one that Nitardy (Beih. Bot. Cent. **32**[2]: 123, *pl.* 5, *fig.* 8. 1914) has reproduced from Corda. The variety is sharply defined through the greater development of the projections.

EUASTROPSIS Lagerheim 1894.

Coenobia free-floating, 2-celled. Cells flattened, approximated at the bases, distal end free and deeply emarginate. Chloroplast single, parietal, with one pyrenoid.

Asexual reproduction by division of cell contents into 2–4–8–16–32 zoospores which are extruded in a vesicle and on becoming quiescent join in pairs to form the characteristic coenobe.

Sexual reproduction unknown.

EUASTROPSIS RICHTERI (Schmidle) Lagerheim. Pl. **49**, Fig. **3**.

Tromsö Museums Aarshefter **17**: 20, *pl.* 1, *figs.* 8–27. 1894.

Cells trapezoidal, outer face deeply emarginate. (Tychoplanktont).
Cells 4.5–25 μ broad, 5–20 μ long. Coenobia 4.5–25 μ broad, 10–40 μ long.

Kegonsa (rrr).

It is quite possible to mistake 2-celled *Pediastrum* colonies for *E. Richteri,* but since such *Pediastrum* colonies are of very rare occurrence the similarity of the two is not apt to lead to confusion.

Family *PLANOSPORACEAE.*

Cells solitary or gregarious; ovoid, pyriform, fusiform, cylindrical, or hemispherical; epiphytic, endophytic, or epizootic. Cells capable of growth but incapable of division after their formation. Generally with a single chloroplast with one or more pyrenoids.

Asexual reproduction by division of cell contents to form many zoospores.

Only one genus, *Characium,* is found in the plankton of the state.

CHARACIUM A. Braun 1849.

Cells sub-spherical, ovoid, pyriform, fusiform, or cylindrical; solitary or gregarious; epiphytic or epizootic; generally attached to substratum by a more or less elongate stipe, rarely sub-sessile. Chloroplast parietal and laminate to cup-shaped in young cells; later expanded to fill entire cell: with one or more pyrenoids. Cells at times with several chloroplasts.

Asexual reproduction by repeated division of cell contents into 8–16–32–64–128 zoospores that are liberated through a lateral or apical pore. Zoospores swarming but a short time and then coming to rest on substratum and developing into vegetative cells.

Microgametes reported but doubtful.

KEY TO THE SPECIES.

Epiphytic upon other algae.
  Cells ovoid.......................................C. STIPITATUM
  Cells lunate.......................................C. CURVATUM
Epizootic upon crustaceans.
  Apices of cells rounded.............................C. CYLINDRICUM
  Apices of cells produced into a fine seta.
    Stipe simple.....................................C. LIMNETICUM
    Stipe ending in filiform rhizoids.......................C. GRACILIPES

CHARACIUM STIPITATUM (Bachmann) Wille. Pl. 49, Figs. 4–7.

in Engler & Prantl, Nat. Pflanzenfam. **1**[2] (Nachtr.); 45. 1909; G. M. Smith Bull. Torr. Bot. Cl. **43**: 472, *pl.* 24, *figs.* 4–6. 1916.

Cells spherical to pyriform, with a delicate hyaline stipe somewhat longer than the cell. Chloroplast single, parietal, cup-shaped, always on the side away from the stipe; pyrenoid single. Zoospores broadly ovoid, biciliate, with a cup-shaped chloroplast at the posterior end (Euplanktont).

Epiphytic upon and embedded within the gelatinous envelope of *Coelosphaerium Naegelianum* Unger.

Cells 5–8 μ broad. Stipe 10–16 μ long.

Amicoy (r), Balsam (rr), Beartrap (rr), Beaverdam (s), Big (rr), Big McKenzie (ss), Clam (ss), Devils (2) (ss), Horseshoe (1) (ss), Horseshoe (2) (rr), Long (2) (ss), Kawaguesaga (ss), North Twin (sss), Pardee (r) Pike (rr), Pokegama (1) (s), Pokegama (2) (r), Poplar (rrr), St. Croix (r) Sand (4) (rr), Sishebogema (rr).

In one case the alga was found upon *Oocystis* but in all others it was inside the gelatinous envelope of *Coelosphaerium Naegelianum* Unger When the alga grows upon *Coelosphaerium* the colony is apt to be infested with 50–100 individuals as a rule. The alga was originally described as a *Chlamydomonas* species but since the sessile phase is the dominant one of the life cycle it is better placed as a *Characium* species Although found at only one station in Europe it is widely distributed in northern Wisconsin.

CHARACIUM CURVATUM G. M. Smith. Pl. 49, Figs. 8–15.

Trans. Wis. Acad. Sci., Arts, & Lett. **19**[1]: 641, *pl.* 11, *figs.* 6–13. 1918.

Cells lunate or recurved, with acute or sub-acute apices. Stipe hyaline, thick, tubular or conical, sometimes irregular in outline. Chloroplast single, parietal, laminate to cup-shaped; with or without a pyrenoid. (Euplanktont).

Epiphytic upon *Coelosphaerium Naegelianum* Unger, *Dictyosphaerium pulchellum* Wood, and *Phaeococcus planctonicus* W. & G. S. West

Cells 3–6 μ broad, 8–18 μ long (without stipe), 13–22 μ long with stipe.

Bear (1) (rr), Birch (2) (cc), Chain (rr), Chetac (aa), Chetek (rr), Clam (sss), East (rr), Horseshoe (1) (ss), Lindy (rr), Mud (1) (rr), South Crab (ss), Wild Goose (rr).

An alga that resembles *C. stipitatum* in the manner of growth but one which is found upon a wider range of genera. It also differs in the curved cells and the broader stipe.

CHARACIUM CYLINDRICUM Lambert. Pl. **49**, Figs. **16–19.**

Rhodora **11**: 70, *pl.* 79, *figs.* 1–2, 7–8, 10–13, 23–25. 1909.

Cells cylindrical, with rounded apices, base tapering slightly into a short stipe without a basal disc. Chloroplasts 1–16, parietal, disciform to laminate, without pyrenoids. (Euplanktont).

On *Diaphanosoma brachyura* and *Branchipus vernalis.*

Cells 8–20 μ broad, 20–430 μ long.

Howey (cc).

The lack of a pyrenoid suggests *Characiopsis* but the bright green color of the chloroplast shows that the alga is one of the Chlorophyceae and so is placed among *Characium* species. Young cells contain a single chloroplast but with growth there is an increase in the number of chloroplasts and all stages between cells with a single chloroplast and those with sixteen may be found on the same crustacean. Lambert holds that this condition is a stage in the division into zoospores but in my opinion it is the normal condition of the adult cell. Another point of interest is the position of the alga upon the host since it is only found on the tail of the crustacean.

CHARACIUM LIMNETICUM Lemmermann. Pl. **49**, Figs. **25–29.**

Bot. Not. **1903**: 81, *pl.* 3, *figs.* 7–10. 1903.

Cells broadly fusiform, lunate, rarely straight; apex produced into a single delicate hyaline seta; base a delicate stipe without basal discs. 1–8 parietal band-shaped chloroplasts, each containing 1–2 pyrenoids. (Euplanktont).

Cells 5–14 μ broad, 25–110 μ long; stipe 5–10 μ long.

On *Diaphanosoma brachyura.*

Boulder (rr), Browns (rr), Delavan (rrr).

Lemmermann describes the chloroplast as single but reports a division of the chloroplast prior to division. I have found that in all normal adult cells of *C. limneticum, C. gracilipes,* and *C. cylindricum* there is more than one chloroplast.

The species is very closely related to *C. gracilipes* but is much smaller, proportionally broader, and has a simple stipe. It is very probable that *Raphidium longissimum* var. *septatum* Chodat (*Ankistrodesmus longissimum* var. *septatum* (Chodat) Brunnthaler) is a cell of *C. limneticum* broken off from the crustacean during collection.

CHARACIUM GRACILIPES Lambert. Pl. **49**, Figs. **20–24.**

Rhodora **11**: 65, *pl.* 79, *figs.* 3–6, 9, 14–22. 1909.

Cells straight or curved with the median portion fusiform; apex tapering into a long seta and base attenuated to a filiform stipe. Chloroplast parietal, band-shaped, 1–32 per cell; with one pyrenoid in each chloroplast. (Euplanktont).

On *Diaphanosoma brachyura.*

Cells 5–14 μ broad, 70–480 μ long.

Howey (cc).

This alga was found on the same organism and at the same time as *C. cylindricum.* It was confined to the forward appendages while *C. cylindricum* was found at the tail only.

Order **ULOTRICHALES.**

Cells usually in a simple or branched thallus, rarely forming disciform thalli one cell in thickness. Thallus generally attached, rarely free-floating; naked, or enclosed by a tough and leathery or copious and gelatinous, homogeneous or lamellated envelope. Cells generally with a single parietal laminate to cup-shaped chloroplast with one or more pyrenoids. Cell shape usually cylindrical with flattened ends, rarely cylindrical with rounded ends, spherical or cubical.

Asexual reproduction by vegetative cell division, by a division of vegetative or special cells to form 2–16 bi- or tetraciliate zoospores, and by aplanospores.

Sexual reproduction either isogamous by a division of cell contents to form bi- or tetraciliate zoogametes; or heterogamous by the fusion of tetraciliate zoogametes of different size or the union of large non-motile oospheres and small biciliate antherozoids.

KEY TO THE FAMILIES.

Cells united to form simple filaments.......................... ULOTRICHACEAE
Cells united to form branching filaments.
    Sexual reproduction by isogamous zoogametes.......... CHAETOPHORACEAE
    Sexual reproduction by heterogamous zoogametes...... APHANOCHAETACEAE

Family *ULOTRICHACEAE.*

Cells united to form simple free-floating or sessile filaments which are enclosed at times by a copious, hyaline, homogeneous or radially fibrillar, gelatinous envelope. Cell shape spherical to cylindrical with rounded or flattened ends. Walls of cells delicate or thick, homogenous or lamellated. Cells containing a nucleus and one parietal, laminate or cup-shaped chloroplast with one or more pyrenoids.

Asexual reproduction by vegetative cell division, aplanospores or biciliate zoospores.

Sexual reproduction isogamous, by a fusion of bi- or tetraciliate zoogametes.

ULOTHRIX Kützing 1833.

Filaments simple, free-floating or with the basal cell modified to form a hold-fast. Cells cylindrical, with flattened end walls, united to form simple filaments that are the same diameter throughout. Wall of cells delicate or thick, homogeneous or lamellated, without a gelatinous outer layer. Chloroplast single, parietal, laminate, with one or more pyrenoids. Cells uninucleate.

Asexual reproduction by the division of cell contents to form 1–32 aplanospores and bi- or tetraciliate zoospores.

Sexual reproduction isogamous, by a fusion of biciliate zoogametes.

KEY TO THE SPECIES.

Filaments 11–45 μ in diameter.................................U. ZONATA
Filaments 5–7 μ in diameter...............................U. SUBCONSTRICTA

ULOTHRIX ZONATA (Weber & Mohr) Kützing. Pl. 50, Fig. 6.

Flora **16**: 519. 1833; Hazen, Mem. Torr. Bot. Cl. **11**: 147, *pl.* 20, *figs.* 1–4. 1902.

Filaments attached or free-floating and forming greenish masses. Cells disciform to cylindrical, ⅓ to 1½ times as long as broad. End walls of cells straight, side walls straight or slightly convex in older cells. Chloroplast forming a broad or narrow parietal annular girdle containing several pyrenoids. Wall of cell thick, generally homogeneous but at times lamellated in structure. (Tychoplanktont).

Mature cells 11–45 μ in diameter.

Mendota (rrr).

ULOTHRIX SUBCONSTRICTA G. S. West. Pl. 50, Figs. 4–5.

Jour. of Bot. **53**: 82, *figs.* 6A–6C. 1915.

Filaments free-floating, narrow, slightly constricted at cross walls of cells, with or without a homogeneous gelatinous envelope. Cells elongate, 2–6 times as long as broad, slightly inflated at the poles. Chloroplast single, parietal, laminate, with irregular edges and occupying the median two thirds of the cell. With or without 1–2 pyrenoids. Cell wall very delicate. (Facultative Planktont).

Cells 5.7–9 μ broad.

The specimens that I have found in our lakes differ somewhat from those described by West since the cells are not slightly dilate at the poles and the chloroplasts usually contain pyrenoids. The Wisconsin organism does not seem sufficiently differentiated, however, to warrant the description of a new variety or species. This alga is of particular interest since it is the only one of the order which has become established as a true member of the plankton.

Family *CHAETOPHORACEAE.*

Cells united to form branching filamentous, discoidal or parenchymatous thalli. Branches of thallus generally attenuated at apices and sometimes ending in long hair-like processes. Thallus naked or enclosed in a copious, hyaline, homogeneous, gelatinous envelope. Cell shape ordinarily cylindrical with flattened ends. Chloroplast single, parietal and laminate to girdle-shaped; with one or more pyrenoids. Nucleus single.

Asexual reproduction by aplanospores and bi- or tetraciliate zoospores.

Sexual reproduction isogamous, by a fusion of biciliate zoogametes.

STIGEOCLONIUM Kützing 1843.

Thallus free-floating or attached, with branches scattered and not arising in dense tufts. Thallus without an appreciable difference in size between the main stem and minor branches; covered with a homogeneous, hyaline, gelatinous envelope. Cells cylindrical to disciform, gradually decreasing in diameter towards the apex and with terminal cells conical or elongated to hair-like unseptate processes. Chloroplast single, parietal; gridle-shaped and filling median portion of older cells; completely filling younger cells. Each chloroplast generally containing several pyrenoids.

Asexual reproduction by aplanospores and tetraciliate zoospores.

Sexual reproduction isogamous, by the fusion of biciliate zoogametes.

STIGEOCLONIUM LUBRICUM (Dillwyn) Kützing. Pl. **50**, Fig. **7**.

Phycol. germ. 189. 1845.

Myxonema lubricum (Dillw.) Fries. Hazen, Mem. Torr. Bot. Cl. **11**: 195, *pl.* 28, *figs.* 1–2. 1902.

Filaments forming dark green tufts up to 30 cm. long. Filaments much branched and branches usually opposite, rarely single. Cells of main filament at base of branches sub-spherical, shorter than other cells.

Branchlets abundant, opposite, scattered in basal portion, dense in distal, only slightly narrower than the branch from which they arise and with terminal cell conical or setiferous. Lower cells of main branch swollen, 2/3–4 times as long as broad, with a girdle-like chloroplast in median portion. (Tychoplanktont).

Main branches 14–16.5 μ broad; branchlets 6–7 μ broad.

Mendota (rrr), Wild Goose (rrr).

Family *APHANOCHAETACEAE*.

Thallus epiphytic, generally upon other algae, creeping in habit and composed of simple or irregularly branching filaments that taper from the center to short, narrow terminal cells that do not end in terminal setae. Central cells of filament spherical to spherical-cylindrical; those towards the end cylindrical with straight or convex sides; terminal cells conical. Cells generally bearing one or more long, hyaline, unseptate setae with swollen bases. Chloropast parietal, completely filling the cell, or gridle-like and in the median portion; generally with more than one pyrenoid.

Asexual reproduction by tetraciliate zoospores or aplanospores.

Sexual reproduction heterogamous, by the fusion of tetraciliate zoogametes that are markedly different in size.

The family contains the single genus *Aphanochaete* whose characters are those of the family.

Aphanochaete vermiculoides Wolle. Pl. 50, Fig. 8.

Freshw. algae of U. S. 119, *pl.* 105, *figs.* 9–10. 1887.

Herposteiron hyalothecae Hansgirg. G. M. Smith, Trans. Wis. Acad. Sci., Arts, & Lett. **18**[2]: 560. 1916.

Filaments but slightly branched and confined to the peripheral portion and twisted around or arched from the gelatinous envelope of *Hyalotheca* filaments. Cells cylindrical, with straight or convex sides and containing a single chloroplast with 1–4 pyrenoids. Setae very sparse, rarely more than five on any filament. (Tychoplanktont).

Cells 4–6 μ broad; 5–12 μ long.

Beaverdam (rr).

All other species of the genus grow directly upon the cells of the substratum but the wide gelatinous envelope of the *Hyalotheca* is in this case sufficiently firm to support the *Aphanochaete.* The alga has the appearance of an arching worm crawling along the filament.

Order **SIPHONOCLADIALES.**

Cells multinucleate, generally cylindrical and united to form simple or branched thalli that are, at times, quite complicated in structure. Chloroplasts disciform, parietal and numerous, or united by strands to form a single reticulate structure. Pyrenoids generally present.

Asexual reproduction by zoospores and aplanospores.

Sexual reproduction isogamous by means of zoogametes; or heterogamous by large non-motile oospheres and small motile antherozoids.

The order is largely marine and only one representative, *Cladophora*, has been found in the plankton.

Family *CLADOPHORACEAE.*

Filaments generally sessile, simple or branched, with septations between the cells complete. Walls of cells generally heavy, homogeneous or lamellose, without a gelatinous envelope. Cells with numerous parietal disciform chloroplasts or a single reticulate chloroplast; multinucleate; with a single large central vacuole.

Asexual reproduction by zoospores and akinetes.

Sexual reproduction isogamous, by means of zoogametes.

CLADOPHORA Kützing 1843.

Filaments branched, with the component cells in a single linear series; lateral branches arising at distal part of main branch cells and generally with an abundantly developed system of branching. Growth of filaments confined to the apical region. Cells cylindrical, several times as long as broad; multinucleate; with numerous parietal disciform chloroplasts or the chloroplasts united by processes to form a single reticulate chloroplast; pyrenoids numerous.

Asexual reproduction by the formation of many biciliate zoospores in any cell of the filament.

Sexual reproduction isogamous, by means of biciliate zoogametes.

CLADOPHORA GLOMERATA (L.) Kützing. Pl. **50**, Figs. **1–3.**

Phycol. gener. 266. 1843; Collins, Tufts Coll. Studies **2**: 350, *pl.* 13, *fig.* 124. 1909.

Branching of filaments fairly dense in lower portion and very dense at the apex. Ultimate branchlets containing several cells, straight or slightly incurved, with ends rounded but not markedly attenuated. Cells of main branches 6–7 times as long as broad, 3–6 times as broad as branchlets. Zoospore formation confined to terminal and subterminal cells. (Tychoplanktont).

Diam. main cells 75–100 μ, branchlets 35–50 μ.

Mendota (rr).

Division **AKONTAE.**

Cells solitary or in filaments, variously shaped, with one or more grass-green chloroplasts that usually contain more than one pyrenoid.

Asexual reproduction ordinarily by vegetative cell division, rarely by aplanospores, never by zoospores.

Sexual reproduction isogamous, always by aplanogametes, never by ciliate zoogametes.

The lack of cilia in every type of reproduction is the great characteristic of the division. It includes but one order, the Conjugales, which forms a very homogeneous though diverse group.

Order **CONJUGALES.**

Cells solitary or in simple filaments; cell shape various, cylindrical without median constrictions or bilaterally symmetrical and usually with a median constriction dividing the cell into two symmetrical halves. Cell wall smooth or variously ornamented, with an inner firm portion and an outer gelatinous portion that is very copious at times. Chloroplasts 1–8, variously shaped, axial or parietal, with one to several pyrenoids. Nucleus single, at the center of the cell.

Asexual reproduction by vegetative cell division or aplanospores.

Sexual reproduction isogamous, by aplanogametes.

KEY TO THE FAMILIES.

Cells cylindrical, without a median constriction, united to form simple filaments.................................... Zygnemaceae

Cells solitary or in filaments, divided into two symmetrical halves, generally with a median constriction............ Desmidiaceae (See Part II)

Family *ZYGNEMACEAE.*

Cells always cylindrical, with ends flattened or replicate, serially arranged to form simple filaments that occasionally have simple lateral outgrowths. Filaments generally free-floating though at times attached by a modification of terminal cell into a basal cell with short rhizoidal processes. Wall of cell firm, homogeneous, with a narrow or copious, homogeneous or radially fibrillar, gelatinous covering. Cell contents consisting of a single nucleus, and one to several chloroplasts that are either stellate or laminate and axial or parietal. Chloroplasts with one to many pyrenoids.

Asexual reproduction by vegetative cell division or metamorphosis of cell contents into a single large aplanospore.

Sexual reproduction isogamous, by metamorphosis of all or a part of the cell contents into an aplanogamete which fuses with an aplanogamete from the same or another filament. Zygote with 2–3 layers, sometimes germinating immediately but usually after a period of rest.

KEY TO THE GENERA.

Chloroplast stellate .... ZYGNEMA
Chloroplast a longitudinal band.
  Chloroplast axial .... MOUGEOTIA
  Chloroplast parietal .... SPIROGYRA

MOUGEOTIA C. A. Agardh 1824.

Cells cylindrical, generally several times as long as broad. Chloroplast single, forming an axial plate extending the whole length of the cell and generally with several pyrenoids.

Conjugation generally scalariform and with the zygote formed in the conjugation tube. Zygote at times occupying the median portion of male and female gametangia as well as conjugation tube. Wall of zygote of two layers the inner smooth, the outer smooth or sculptured, hyaline or colored.

MOUGEOTIA VIRIDIS (Kützing) Wittrock. Pl. 51, Fig. 1.

Bih. Kgl. Sv. Vet.-Ak. Handl. **1**, No. **1**: 39. 1872.

Staurospermum viride Kützing. Cooke, Brit. Freshw. Algae 107, *pl.* 44, *figs.* 2a–2c. 1882.

Cells 4–10 times as long as broad. Zygospores normally bounded by four cells, quadrangular, with concave sides and truncate corners. (Tychoplanktont).

Cells 6–8 μ broad. Zygotes 22–32 μ broad.

Okauchee (rrr).

ZYGNEMA C. A. Agardh 1817.

Cells cylindrical, rarely more than twice as broad as long. Filaments with a terminal cell at times modified to form a holdfast. Cells with one nucleus and two axial stellate chloroplasts each of which contain a single pyrenoid. Conjugation lateral or scalariform, with a formation of zygospores in the conjugation tube or the gametangia. Zygotes spherical to ellipsoid, wall with three layers.

ZYGNEMA PECTINATUM (Vaucher) C. A. Agardh. Pl. **51**, Fig. **2**.

Synop. algarum Scand. 102. 1817; DeBary, Unters. u. d. Fam. d. Conjugaten. 77, *pl.* 1, *figs.* 15–19. 1858.

Vegetative cells 1–3 times as broad as long. Zygospore formed in conjugation tube, with rounded ends. Middle layer of zygote wall brown and scrobiculate. (Tychoplanktont).

Vegetative cells 30–40 μ broad. Zygote 35–40 μ broad.

Oconomowoc (rrr).

SPIROGYRA Link 1820.

Cells cylindrical, with end walls a flat plane or replicate; united to form simple filaments that may be attached by a modification of the terminal cell into holdfasts. Length of cells slightly greater than to several times the breadth. Chloroplasts one to several, parietal, in long ribbons that extend from end to end in the cell and generally make several spiral turns. Nucleus single, moored in the center of the central vacuole by cytoplasmic strands running to the pyrenoids. Chloroplasts with several pyrenoids that are generally in a single median row.

Conjugation scalariform or lateral, with zygote formation in gametangia and never in the tube. Zygote spherical, ellipsoid, or cylindrical with rounded ends. Middle layer of zygote hyaline or colored, smooth or sculptured.

SPIROGYRA PORTICALIS (Müller) Cleve. Pl. **51**, Fig. **3**.

Nova Acta Reg. Soc. Sci. Upsaliensis **3**, Ser., **6**: No. **11**, 22, *pl.* 5, *figs.* 8–13. 1868.

Length of vegetative cells 2–6 times the breadth. Cells with plane cross walls and containing a single chloroplast that makes 3–5 turns and contains numerous conspicuous pyrenoids. Gametangia but little swollen when zygospore is mature. Zygospores broadly ovoid to sub-cylindrical, with a yellow membrane when mature. (Tychoplanktont).

Diam. cells 30–50 μ. Diam. zygospores 40–50 μ.

Mendota (rrr).

PLATE 1.

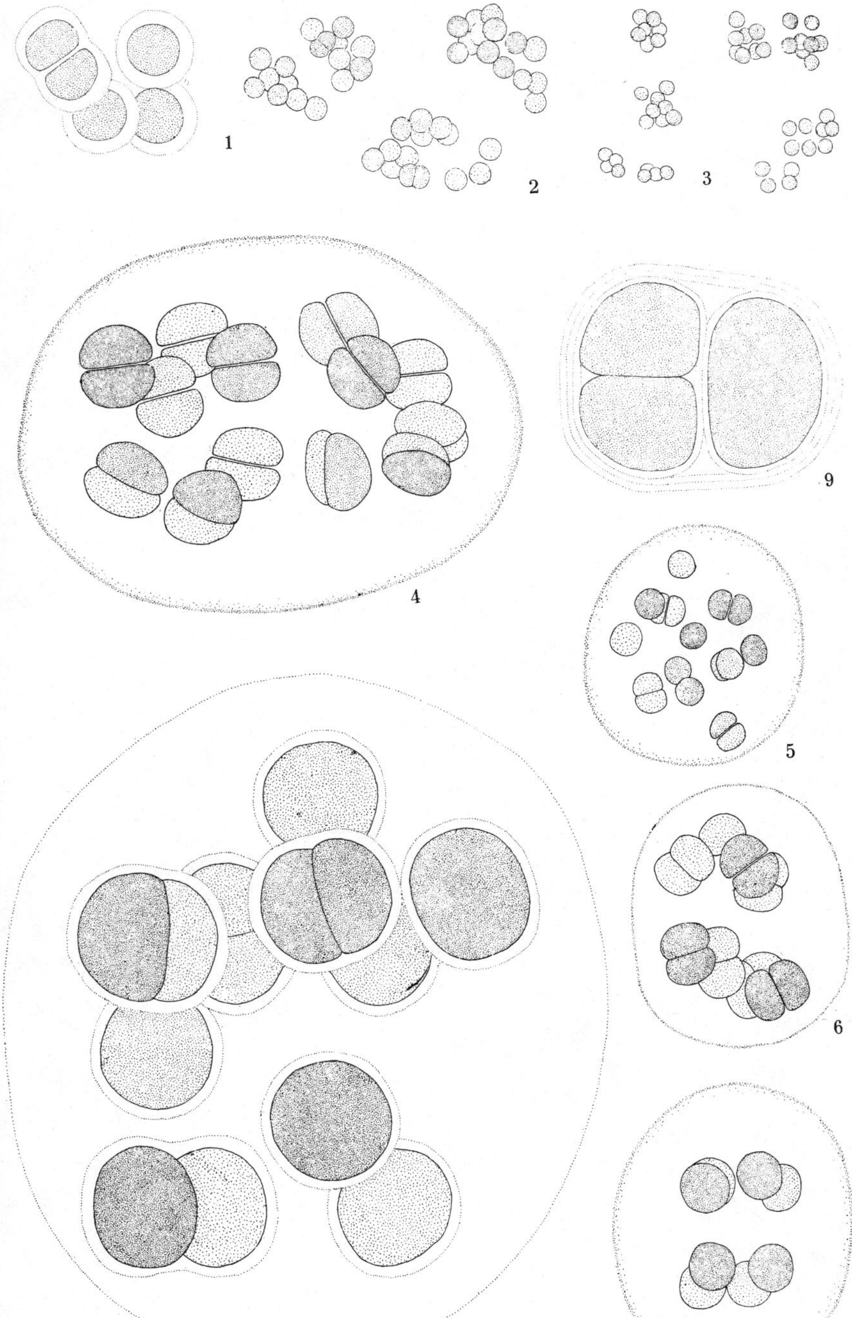

H—ALGAE

COCKAYNE, BOSTON

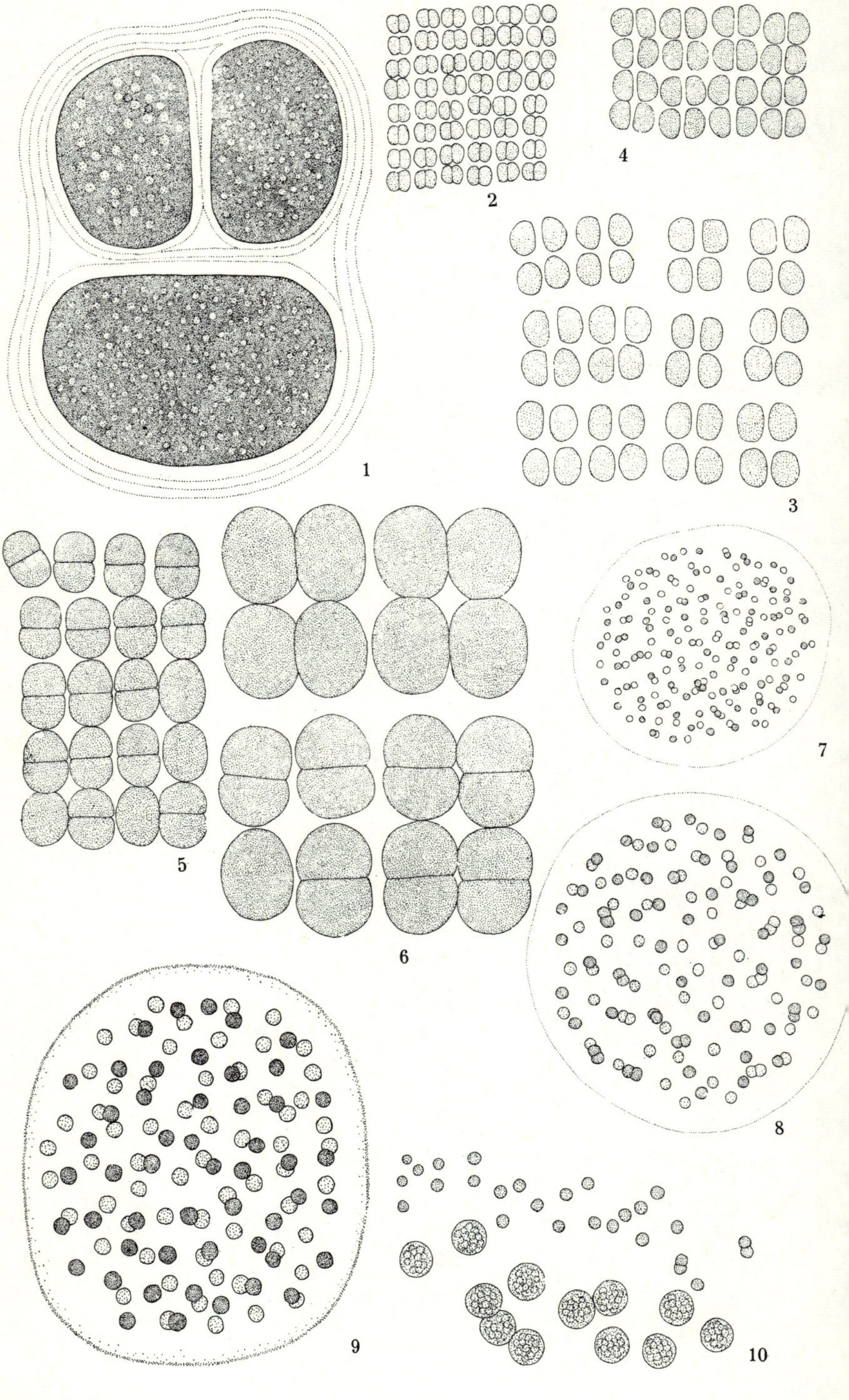

COCKAYNE, BOSTON

PLATE 2.

PLATE 3.

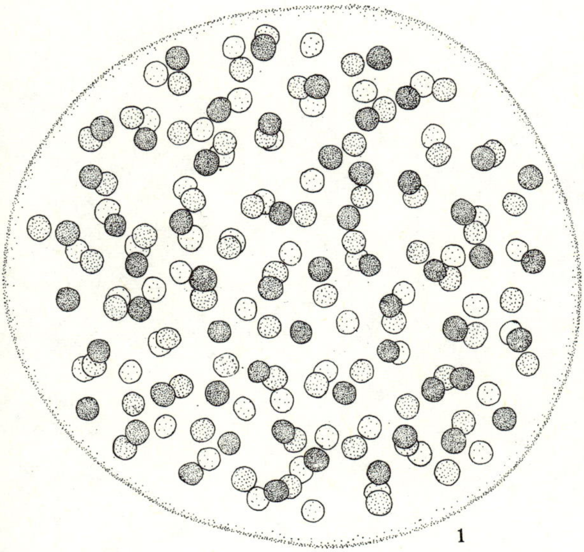

1

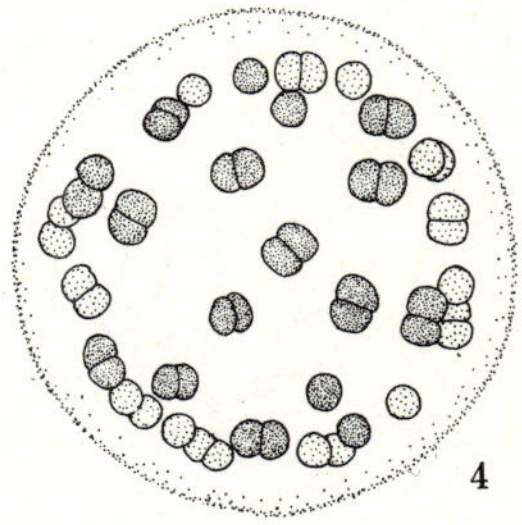

4

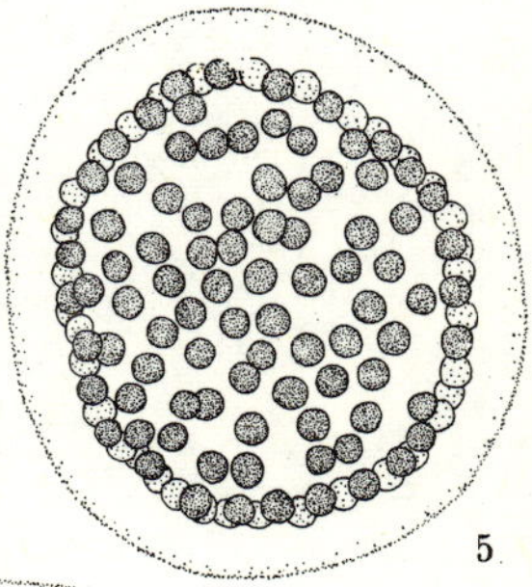

5

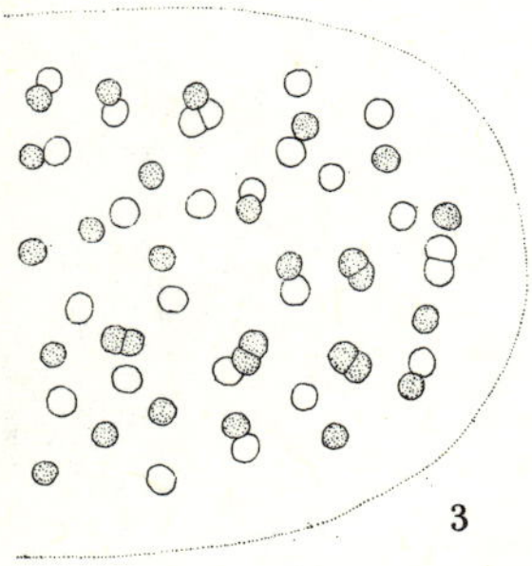

3

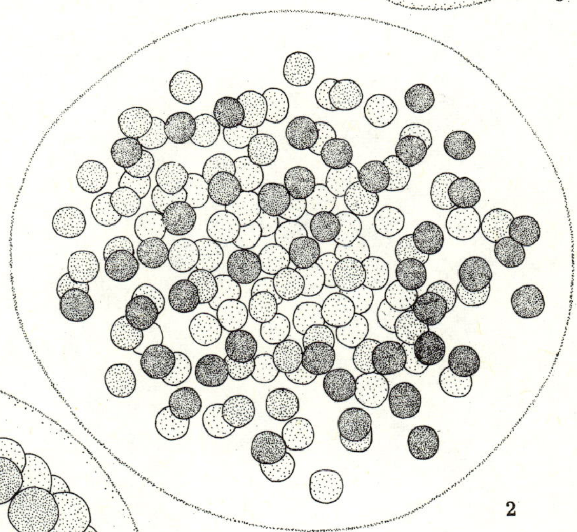

2

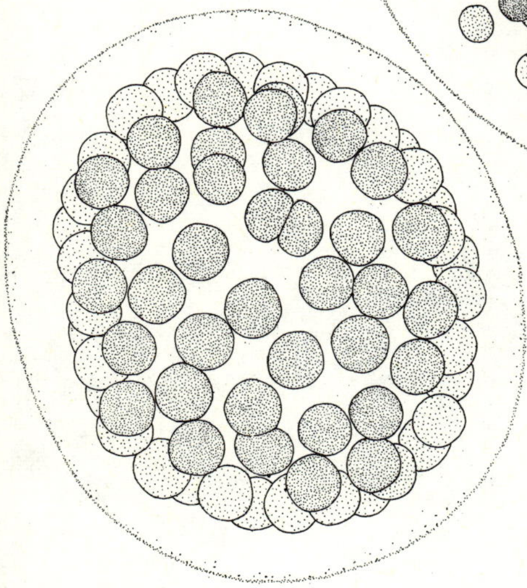

7

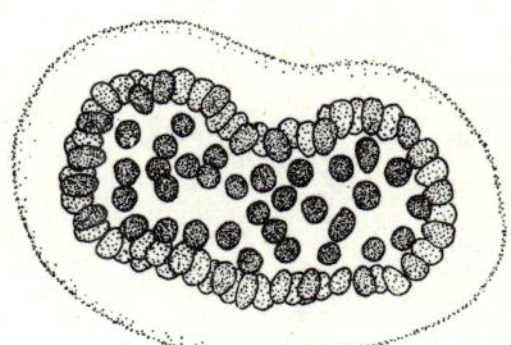

6

COCKAYNE, BOSTON

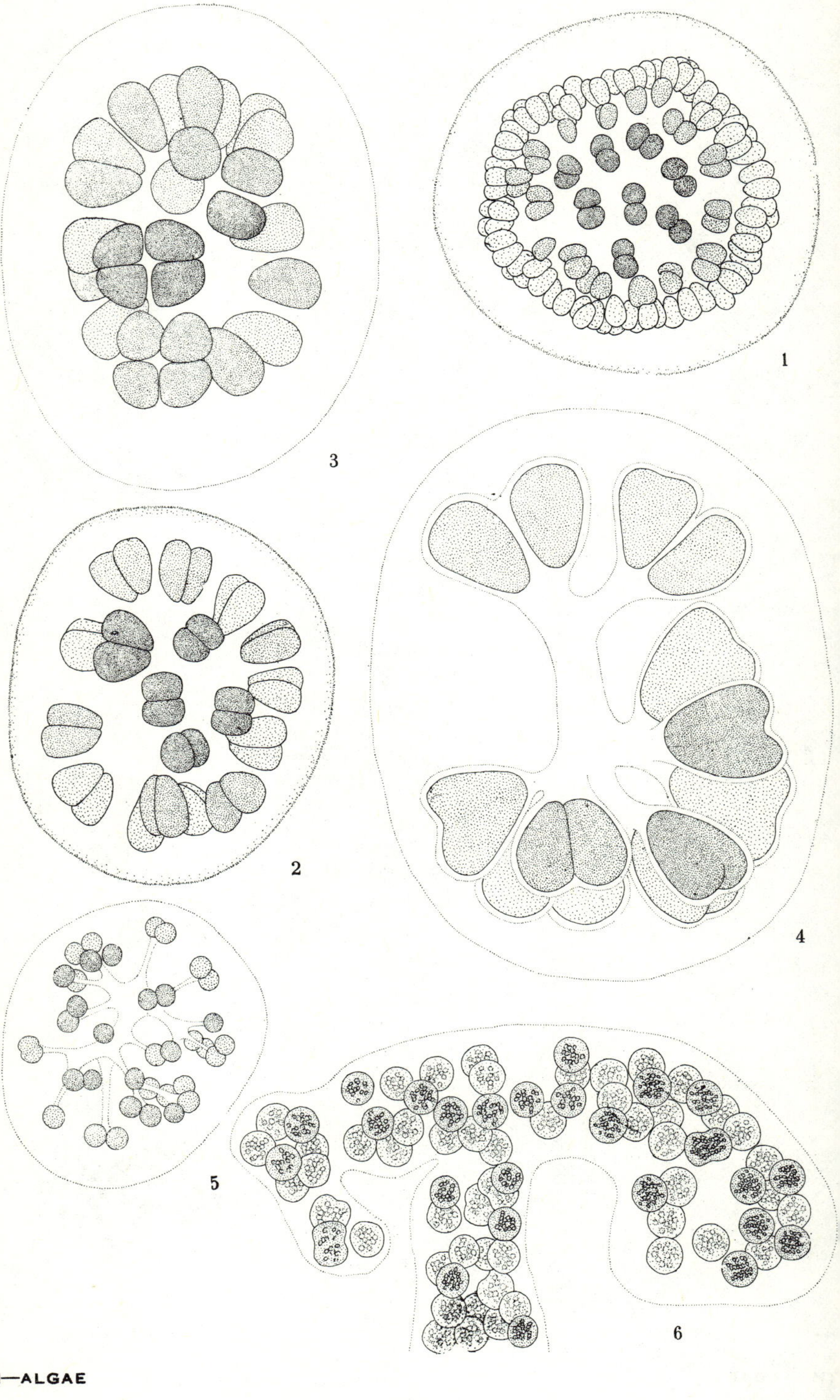

COCKAYNE, BOSTON

PLATE 4.

		Page
Fig. 1.	*Coelosphaerium Naegelianum* (x 825)	35
Figs, 2–3.	*Gomphosphaeria aponina* (x 825)	37
Fig. 4.	*Gomphosphaeria aponina* var. *cordiformis* (x 825)	37
Fig. 5.	*Gomphosphaeria lacustris* (x 825)	36
Fig. 6.	*Microcystis aeruginosa* var. *major* (x 825)	40

PLATE 5.

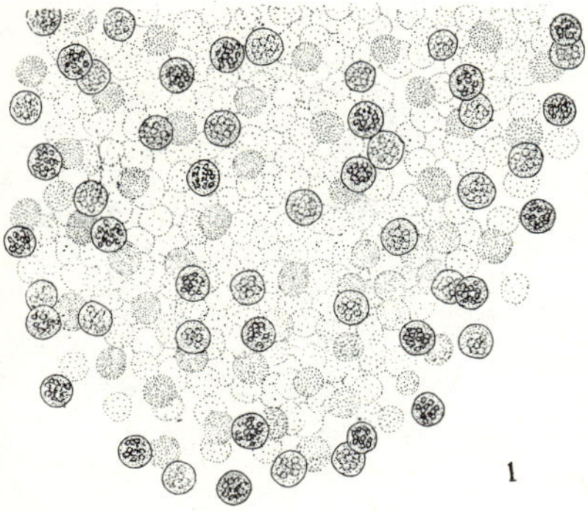

1

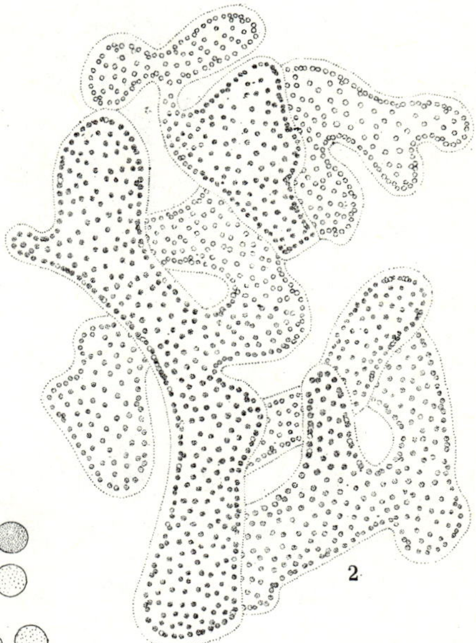

2

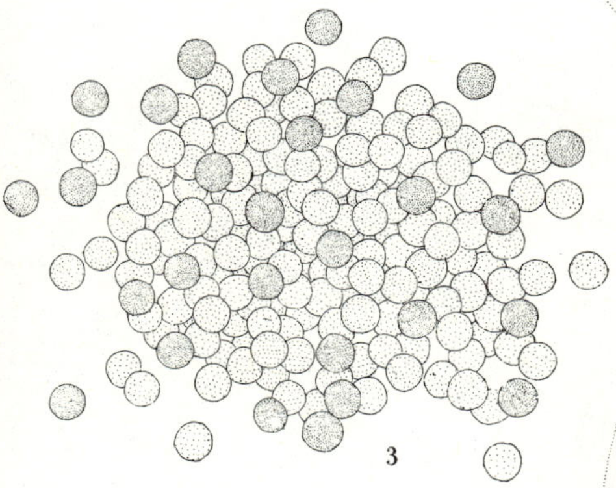

3

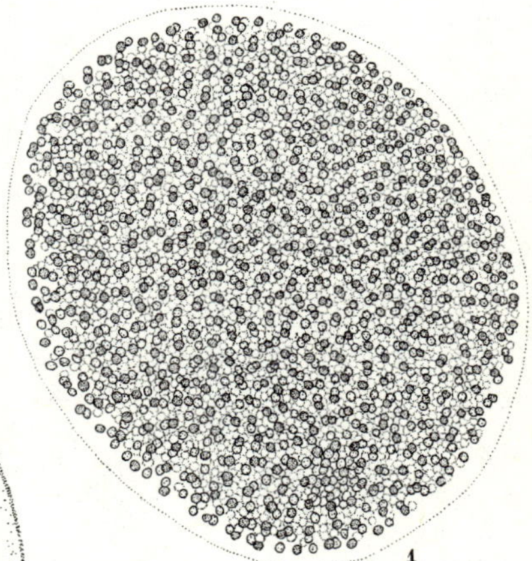

4

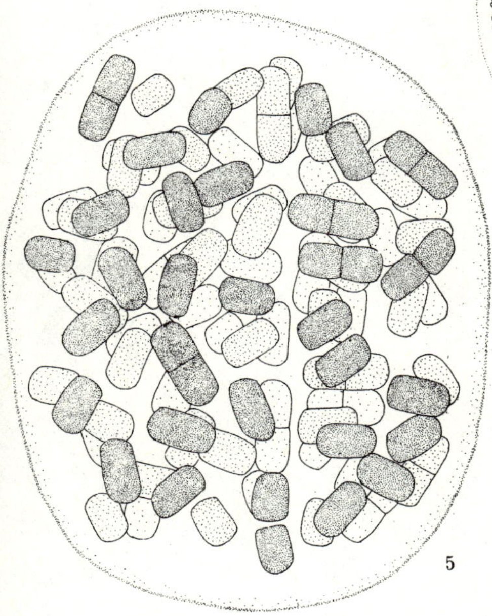

5

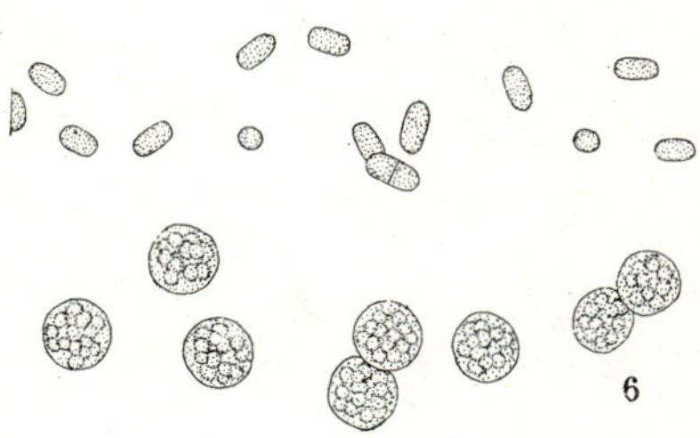

6

H—ALGAE

COCKAYNE, BOSTON

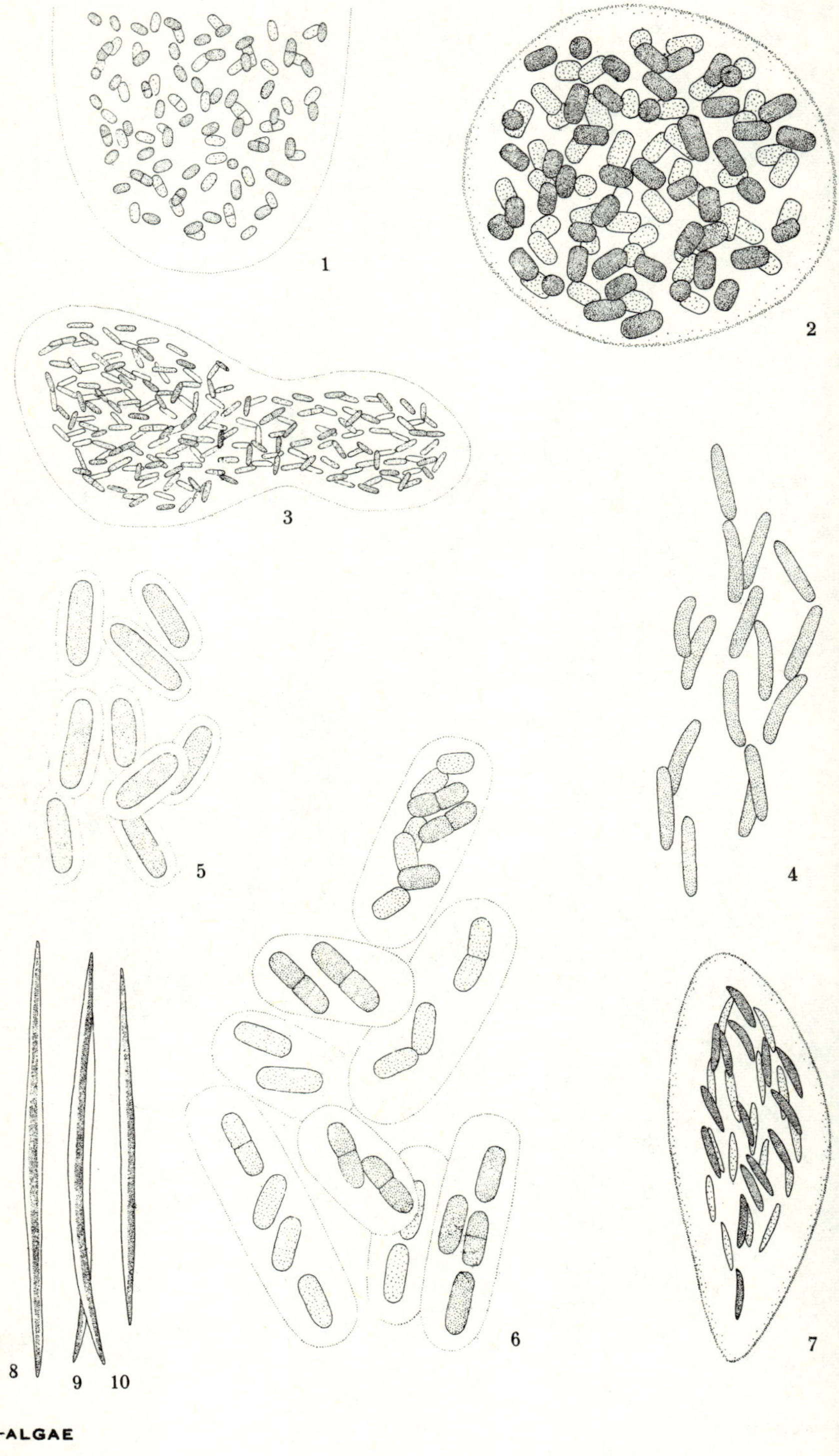

SMITH—ALGAE

COCKAYNE, BOSTON

PLATE 6.

PLATE 7.

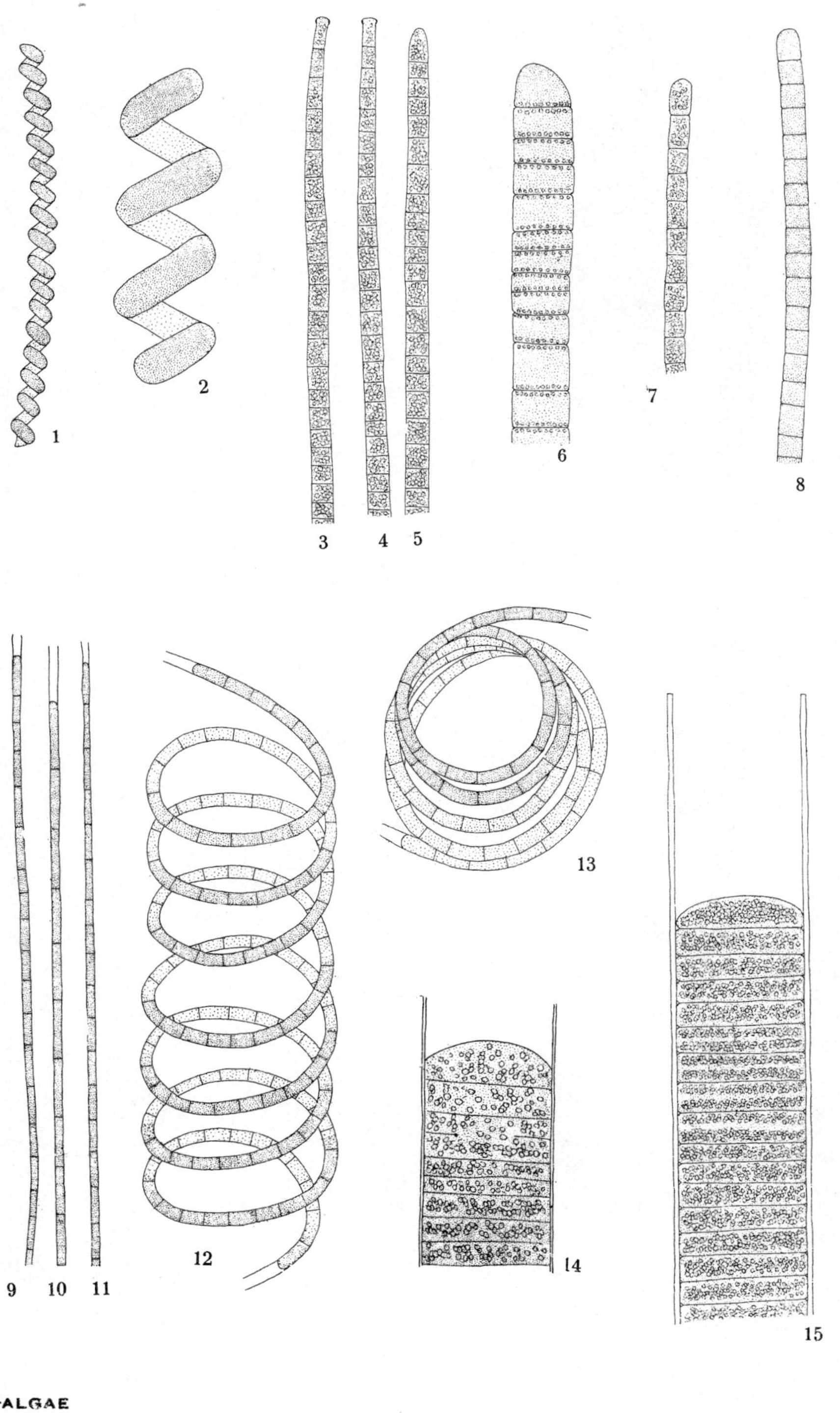

'H—ALGAE

COCKAYNE, BOSTON

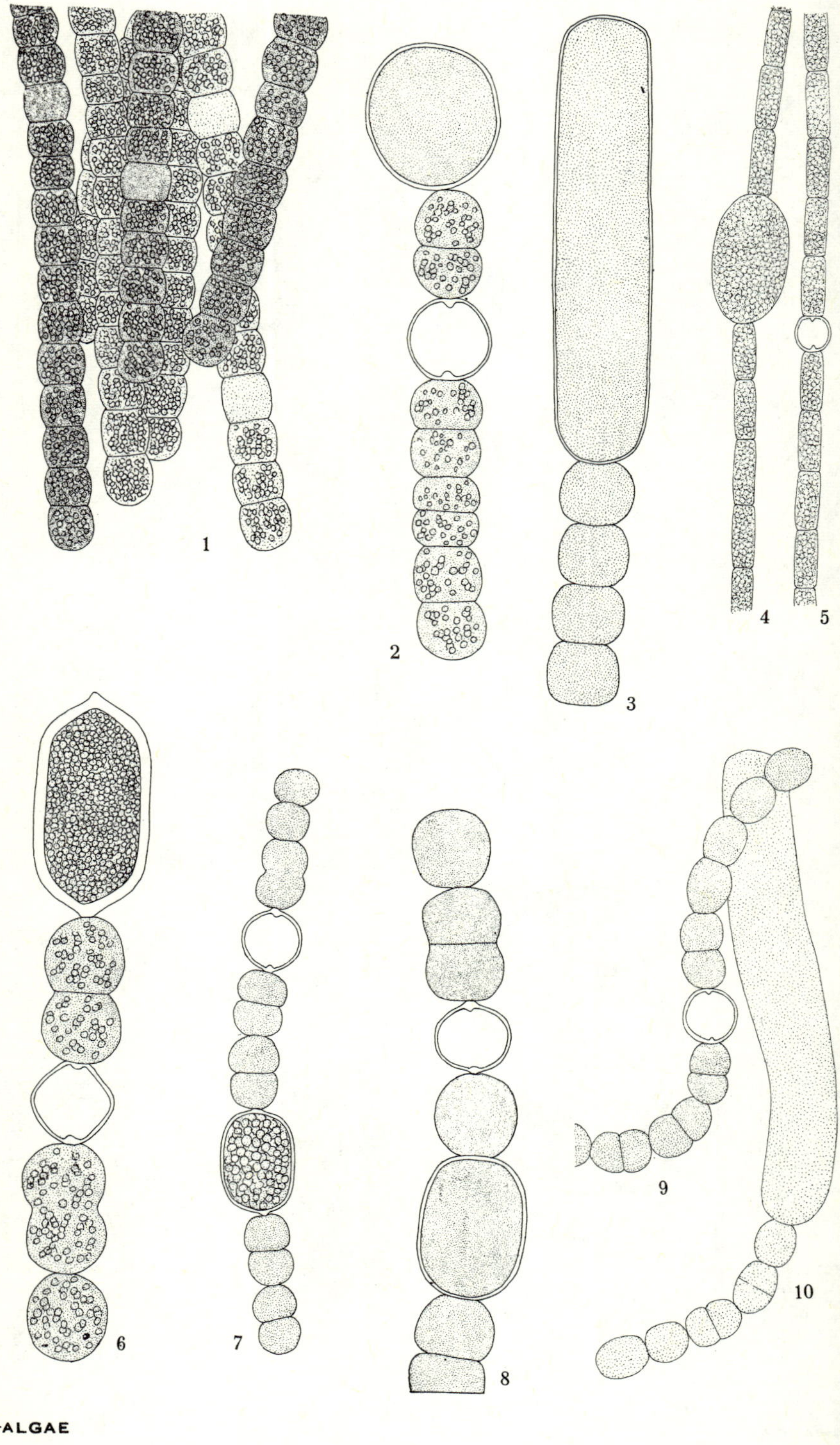

SMITH—ALGAE

COCKAYNE, BOSTON

PLATE 8.

		Page
Fig. 1.	*Trichodesmium lacustre* (x 1000)	54
Fig. 2.	*Anabaena plantonica* (x 825)	56
Fig. 3.	*Anabaena Bornetiana* (x 825)	58
Figs. 4–5.	*Anabaena Levanderi* (x 825)	58
Fig. 6.	*Anabaena macrospora* var. *robusta* (x 825)	57
Fig. 7.	*Anabaena affinis* (x 825)	57
Fig. 8.	*Anabaena limnetica* (x 825)	57
Figs. 9–10.	*Anabaena augstumalis* var. *marchica* (x 825)	58

PLATE 9.

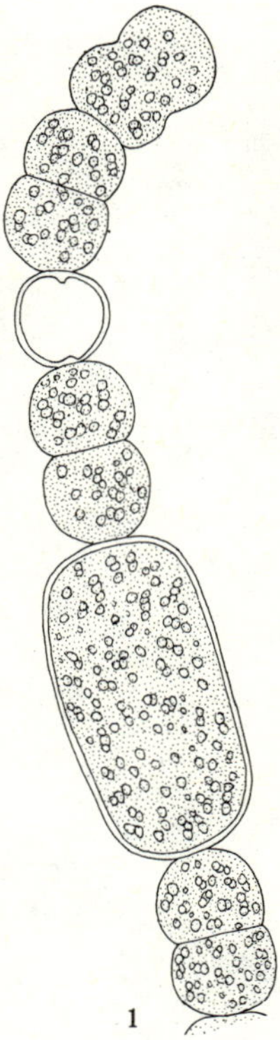

1

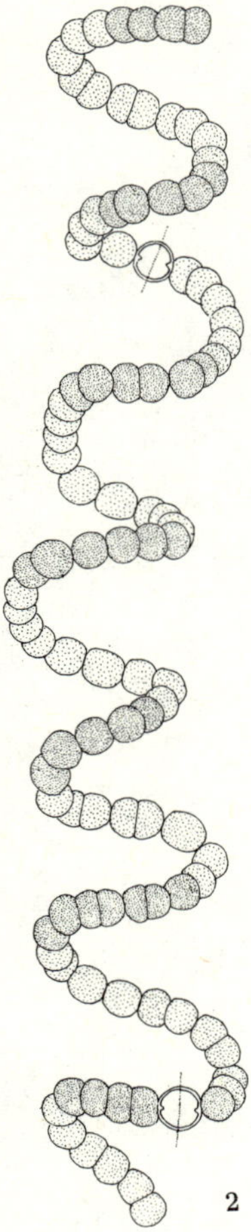

2

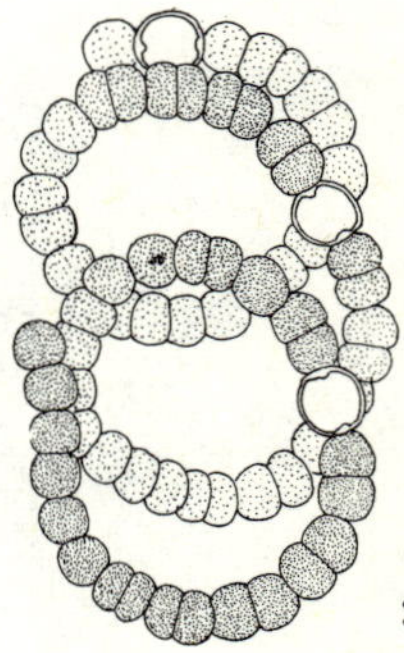

3

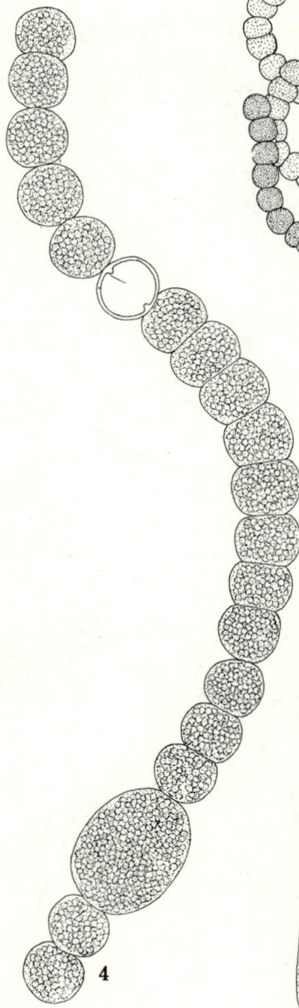

4

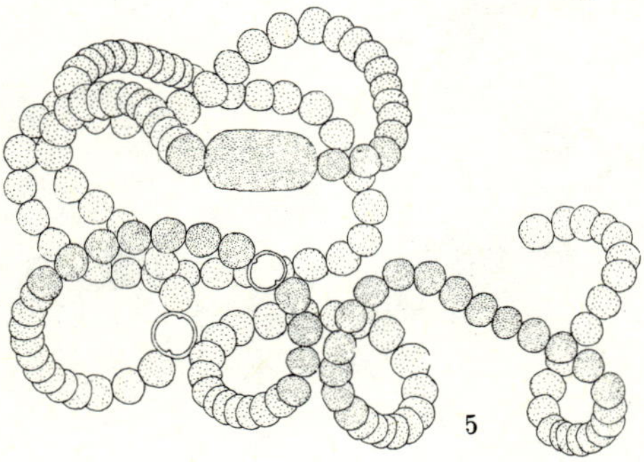

5

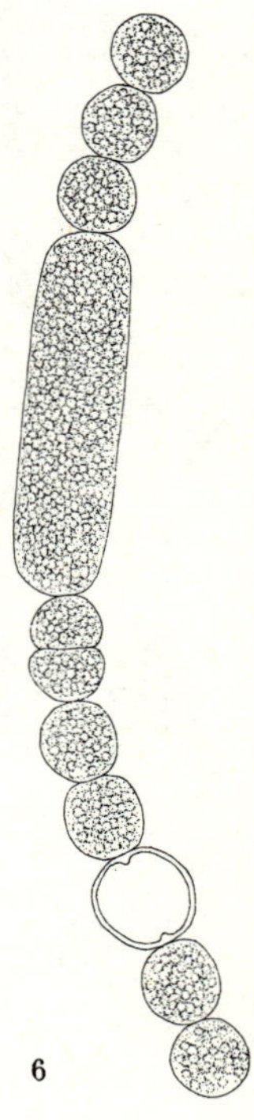

6

COCKAYNE, BOSTON

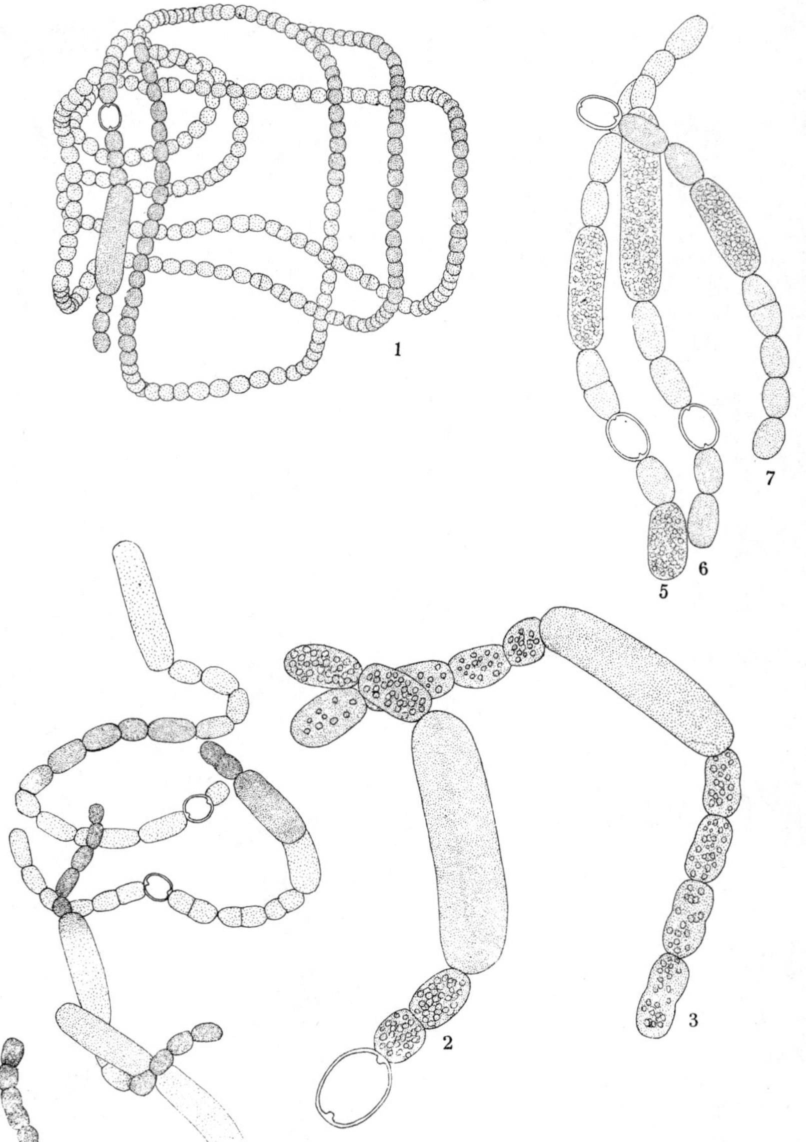

SMITH—ALGAE

COCKAYNE, BOSTON

PLATE 10.

			Page
Fig.	1.	*Anabaena circinalis* var. *macrospora* (x 400)	60
Figs.	2–4.	*Anabaena flos-aquae* (Figs. 2–3, x 825; fig. 4, x 400)	60
Figs.	5–7.	*Anabaena flos-aquae* var. *Treleasii* (x 825)	60
Fig.	8.	*Anabaena Lemmermanni* (x 825)	61

PLATE 11.

		Page
Fig. 1.	*Anabaena Lemmermanni* (x 400)	61
Figs. 2–4.	*Aphanizomenon flos-aquae* (Fig. 2, x 400; figs. 3–4, x 1000) ..	61
Figs. 5–6.	*Gloeotrichia echinulata* (x 400)	63

H—ALGAE

COCKAYNE, BOSTON

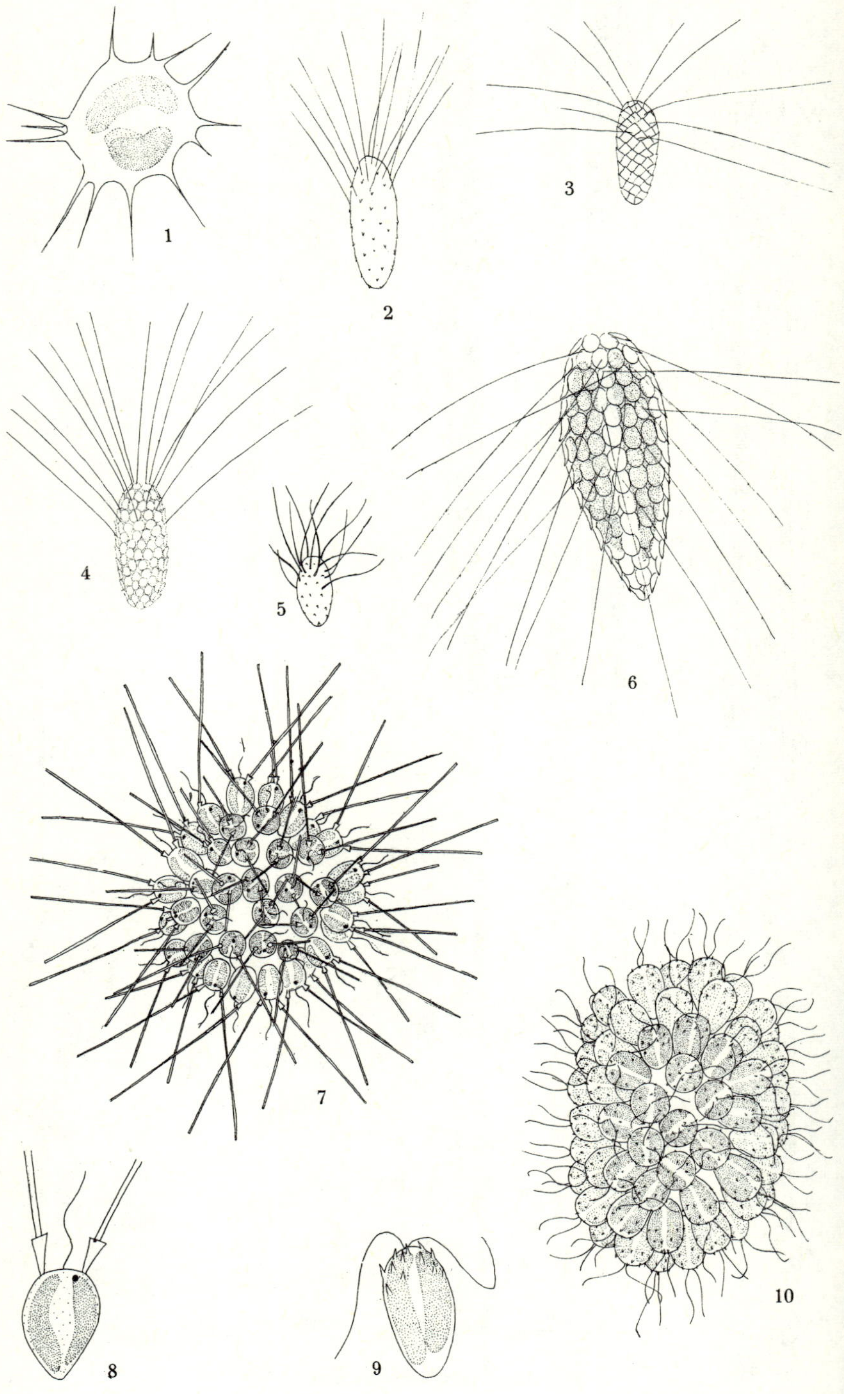

COCKAYNE, BOSTON

PLATE 12.

PLATE 13.

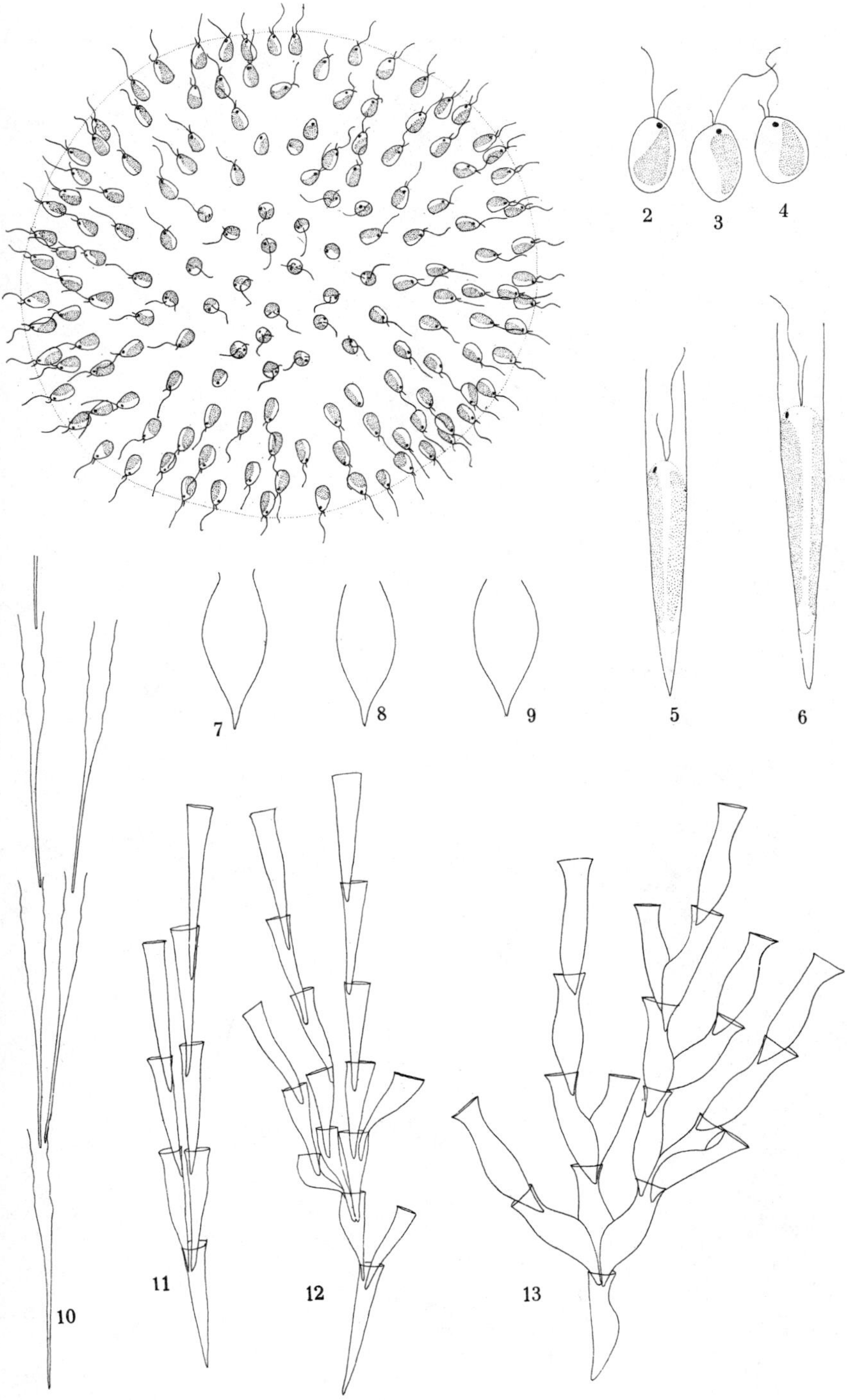

—ALGAE

COCKAYNE, BOSTON

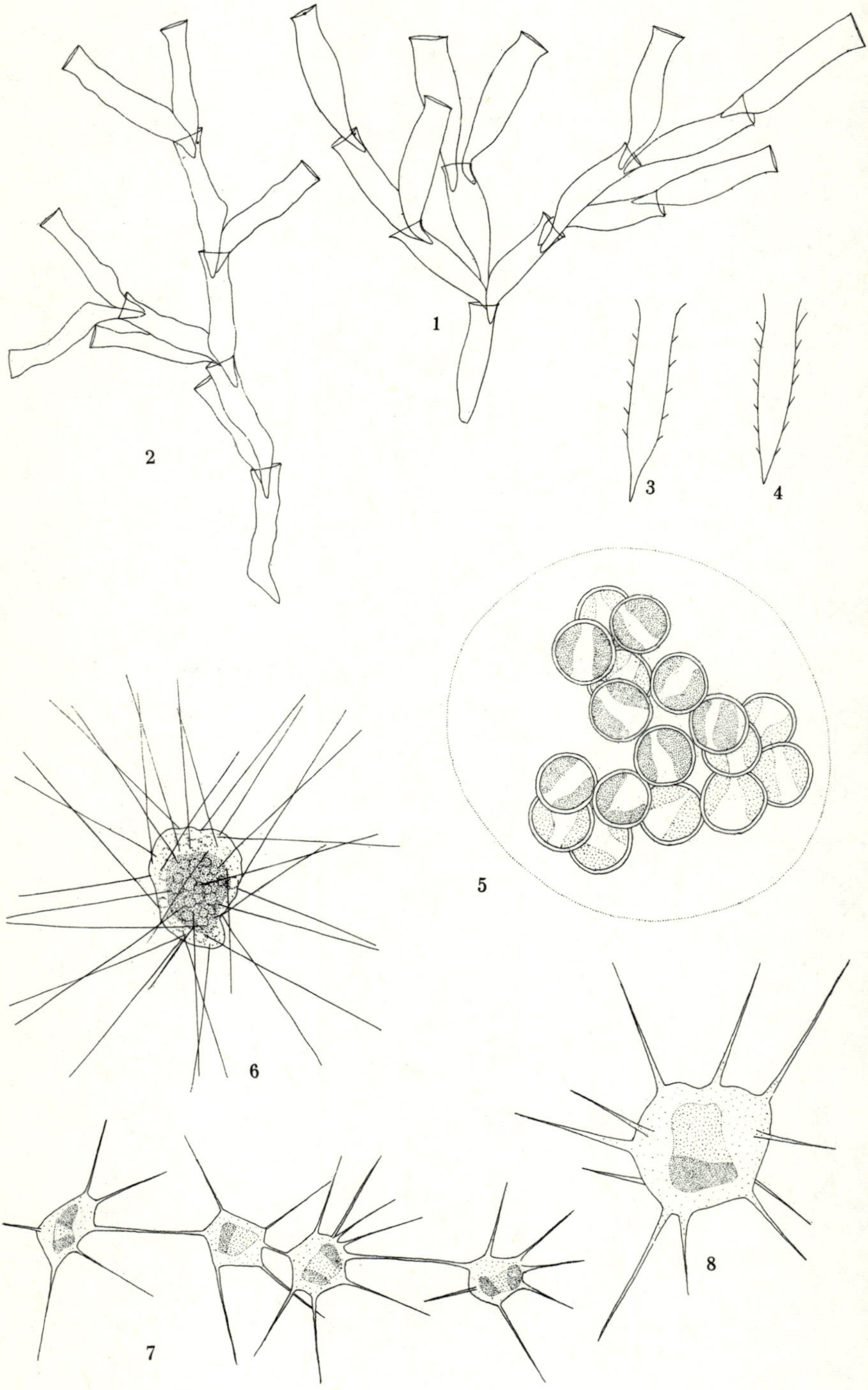

SMITH—ALGAE

COCKAYNE, BOSTON

PLATE 14.

PLATE 15.

		Page
Figs. 1–3.	*Chlorochromonas minuta* (x 1000)	80
Fig. 4.	*Peroniella planctonica* (x 1000)	81
Fig. 5.	*Botryococcus Braunii* (x 400)	84
Figs. 6–7.	*Botryococcus protuberans* var. *minor* (Fig. 6 x 500, fig. 7 x 1000)	85
Fig. 8.	*Botryococcus sudeticus* (x 400)	85
Fig. 9.	*Chlorobotrys limneticus* (x 1000)	82
Fig. 10.	*Chlorobotrys regularis* (x 1000)	82
Fig. 11.	*Ophiocytium parvulum* (x 1000)	86
Figs. 12–13.	*Ophiocytium capitatum* (x 1000)	86
Figs. 14–16.	*Ophiocytium capitatum* var. *longispinum* (x 1000)	86
Figs. 17–18.	*Tribonema minus* (x 1000)	87

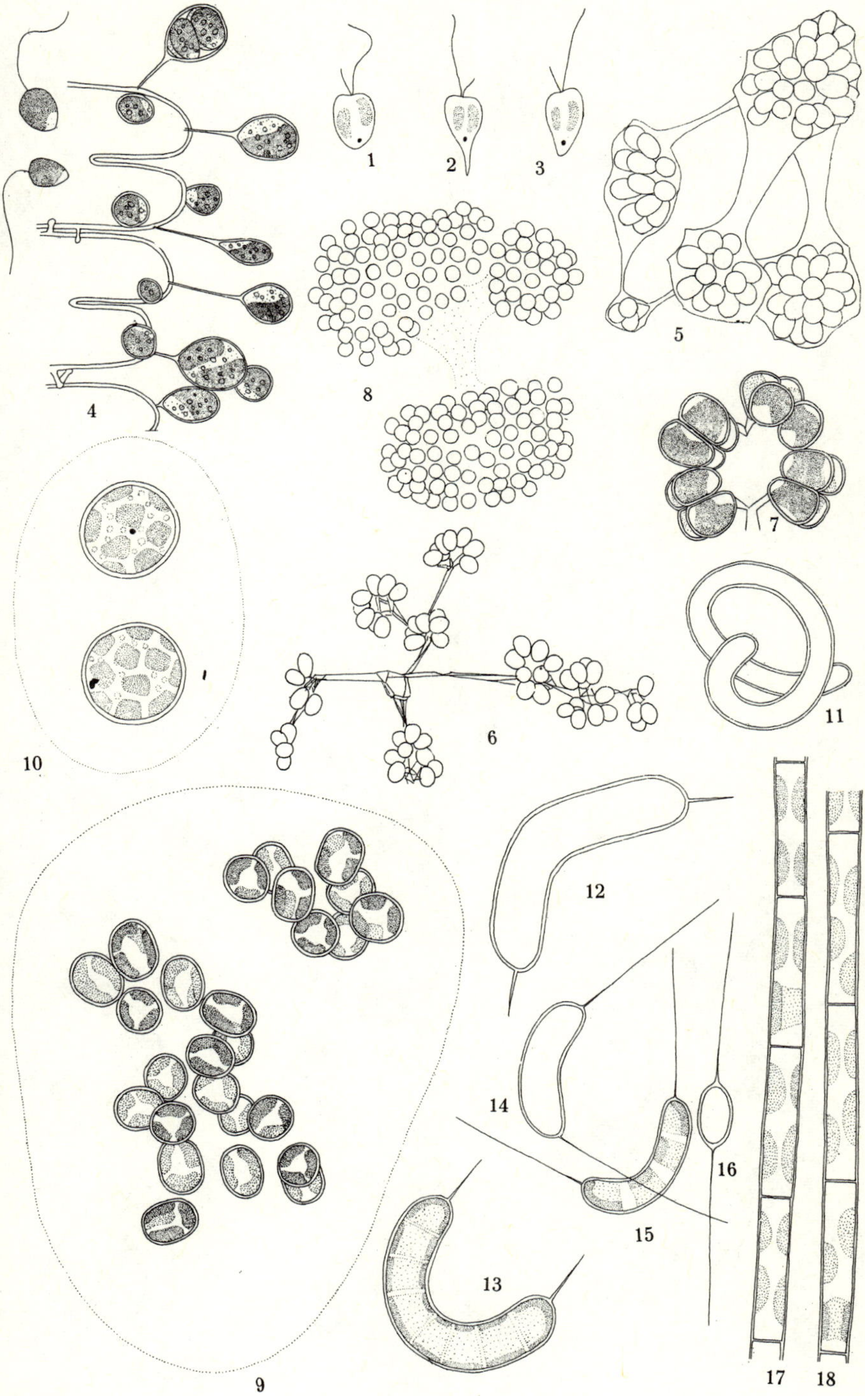

H—ALGAE

COCKAYNE, BOSTON

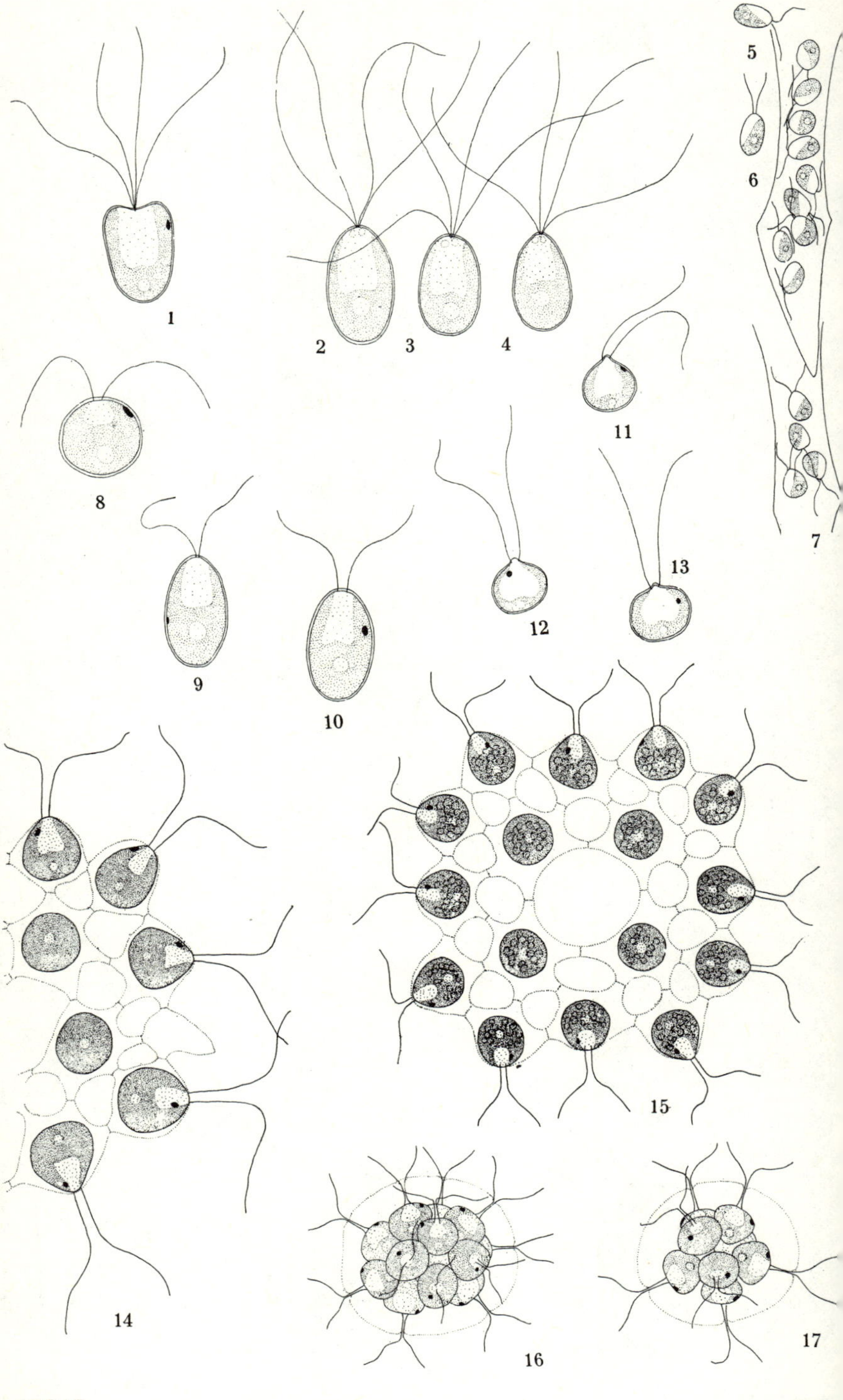

SMITH—ALGAE

COCKAYNE, BOSTON

PLATE 16.

		Page
Fig. 1.	*Carteria cordiformis* (x 1000)	92
Figs. 2–4.	*Carteria Klebsii* (x 1000)	93
Figs. 5–7.	*Chlamydomonas dinobryoni* (x 1000)	91
Fig. 8.	*Chlamydomonas globosa* (x 1000)	90
Figs. 9–10.	*Chlamydomonas Snowii* (x 1000)	91
Figs. 11–13.	*Chlamydomonas epiphytica* (x 1000)	91
Figs. 14–15.	*Gonium pectorale* (x 500)	94
Figs. 16–17.	*Pandorina morum* (x 400)	95

PLATE 17.

	Page
Figs. 1–3. *Pleodorina californica* (x 400)	97

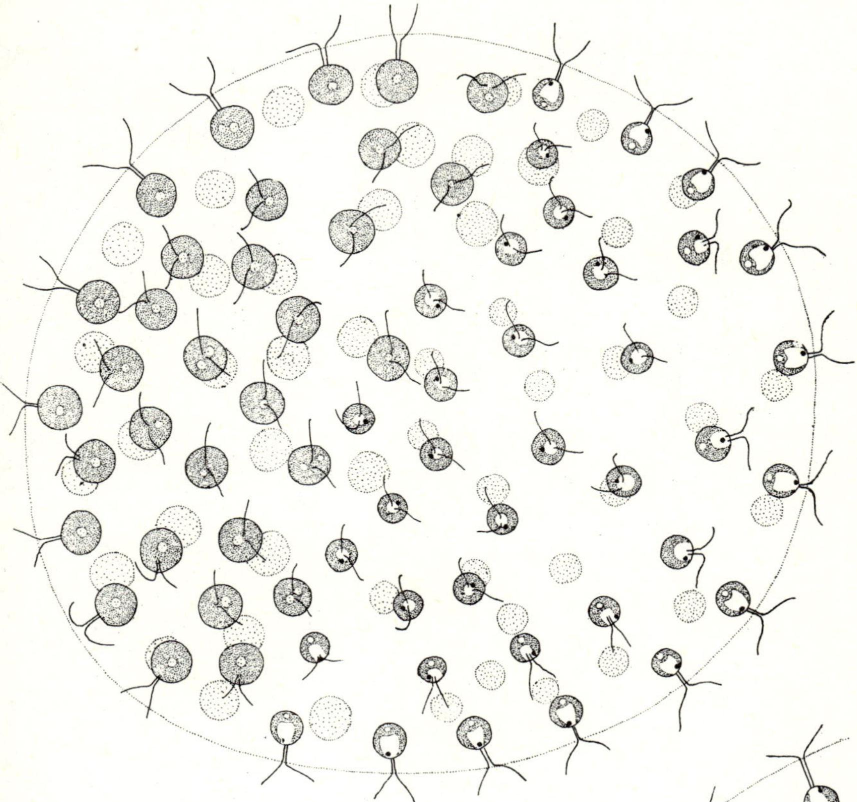

2

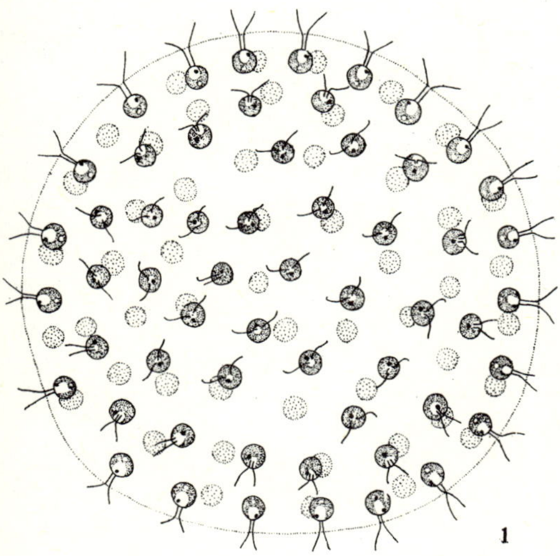

1

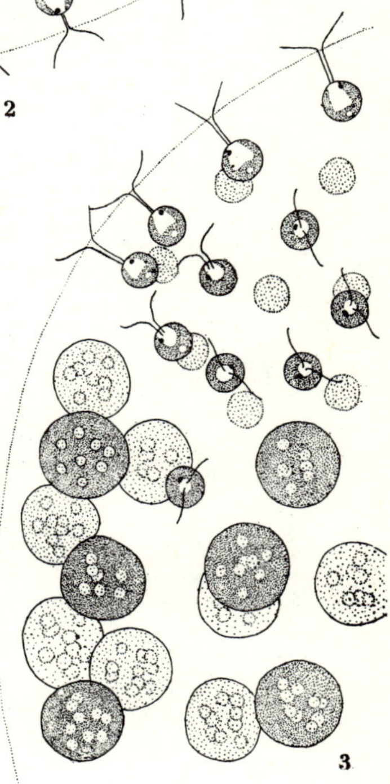

3

COCKAYNE, BOSTON

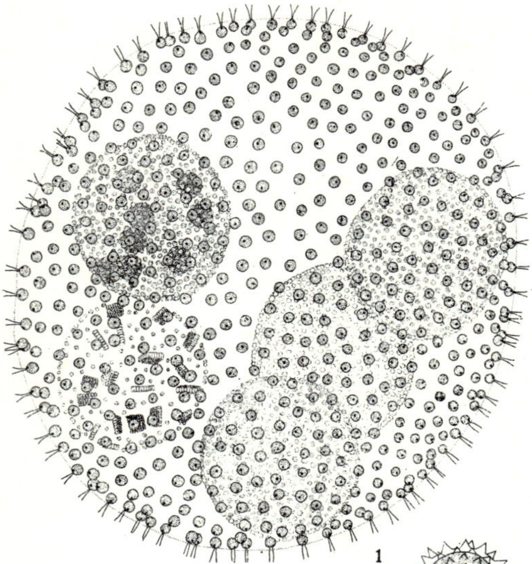

1

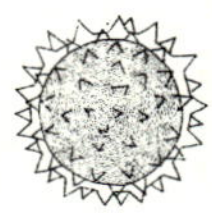

4

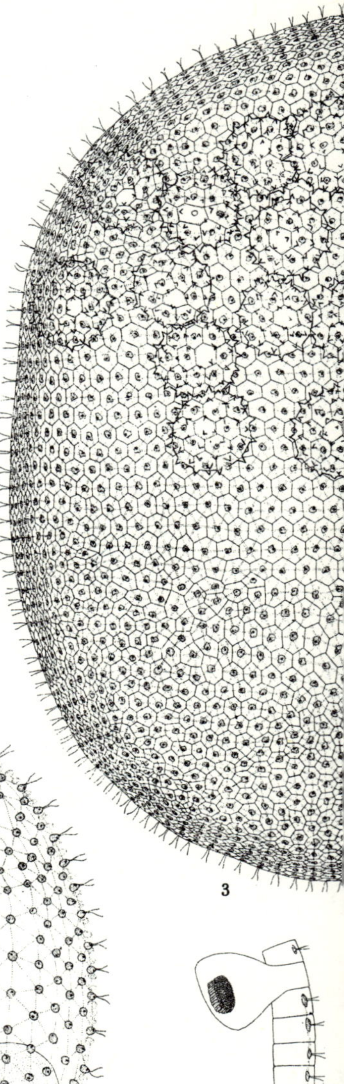

3

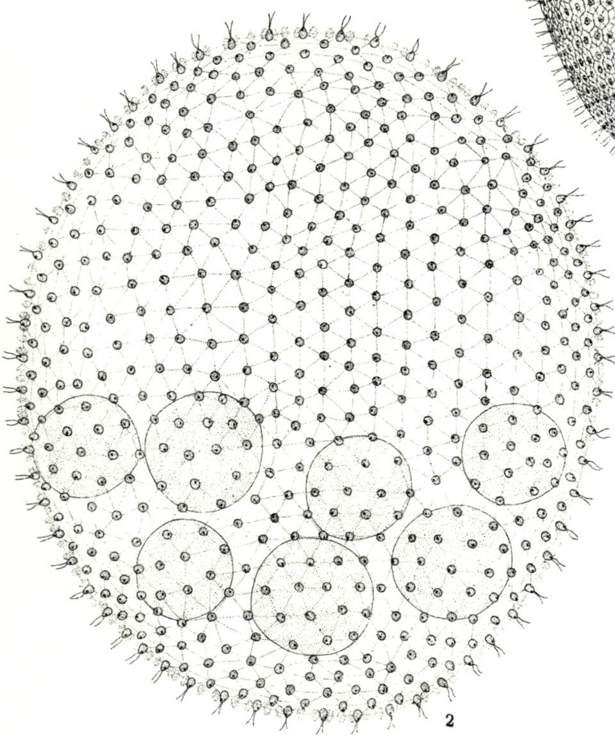

2

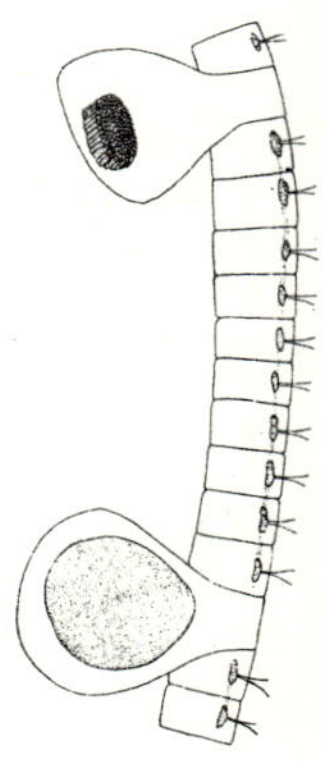

5

COCKAYNE, BOSTON

PLATE 18.

PLATE 19.

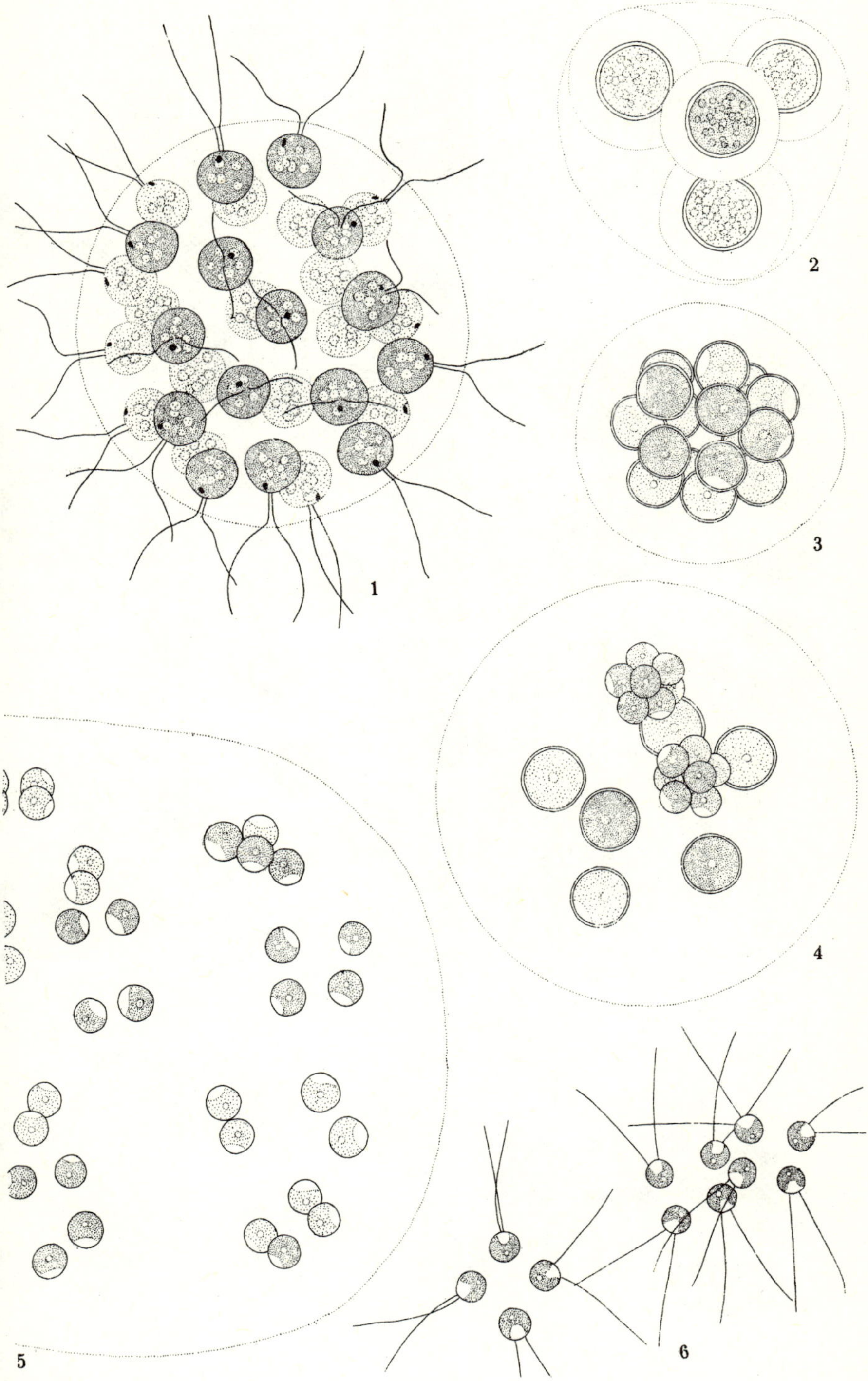

1 2 3 4 5 6

COCKAYNE, BOSTON

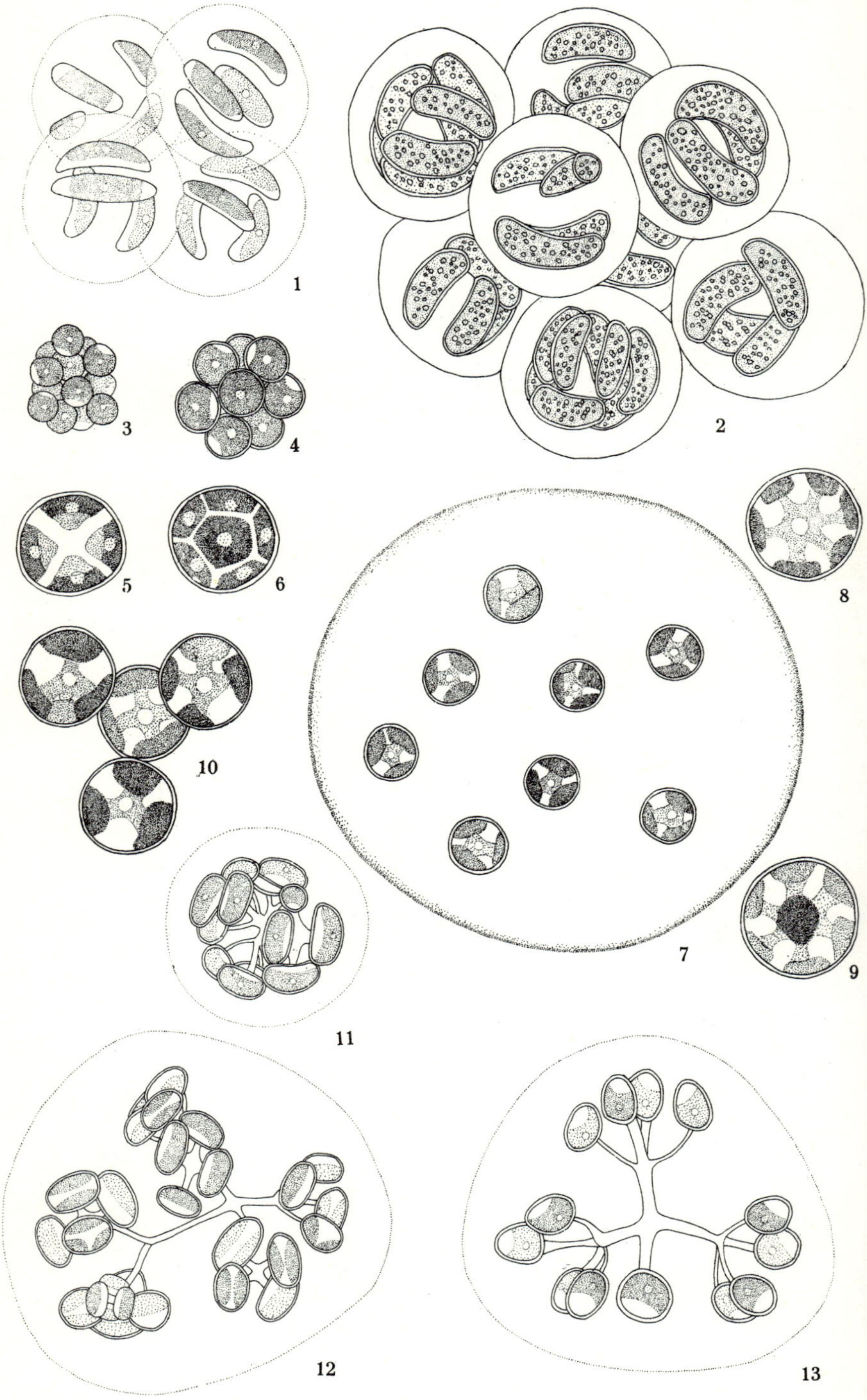

SMITH—ALGAE

COCKAYNE, BOSTON

PLATE 20.

PLATE 21.

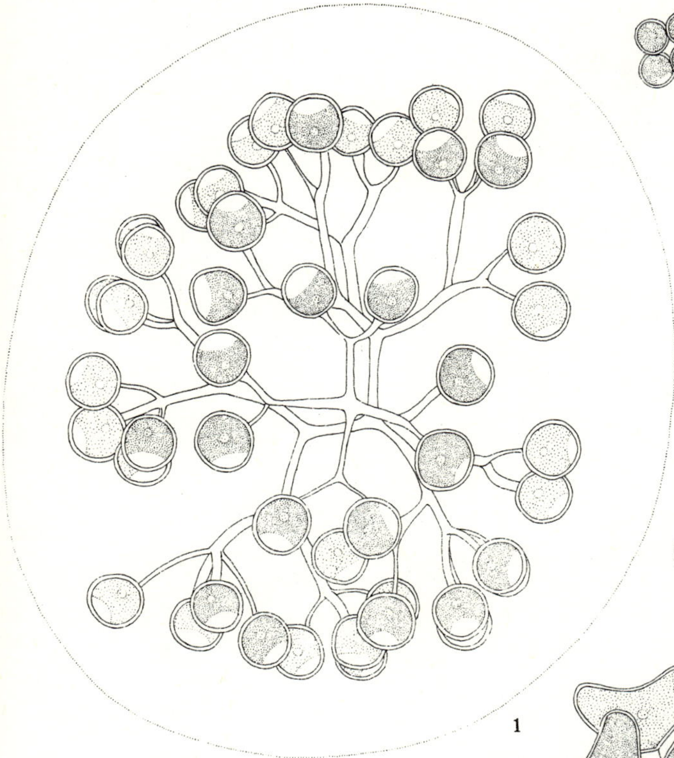

1

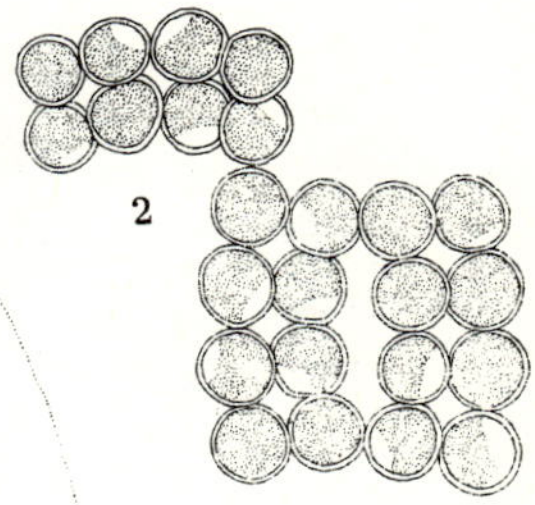

2

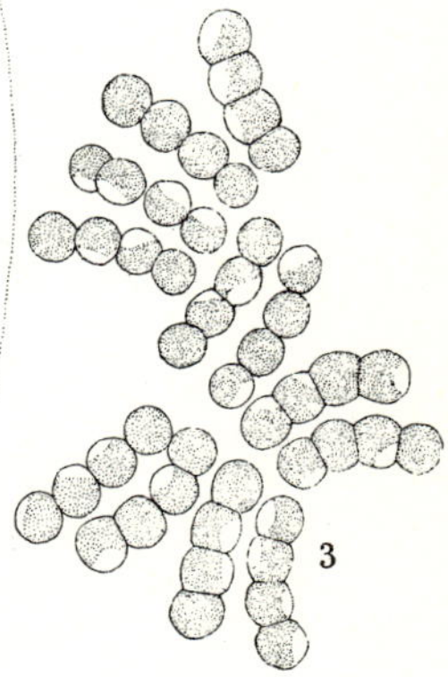

3

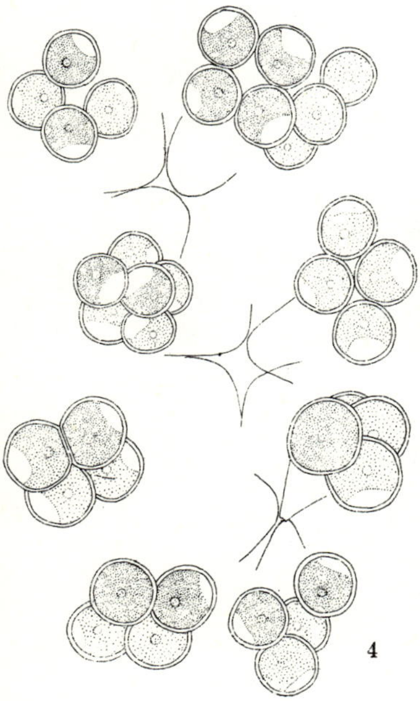

4

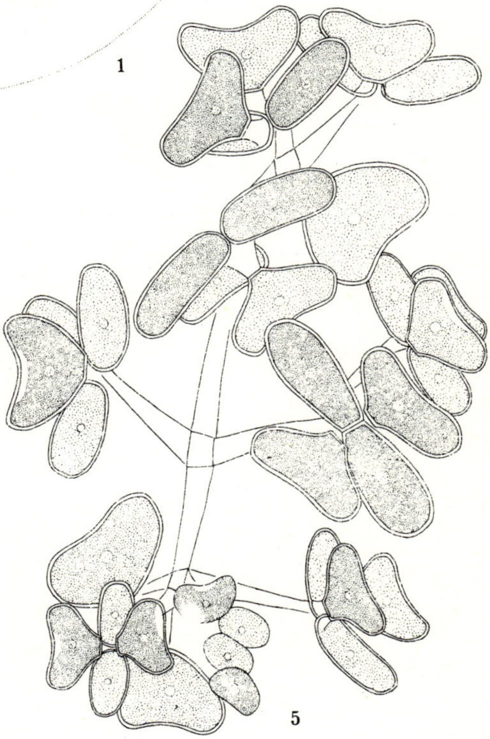

5

ITH—ALGAE

COCKAYNE, BOSTON

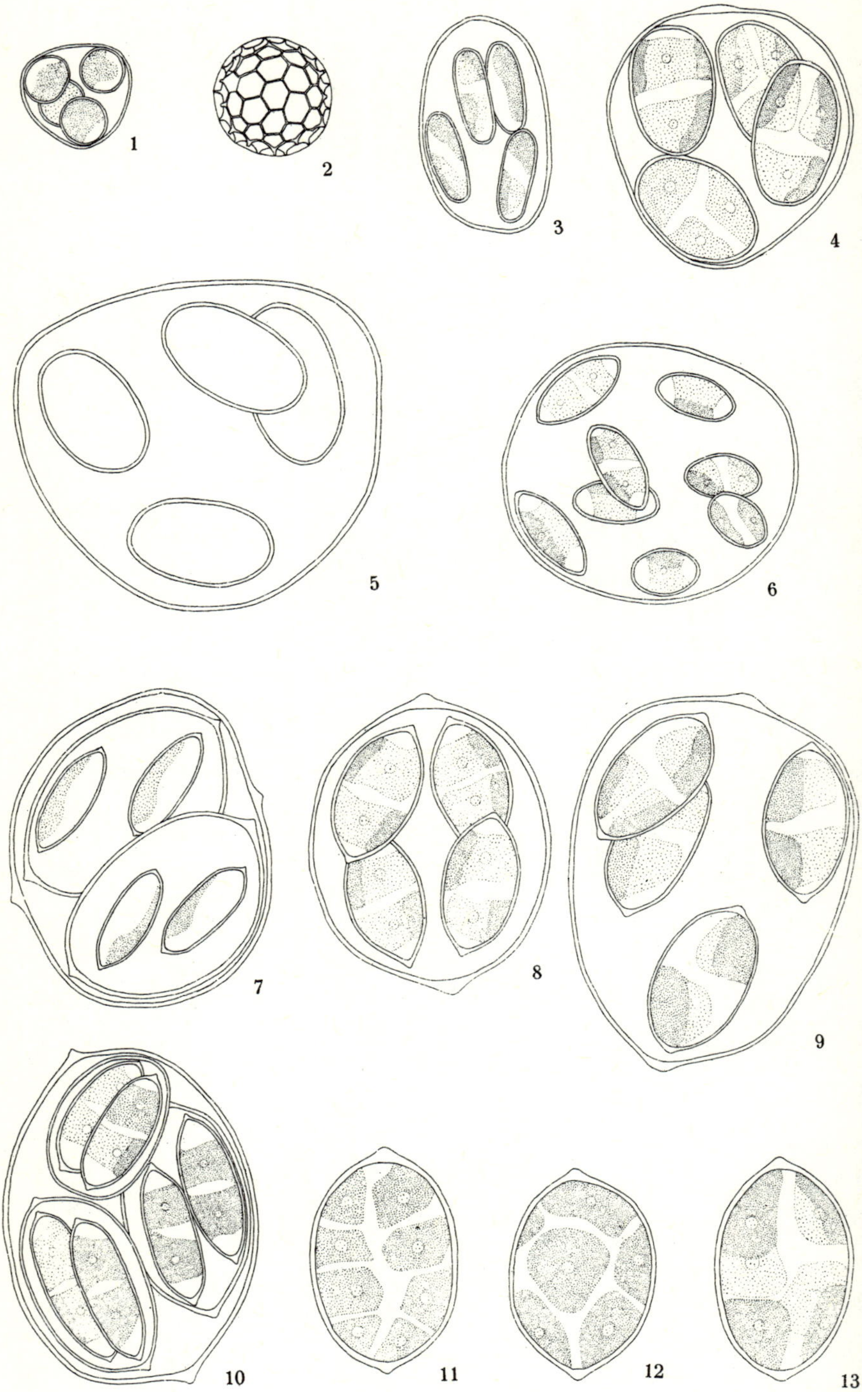

SMITH—ALGAE

COCKAYNE, BOSTON

PLATE 22.

PLATE 23.

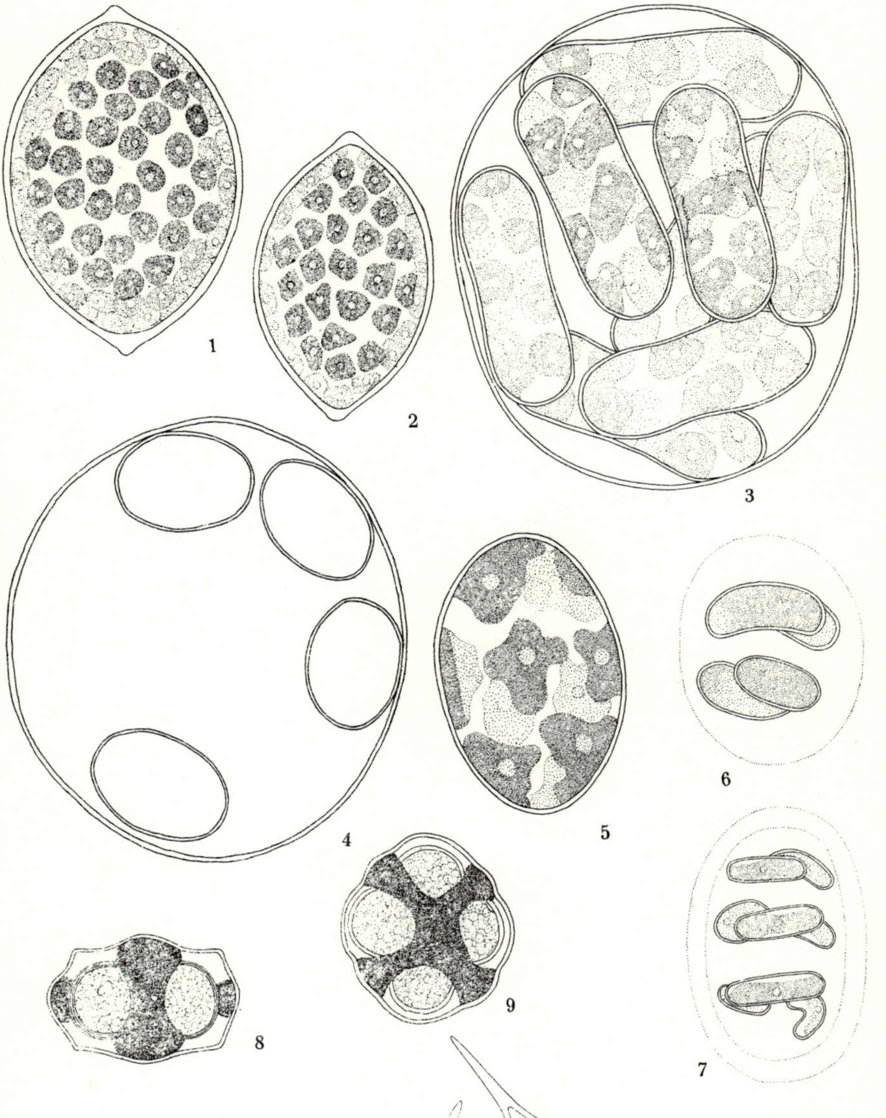

TH—ALGAE

COCKAYNE, BOSTON

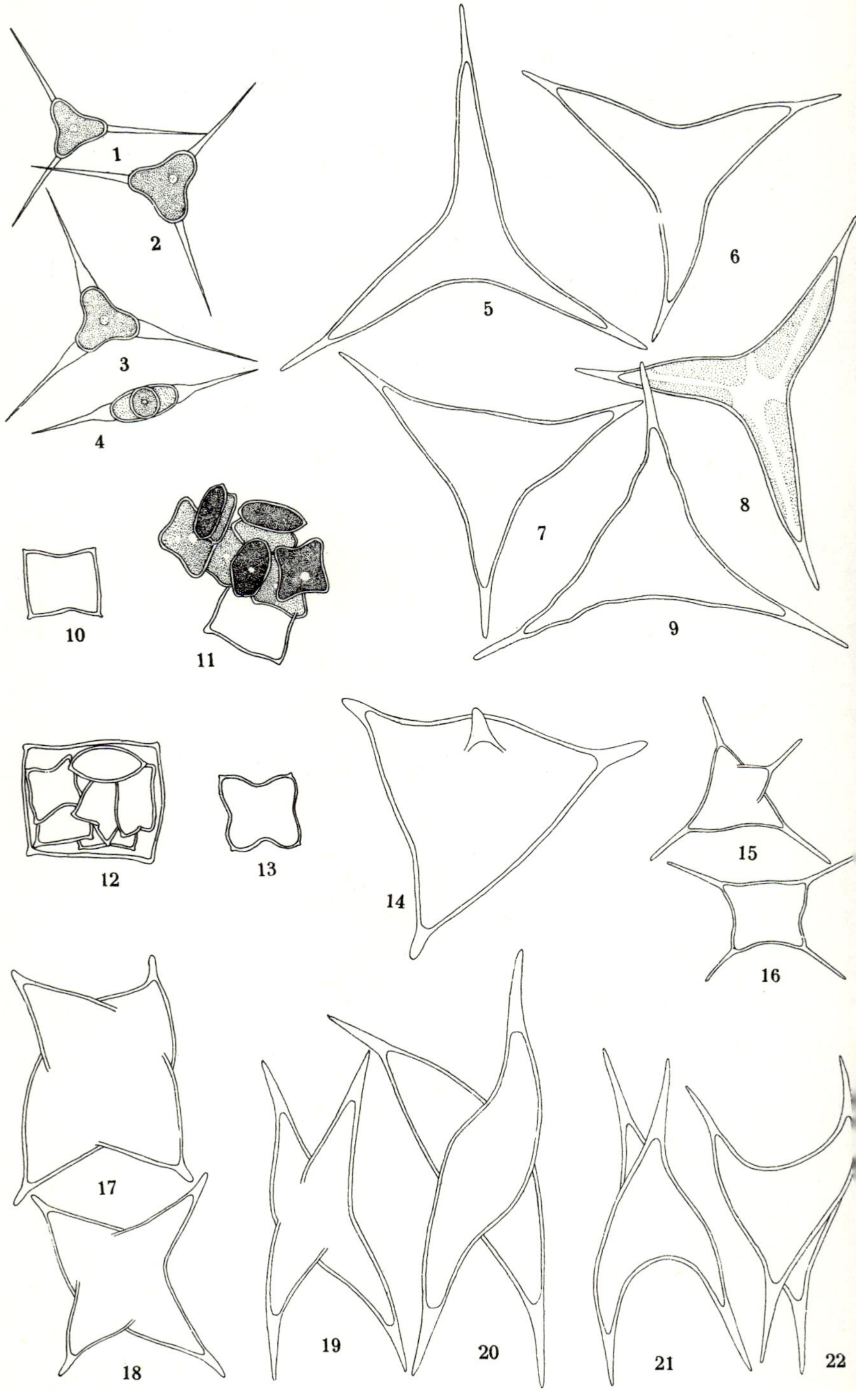

COCKAYNE, BOSTON

PLATE 24.

PLATE 25.

		Page
Fig. 1.	*Tetraëdron victorieae* var. *major* (x 1000)	119
Figs. 2–3.	*Tetraëdron verrucosum* (x 1000)	119
Figs. 4–7.	*Tetraëdron caudatum* (x 1000)	120
Figs. 8–12.	*Tetraëdron caudatum* var. *longispinum* (x 1000)	120
Figs. 13–17.	*Tetraëdron pentaëdricum* (x 1000)	120
Fig. 18.	*Tetraëdron hastatum* (x 1000)	121
Figs. 19–21.	*Tetraëdron hastatum* var. *palatinum* (x 1000)	121
Figs. 22–24.	*Tetraëdron constrictum* (x 1000)	122

25

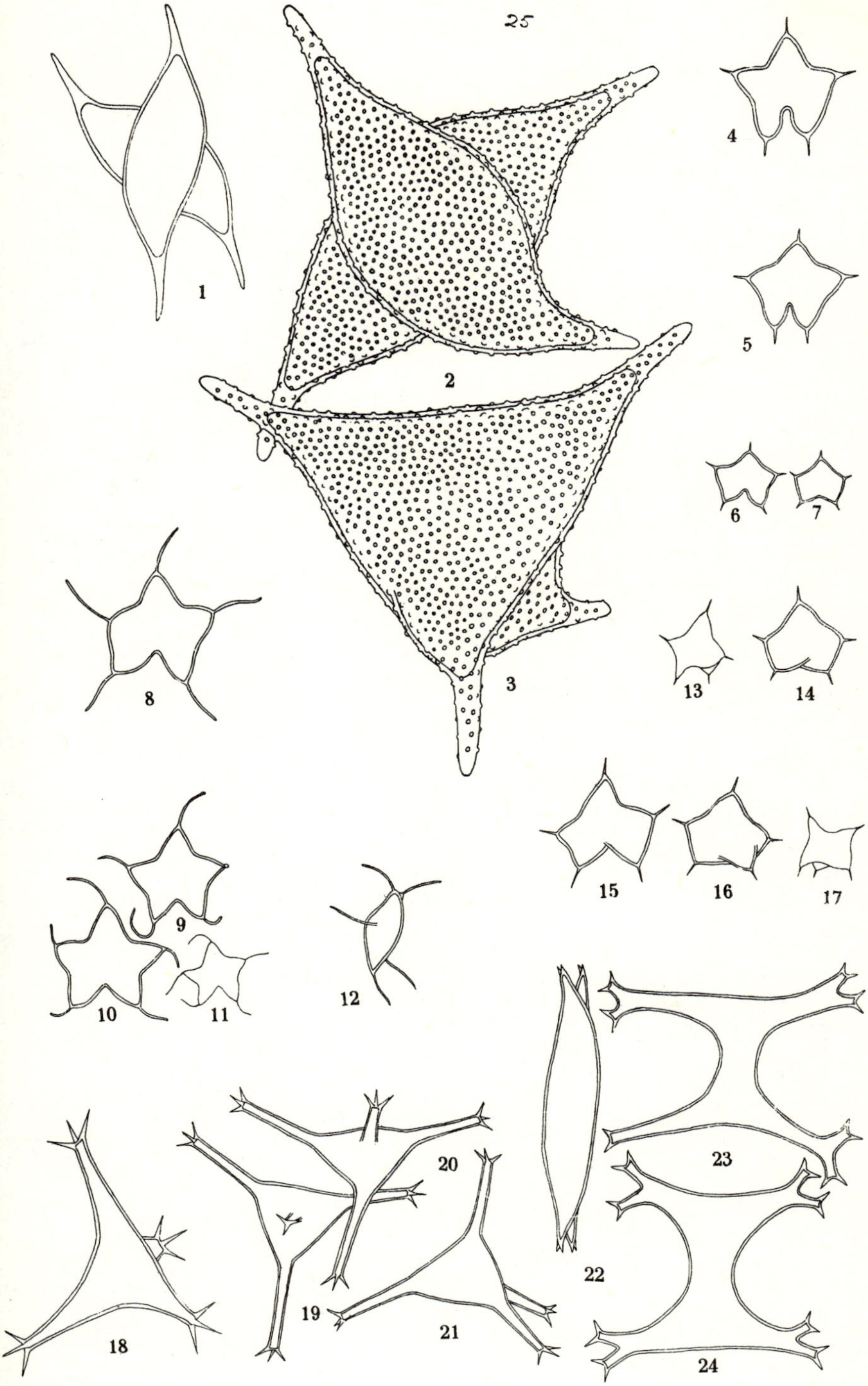

TH—ALGAE

COCKAYNE, BOSTON

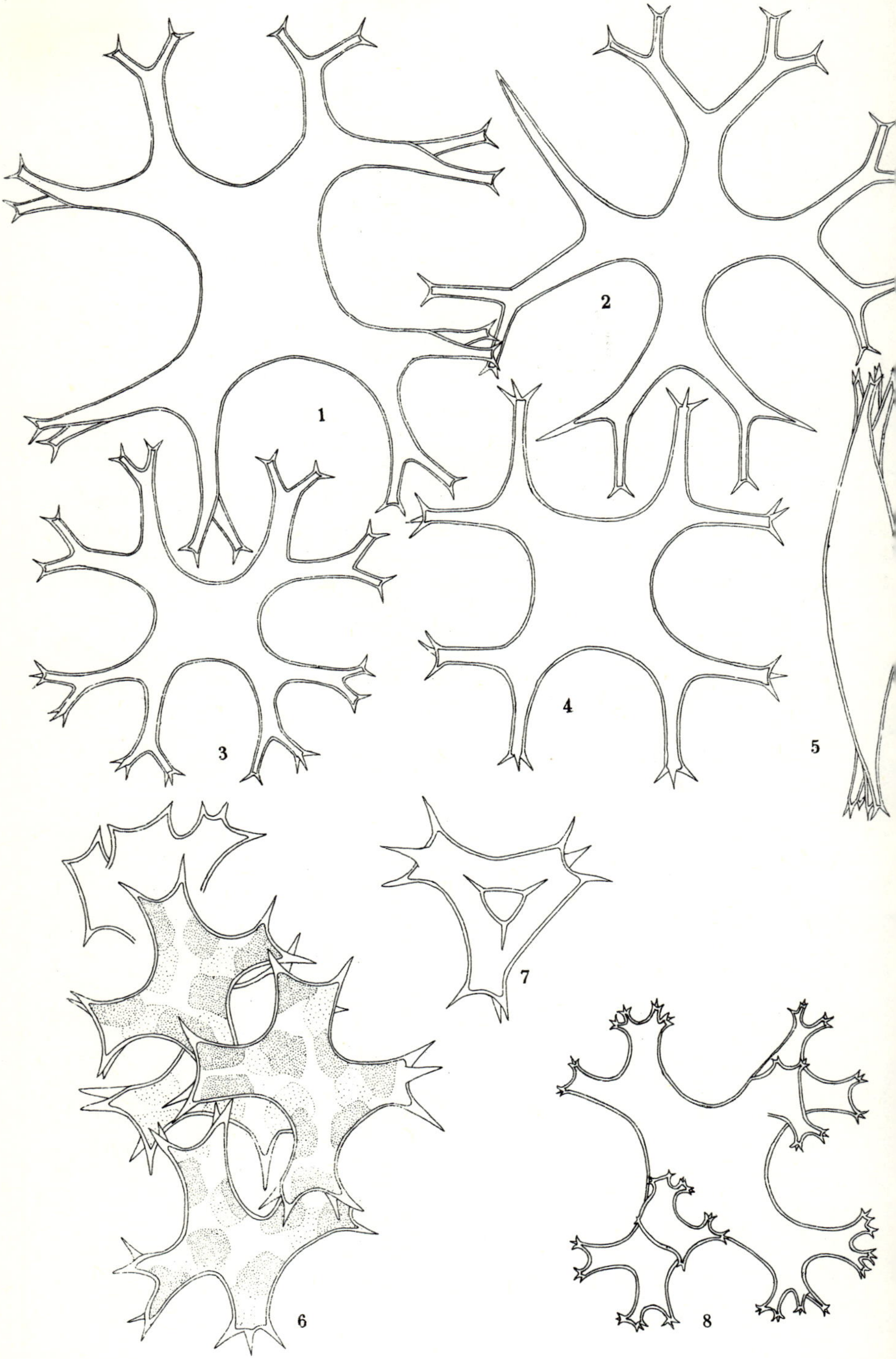

SMITH—ALGAE

COCKAYNE, BOSTON

PLATE 26.

PLATE 27.

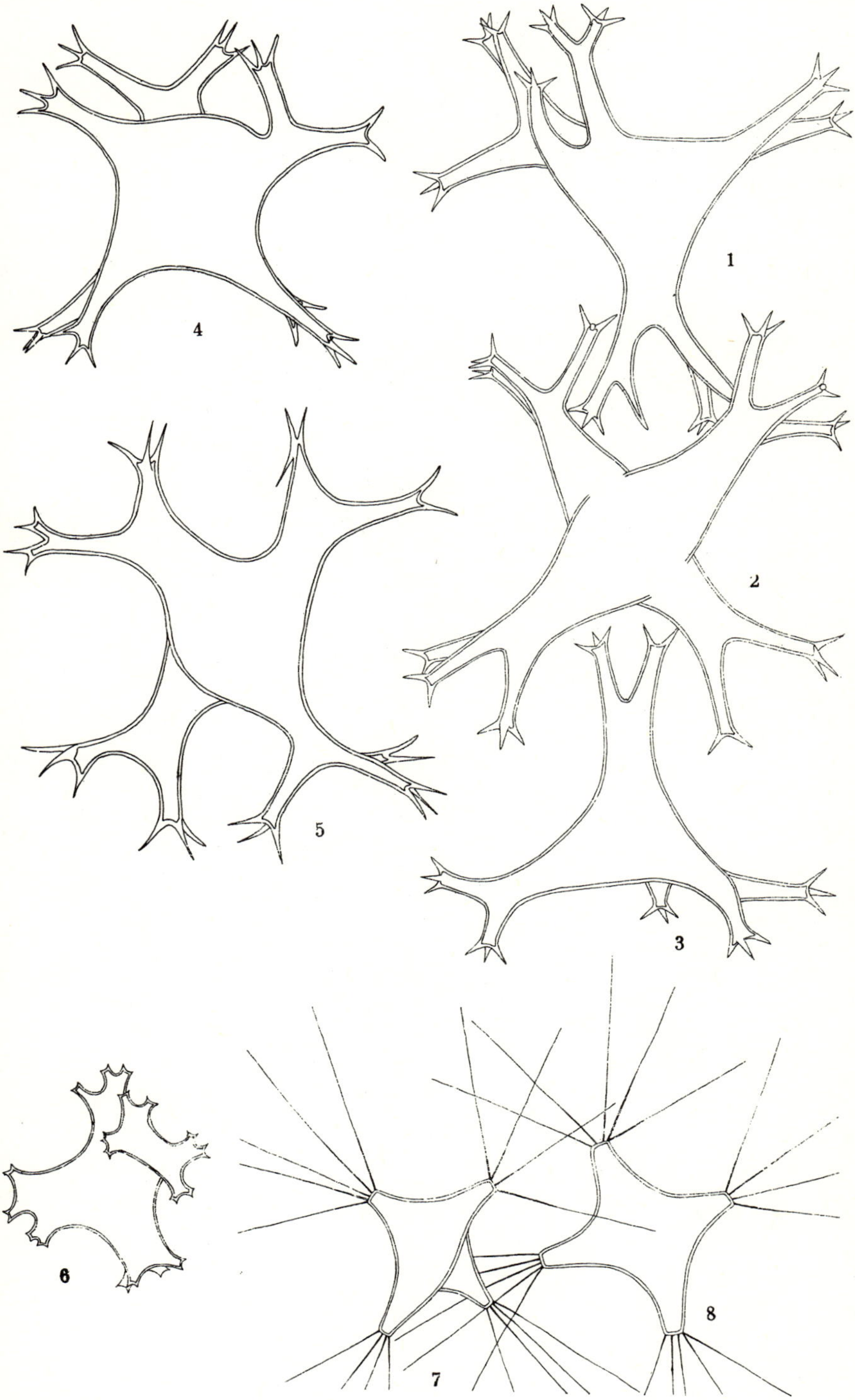

H—ALGAE

COCKAYNE, BOSTON

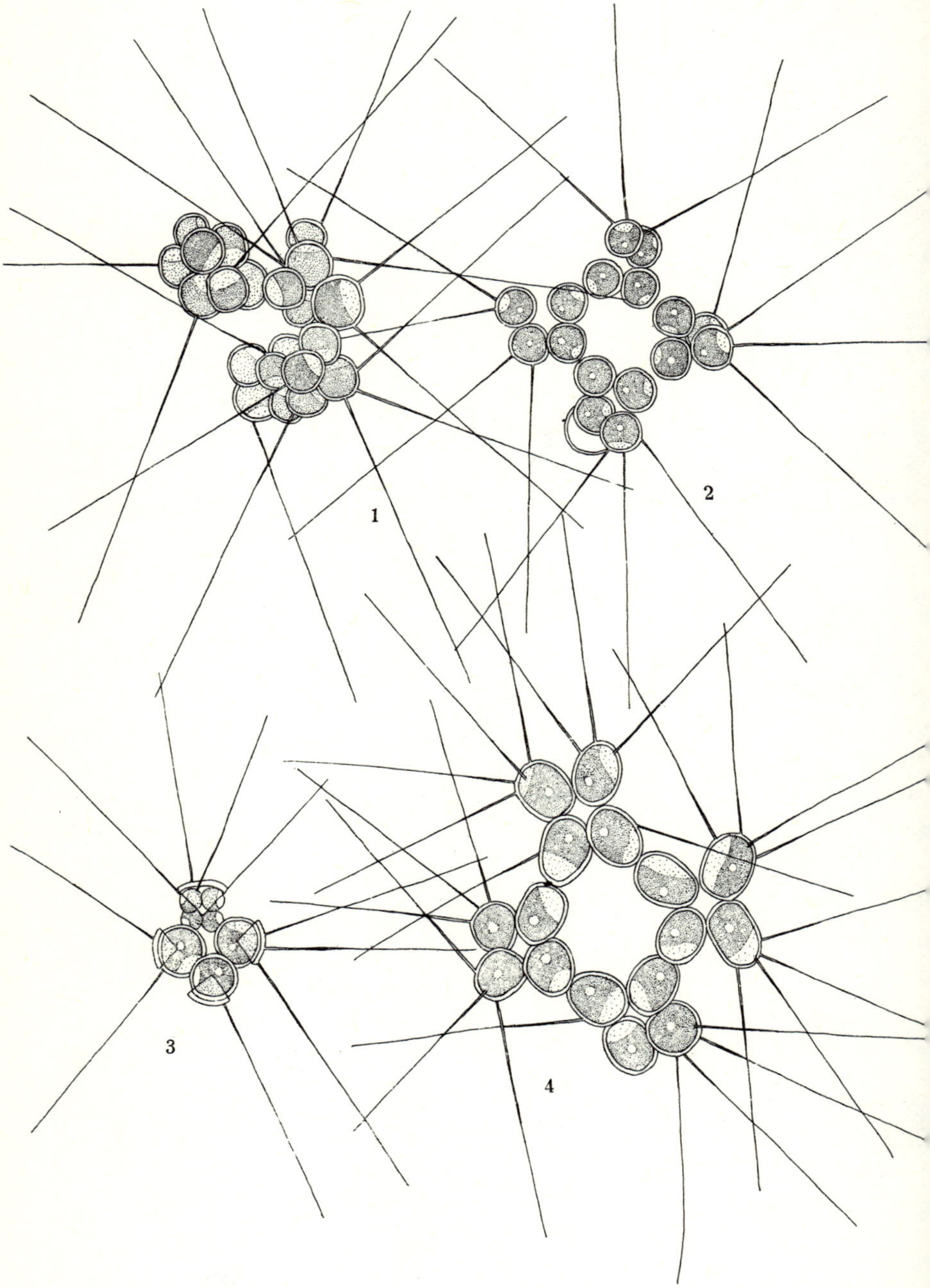

COCKAYNE, BOSTON

PLATE 28.

	Page
Figs. 1–3. ***Micractinium pusillum*** **(x 1000)**........................	**125**
Fig. 4. ***Micractinium quadrisetum*** **(x 1000)**......................	**126**

PLATE 29.

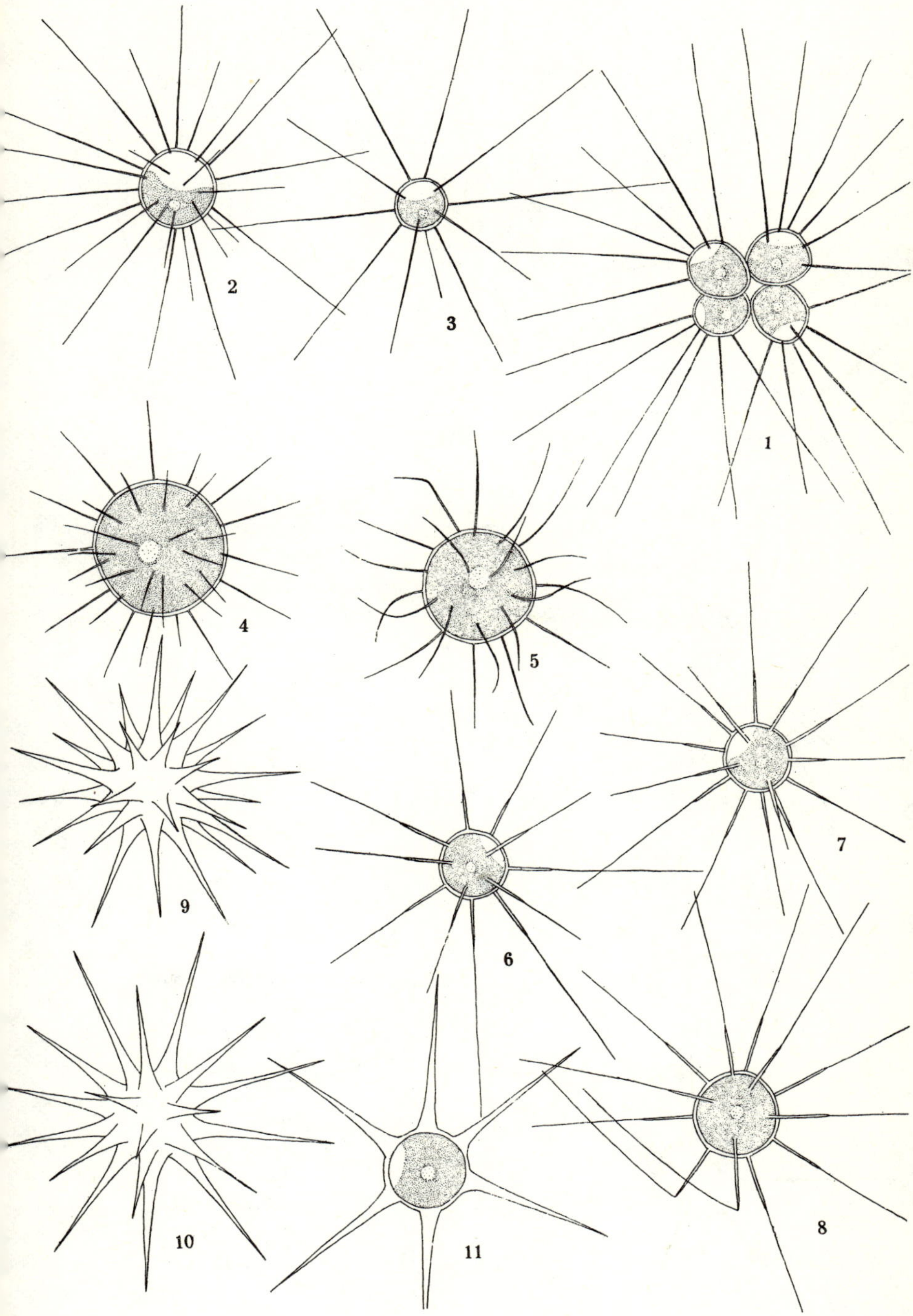

H—ALGAE

COCKAYNE, BOSTON

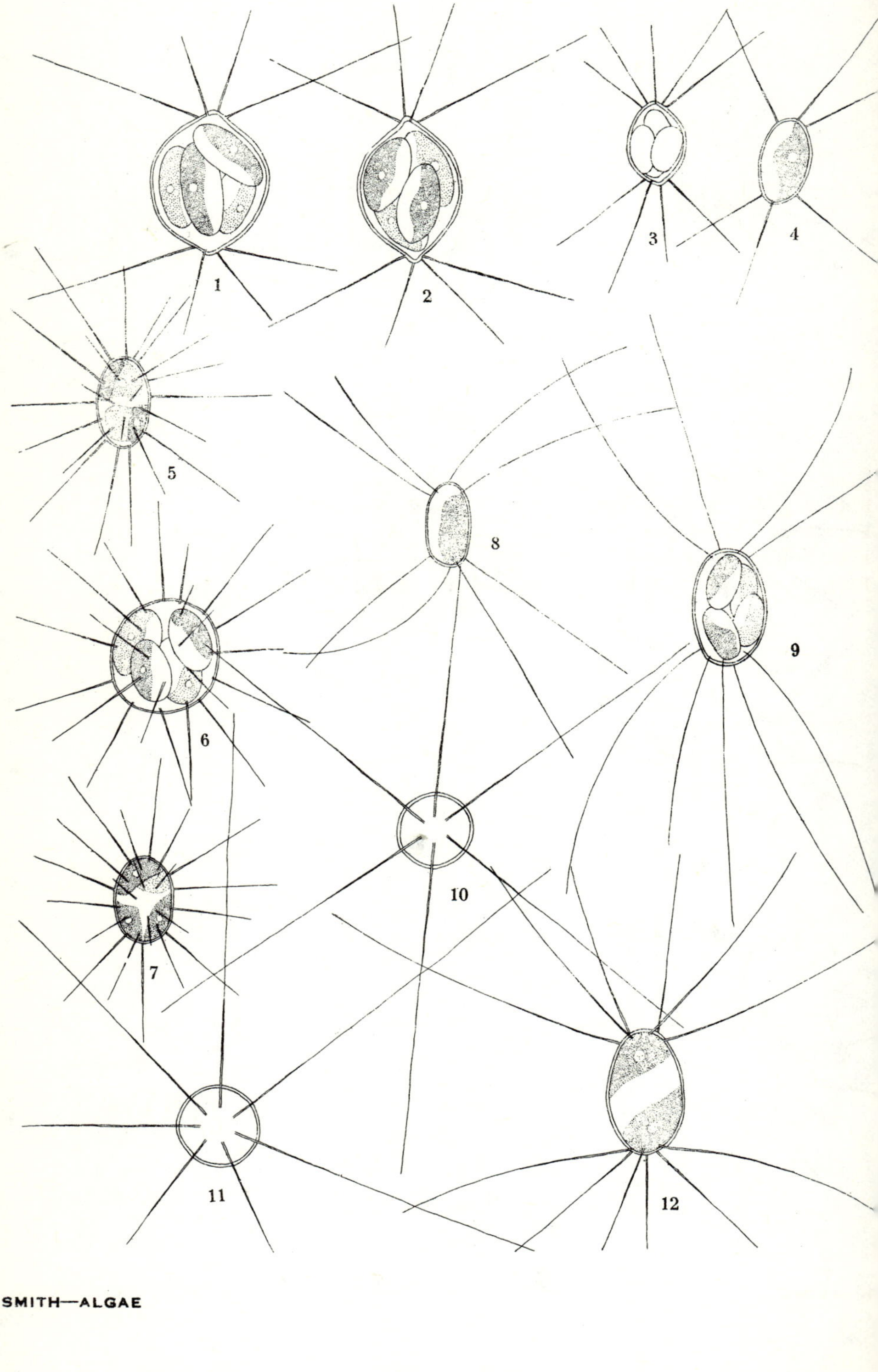

COCKAYNE, BOSTON

PLATE 30.

		Page
Figs. 1–2.	*Lagerheimia citriformis* (x 1000)	130
Figs. 3–4.	*Lagerheimia subsala* (x 1000)	130
Figs. 5–7.	*Lagerheimia Droescheri* (x 1000)	131
Figs. 8–9.	*Lagerheimia longiseta* (x 1000)	130
Figs. 10–12.	*Lagerheimia longiseta* var. *major* (x 1000)	130

PLATE 31.

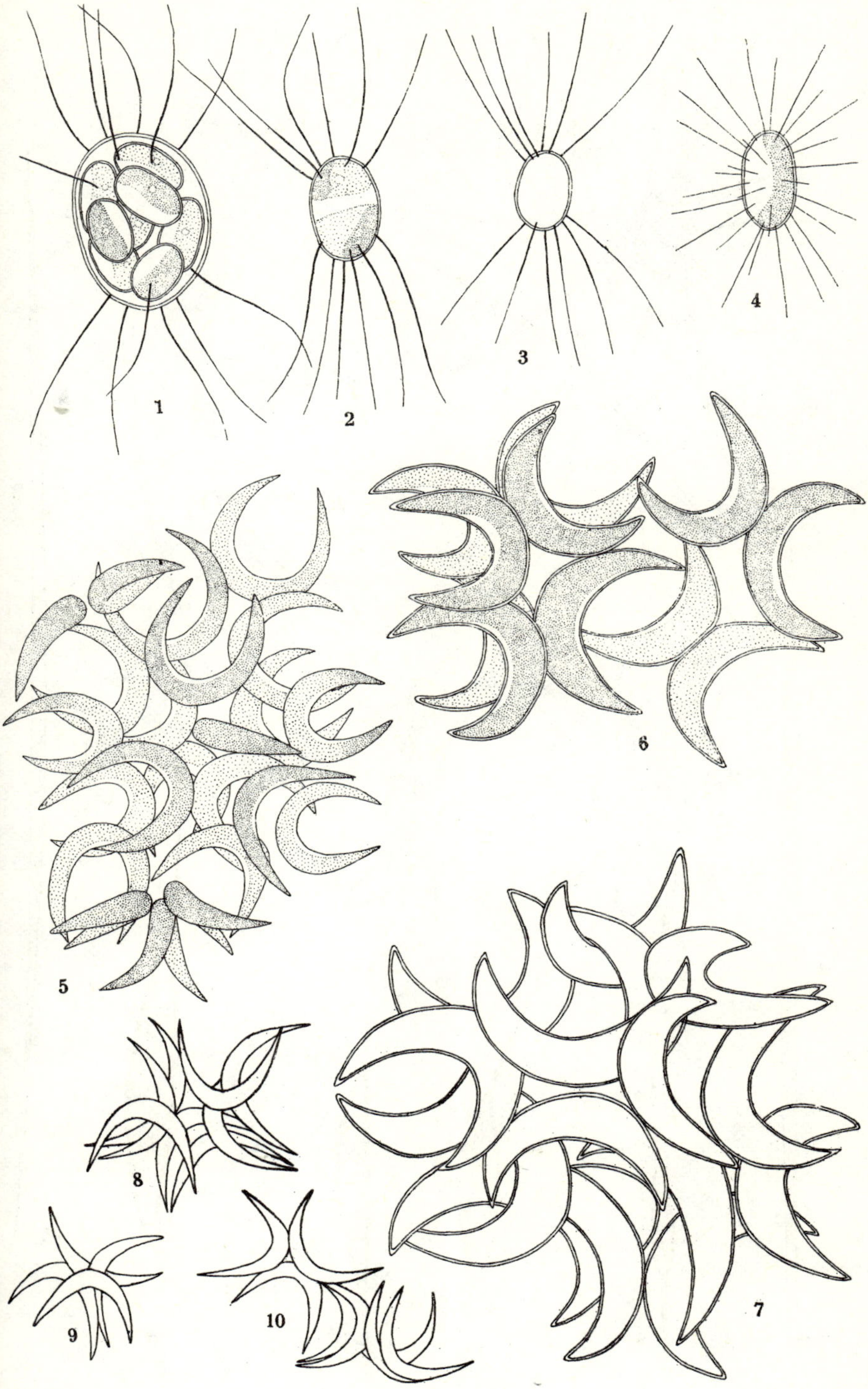

H—ALGAE

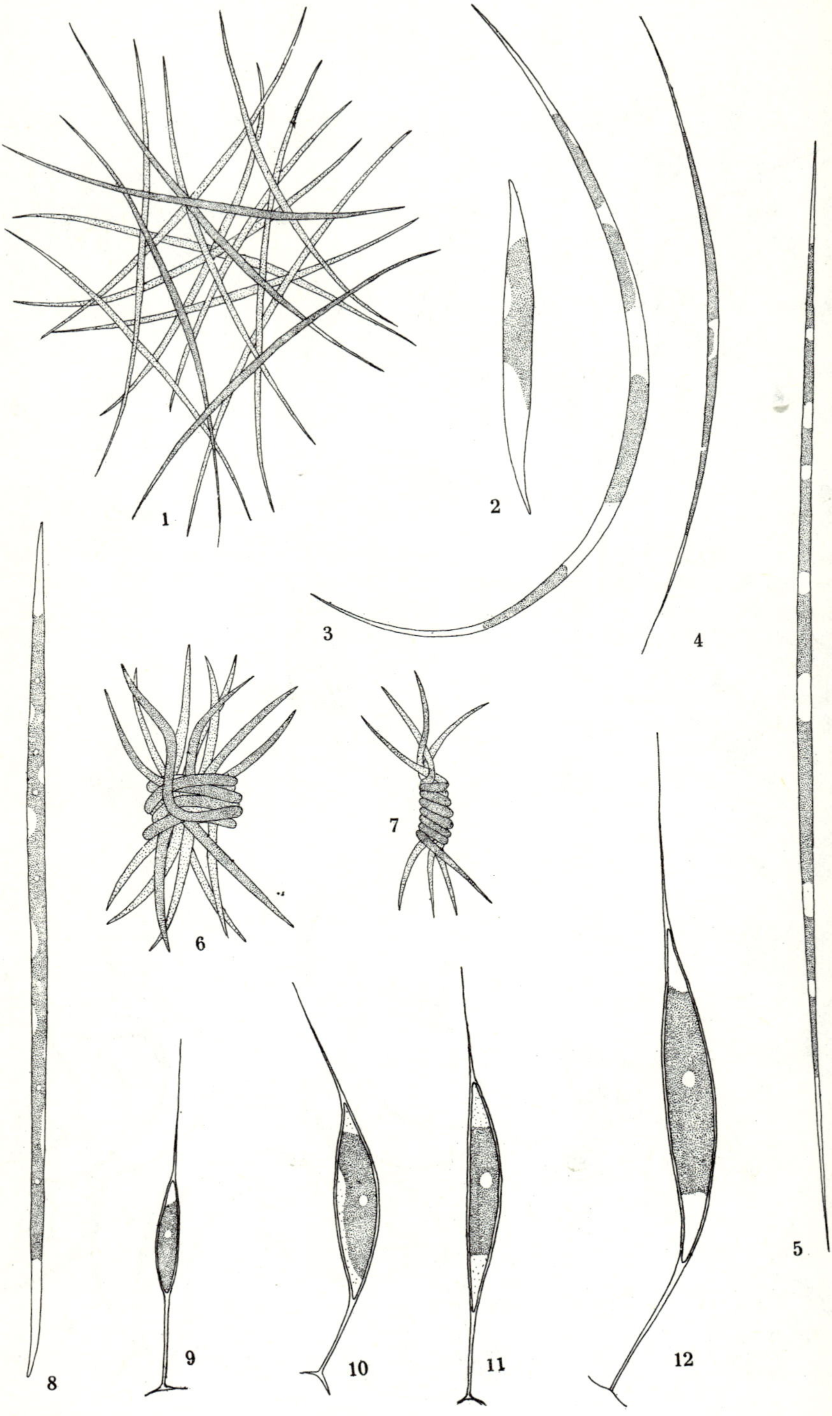

SMITH—ALGAE

COCKAYNE, BOSTON

PLATE 32.

PLATE 33.

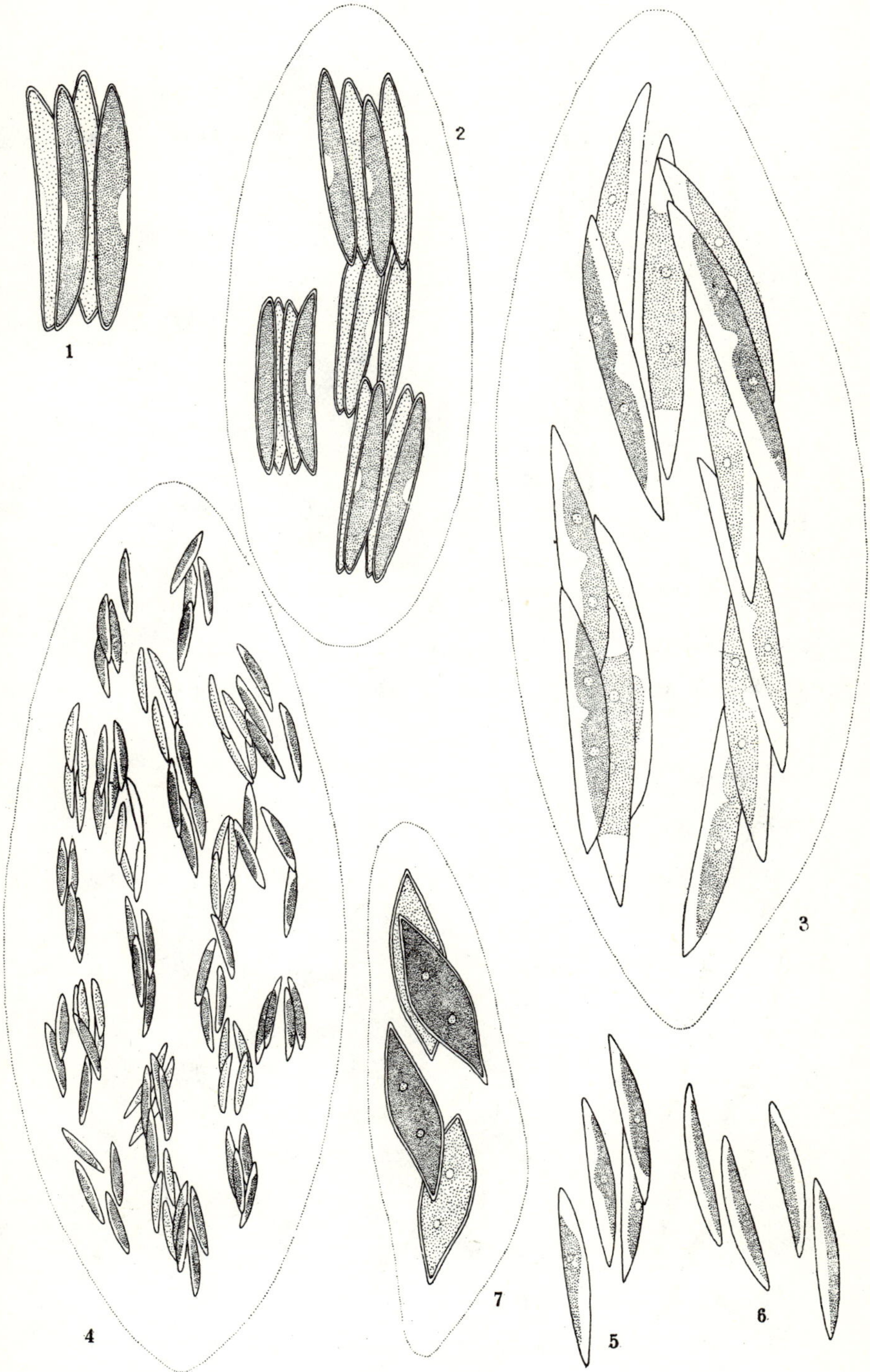

ITH—ALGAE

COCKAYNE, BOSTON

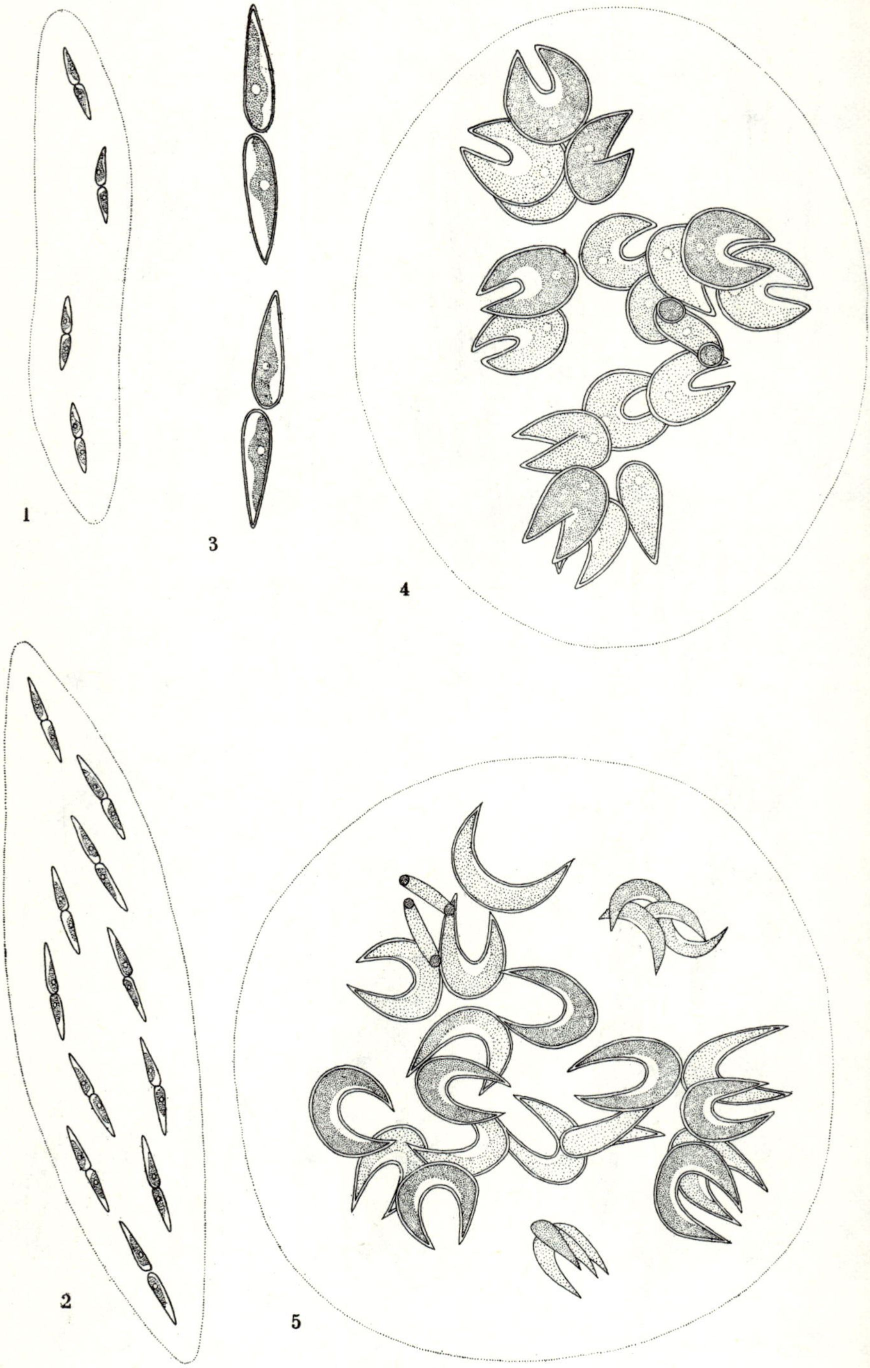

COCKAYNE, BOSTON

PLATE 34.

PLATE 35.

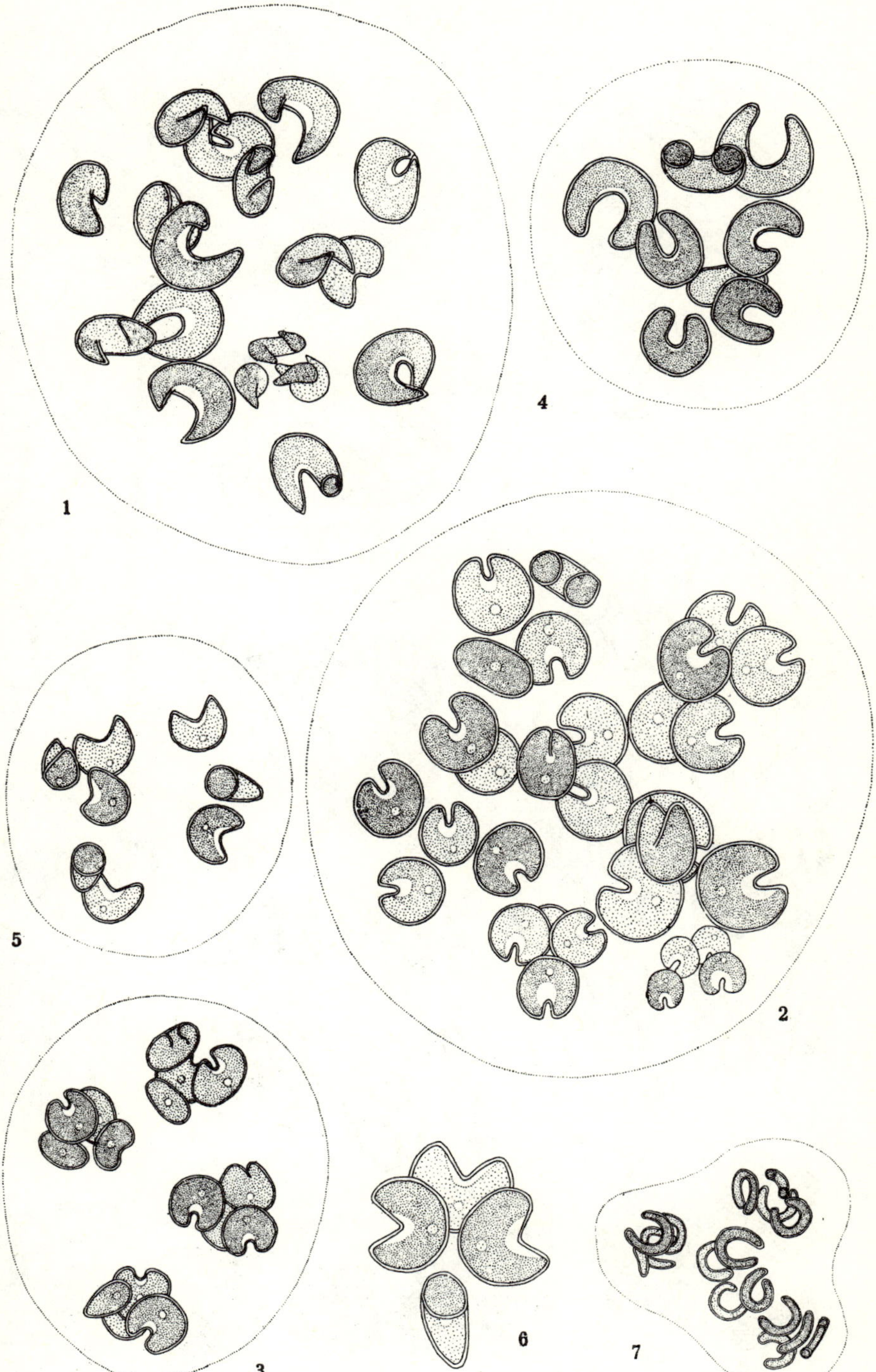

COCKAYNE, BOSTON

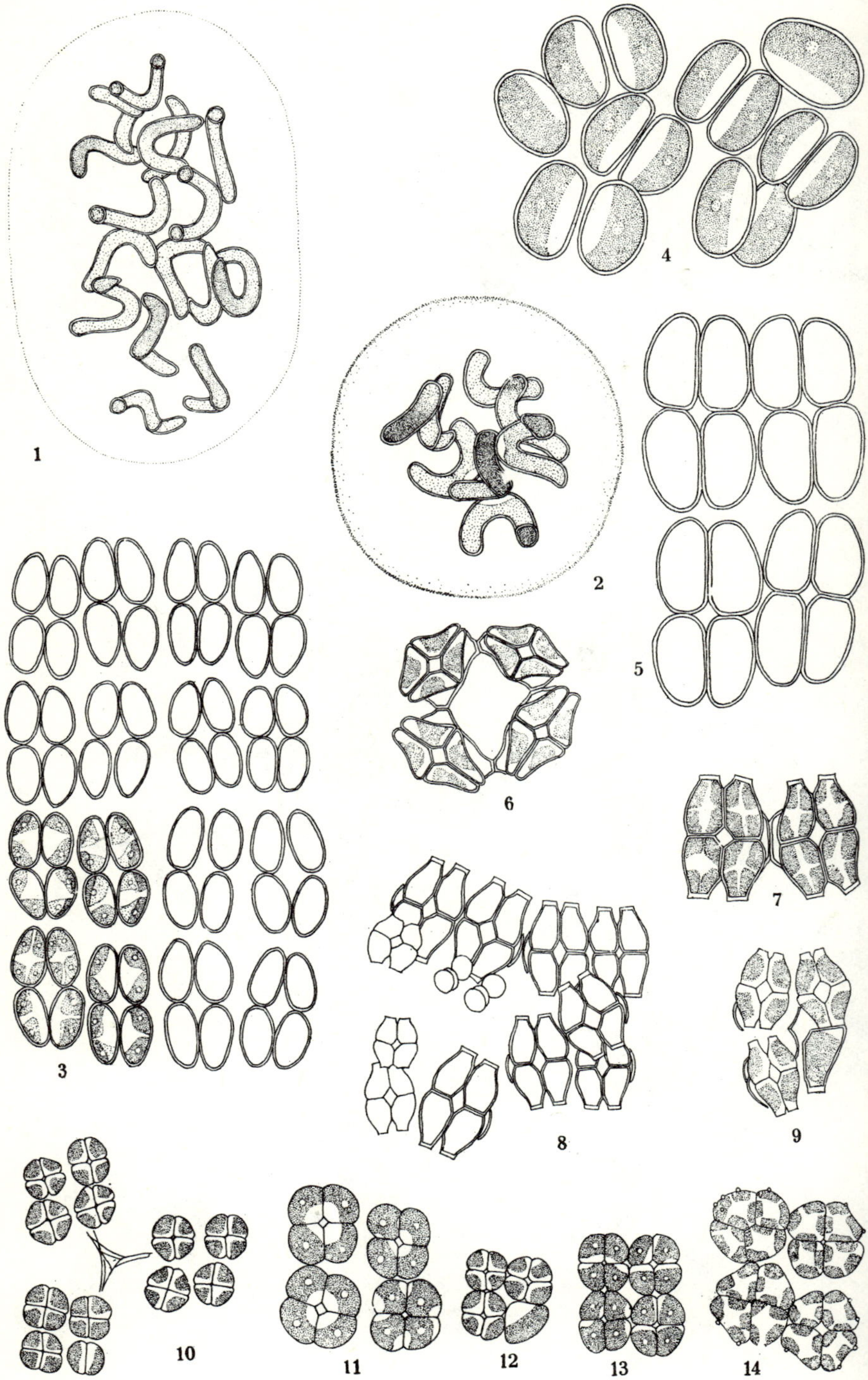

COCKAYNE, BOSTON

PLATE 36.

		Page
Figs. 1–2.	*Kirchneriella elongata* (x 1000)	143
Fig. 3.	*Crucigenia rectangularis* (x 1000)	144
Figs. 4–5.	*Crucigenia irregularis* (x 1000)	145
Fig. 6.	*Crucigenia crucifera* (x 1000)	145
Figs. 7–9.	*Crucigenia truncata* (x 1000)	146
Figs. 10–14.	*Crucigenia quadrata* (x 1000)	147

PLATE 37.

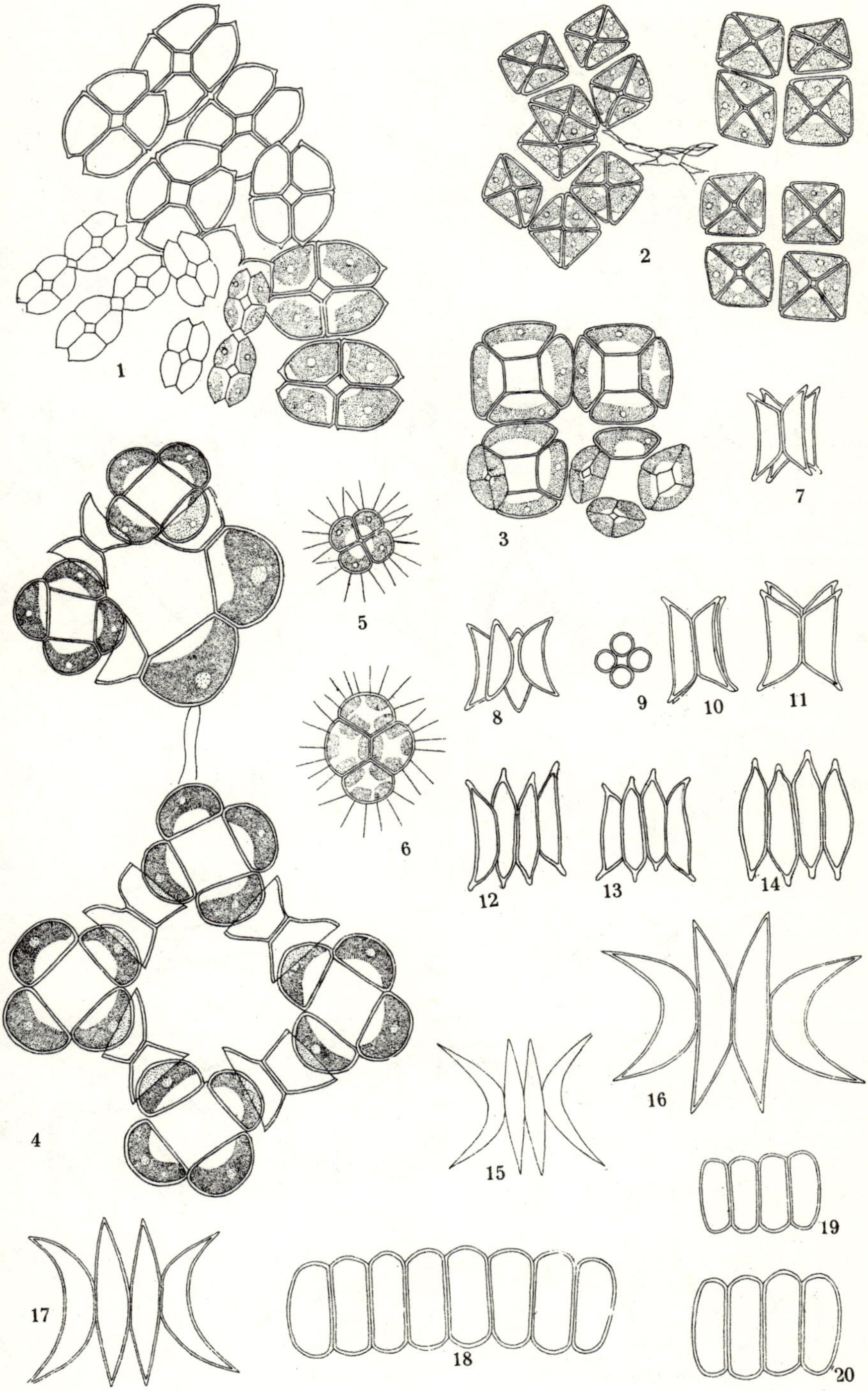

'H—ALGAE

COCKAYNE, BOSTON

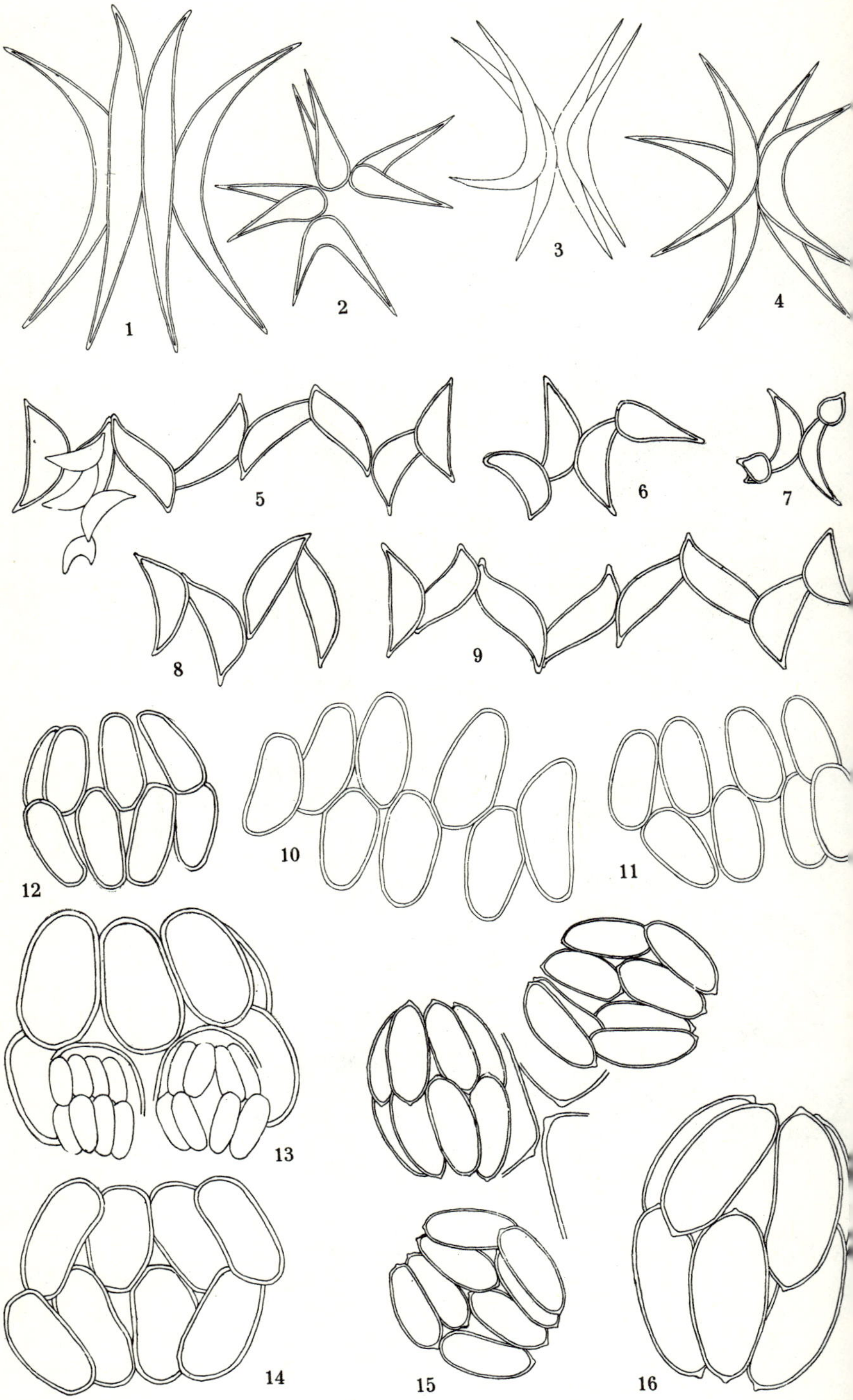

SMITH—ALGAE

COCKAYNE, BOSTON

PLATE 38.

		Page
Figs. 1–4.	*Scenedesmus acuminatus* (x 1000)	152
Figs. 5–9.	*Scenedesmus Bernardii* (x 1000)	152
Figs. 10–11.	*Scenedesmus bijuga* var. *alternans* (x 1000)	153
Figs. 12–14.	*Scenedesmus arcuatus* (x 1000)	153
Figs. 15–16.	*Scenedesmus arcuatus* var. *capitatus* (x 1000)	153

PLATE 39.

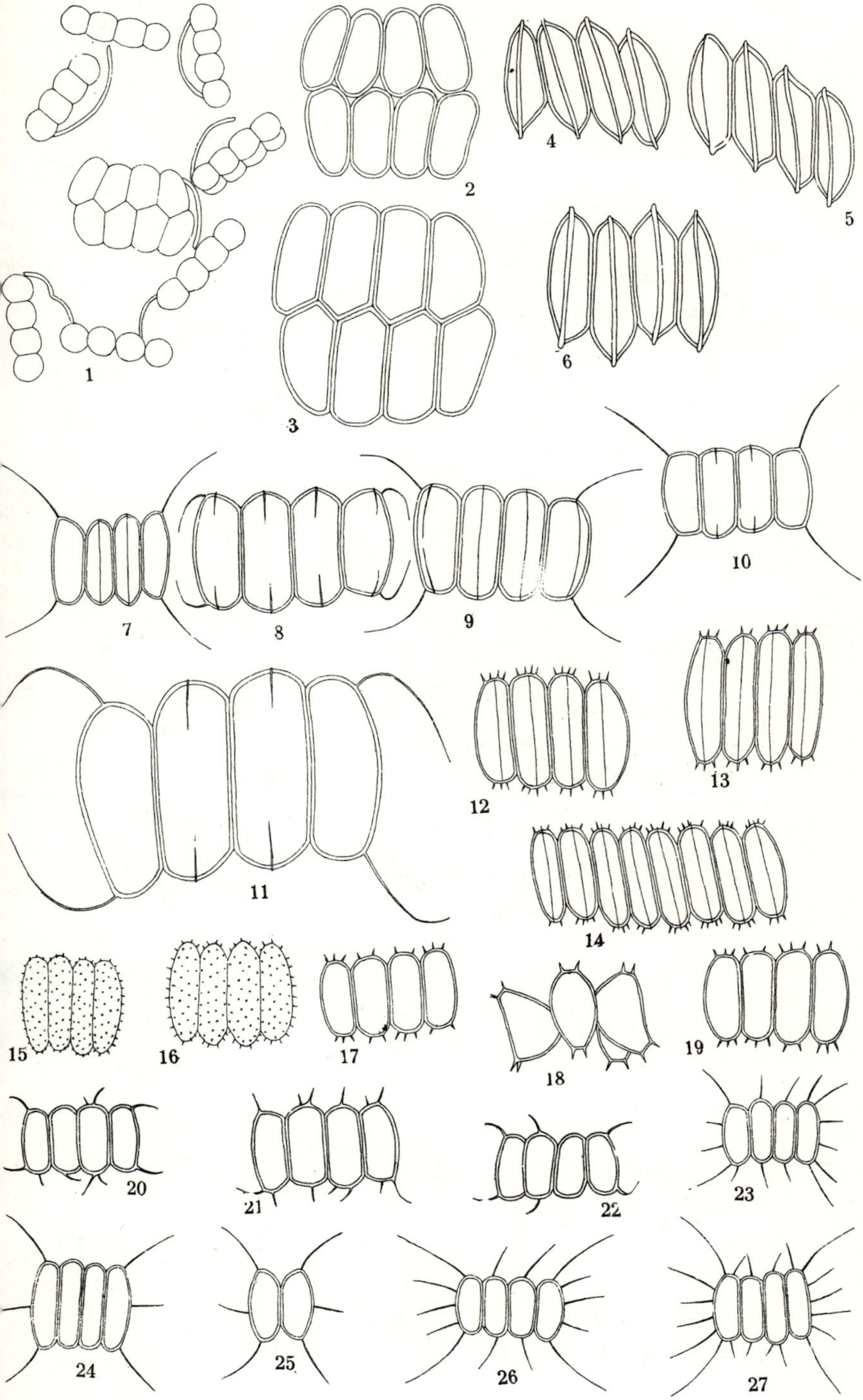

—ALGAE

COCKAYNE, BOSTON

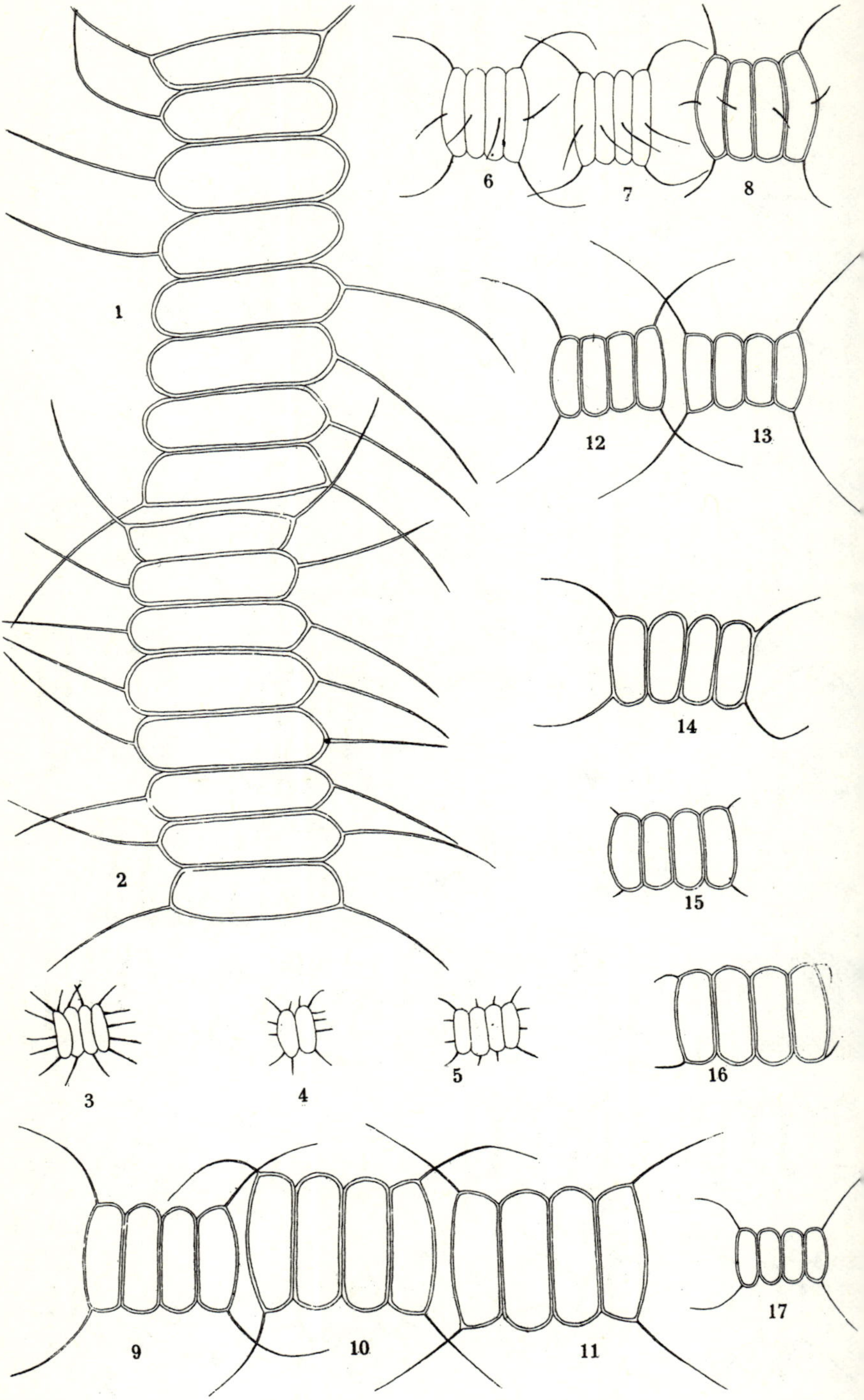

COCKAYNE, BOSTON

PLATE 40.

PLATE 41.

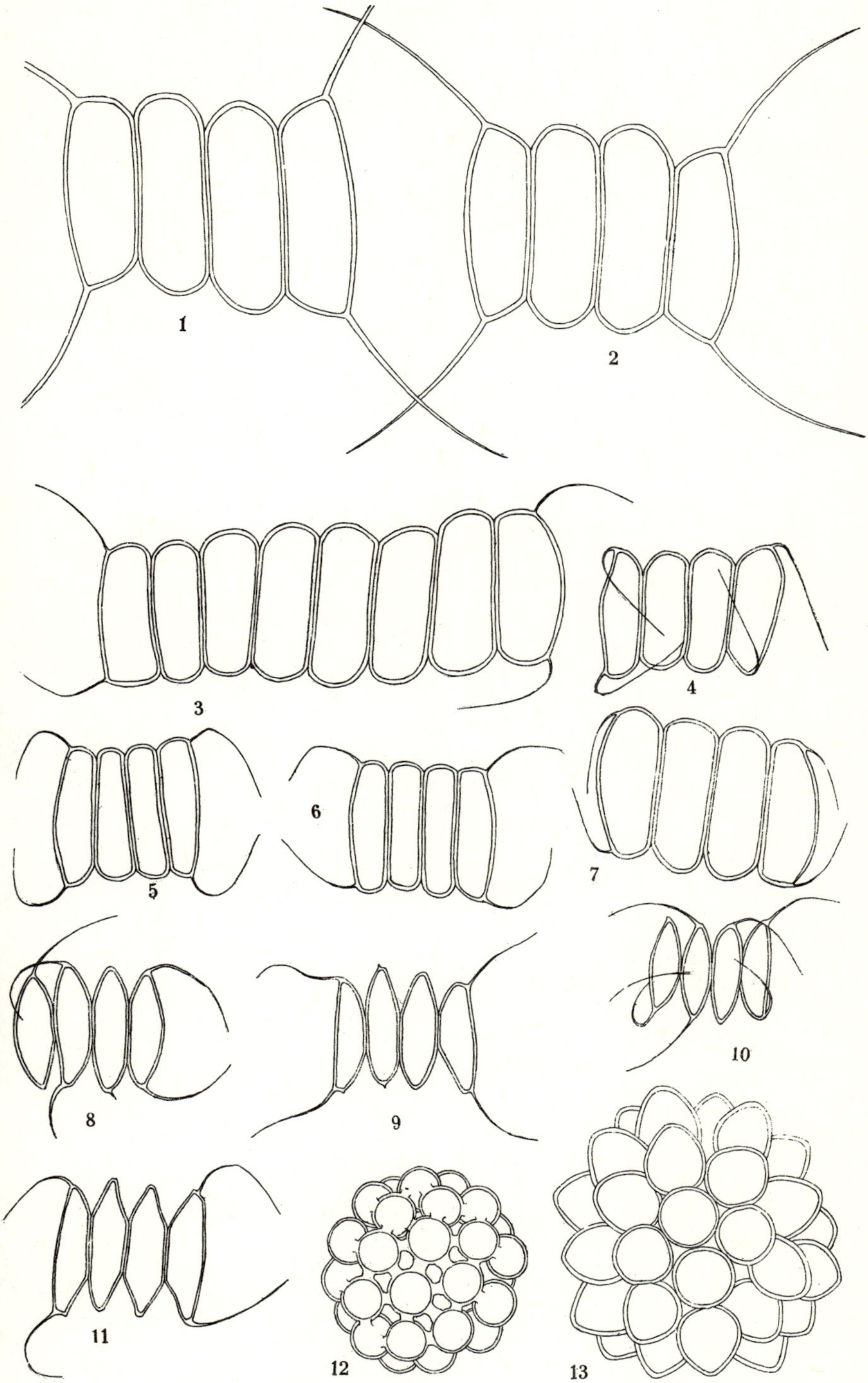

—ALGAE

COCKAYNE, BOSTON

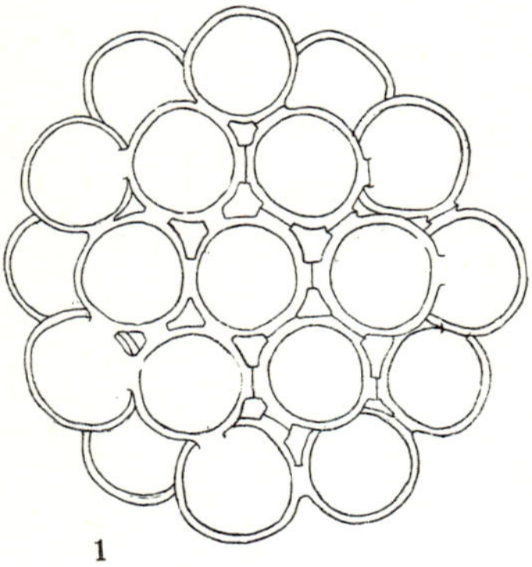
1

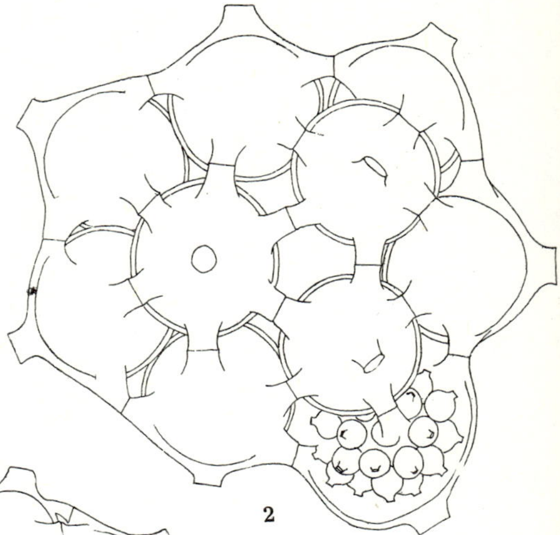
2

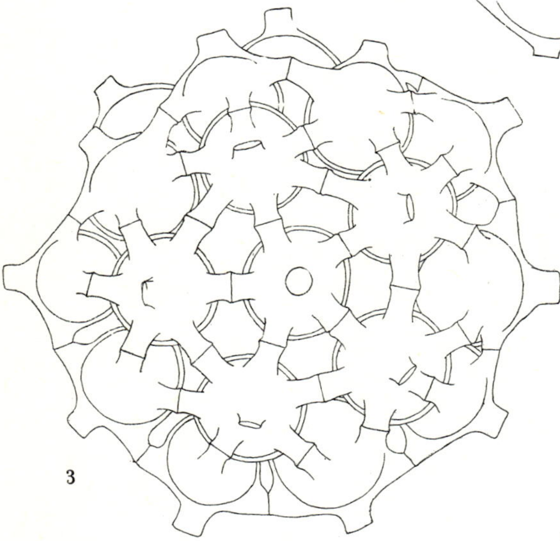
3

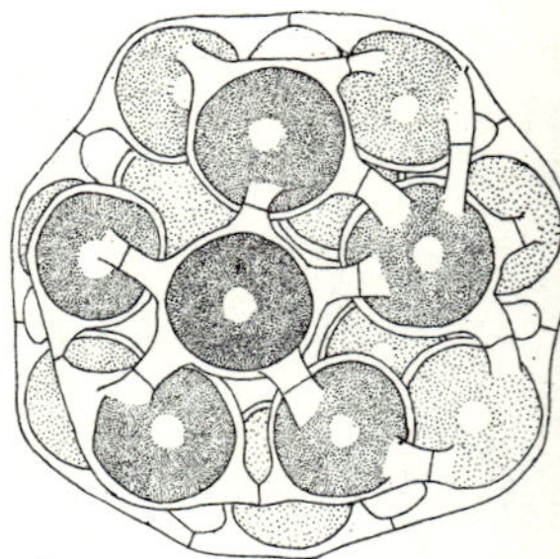
4

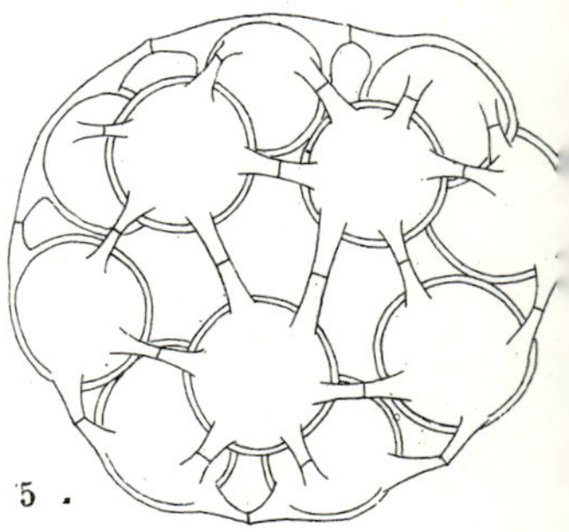
5

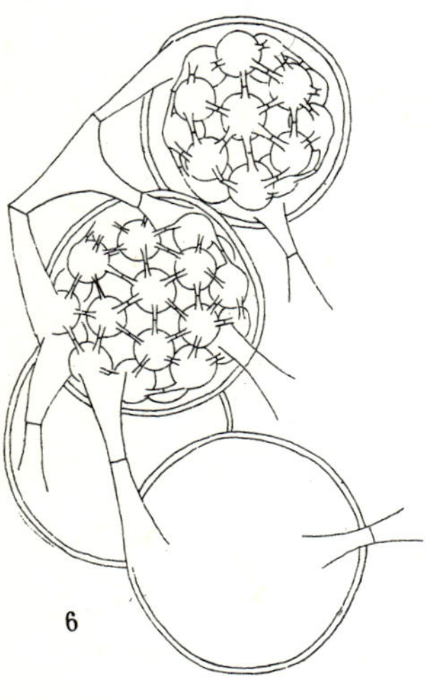
6

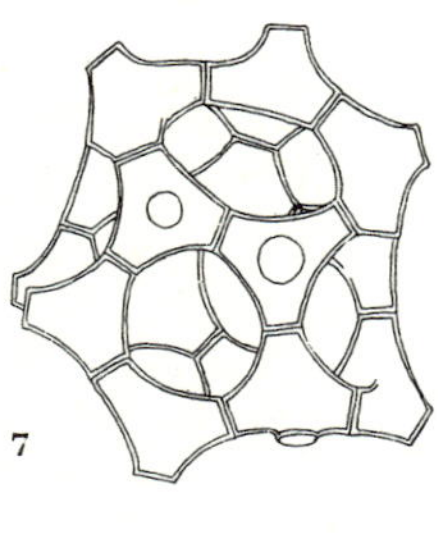
7

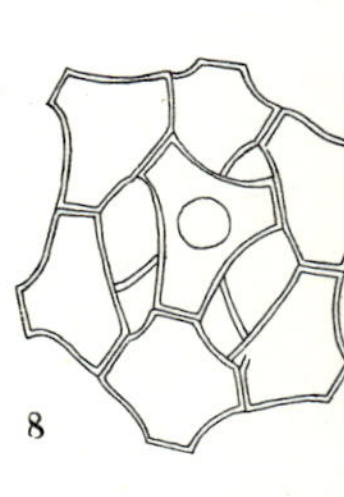
8

COCKAYNE, BOSTON

PLATE 42.

		Page
Fig. 1.	*Coelastrum microporum* (x 825)	160
Figs. 2–3.	*Coelastrum cambricum* (x 825)	161
Figs. 4–6.	*Coelastrum reticulatum* (x 825)	161
Figs. 7–8.	*Coelastrum proboscideum* (x 825)	162

PLATE 43.

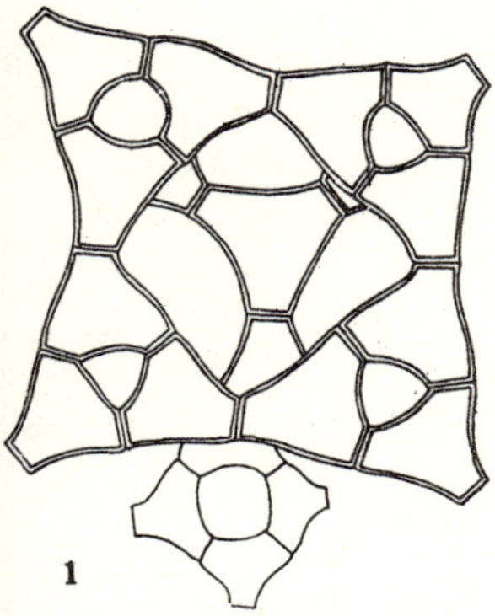

1

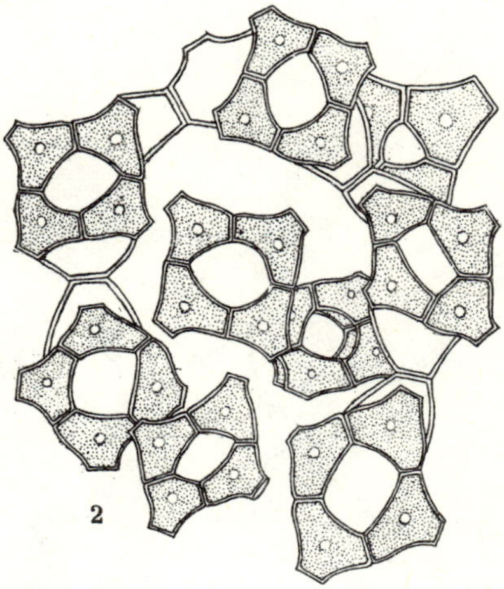

2

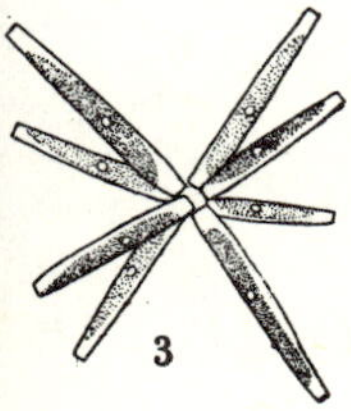

3

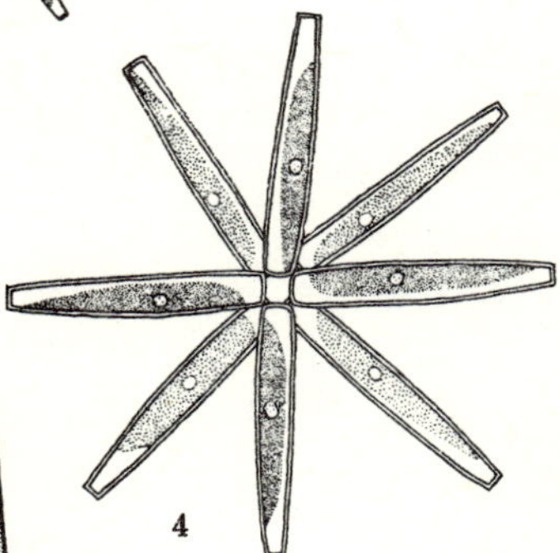

4

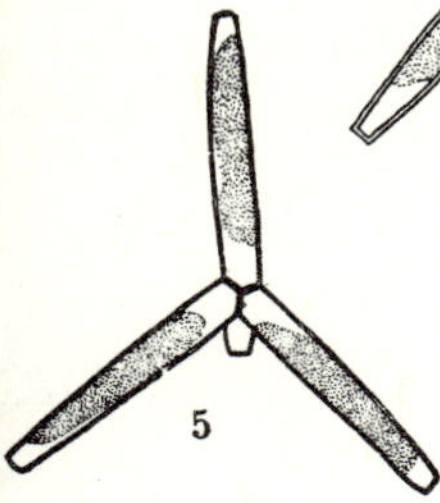

5

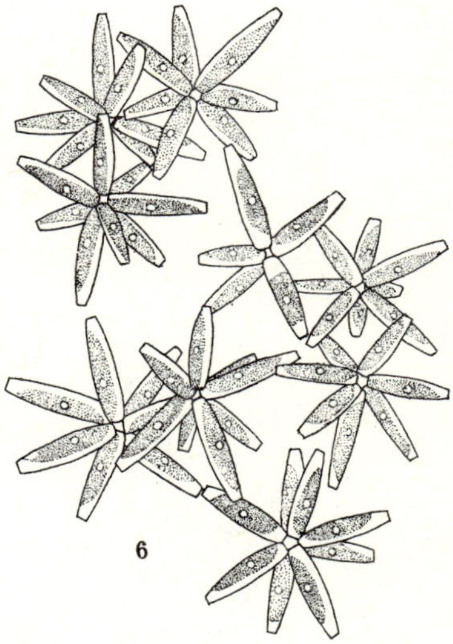

6

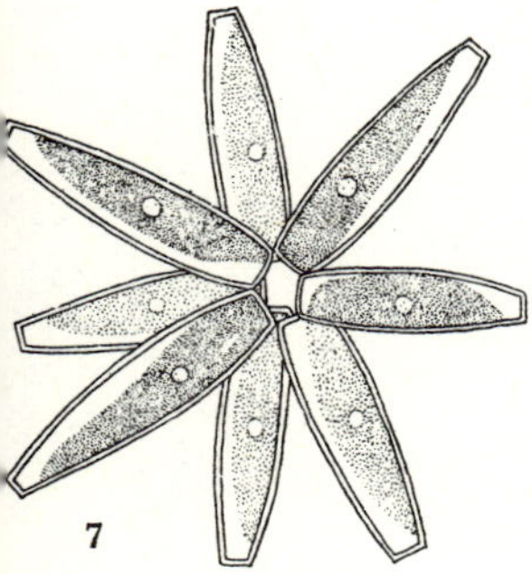

7

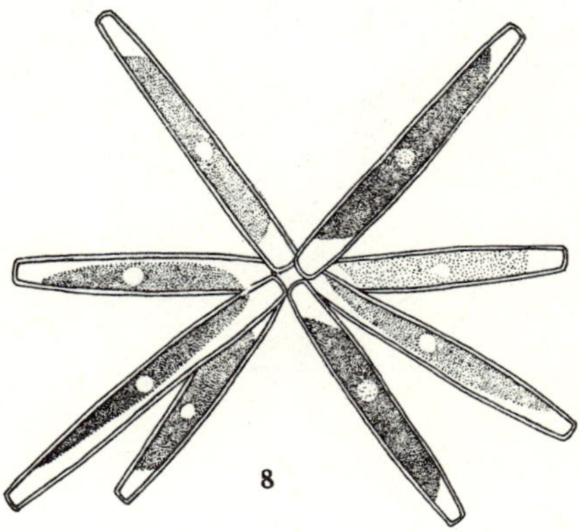

8

ALGAE

COCKAYNE, BOSTON

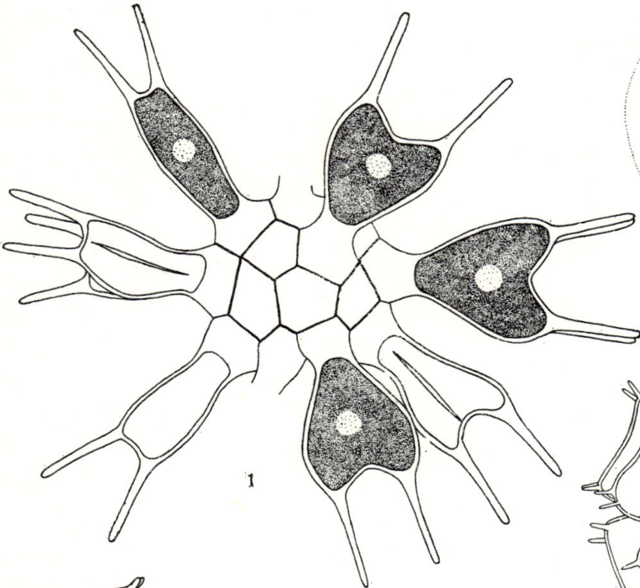

1

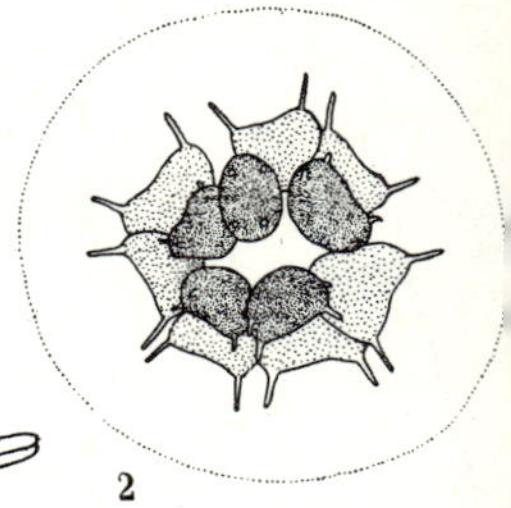

2

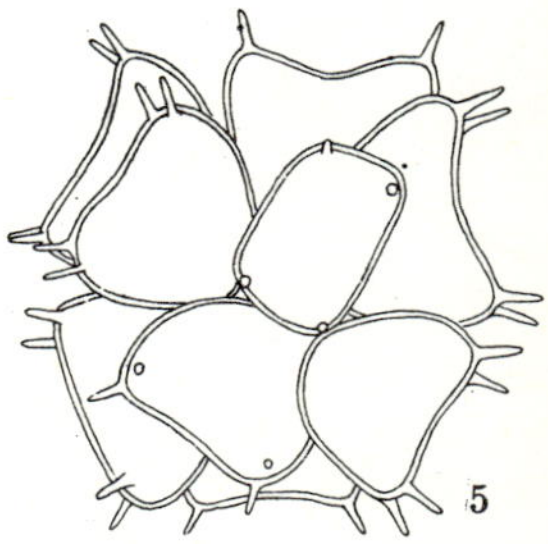

5

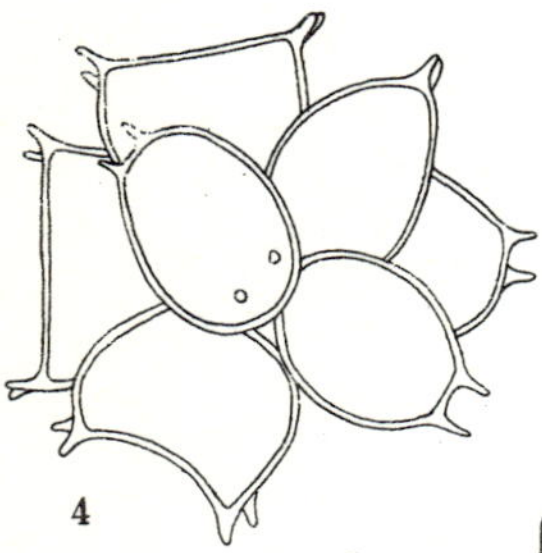

4

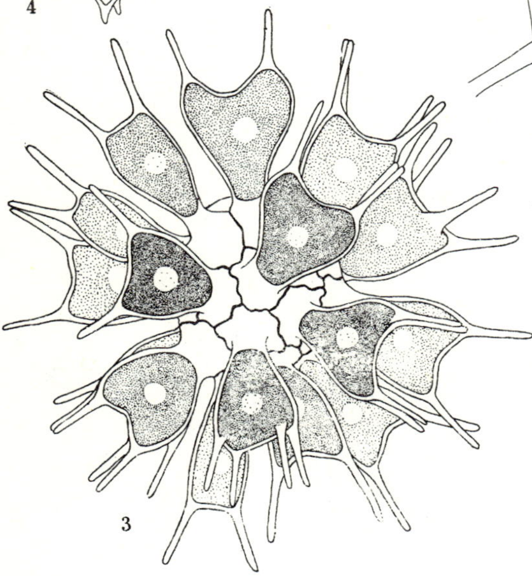

3

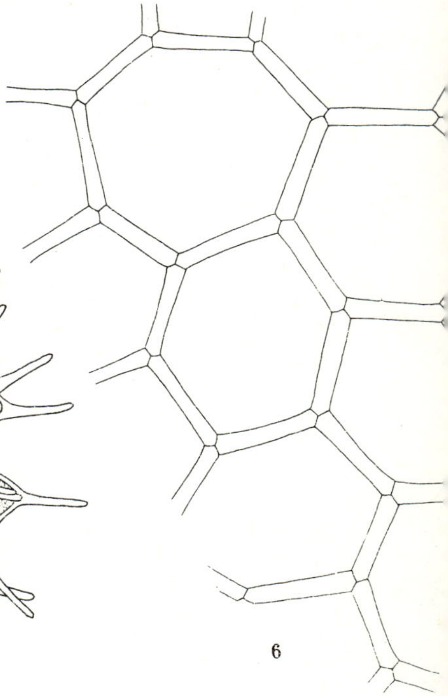

6

SMITH—ALGAE

COCKAYNE, BOSTON

PLATE 44.

PLATE 45.

		Page
Fig. 1.	*Hydrodictyon reticulatum* (x 266)........................	166
Figs. 2–6.	*Pediastrum simplex* var. *duodenarium* (x 333)...........	167
Fig. 7.	*Pediastrum integrum* (x 333)..........................	168
Figs. 8–10.	*Pediastrum integrum* var. *priva* (x 666)................	169
Fig. 11.	*Pediastrum araneosum* (x 333).........................	168
Figs. 12–13.	*Pediastrum araneosum* var. *rugulosum* (Fig 12, x 333, fig. 13, x 666)..	168

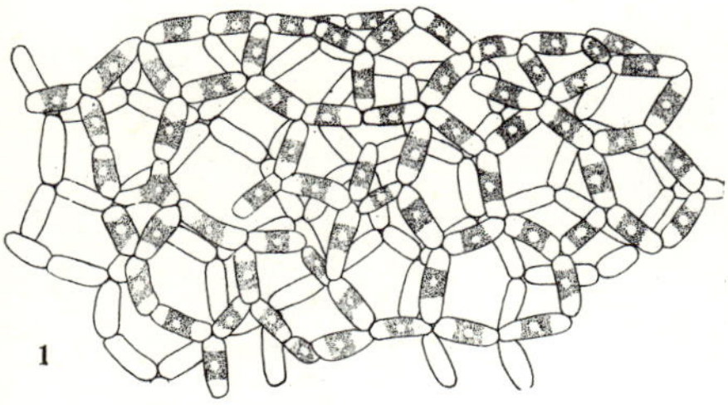
1

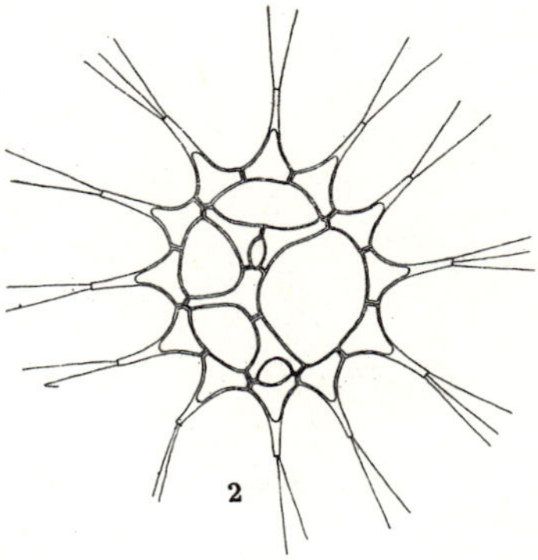
2

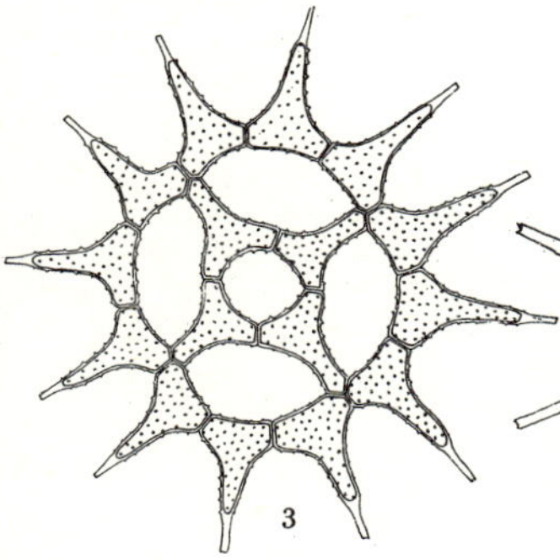
3

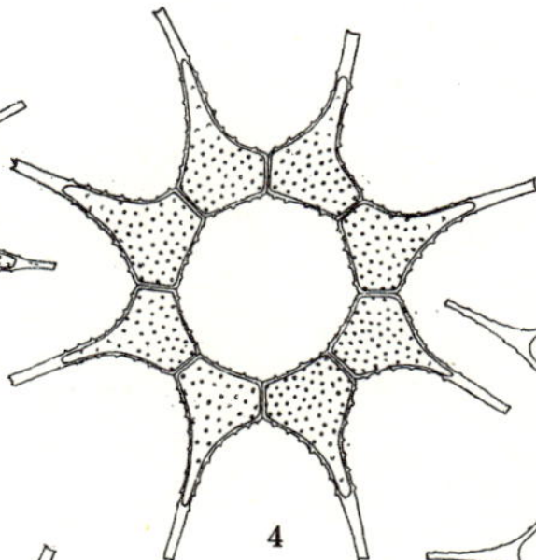
4

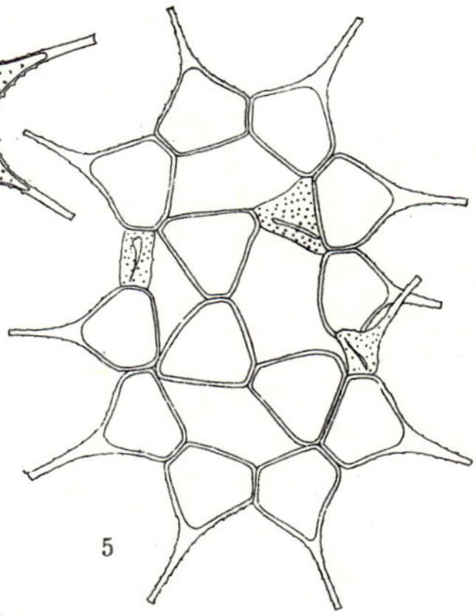
5

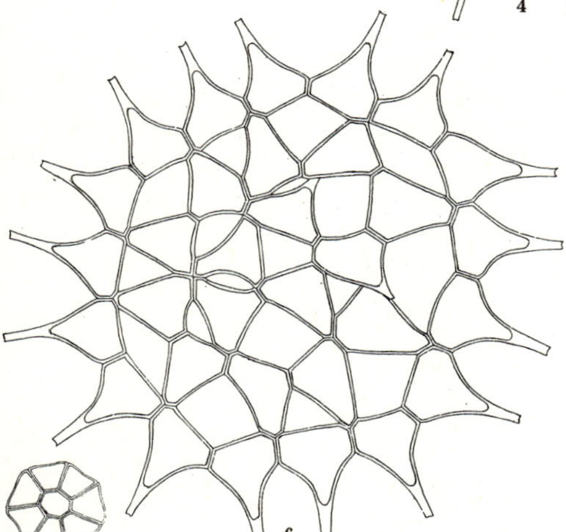
6

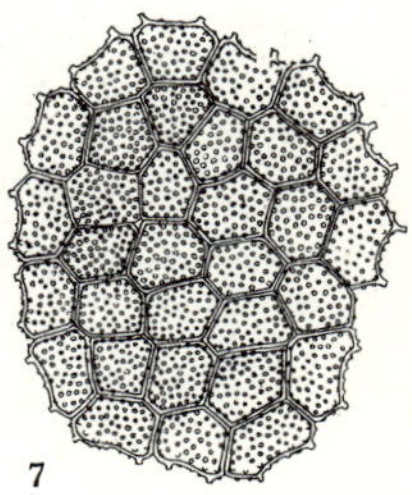
7

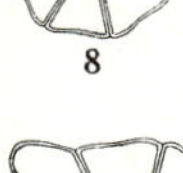
8

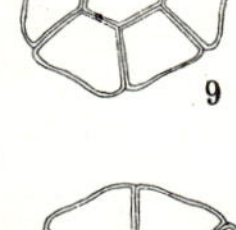
9

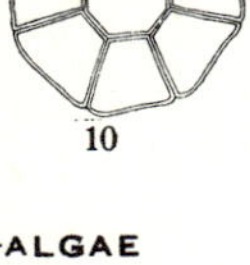
10

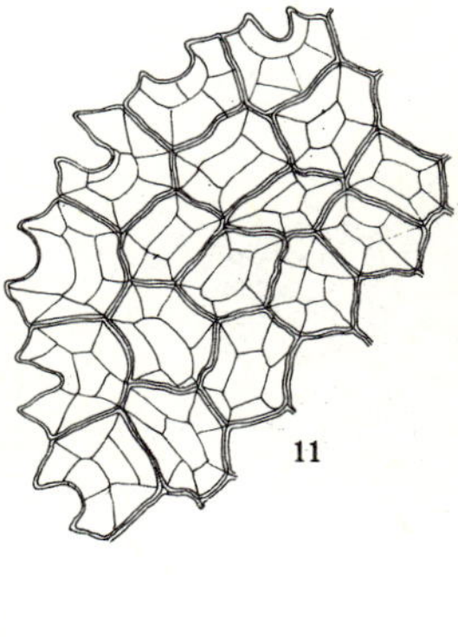
11

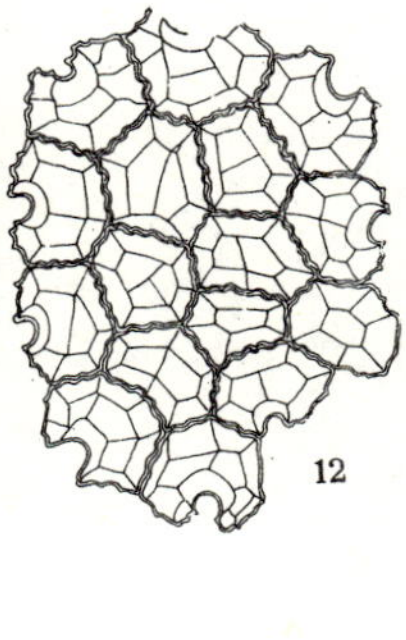
12

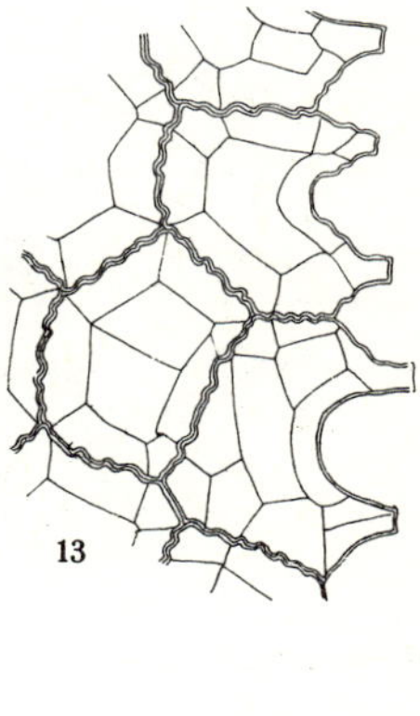
13

COCKAYNE, BOSTON

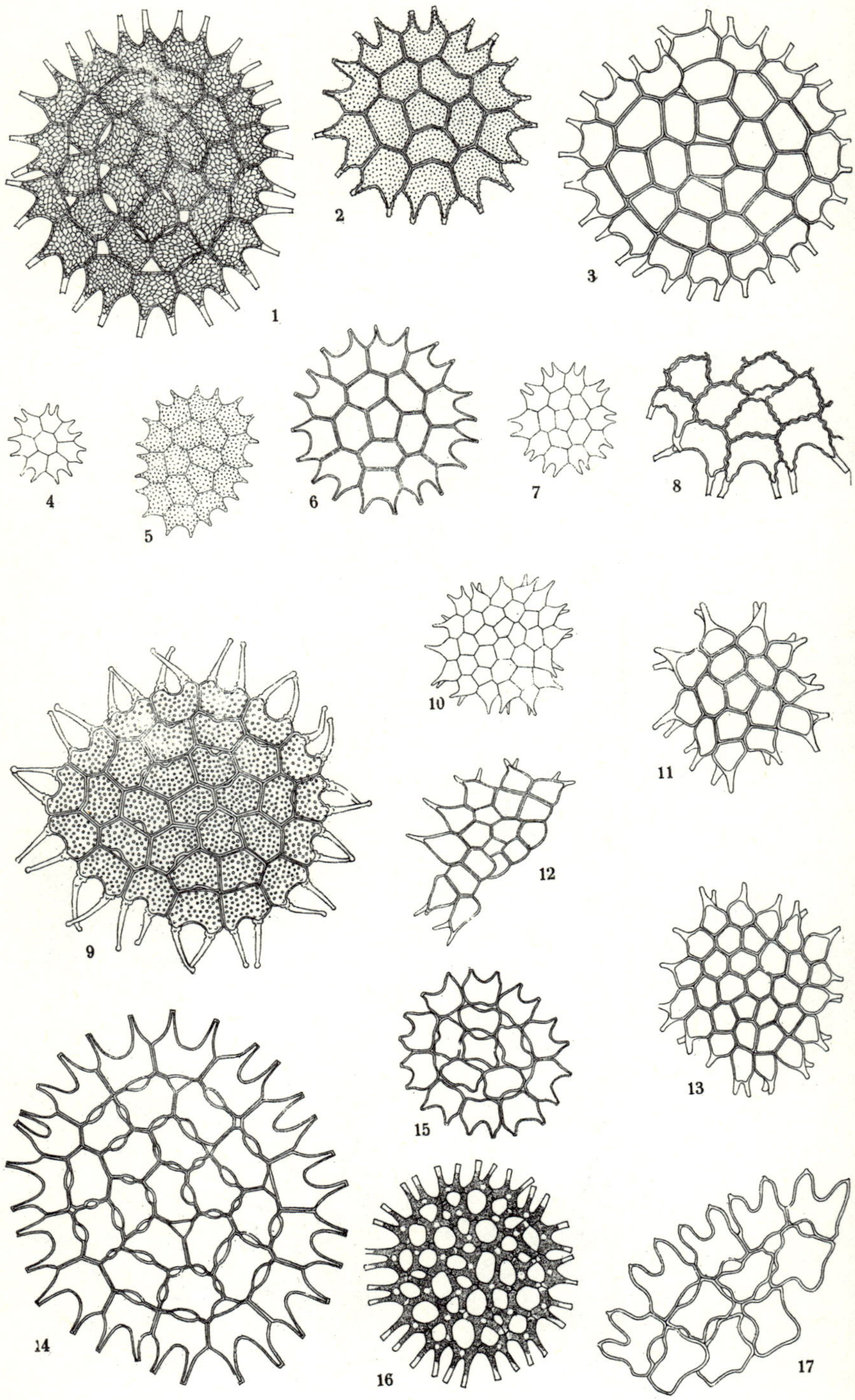

COCKAYNE, BOSTON

PLATE 46.

PLATE 47.

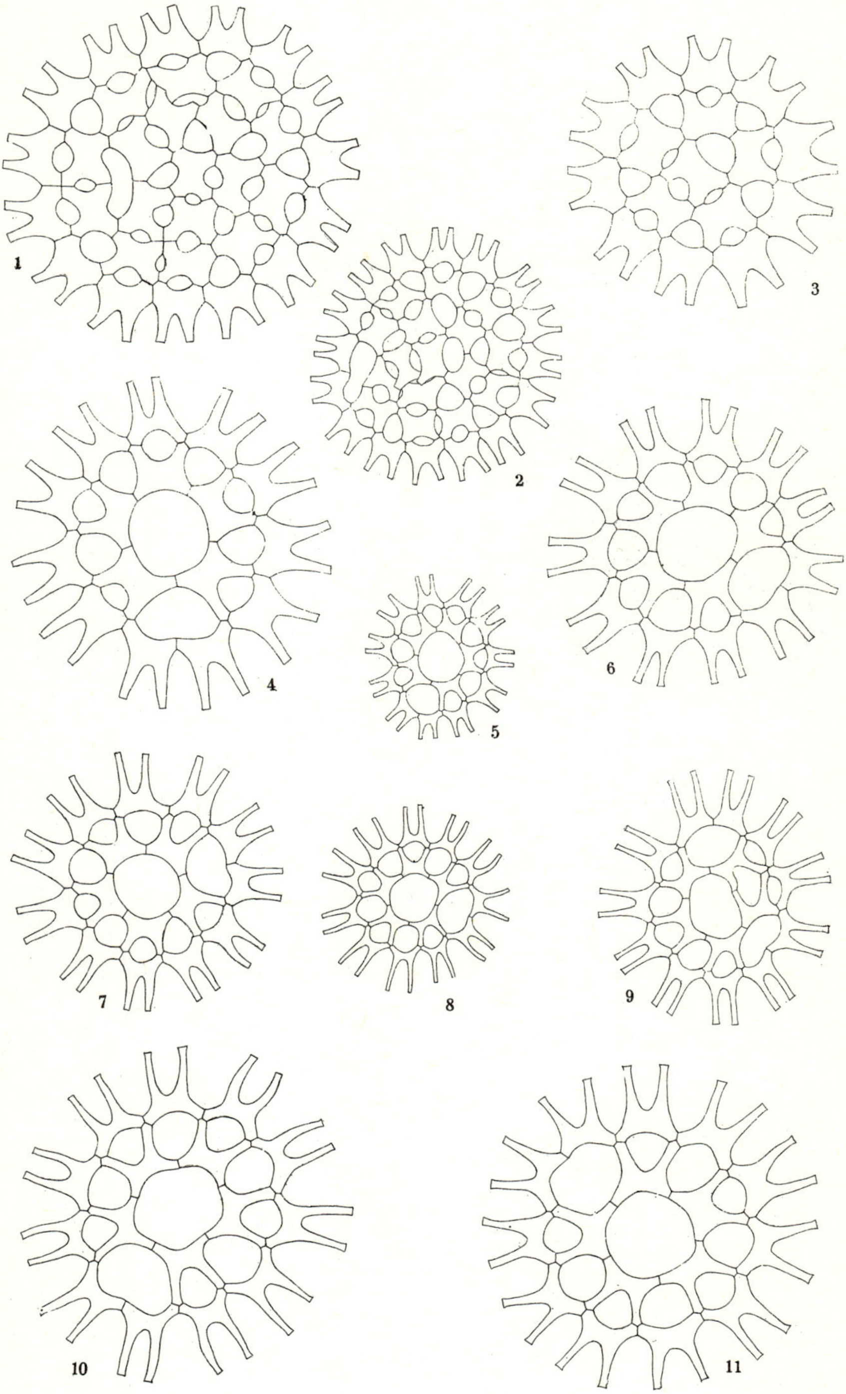

H—ALGAE

COCKAYNE, BOSTON

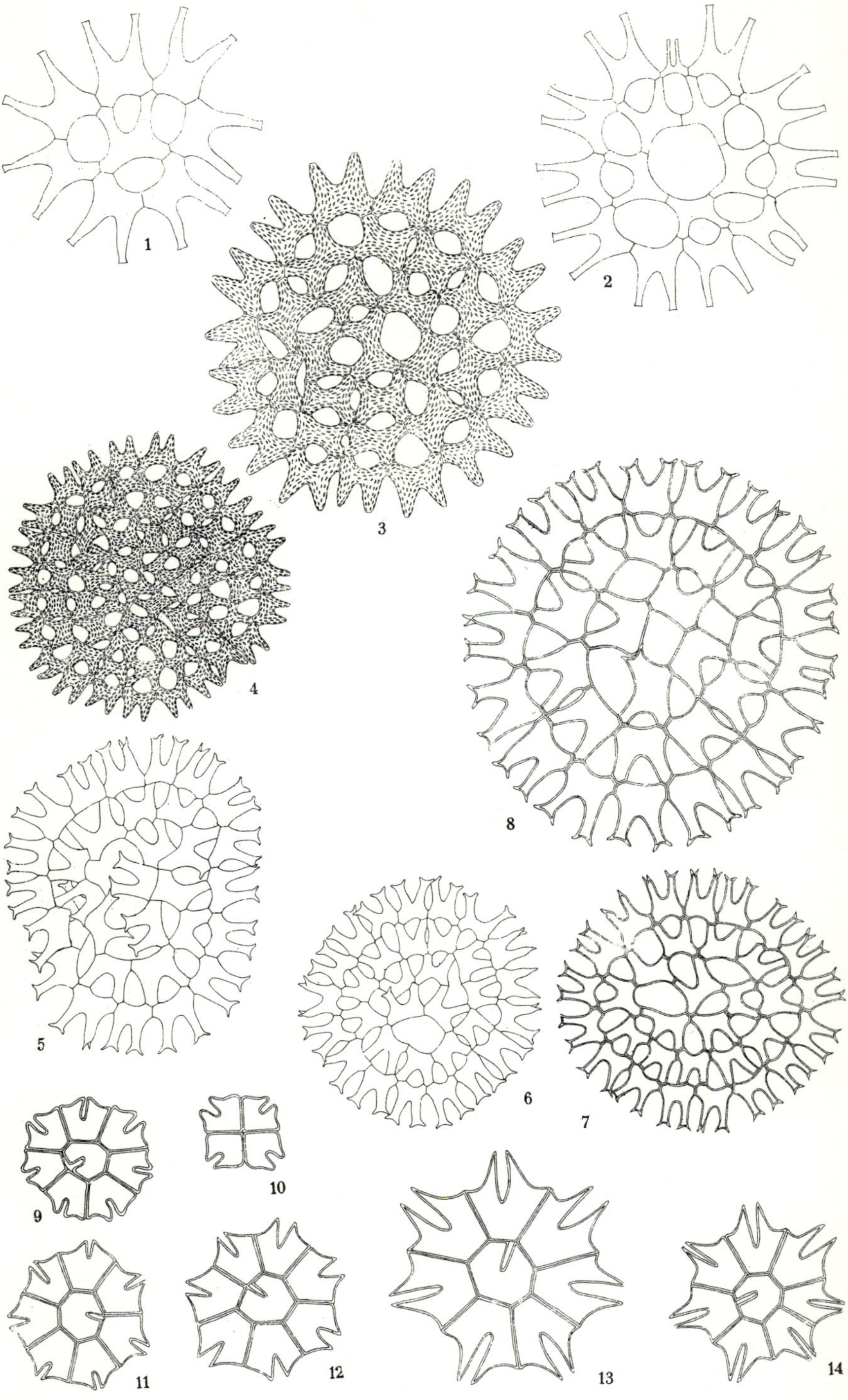

COCKAYNE, BOSTON

PLATE 48.

		Page
Figs. 1–2.	*Pediastrum duplex* var. *gracillimum* (x 333)	172
Figs. 3–4.	*Pediastrum duplex* var. *cohaerens* (x 333)	173
Figs. 5–8.	*Pediastrum biradiatum* (x 333)	173
Figs. 9–12.	*Pediastrum tetras* (x 666)	173
Figs. 13–14.	*Pediastrum tetras* var. *tetraödon* (x 666)	174

PLATE 49.

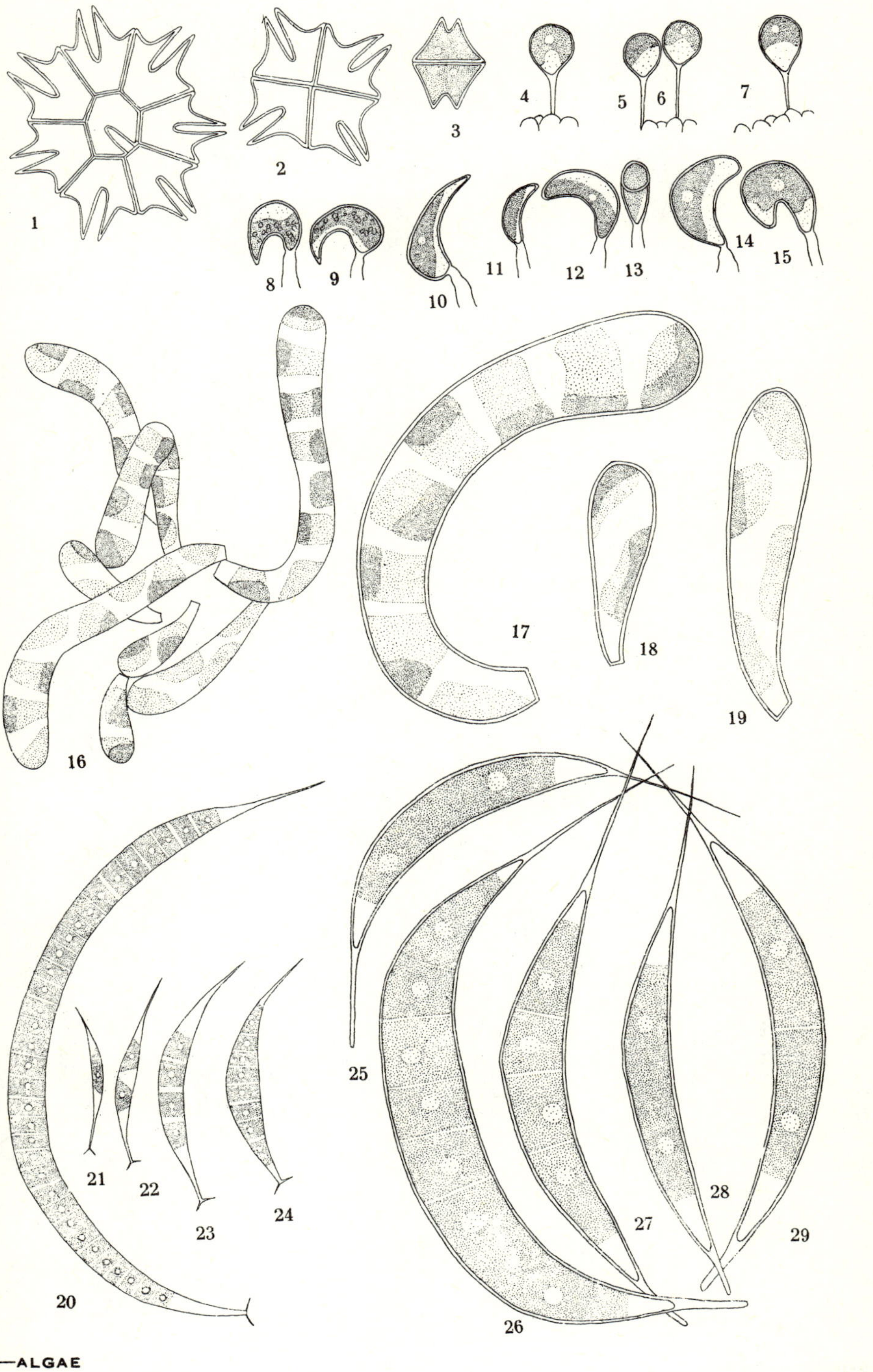

COCKAYNE, BOSTON

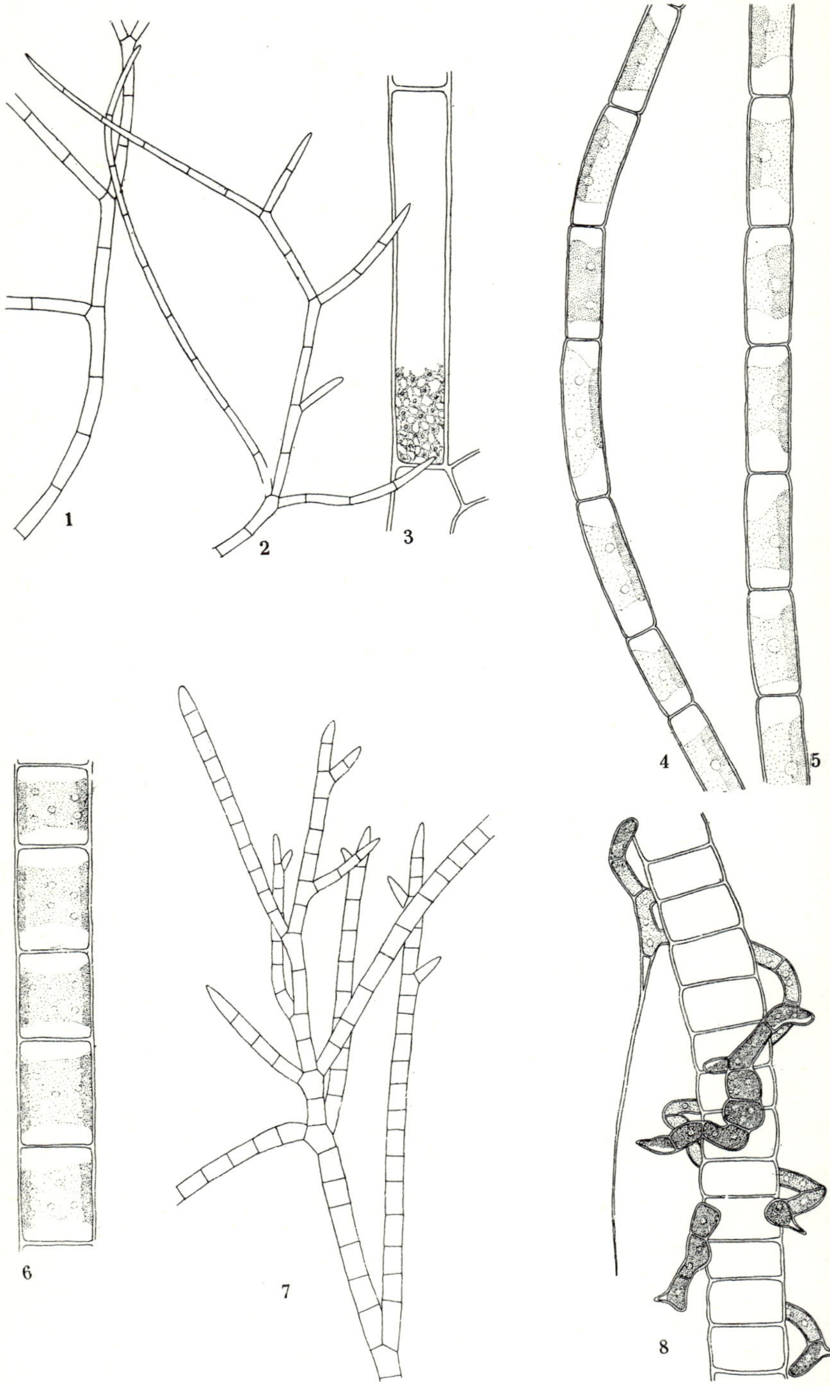

COCKAYNE, BOSTON

PLATE 50.

		Page
Figs. 1–3.	*Cladophora glomerata* (Figs. 1–2, x 50, fig. 3, x 300)	182
Figs. 4–5.	*Ulothrix subconstricta* (x 1000)	179
Fig. 6.	*Ulothrix zonata* (x 400)	179
Fig. 7.	*Stigeoclonium lubricum* (x 300)	180
Fig. 8.	*Aphanochaete vermiculoides* (x 400)	181

PLATE 51.

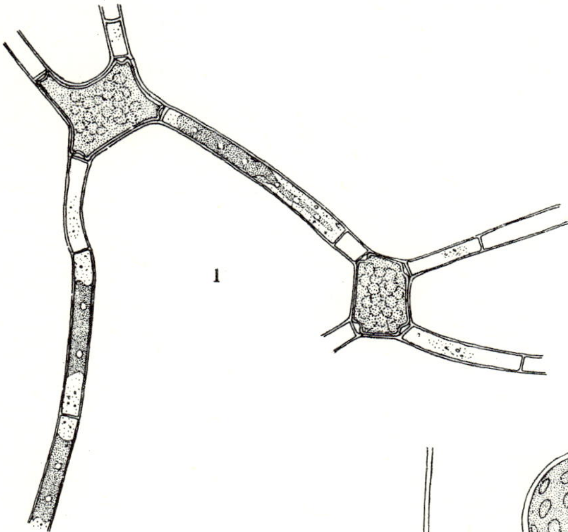

1

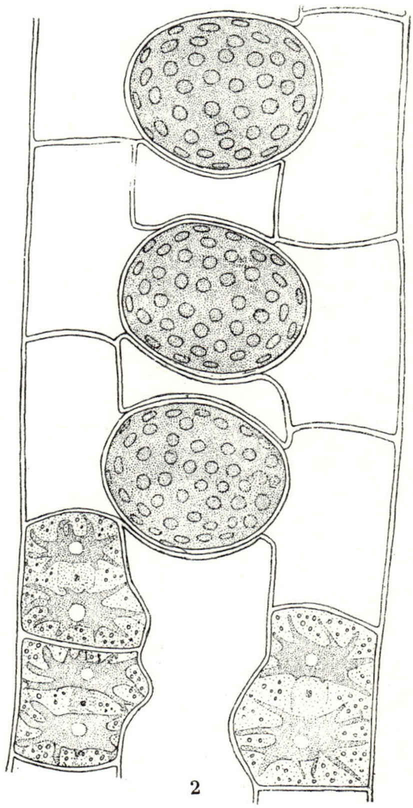

2

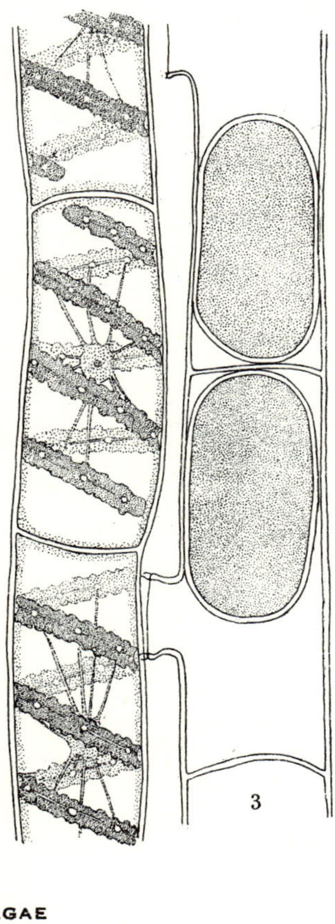

3

—ALGAE

COCKAYNE, BOSTON

INDEX

BULLETIN OF THE UNIVERSITY OF WISCONSIN

SERIAL NO. 1270 GENERAL SERIES NO. 1048

Phytoplankton of the Inland Lakes of Wisconsin

Part II
Desmidiaceae

BY
GILBERT MORGAN SMITH

This Bulletin Constitutes Part II of Bulletin 57 of the Wisconsin Geological and Natural History Survey

MADISON
PUBLISHED BY THE UNIVERSITY
1924

FOREWORD

In 1920 the Wisconsin Geological and Natural History Survey issued the first part of a report on the plankton algae of the inland lakes of Wisconsin[1] by Dr. Gilbert M. Smith. This report included all of the plankton algae, except the desmids and diatoms.

The present volume comprises the second part of this report, and includes the Desmidiaceae. A third part, including the Bacillarieae, is in preparation and will complete the investigations of the algae of Wisconsin lakes. The material has been sent for study and report to Mr. Fr. Meister, of Zürich, Switzerland.

The present report is published as one of the scientific studies of the University of Wisconsin as well as a bulletin of the Survey. The arrangement is made in accordance with Sec. 36.26 of the Wisconsin Statutes. It is especially appropriate that this particular bulletin should be printed by the University since Dr. Smith is associate professor in the department of botany and the report has been prepared in connection with the work of the department. The report is listed as Bulletin 57, Part 2, in the publications of the Survey, and as No. 1270 in the University Bulletins.

E. A. Birge,
President, University of Wisconsin,
President, Commissioners, Wisconsin Geological and Natural History Survey.

Madison, May 12, 1924.

---

[1]Phytoplankton of the Inland Lakes of Wisconsin. Part I. Wisconsin Geol. & Nat. Survey. Bull. No. 57; Sci. series, No. 12. Pp. 1–243; Pl. 1–51. Madison. 1920.

PREFACE.

Part 1 of this bulletin, which covers the plankton algae of Wisconsin, exclusive of the Desmidiaceae and Bacillarieae, appeared in 1920. Since this account of the plankton desmids is a continuation of the same investigation and is based upon the same collections, it is unnecessary to repeat the data given on pages 13–19 of Part 1, which describes the location and morphometry of the various lakes from which plankton samples were collected.

The method of treatment of the various species is the same as that followed in the first section. In addition to the citation of the original description of the alga there has been a liberal citation of articles that contain characteristic figures. A list of all stations for each alga is given, following its description, and the same method has been used for designating its frequency at each station as was used in the first part.

Since all drawings have been made from preparations the vertical and front views are from different individuals. For this reason each drawing has been given a different number. All of the figures of the plates have been drawn from Wisconsin specimens and all dimensions, unless expressly stated, are those of cells as found in the plankton of Wisconsin lakes. Since only varieties of a number of species are represented in the local flora descriptions of the type are taken from other sources. Where illustrations have been drawn of these types they are given as text figures and with an indication of the source of the material.

The difficulties involved in the determination of the species of Desmidiaceae are well known to all students of the family. In the preparation of this section the Wisconsin specimens have been compared with species found in the plankton of various European lakes and the opportunity of studying foreign plankton Desmidiaceae has added greatly to the certainty of determination of members of the local flora.

In the course of this investigation, through the kindness of Professor Yapp of the University of Birmingham, I have had free access to the collections and drawings of the late Professor G. S. West. While studying the material at Birmingham Dr. Nellie Carter placed at my disposal her material for the completion of W. & G. S. West's British Desmidiaceae. Dr. A. B. Rendle furnished facilities for examining the herbarium of the British Museum of Natural History and for consulting the rich library of that institution. The privilege of studying the Huitfeld–Kaas collections from Norwegian lakes was granted by Professor Wille of Christiania. Mr. Karl Strom supplemented these collections with his own from other Norwegian lakes and made determinations of certain puzzling forms. I am under especial obligations to Dr. O. Borge of Stockholm, Sweden, and to Charles Bullard of Cambridge, Massachusetts. Dr. Borge devoted a large portion of his time during my stay in Stockholm to checking over the determination of a majority of the Wisconsin Desmids and comparing them with his extensive iconography. Besides demonstrating a method of making mounts of individual Desmids, Mr. Bullard has allowed me to consult his preparations and his unpublished drawings of American Desmidiaceae and has placed at my disposal his wide experience with the group. Although the investigators named above have given assistance in determination of doubtful species, I assume full responsibility for all errors of determination.

The University of Wisconsin,
Department of Botany,
March, 1922.

Family *DESMIDIACEAE.*

Cells solitary or united in colonies. Naked or enclosed by a hyaline gelatinous envelope that is homogeneous or radially fibrillar. Colonies filamentous or growing in three dimensions. Shape of the cells exceedingly variable, generally with a pronounced constriction in the median portion dividing them into two similar semicells. Semicells always with a bilateral or a radial symmetry. Cells frequently with the walls having a characteristic ornamentation. Nucleus single and lying at the median constriction. Chloroplasts of each semicell generally solitary and axial, with one or more pyrenoids; more rarely with more than one axial or parietal chloroplast in each semicell or extending throughout the entire length of the cell.

Asexual reproduction by vegetative cell division, the divisions taking place at the isthmus and new semicells being formed at the point of division. Aplanospore formation known in a few instances.

Sexual reproduction isogamous, by aplanogametes.

KEY TO THE GENERA.

Cells not united in colonies.

Cell length at least twice the breadth.
  Without a median constriction.
    Chloroplast ribbon-like.
      Ribbon axial and straight........................GONATOZYGON (p. 4)
      Ribbon parietal and spiral.......................SPIROTAENIA (p.p.) (p. 5)
    Chloroplast not ribbon-like.
      Cells straight.
        Longitudinal ridges of chloroplast notched............NETRIUM (p. 6)
        Longitudinal ridges of chloroplast unnotched.
          Ridges straight.............................PENIUM (p.p.) (p. 7)
          Ridges spirally twisted...................SPIROTAENIA (p.p.) (p. 5)
      Cells curved and generally strongly attenuated at poles..CLOSTERIUM (p. 8)
  With a median constriction.
    Cell apices incised..................................TETMEMORUS (p. 20)
    Cell apices not incised.
      Bases of semicells not swollen....................PENIUM (p.p.) (p. 7)
      Bases of semicells swollen.
        Semicell bases with vertical plications...............DOCIDIUM (p. 17)
        Semicell bases without plications.
          Wall with transverse rings of verrucae..........TRIPLOCERAS (p. 18)
          Wall without transverse rings of verrucae....PLEUROTAENIUM (p. 13)
Cell length not over twice the breadth.
  Cells compressed.
    Apex of semicells incised.
      Lateral margins not deeply incised..................EUASTRUM (p. 21)
      Lateral margins deeply incised.............MICRASTERIAS (p.p.) (p. 40)

Apex of semicells not incised.
Lateral margins deeply incised..............MICRASTERIAS (p.p.) (p. 40)
Lateral margins not incised.
With spines.
Cell face with central protuberance.......XANTHIDIUM (p.p.) (p. 55)
Cell face without a central protuberance......ARTHRODESMUS (p. 125)
Without spines.
Angles continued in processes...........STAURASTRUM (p.p.) (p. 62)
Angles not continued in processes..............COSMARIUM (p. 28)
Cells radially symmetrical.
Face with a central protuberance..............XANTHIDIUM (p.p.) (p. 55)
Face without a central protuberance...........STAURASTRUM (p.p.) (p. 62)

Cells united to form permanent colonies.

Colonies filamentous.
Cell length several times the breadth...........PLEUROTAENIUM (p.p.) (p. 13)
Cell length never double the breadth.
Cells united by special apical processes.
Processes interlocking..........................SPHAEROZOSMA (p. 136)
Processes overlapping adjacent cells...............ONYCHONEMA (p. 135)
Cells not united by apical processes.
Without a girdle-like thickening at point of division.
Deeply constricted..........................SPONDYLOSIUM (p. 139)
Not deeply constricted.........................HYALOTHECA (p. 141)
With a girdle-like thickening at point of division.
Length greater than breadth....................GYMNOZYGA (p. 146)
Breadth greater than length.....................DESMIDIUM (p. 143)
Colonies not filamentous..............................COSMOCLADIUM (p. 37)

GONATOZYGON De Bary 1856.

Cells cylindrical, length many times the breadth, without a trace of a median constriction; sides of cells parallel, except near apices where they may be either slightly dilated or convergent; apices truncate, with angles only slightly rounded, at times subcapitate. Cell wall smooth, finely granulate, or clothed with minute spicules or spines. Chloroplasts axial ribbon-like strands extending from pole to pole or from each pole to middle of cell; pyrenoids fairly numerous and at regular intervals along the chloroplast. Cell division at any point in the cell wall, the cells frequently remaining attached at apices and forming temporary colonies of variable length. Colonies readily separating into component cells on being disturbed and always separating prior to conjugation.

Zygospores spherical and with smooth walls.

KEY TO THE SPECIES.

Cell wall granulate.................................(1) *G. monotaenium*
Cell wall with spines.
Spines the same diameter at base and apex...................(2) *G. pilosum*
Spines broader at bases..................................(3) *G. aculeatum*

1. Gonatozygon monotaenium De Bary. Pl. **52**, Fig. **1**.

in Rabenhorst, Die Algen Sachsens, Dec. 53; No. 536. 1856.

Gonatozygon Ralfsii De Bary, Unters. u. d. Fam. d. Conj. 76, *pl.* 4, *figs.* 23–25. 1858.

Cells usually quite long, cylindrical, length 25–30 times the breadth, apices somewhat dilated and truncate, angles slightly rounded. Cell wall minutely granulate except at extreme ends of cells. Chloroplasts axial and with about ten pyrenoids. (Tychoplanktont.)

Zygospores spherical, with smooth walls (De Bary).

Length 370–408 μ; breadth of median portion 8–11.5 μ, breadth apices 11–15 μ.

Diameter zygospores 25–28 μ (De Bary).

Crab (rrr).

2. Gonatozygon pilosum Wolle. Pl. **52**, Fig. **2**.

Bull. Torr. Bot. Cl. 9: 27, *pl.* 13, *fig.* 16. 1882; W. & G. S. West, Monogr. Brit. Desm. 1: 34, *pl.* 1, *figs.* 19–20. 1904.

Length about eighteen times the breadth; cell apices very slightly dilated, truncate, with angles slightly rounded. Cell wall very densely covered with very short, hair-like spicules. Chloroplasts axial, with about six pyrenoids. (Tychoplanktont.)

Zygospores unknown.

Length 142 μ; breadth without spines 7.5 μ; length spines 1–1.5 μ.

Rudolph (rrr).

3. Gonatozygon aculeatum Hastings. Pl. **52**, Fig. **3**.

Am. Mo. Micr. Jour. 13: 29 (with text figure). 1892; W. & G. S. West, Trans. Linn. Soc. 2nd. Ser. Bot. 5: 230, *pl.* 12, *figs.* 1–2. 1896.

Cells fairly long, cylindrical, length 15–20 times the breadth (without spines); apices somewhat dilated and truncate, apical angles very slightly rounded. Cell wall densely clothed with small, sharp, erect spines; truncate ends without spines. Chloroplasts with 6–9 pyrenoids. (Tychoplanktont.)

Zygospores unknown.

Length 211–212 μ; breadth of median portion 12.5 μ; breadth of apices 15–22 μ; length of spines 4.5–9.5 μ.

Bear (rrr), Dummy (rrr).

SPIROTAENIA De Brébisson 1848.

Cells of variable size, length at least five times the breadth, neither constricted nor compressed; cells usually cylindrical and with parallel sides, more rarely with sides attenuated or fusiform; poles commonly broadly rounded, infrequently subacute. Cell wall smooth and without pores. Chloroplast a fairly broad, parietal, spirally twisted ribbon extending from pole to pole; or axial and with longitudinal blades that are spirally twisted; in either case with several pyrenoids. Cell wall with no fixed point at which division takes place.

Zygospores generally spherical; with smooth, areolate, or short-spined walls.

The investigations of Lütkemüller show that the segmentation or lack of segmentation of the wall, a feature that is associated with the presence or absence of pores, is a fundamental point in the classification of the family. Most genera belong to the first type; *Gonatozygon, Spirotaenia* and *Netrium* are the only genera with unsegmented walls found in the Wisconsin plankton.

SPIROTAENIA CONDENSATA De Brébisson. Pl. **52**, Fig. **4**.

in Ralfs, Brit. Desm. 179, *pl.* 34, *figs.* 1A–1C. 1848; Archer, Quart. Jour. Micr. Sci. N.S. 7: 186, *pl.* 8, *figs.* 5–11. 1867.

Cells relatively large, straight or slightly curved, length 5–12 times the breadth; cell shape cylindrical, not attenuated towards the apices, poles broadly rounded. Cell wall smooth. Chloroplast a broad spiral ribbon, extending from pole to pole and almost completely covering the interior of the cell wall; pyrenoids fairly numerous, at least one every turn. (Tychoplanktont.)

Zygospores spherical; wall with a thin, inner, brownish layer and a thick, colorless, areolate, outer layer (Archer).

Length 150 μ; breadth 25 μ.

Diameter zygospores 60 μ (Archer).

Devils (rrr).

NETRIUM Nägeli 1849 (emend. Lütkemüller 1902).

Cells generally large, length at least three times the breadth, neither constricted in the median portion nor compressed; cell shape cylindrical to naviculoid, with lateral margins gradually attenuated to truncate apices, or parallel and then sharply convergent in the apical region. Cell wall always smooth and without pores. Chloroplasts axial, usually one in each half-cell, with 6–12 longitudinal, radiating plates that have strongly notched edges; pyrenoids always axial, usually a single long rod, but at times with the rod fragmenting to form an axial row of spherical to irregularly shaped pyrenoids. Cell division at no definite point.

Zygospores spherical, with smooth, thick walls.

W. & G. S. West (Monogr. Brit. Desm. **1**: 63) give the reasons for adopting Lütkemüller's separation of *Netrium* from *Penium* on the basis of the structure of the cell wall.

NETRIUM DIGITUS (Ehrenberg) Itzigsohn & Rothe. Pl. **52**, Fig. **5**.

in Rabenhorst, Die Algen Sachsens Dec. **50**: No. 508. 1856; W. & G. S. West, Monogr. Brit. Desm. **1**: 64, *pl.* 6, *figs.* 14–16. 1904; Carter, Ann. of Bot. 33: 227, *pl.* 14, *figs.* 13–15. 1919.

Cells usually large, length 3–5 times the breadth, neither constricted nor compressed; cell shape broadly naviculoid, with lateral margins convex and attenuated to truncate apices with rounded angles. Cell wall smooth. Chloroplasts axial, one in each half-cell, usually with eight radiating longitudinal plates that end at the cell wall and have deep notches that form rectangular teeth; pyrenoids usually a single elongate rod in the axial portion of the chloroplast. (Tychoplanktont.)

Zygospores spherical, with smooth thick walls (W. & G. S. West).

Length 177–230 μ; breadth of median portion 48 μ; breadth at apices 12.5–23 μ.

Diameter of zygospores 73.6 μ; thickness of zygospore wall 3.2 μ (W. & G. S. West).

Clear Crooked (rrr), Devils (1) (rrr), Devils (2) (rrr), Speese (rrr), Squirrel (rrr).

Exceedingly variable in size and relative proportions of length and breadth. The striking axial chloroplast with notched longitudinal rays is the most prominent feature of the alga.

Penium De Brébisson 1844.

Cells of variable length, length usually several times the breadth, generally straight, with median portion somewhat constricted or without a median constriction; sinus, when present, only a slight depression. Cells not compressed, shape generally cylindrical, at times fusiform to subelliptic, frequently somewhat attenuated at poles, apices usually truncate and with rounded angles. Point of cell division definite. Cell wall colorless to brownish, smooth to punctate or granulate, at times with the granules arranged in definite longitudinal series. Chloroplasts axial, with longitudinal radiating plates that extend to the cell wall; pyrenoids generally in an axial row at center of chloroplasts.

Zygospores spherical to rectangular, walls usually smooth.

KEY TO THE SPECIES.

Cell wall with rows of granules........................(1) ***P. margaritaceum***
Cell wall without granules ....(2) ***P. minutum***

1. Penium margaritaceum (Ehrenberg) De Brébisson. Pl. **52**, Fig. **6**.

in Ralfs, Brit. Desm. 149, *pl.* 25, *figs.* 1A–1C; *pl.* 33, *fig.* 3. 1848.

Cells of medium size, length about ten times the breadth, slightly constricted in median portion; cell shape cylindrical, with lateral margins very slightly attenuated at poles and apices broadly rounded. Cell wall with transverse lines at points of division; ornamented with vertical rows of minute granules, older portions of wall brownish. Chloroplasts with a fairly massive axial portion from which about ten longitudinal ridges radiate to the cell walls; pyrenoids fairly numerous and forming an axial row in the central portion of chloroplast. (Tychoplanktont.)

Zygospores spherical, with smooth walls (Ralfs).

Length 160 μ; maximum breadth 17.5 μ; breadth at apices 10 μ.

Diameter of zygospores 46 μ (Ralfs).

Lynx (rrr).

2. Penium minutum (Ralfs) Cleve. Pl. 52, Fig. 7.

Öfvers. Kgl. Sv. Vet.-Ak. Förh. 20: 493. 1864.
Docidium minutum Ralfs, Brit. Desm. 158, *pl.* 26, *figs.* 5A–5B. 1848.

Cells of medium size, length about ten times the breadth, slightly constricted, sinus a small V-shaped depression, isthmus very broad; semicells cylindrical, with lateral margins very slightly attenuated towards the apices, bases of semicells not inflated, apices truncate, with angles somewhat rounded. Vertical view circular. Cell wall smooth or finely punctate. Chloroplast axial, with an axial row of about six pyrenoids. (Tychoplanktont.)

Zygospores unknown.

Length 116 μ; breadth semicell base 10 μ, semicell apex 6.25 μ; breadth isthmus 7.5 μ.

Lindy (rrr).

The specimens observed were somewhat narrower than the typical *P. minutum* and approached the variety *gracile.* The lack of a basal swelling of the semicell, as well as a lack of apical or basal ornamentation, shows that the species cannot be referred to either *Docidium* or *Pleurotaenium.*

CLOSTERIUM Nitzsch 1817.

Cells greatly elongated and always, usually markedly, attenuated; in most cases curved, rarely straight; without a median constriction. Cell wall colorless or yellow to brown in color; smooth or with longitudinal striae that are usually ridges but may be series of granules; commonly with one or more transverse lines at middle of cell or at different points along the cell. Cells with a single chloroplast in each semicell that is either entire or with longitudinal ridges radiating from a comparatively slender central axis; generally with a single axial row of pyrenoids, more rarely with pyrenoids scattered throughout chloroplast. Cell apices hyaline in portion not occupied by chloroplasts and with a conspicuous vacuole that contains one or more granules that show a Brownian movement.

Zygospores spherical, ellipsoid, ovoid, or quadrangular; with a heavy, smooth or scrobiculate wall that is without processes.

KEY TO THE SPECIES.

Cells with a median girdle .... (1) *Cl. subtruncatum*
Cells without a median girdle.
  Poles not sharply attenuated.
    Curvature pronounced .... (2) *Cl. Venus*
    Curvature not pronounced.
      Inner margin tumid .... (3) *Cl. moniliferum*
      Inner margin not tumid .... (4) *Cl. acerosum*

Poles sharply attenuated.
  Wall longitudinally striated.
    Poles truncate....................................(8) *Cl. Ralfsii*
    Poles not truncate.
      Attenuated poles very narrow..................(9) *Cl. Kuetzingianum*
      Attenuated poles fairly broad.......................(10) *Cl. decorum*
  Wall smooth.
    Apices rounded.......................................(5) *Cl. gracile*
    Apices acute.
      Length at least 75 times the breadth..................(6) *Cl. aciculare*
      Length less than 35 times the breadth..................(7) *Cl. acutum*

1. Closterium subtruncatum W. & G. S. West. Pl. 52, Fig. 8.

Jour. Linn. Soc. Bot. 33: 159, *pl.* 8, *fig.* 4. 1897; Borge, Ark. för Botanik 1: 78, *pl.* 1, *fig.* 14. 1903.

Cells of medium size, length 9–12 times the breadth, curvature moderate, arc of curvature of outer margin about 65°; inner margin concave; somewhat attenuated towards the poles which are broadly truncate and slightly tumid in polar region of outer margin. Cell wall slightly thicker at poles, colorless to brownish yellow; with a median girdle and with numerous longitudinal striae. Chloroplast with an axial row of about eight pyrenoids. (Tychoplanktont.)

Zygospores unknown.

Distance between apices 225–300 μ; breadth of median portion 20–26 μ; about 12 striae in 10 μ.

Clam (rrr), Fish (rrr).

2. Closterium Venus Kützing. Pl. 52, Fig. 9.

Phycol. Germ. 130. 1845; Ralfs, Brit. Desm. 220, *pl.* 35, *fig.* 12. 1848; W. & G. S. West, Monogr. Brit. Desm. 1: 137, *pl.* 15, *figs.* 15–20. 1904.

Cells small, distance between apices 8–10 times the breadth; strongly curved, the outer margin making an arc of 150°–165°, inner margin with a curvature resembling that of the outer; gradually attenuated to subacute apices. Cell wall colorless, rarely colored; without a median girdle; smooth. Terminal vacuoles large. (Tychoplanktont.)

Zygospores oblong, ends concave and sides tumid in median portion; angles rounded; spore wall smooth. Halves of zygospore at times cruciately twisted (W. & G. S. West).

Distance between apices 70 μ; breadth 8 μ.

Length of zygospore 23–28.5 μ; breadth of zygospore 18–22 μ (W. & G. S. West).

Monona (rrr).

3. Closterium moniliferum (Bory) Ehrenberg. Pl. 52, Fig. 10.

Die Infusionsthierchen 91, *pl.* 5, *figs.* 16[1]–16[7]. 1838; Ralfs, Brit. Desm. 166, *pl.* 28, *figs.* 3A–3B. 1848; Lundell, Nova Acta Reg. Soc. Sci. Upsaliae, 3 Ser. 8: 80, *pl.* 5, *fig.* 14. 1871.

Cells fairly large, length 6–8 times the breadth, curvature of outer margin conspicuous, inner margin with central portion tumid and apical portions with approximately the same curvature as the outer margin; apices bluntly rounded. Cell wall colorless, without a median girdle; smooth. Chloroplasts with 6–8 longitudinal ridges and an axial row of 5–9 pyrenoids that vary greatly in size. Terminal vacuoles conspicuous. (Tychoplanktont.)

Zygospores ovoid, smooth; enclosed in a laminated gelatinous envelope (Lundell).

Distance between apices 244–252 μ; breadth 46–50 μ.

Monona (rr).

4. CLOSTERIUM ACEROSUM (Shrank) Ehrenberg. Pl. **53**, **Fig. 1.**

Symbolae Physicae seu Icones et Descriptiones etc. *pl.* 2, *fig.* 9. 1828 (*sec.* Nordst. Index Desm.); Ralfs, Brit. Desm. 164, *pl.* 27, *figs.* 2A–2L. 1848.

Cells fairly large, length 10–13 times the breadth; median portion straight and apical region slightly curved; outer margin curved but inner margin straight; gradually attenuated to subacute or rounded apices. Cell wall colorless, rarely brownish in older specimens; without a median girdle; smooth, or in rare cases with delicate striae. Chloroplast with two longitudinal ridges in front view of cell and an axial row of 10–12 pyrenoids. Terminal vacuoles with numerous granules. (Tychoplanktont.)

Zygospores spherical and smooth-walled (Ralfs).

Length 288–328 μ; breadth 19–22.5 μ.

Diameter of zygospore 63 μ (Ralfs).

Camp (rr), Center (rrr).

5. CLOSTERIUM GRACILE De Brébisson.

in Chevalier, Des microscopes et de leur usage 272. 1839; Mém. de la Soc. imp. d. Sc. nat. de Cherbourg **4**: 155, *pl.* 2, *fig.* 45. 1856; W. & G. S. West, Monogr. Brit. Desm. **1**: 166, *pl.* 21, *figs.* 8–12. 1904.

Cells fairly small, acicular, very delicate, length about thirty times the breadth; median portion straight and with sides parallel, terminal portions gracefully attenuated to blunt apices. Cell wall colorless; without a median girdle; smooth. Chloroplast with 5–7 pyrenoids.

Fig. 1. *Closterium gracile* De Brébisson (x 610). From a herbarium specimen in the British Museum of Natural History collected by De Brébisson.

Zygospores spherical to subquadrate; with a smooth wall.

Length 130–190 μ, breadth median portion 3.4–6 μ; breath apices 1.2–2.4 μ (W. & G. S. West).

Diameter of zygospore 20–27.5 μ (W. & G. S. West).

Not found in the plankton of Wisconsin lakes.

var. ELONGATUM W. & G. S. West. Pl. **53,** Fig. **2.**

Monogr. Brit. Desm. 1: 168, *pl.* 21, *figs.* 14–16. 1904.

Cells very much longer and with curvature confined to the apical regions. (Tychoplanktont.)

Length 418–480 μ; maximum breadth 8 μ; breadth near apices 5–8 μ.

Little Bass (rrr), Pokegama (rr).

6. CLOSTERIUM ACICULARE T. West.

Trans. Micr. Soc. London N. S. 8: 153, *pl.* 7, *fig.* 16. 1860 (in Quart. Jour. Micr. Sci. 8: 1860).

Cells greatly elongated, length many times the breadth; median portion narrow and straight, terminal portions curved; very gradually attenuated to acutely rounded apices. Cell wall colorless; without a median girdle; smooth. Chloroplast with an axial row of 6–8 pyrenoids.

Zygospores unknown.

Length 500 μ; greatest breadth 5 μ (T. West).

Not found in the plankton of Wisconsin lakes.

var. SUBPRONUM W. & G. S. West. Pl. **53,** Fig. **3.**

Monogr. Brit. Desm. 1: 175, *pl.* 23, *figs.* 4–5. 1904.

Cells longer and with attenuation of terminal portion more delicate. Chloroplast extending half way from cell center to cell apex. (Facultative planktont.)

Length 540–610 μ; breadth 7.5 μ.

Browns (r), Green (r), Mendota (r).

7. CLOSTERIUM ACUTUM (Lyngbye) De Brébisson. Pl. **53,** Fig. **4.**

in Ralfs, Brit. Desm. 177, *pl.* 30, *figs.* 5A–5E; *pl.* 34, *fig.* 5. 1848.

Cells small, generally curved, rarely straight; very gradually attenuated to subacute or rounded apices. Cell wall colorless; without a median girdle. Chloroplasts without longitudinal ridges and with an axial row of 2–4 pyrenoids. (Tychoplanktont.)

Zygospores oblong, with ends very deeply emarginate and sides slightly convex; angles rounded and forming projections that lie parallel to the long axis of the zygospore (Ralfs).

Length 177 μ; breadth 3.75 μ.

Length of zygospore 43 μ; breadth of zygospore 13 μ (Ralfs).

Green (rrr).

8. CLOSTERIUM RALFSII De Brébisson.

in Jenner, Flora of Tunbridge Wells X. 1845 (Ref. Nordst. Index Desm.); in Ralfs, Brit. Desm. 174, *pl.* 30, *figs.* 2A–2B. 1848.

Cells large, length 6–8 times the breadth, somewhat curved, arc of curvature of outer margin about 45°, inner margin tumid in median portion; more sharply attenuated at extremities and with apices bluntly truncate. Cell wall yellow-brown, with apical portion a deeper shade; without a median girdle, with numerous fine striae extending to the cell apex. Chloroplast ridged (?), with an axial row of 5–6 pyrenoids.

Zygospores unknown.

Length 317 μ; breadth 54 μ (Ralfs.)

Not found in the plankton of Wisconsin lakes.

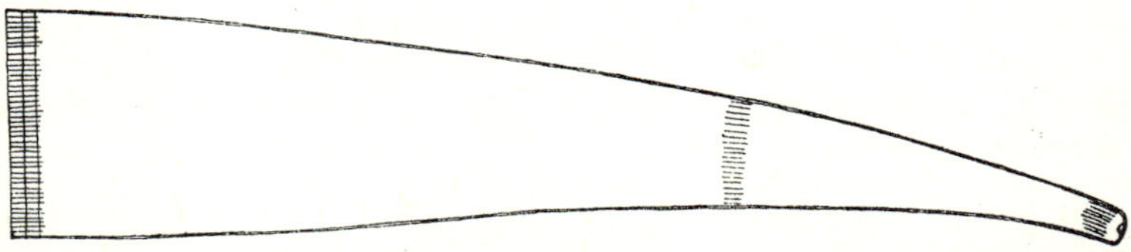

Fig. 2. *Closterium Ralfsii* De Brébisson (x 350). From a herbarium specimen in the British Museum of Natural History collected by De Brébisson.

var. HYBRIDUM Rabenhorst. Pl. **53,** Fig. **5.**

Kryptogamenfl. Sachs. Abt. 1: 174. 1863; W. & G. S. West, Monogr. Brit. Desm. 1: 183, *pl.* 24, *figs.* 8–13. 1904.

Cells narrower than the foregoing and with inflation of the inner margin less prominent. (Tychoplanktont.)

Zygospores double; subovoid, with smooth walls (W. & G. S. West).

Length 430 μ; breadth in median portion 38 μ, at apices 9 μ; 10 striae in 17.5 μ.

Diameter of zygospores 51.5–80 μ (W. & G. S. West).

Found (rrr), Jag (rrr).

9. CLOSTERIUM KUETZINGII De Brébisson. Pl. **53,** Fig. **6.**

Mem. d. l. Soc. imp. d. Sc. nat. de Cherbourg 4: 156, *pl.* 2, *fig.* 40. 1856; Raciborski, Sprawozd. Kom. fizyjogr. Akad. um. w Krakowie 19: 8, *pl.* 1, *fig.* 2. 1884; W. & G. S. West, Monogr. Brit. Desm. 1: 186, *pl.* 25, *figs.* 6–11. 1904.

Cells long, of medium breadth; straight, median portion naviculoid and ends continued in processes with parallel sides and rounded apices; terminal portion of processes somewhat curved. Cell wall colorless to straw colored, without a median girdle but longitudinally striated; striae delicate, 15–20 visible across face of cell, extending to base of terminal processes. Chloroplast occupying striated portion of cell, without longitudinal ridges, with an axial row of 4–6 pyrenoids. Terminal vacuoles conspicuous. (Tychoplanktont.)

Zygospores subrectangular, with angles truncate and sides between angles concave (Raciborski, W. & G. S. West).

Cells 367–398 μ long; 18–20 μ broad in median portion, 2.5 μ broad at apices; 10 striae in 10 μ.

Greatest breadth of zygospore 48–51 μ, shortest breadth 35–36 μ (W. & G. S. West).

Carson (rrr), Crab (rrr), Meronk (rrr).

10. Closterium decorum De Brébisson. Pl. 53, Fig. 7.

Mem. d. l. Soc. imp. d. Sc. nat. de Cherbourg 4: 151, *pl.* 2, *fig.* 39. 1856; Kirchner, Kryptogamenfl. v. Schles. 2[1]: 139. 1878; W. & G. S. West, Monogr. Brit. Desm. 1: 184, *pl.* 17, *figs.* 7–8; *pl.* 25, *figs.* 1–3. 1904.

Cells fairly large, length 12–15 times the breadth, median portion straight, terminal portion curved; median area narrowly fusiform and attenuated towards the apices, terminal portions with subparallel sides but fairly stout and with apices truncate. Cell wall yellowish-brown; without a median girdle but longitudinally striated, the striae extending to cell apices; 15–18 striae visible across the face of the median portion of the cell. Chloroplasts ridged and with an axial row of 7–10 pyrenoids. (Tychoplanktont.)

Zygospores angular (Kirchner); subspherical and smooth-walled (W. & G. S. West).

Cells 450 μ long; 27.5 μ broad in median portion, 5 μ broad at apices; 10 striae in 14 μ.

Diamond (rrr).

PLEUROTAENIUM Nägeli 1849.

Cells usually quite large, length several times the breadth, median constriction well defined but not deep, cells always straight; semicells never compressed, commonly cylindrical and somewhat attenuated towards the poles, very rarely ovate; poles always truncate, usually with angles more or less rounded and, except in rare cases, with a ring of mamillate to conical tubercules; bases of semicells inflated, at times with lateral margins above the inflations undulate. Cell walls rarely smooth, usually finely to coarsely punctate, never with vertical plications on swollen bases of semicells; walls at times of uneven thickness, the thinner portions producing a somewhat definite pattern. Chloroplasts usually numerous and parietal, in the form of straight to undulate bands that extend the length of the semicell and contain numerous pyrenoids; rarely axial and with longitudinal plates and an axial row of pyrenoids. Parietal chloroplasts at times fragmenting into small rhomboidal masses each containing a single pyrenoid.

Zygospore formation rare. Zygospores spherical and with smooth walls.

KEY TO THE SPECIES.

Lateral walls the same thickness.
- Poles without tubercules.................................(1) *P. Trabecula*
- Poles with tubercules.
  - Walls with whorls of tumors.................................(5) *P. nodosum*
  - Walls without whorls of tumors.
    - Strongly attenuated at poles.................................(4) *P. truncatum*
    - Not strongly attenuated at poles.
      - Sides undulate above basal inflation.................(3) *P. Ehrenbergii*
      - Sides not undulate above inflation.................(2) *S. subcoronulatum*

Lateral walls with thin areas.................................(6) *P. Trochiscum*

1. Pleurotaenium Trabecula (Ehrenberg) Nägeli.

Gatt. einz. Algen 104, *pl.* 6, *fig.* A. 1849; W. & G. S. West, Monogr. Brit. Desm. 1: 209, *pl.* 30, *figs.* 11–13. 1904.

Cells fairly large, length 11–15 times the breadth, median constriction well defined; semicells cylindrical, very slightly attenuated towards the poles, with a well defined basal inflation and a single undulation above the basal inflation; apices truncate, with well rounded angles, smooth and without tubercules. Cell wall punctate.

Zygospores ellipsoidal, with smooth walls (W. & G. S. West).

Cells 390–664 μ long; 26–46 μ broad at base of semicells, 16–32 μ at apices.

Zygospores 70 μ long; 48 μ broad (W. & G. S. West).

Not found in the plankton of Wisconsin lakes.

var. rectum (Delponte) W. & G. S. West. Pl. 54, Fig. 1.

Monogr. Brit. Desm. 1: 212, *pl.* 30, *figs.* 9–10. 1904.

Cells smaller, length generally proportionally greater; semicells without undulations above the basal inflations. (Tychoplanktont.)

Cells 284 μ long; 20 μ broad at base, 12.5 μ broad at apex; isthmus 15 μ broad.

Lindy (rrr).

2. Pleurotaenium subcoronulatum (Turner) W. & G. S. West.

Trans. Linn. Soc. 2nd. Ser. Bot. 5: 44, *pl.* 5, *fig.* 33. 1895.

***Docidium subcoronulatum* Turner, Kgl. Sv. Vet.–Ak. Handl. 25, No. 5: 29, *pl.* 3, *fig.* 1. 1893.**

Cells fairly large, length about 12 times the breadth, united to form filaments, median constriction fairly well marked; semicells cylindrical, with bases inflated and lateral margins slightly undulate in the lower third of the semicell, lateral margins very slightly attenuated towards the apices; apices dilated, truncate, with rounded angles and a transverse ring of elliptic tubercules (about 10 visible in front view of cell). Cell wall smooth or finely punctate.

Zygospores unknown.

Cells 407–453 μ long; 29–37 μ broad at base of semicells, 25–29 μ broad at apices (W. & G. S. West).

Not found in the plankton of Wisconsin lakes.

var. detum W. & G. S. West. Pl. 54, Fig. 2–4.

Trans. Linn. Soc. 2nd. Ser. Bot. 5: 235, *pl.* 13, *figs.* 2–3. 1896; G. S. West, Jour. Linn. Soc. Bot. 38: 110. 1907.

Cells with a distinct ligature-like constriction just below the apex that gives the apex a distinctly dilated appearance. Cells united to form filaments. (Facultative planktont.)

Cells 385–567 μ long; 24–30 μ broad at base of semicells, 16–22 μ at apical constriction, 24–28 μ at apex (with tubercules); isthmus 23–24 μ broad.

Bass (1) (rrr), Bass (3) (rrr), Diamond (rrr), Howey (sss), Red Bass (rrr).

The ligature-like constriction lies about a tenth of the distance from the apex to the semi-cell base end and is the chief distinction between the variety and the type. This constriction is variable in depth and is much more prominent in some specimens than in others. The interlocking of the apical tubercules is a very constant feature and the cells are united in filaments that have a considerable degree of permanence, in fact if a break occurs in a filament it is more commonly at the isthmus than at the point of contact between adjacent cells. This union of cells to form filaments is of frequent occurrence in the tropics but is very rare in temperate regions. The alga was found in sufficient abundance to warrant the assumption that it can multiply under plankton conditions.

3. PLEUROTAENIUM EHRENBERGII (De Brébisson) De Bary. Pl. **54**, Figs. **5–8**.

Unters. u. d. Fam. d. Conj. 75. 1858; W. & G. S. West, Monogr. Brit. Desm. 1: 205, *pl.* 29, *figs.* 9–11; *pl.* 30, *fig.* 1. 1904.

Cells fairly large, length 15–30 times the breadth at the base, slightly constricted, sinus a shallow depression; semicells cylindrical, with lateral margins slightly attenuated, the convergence being more pronounced in the apical region; basal inflation conspicuous and with a single undulation above it that looks like a second basal inflation; apex truncate and with angles slightly rounded, apex with a subapical ring of 8–9 (5 generally visible in front view) conical papillae with very blunt apices. Chloroplasts parietal, numerous, forming undulate longitudinal bands; with several pyrenoids. (Tychoplanktont.)

Zygospores spherical to subspherical, with smooth walls (W. & G. S. West).

Cells 350–650 μ long; 25–35 μ broad at base of semicells, 15–24 μ at apices; isthmus 23–30 μ broad.

Dutchmans (rrr), Little Doctor (rrr), Meta (rrr), Muskallonge (rrr), Oak (rrr), Sand (rrr), Speese (rrr).

The organism is subject to considerable variation in size, in relative proportions between length and breadth as well as in the apical attenuation. In typical specimens there is but a single undulation above the basal inflation but it is not at all uncommon to have more than one of these inflations. Specimens like those shown in Figures **7–8**, with the undulations extending half way to the apex from the bases of the semicells, approach the variety *undulatum*.

var. UNDULATUM Schaarschmidt. Pl. **54**, Fig. **9**.

Magyar. Tudom. Akad. Math. s. Természettud. Közelémenyek 18: 278, *pl.* 1, *fig.* 21. 1882 (Ref. Nordst. Index Desm.); W. & G. S. West, Trans. Linn. Soc. 2nd. Ser. Bot. 6: 145, *pl.* 18, *fig.* 28. 1902.

Lateral margins gently undulate from basal inflation to the semicell apex. (Tychoplanktont).

Cells 355 μ long; 24 μ broad at semicell base, 16.5 μ at apex; isthmus 22 μ broad.

Speese (rrr).

The dimensions of the variety as found in Wisconsin lakes agree very closely with those that W. & G. S. West give for their Ceylon specimens.

var. ELONGATUM W. West. Pl. **54,** Fig. **10.**

Jour. Linn. Soc. Bot. **29:** 119. 1892; W. & G. S. West, Monogr. Brit. Desm. **1:** 207, *pl.* 30, *fig.* 3. 1904.

Length about 25 times the breadth; cells somewhat narrower than the type. (Tychoplanktont.)

Cells 560 μ long; 24 μ broad at base of semicells, 17 μ at apex; isthmus 20 μ broad.

Jenny (rrr).

4. PLEUROTAENIUM TRUNCATUM (De Brébisson) Nägeli. Pl. **55,** Figs. **1–2.**

Gatt. einz. Algen 104. 1849; W. & G. S. West, Monogr. Brit. Desm. **1:** 203, *pl.* 29, *figs.* 3–4. 1904.

Cells large, length about 8 times the greatest breadth, median constriction well defined; semicells cylindrical in lower half and then attenuated to a truncate apex that is about half the breadth of the base of the semicells, semicell bases with a single, distinct, basal inflation; lateral margins convex, apex slightly tumid and with rounded lateral angles; apical tubercules small flattened elevations, about seven visible in front view of cell. Chloroplasts numerous, about eight visible in front view of cells; parietal, slightly undulate, longitudinal bands with numerous pyrenoids. (Tychoplanktont.)

Zygospores unknown.

Cells 535–577 μ long; 70–72.5 μ broad at base of semicells, 32.5–77 μ at apex; isthmus 60–64 μ broad.

Carroll (rrr), Dutchmans (rrr).

In one specimen the lateral margins possessed two undulations above the swollen bases of the semicells. Unfortunately no other cells were collected from this station so that it is impossible to state whether or not this undulation in the lower part of the semicell is a constant feature.

5. PLEUROTAENIUM NODOSUM (Bailey) Lundell. Pl. **54,** Fig. **11.**

Nova Acta Reg. Soc. Sci. Upsaliae 3 Ser., **8:** 90. 1871; W. & G. S. West, Monogr. Brit. Desm. **1:** 214, *pl.* 31, *figs.* 3–6. 1904.

Cells large, length about 8 times the breadth, median constriction conspicuous; semicells cylindrical, with lateral margins convergent towards a truncate apex and with four tumors on each margin; with four transverse whorls of mamillate tumors, one whorl at semicell base and the other three equidistant between basal whorl and apex;

polar region dilated, with apex slightly tumid and angles rounded; pole ornamented with a subapical whorl of conspicuous, conical, upwardly curved teeth (5 generally visible in front view of cell) that do not extend beyond the apex. Cell wall smooth or punctate. Front view with about four longitudinal, parietal, slightly undulate chloroplasts that contain numerous pyrenoids. (Tychoplanktont.)

Zygospores unknown.

Cells 337–410 μ long; 47.5–50 μ broad at inferior whorl, 37–42.5 μ at superior whorl, 25–27.5 μ at apex (without spines); breadth isthmus 30–32.5 μ.

Little Doctor (rrr), Tank (rrr).

Sharply differentiated from the other species by the whorls of mamillate protuberances.

6. Pleurotaenium Trochiscum W. & G. S. West.

Trans. Linn. Soc. 2nd. Ser. Bot. 5: 235, *pl.* 13, *figs.* 4–5. 1896; *ibid.* 6: 142, *pl.* 19, *figs.* 7–8. 1902.

Cells of medium size, length 10–16 times the breadth, constriction well marked; semicells cylindrical, with sides slightly attenuated towards the poles, basal inflation pronounced; poles truncate, flattened, with rectangular angles, and without polar tubercules. Cell wall with 12–15 transverse rings of irregular to quadrangular areas where the wall is not internally thickened; thin areas on basal inflation much smaller and more irregular, thin areas of polar region irregularly elongate.

Zygospores unknown.

Cells 265–468 μ long; 20–42 μ broad at base of semicells, 21–29 μ at apex. (W. & G. S. West.)

Not found in the plankton of Wisconsin lakes.

Var. **tuberculatum** var. nov. Pl. **55,** Fig. **3.**

Apices with a ring of conical to rounded tubercules (about 5 visible in front view of cell). About 5 parietal, longitudinal, ribbon-like chloroplasts with fairly smooth outline visible in front view; pyrenoids numerous. (Tychoplanktont.)

Cells 384–420 μ long; 32–34 μ broad at base of semicells, 23–27.5 μ at apex; isthmus 24–26.5 μ broad.

Little Doctor (rrr), Sunday (rrr).

In the scanty material observed the apical tubercules so commonly found in the genus were well defined. Since such careful investigators as W. & G. S. West have not described them it is evident that the Wisconsin plants differ from the type and since this presence or absence of apical tubercules is such an important feature of the genus these cells are considered a distinct variety.

DOCIDIUM De Brébisson 1844 (emend. Lundell 1871).

Cells solitary, relatively small, straight, length several times the breadth, cylindrical; median constriction shallow, sinus always widely open, isthmus breadth practically the same as the breadth of the cell;

semicells with straight or undulate sides and with truncate apices that are without ornamentation, inflated bases of semicells always with short vertical plications that frequently have a granule at the base of each plication. Cell wall smooth, finely punctate, or with delicate vertical striae. Chloroplasts axial and with irregular longitudinal ridges; with 6–8 pyrenoids arranged in an axial series.

Zygospores unknown.

In his British Desmidieae Ralfs includes both *Pleurotaenium* and *Docidium* in the same genus. Lundell was the first to differentiate clearly between the two and this characterization of the two genera has been followed by W. & G. S. in their Monograph. *Docidium* differs from *Pleurotaenium* in the vertical plications at the bases of the semicells, the lack of ornamentation at the apices of the semicells, and in the axial chloroplasts.

Docidium undulatum Bailey. Pl. **55**, Fig. **4**.

Smithsonian Cont. to Knowl. 2, Art. 8: 36, *pl.* 1, *fig.* 2, **1851** (sep.); W. & G. S. West, Monogr. Brit. Desm. 1: 194, *pl.* 27, *figs.* 7–10. 1904.

Length about 15 times the breadth, cells very slightly constricted, sinus widely open; semicells with sides strongly undulate (6–8 undulations between isthmus and apex) and apex narrowly obversely trapezoidal, angles of apex rounded and top flattened; basal plications ending in a small granule on the end towards the isthmus, about six visible in front view of semicell. Cell wall smooth. Chloroplast with an axial row of about eight pyrenoids. (Tychoplanktont.)

Zygospores unknown.

Cells 205 μ long; 15 μ broad at basal inflation, 12.5 μ at apex.

Diamond (rrr).

TRIPLOCERAS Bailey 1851.

Cells large, length several times the breadth, median constriction well defined but shallow; semicells cylindrical, always straight, somewhat attenuated from bases to poles, never compressed, with lateral margins strongly undulate; undulations caused by transverse whorls of mamillate protuberances that terminate in a single spine or broadly emarginate verrucae; poles flattened and with two long, upturned, divergent, obliquely disposed processes with truncate ends that bear 2–3 short sharp spines (front view of polar processes giving the cell apices a deeply incised appearance); apex with a protuberance midway between bases of processes that bears one or two erect spines. Cell wall, except for verrucae, smooth. Chloroplasts axial, with longitudinal plates that extend from semicell base to apex; pyrenoids in an axial row.

Zygospores unknown.

A highly differentiated genus that shows but little in common with the other genera that have straight elongate cells. The apical processes and transverse rows of verrucae on the cell wall are not found in related genera.

KEY TO THE SPECIES.

Protuberances with simple spines..............................(1) *T. gracile*
Protuberances with emarginate verrucae....................(2) *T. verticillatum*

1. Triploceras gracile Bailey. Pl. 55, Figs. 5–6.

Smithsonian Cont. to Knowl. 2, Art. 8: 38, *pl.* 1, *fig.* 10. 1851 (sep); G. S. West, Jour. Linn. Sic. Bot. 39: 54, text figs. 7A–7B. 1909.

Cells large, length 12–20 times the breadth; median constriction fairly well defined; semicells cylindrical, somewhat attenuated at poles, without line of lateral margins strongly undulate, undulate outline caused by transverse whorls of mamillate protuberances that terminate in acute, horizontally disposed, spinate verrucae in inferior portion of semicell and upwardly disposed spinate verrucae in superior portion; semicell apices flattened and with two fairly long, divergent, obliquely disposed, truncate processes terminating in two or three short spines; apex also with a single (rarely two) mamillate protuberance arising midway between the bases of the polar processes and terminating in an acute erect spine. Vertical view of pole elliptic, with poles terminating in spines and each lateral margin with a single, alternately disposed, truncate process. Cell wall without pores. Chloroplast axial, with 6–8 straight delicate longitudinal ridges extending from semicell base to apex; pyrenoids fairly numerous and forming an axial row. (Tychoplanktont.)

Zygospores unknown.

Cells 368–550 μ long; 27.5–31.5 μ broad at basal whorl (with verrucae), 18–25 μ broad at apical whorl (with verrucae), and 10–12.5 μ broad just above apical whorl. Distance between ends of apical processes 30–36 μ.

Crooked (rrr), Devils (rrr), Jag (rrr), Little Doctor (rrr), Rudolph (rrr).

The great variability of this species is well known and there has always been a conflict of opinion as to the limits of variation in the type. G. S. West's discussion of the type and its varieties is so thoroughgoing that further remarks are superfluous.

2. Triploceras verticillatum Bailey. Pl. 55, Figs. 7–9.

Smithsonian Cont. to Knowl. 2, Art. 8: 37, *pl.* 1, *fig.* 9. 1851 (sep.).

Docidium verticillatum Bailey, in Ralfs, Brit. Desm. 218, *pl.* 35, *figs.* 9A–9E. 1848; Wolle, Freshw. Algae of U. S. 27, *pl.* 61, *fig.* 18. 1887.

Cells large, stouter than the preceding species, length 8–12 times the breadth, median constriction well defined; semicells cylindrical, somewhat attenuated towards the poles, with outline of lateral margins strongly undulate, with 12–22 transverse whorls of mamillate protuberances that terminate in broad emarginate verrucae (triangular in front view), protuberances in superior portion of semicell at times terminating in simple spines; apices truncate, with two long, divergent, obliquely disposed processes that terminate in two long spines; semicells apices also with a mamillate protuberance bearing two erect spines that lies midway between the bases of the apical processes. Cell wall smooth. Chloroplast axial, with longitudinal plates extending from semicell base to apex; pyrenoids forming a median row and fairly numerous. (Tychoplanktont.)

Zygospores unknown.

Cells 480–504 μ long; 30–41 μ broad at basal whorl (with verrucae), 25–28 μ broad at apical whorl (with verrucae), 13–21 μ broad just above apical whorl. Distance between ends of apical processes 31–48 μ.

Des Moines (rrr), Little Doctor (rrr).

A much rarer species than *T. gracile* and differing in the nature of the transverse whorls of verrucae. Figure 9 is a transition form between the two in which the base of the semicell is typical *verticillatum* and the apex *gracile*.

TETMEMORUS Ralfs 1844.

Cells usually fairly large, length 2–8 times the breadth, median constriction slight to fairly deep but always well defined, sinus always open; semicells straight, cylindrical to subfusiform, with poles somewhat compressed and having a very conspicuous vertical incision, incision generally linear and with a slightly inflated apex. Cell wall smooth, punctate, or minutely scrobiculate. Chloroplasts axial and with 8–10 vertical radiating plates that are simple or bifurcate at the point where they reach the cell wall; pyrenoids fairly numerous, spherical to bacillar and arranged in a linear series along the central axis.

Zygospores spherical to rectangular, with smooth walls.

The elongate cells with apical incisions separate this genus from all others found in Wisconsin lakes.

TETMEMORUS BREBISSONII (Meneghini) Ralfs. Pl. **55**, Fig. **10**.

Ann. & Mag. of Nat. Hist. 14: 257, *pl.* 8, *figs.* 1A–1D. 1844; Nordstedt, Lunds Univ. Arsskr. 9: 39. 1873.

Cells large, length 5–6 times the greatest breadth, fairly deeply constricted, sinus with a subacute apex and outer margin widely open; semicells cylindrical, somewhat attenuated towards the poles, basal angles very broadly rounded; apex rounded, with a very deep linear incision in the median portion. Lateral view of semicells compressed in the polar region and giving the whole cell a subfusiform appearance. Cell wall slightly thickened near the outer margins of the polar incisions; entire wall covered with numerous punctae arranged in vertical rows running from semicell base to apex. Chloroplasts axial, with about eight vertical radiating plates that terminate in broad parietal forks at the cell wall; pyrenoids axial and variable in shape. (Tychoplanktont.)

Zygospores spherical, with smooth walls (Nordstedt).

Cells 226–264 μ long; 39–45 μ broad at base, 25–31 μ at apex, 21–27.5 μ at isthmus; apical incision 10 μ deep.

Zygospores 80 μ in diameter (Nordstedt).

Duck (rrr), Tank (rrr).

EUASTRUM Ehrenberg 1832 (emend. Ralfs 1844).

Cells of variable size, length generally about twice the breadth, deeply constricted, sinus generally linear and with apex inflated and outer portion open; semicells usually truncate-pyramidate, incised to form three or five lobes, polar lobe always with an apical incision which is usually deep but which may be reduced to a shallow depression; incisions forming lateral lobes frequently greatly reduced. Cell wall smooth, punctate or granulate, sometimes with spines on outer margin; with one or more protuberances at base of semicells and at times with other protuberances at center of cell face or at poles; in certain species with small rounded or elongate verrucae definitely arranged on the cell wall. Vertical view generally elliptic, with poles rounded and definite protuberances at sides. Lateral view of semicells narrowly truncate-pyramidate, with basal inflations fairly prominent. Chloroplasts of larger species axial and with two lobes to each cell face that end in massive parietal laminae; chloroplasts of smaller species axial and with simple lobes; in one case with two chloroplasts in each semicell; pyrenoids single to numerous.

Zygospores spherical to ellipsoidal; with smooth, mamillate or short-spined walls.

The compressed cells in which the length is not more than about twice the breadth show that this genus is related to *Cosmarium,* but the apical incision distinguishes it from *Cosmarium.* This apical incision may be reduced to a shallow depression so that the distinction between the two genera is very finely drawn. The apical incision is, however, associated with the presence of one or more protuberances at the base of the semicells. In the apical incision it shows an affinity with certain elongate genera, as *Tetmemorus* and *Ichthyocercus,* but in these genera the cells are not conspicuously compressed and the length is several times the breadth.

KEY TO THE SPECIES.

Polar lobe incised.
- Semicell base with more than one protuberance.
  - Three-lobed semicells.................................(1) *E. crassum*
  - Five-lobed semicells.
    - Polar lobe rectangular.................................(2) *E. didelta*
    - Polar lobe not rectangular.
      - Protuberances of the same size.........................(3) *E. oblongum*
      - Protuberances not of the same size.
        - Two small protuberances above isthmus...............(4) *E. affine*
        - One large protuberance above isthmus...............(5) *E. pinnatum*
- Semicell base with one protuberance.
  - Polar incision ovate.................................(6) *E. Glaviovii*
  - Polar incision not ovate.
    - Apex flattened.................................(7) *E. pulchellum*
    - Apex elevated.
      - Lateral lobes with spines.........................(9) *E. divaricatum*
      - Lateral lobes without spines.........................(8) *E. elegans*

Polar lobe emarginate.
- Lateral lobes cuneate.................................(11) *E. verrucosum*
- Lateral lobes subquadrate.................................(10) *E. gemmatum*

1. Euastrum crassum (De Brébisson) Kützing. Pl. **56**, Fig. **1**.

Phycol. Germ. 135. 1845; Ralfs, Brit. Desm. 81, *pl.* 11, *figs.* 3A–3E. 1848.

Cells very large, length about twice the breadth, deeply constricted, sinus narrowly linear, with apex slightly inflated and outer margin closed, isthmus narrow; semicells subrectangular, three-lobed, the polar lobe much smaller than the lateral lobes, polar lobe broadly cuneate, with apex tumid, angles broadly rounded; polar incision deep, sublinear, and somewhat expanded at the apex; incisions forming lobes open, with rounded apices and parallel sides; lateral lobes very broad, margins slightly concave, superior and inferior angles rounded. Cell wall finely punctate, conspicuously thickened at the angles of the lobes, with three prominent hemispherical protuberances across the base of the semicell, one above the isthmus and one near each basal angle. Vertical view broadly oval, with a tumid inflation at each pole and three less pronounced inflations on each side. Lateral view of semicells narrowly pyramidate, with apices truncate and angles somewhat rounded. Chloroplast an axial plate with four radiating arms, two to each cell face, that form parietal masses against the cell wall; pyrenoids fairly numerous and mostly in parietal portions of chloroplast. (Tychoplanktont.)

Zygospores unknown.

Cells 230 μ long; 80 μ broad at base of semicells, 62.5 μ at apex, 45 μ at apical lobe; 30 μ at isthmus; apical incision 11.5 μ deep.

Meta (rrr), Speese (rrr).

2. Euastrum didelta (Turpin) Ralfs. Pl. **56**, Fig. **2**.

Ann. & Mag. of Nat. Hist. 14: 190, *pl.* 7, *figs.* 2A–2B. 1844; W. & G. S. West, Monogr. Brit. Desm. 2: 15, *pl.* 35, *figs.* 3–7. 1905; Carter, Ann. of Bot. 33: 241, *pl.* 16, *figs.* 63–65. 1919.

Cells fairly large, length about twice the breadth, very deeply constricted, sinus linear, with apical portion inflated and outer portion widely open, isthmus narrow; semicells narrowly truncate-pyramidate, five-lobed, the lateral lobes at times reduced to strong undulations; polar lobe vertically oblong, with angles rounded, polar incision fairly deep and linear but with outer portion slightly expanded; lateral incisions very shallow and broadly rounded; lateral lobes very broadly rounded. Cell wall smooth or punctate, with five hemispherical protuberances on each semicell, a transverse row of three at the base of the semicell, one just above the isthmus and one midway between the isthmus and each basal angle; with a second transverse row of two protuberances just above and alternating with those of the basal row. Vertical view elliptic, poles bearing a single protuberance and sides each with five protuberances. Lateral view of semicells truncate-pyramidate and with angles rounded. Chloroplast axial and with two processes extending to each cell face; pyrenoids confined to processes. (Tychoplanktont.)

Zygospores spherical, clothed with very short papillae that are frequently curved (W. & G. S. West).

Cells 97.5 μ long, 52.5 μ broad at base of semicells, 24 μ broad at apical lobe, 17.5 μ at isthmus; apical incision 6 μ deep.

Zygospores 73.6–86 μ in diameter without papillae; papillae 9–13 μ long (W. & G. S. West).

Big. St. Germain (rrr), Fish (rrr), Meta (rrr).

3. EUASTRUM OBLONGUM (Greville) Ralfs. Pl. **56**, Fig. **3**.

Ann. & Mag. of Nat. Hist. **14**: 189, *pl.* 6, *figs.* 4A–4C. 1844; Brit. Desm. 80, *pl.* 12, *figs.* A–G. 1848.

Cells large, length about twice the breadth, deeply constricted, sinus linear, with apical portion somewhat inflated and outer margin widely open, isthmus narrow; semicells subrectangular, distinctly five-lobed, polar lobe broadly cuneate, with apex strongly tumid, angles broadly rounded and polar incision deep and narrowly V-shaped; interlobular incisions deep, open and with rounded apices; superior lateral lobes subquadrate, somewhat smaller than the inferior lateral lobes and at an angle with the long axis of the semicell, margins somewhat retuse; inferior lobes subquadrate, with rounded angles and concave margins. Cell wall punctate and with a large protuberance just above the isthmus and one just within the margins of both the inferior and superior lateral lobes. Vertical view broadly elliptic, with poles rounded and four protuberances on each side of the cell. Lateral view narrowly elongate-pyramidate, with bases parallel for a short distance and then converging to a truncate apex. Chloroplast an axial plate with four lobes, two extending to each cell face and forming parietal masses. (Tychoplanktont.)

Zygospores spherical and densely covered with short mamillate-conical protuberances (Ralfs).

Cells 152 μ long; 70 μ broad at basal lobes of semicells, 64 μ broad at median lobes, 41 μ broad at apical lobe, 21 μ broad at isthmus; apical incision 10 μ deep.

Zygospores 148 μ in diameter; papillae 25 μ long (Ralfs).

Mercer (rrr).

4. EUASTRUM AFFINE Ralfs. Pl. **56**, Fig. **4**.

Ann. & Mag. of Nat. Hist. **14**: 191, *pl.* 7, *figs.* 3A–3B. 1844; Brit. Desm. 82, *pl.* 13, *figs.* 3A–3E. 1848.

Cells fairly large, length nearly twice the breadth, very deeply constricted, sinus linear, with apex slightly inflated and outer margin somewhat open, isthmus narrow; semicells truncate-pyramidate, distinctly five-lobed, polar lobe subincudate, with apex tumid, angles broadly rounded, polar incision deep and with a slightly expanded apex; incisions forming lateral lobes fairly deep, broadly rounded; superior lateral lobes mamillate, about equidistant between semicell apex and isthmus; inferior lateral lobes subquadrate, with angles rounded and lateral margins strongly emarginate. Cell wall punctate, with a transverse row of rounded protuberances at base of semicell, two small protuberances just above the isthmus and a larger protuberance at each basal angle, with a second transverse row of two protuberances just above the smaller ones of the inferior row. Vertical view elliptic, with poles rounded and four protuberances on each side. Lateral view oblong-pyramidate, with basal portion subquadrate and sides converging to a quadrate apex, angles somewhat rounded. Chloroplast with a central axial plate from which radiate four parietal plates, two to each cell face; pyrenoids 5–8, all in parietal plates. (Tychoplanktont.)

Zygospores unknown.

Cells 104–110 μ long; 58–61 μ broad at basal lobes of semicells, 40–42.5 μ at median lobes, 28–31 μ at apical lobe, 17.5 μ at isthmus; apical incisions 5–7 μ deep.

Little Doctor (rrr), Lost (rrr).

Differing from *E. didelta* in the deeper incisions forming the lobes and in both number and arrangement of the protuberances on the cell wall.

5. Euastrum pinnatum Ralfs. Pl. **56**, Fig. **5**.

Brit. Desm. 81, *pl.* 13, *figs.* 1A–1E. 1848; W. & G. S. West, Monogr. Brit. Desm. 2: 10, *pl.* 34, *figs.* 3–6. 1905.

Cells fairly large, length about twice the breadth, deeply constricted, sinus narrowly V-shaped to linear, with apex expanded, isthmus narrow; semicells truncate-pyramidate, distinctly five-lobed, polar lobe anvil-shaped, with apex concave, angles broadly rounded and polar incision deep and expanded towards the apex; incisions forming lobes deep but widely open and with rounded apices; superior lateral lobes elongate-mamillate, with superior margin at right angles to inferior margin and forming an angle of 45 degrees with the long axis of the semicell; inferior lateral lobes subquadrate, with margins strongly retuse and angles rounded. Cell wall punctate, with a large protuberance just above the isthmus and one within each lateral lobe as well as with two smaller protuberances at bases of lateral lobular incisions. Vertical view elliptic, with poles mamillate and each side with three protuberances. Lateral view of semicells narrowly truncate-pyramidate, with lateral margins concave just below the apex and angles rounded. (Tychoplanktont.)

Zygospores unknown.

Cells 137.5 μ long; 67 μ broad at basal lobes of semicells, 57 μ broad at median lobes, 35 μ at apical lobes, 19 μ at isthmus; apical incision 6.5 μ deep.

Crooked (rrr), Found (rrr).

6. Euastrum Glaziovii Børgesen. Pl. **56**, Fig. **6**.

Vidensk. Medd. f. d. naturhist. Forening i Kiöbenhavn **1890**: 36, *pl.* 3, *fig.* 23. 1890.

Cells somewhat under medium size, length about twice the breadth, very deeply constricted, sinus linear, with apex somewhat inflated and outer margin widely open, isthmus very narrow; semicells distinctly three-lobed, polar lobe almost anvil-shaped, with angles somewhat rounded and each bearing a stout horizontally disposed spine, apex strongly elevated, polar incision very deep, obversely ovate and with outer margin almost closed, lateral margins of polar lobe with a small spine at the base; incisions between lobes widely open and with broadly rounded apices; lateral lobes subquadrate, with margins retuse in the median portion, angles rounded and each bearing a pair of short sharp divergent spines. Cell wall smooth, thickened at the polar angles and at the angles on either side of the apical incision, with a central protuberance above the isthmus and scattered granules on either side of the polar incision and at the base of the semicells. Vertical view elliptic, with poles strongly truncate and a protuberance in the middle of the sides. Lateral view of semicells ovate, with a small protuberance at the bases. (Tychoplanktont.)

Zygospores unknown.

Cells 63 μ long; without spines 37 μ broad at base of semicells, 22 μ at apex; with spines 40 μ broad at base of semicell, 35 μ at apex; 8 μ broad at isthmus; apical incision 7.5 μ deep.

Little Doctor (rrr).

7. Euastrum pulchellum De Brébisson. Pl. **56**, Fig. **7**.

Mem. d. l. Soc. imp. d. sc. nat. de Cherbourg 4: 124, *pl.* 1, *fig.* 5. 1856; W. & G. S. West, Monogr. Brit. Desm. 2: 46, *pl.* 38, *figs.* 14–15. 1905.

Cells small, length about one and a quarter times the breadth, deeply constricted, sinus linear, with outer portion slightly open, isthmus narrow; semicells truncate-pyramidate, the division into three lobes very poorly defined, polar lobe transversely oblong, with apex slightly concave and polar incision quite broad, angles acute and continued in short widely divergent spines; incisions forming lobes broad, very shallow depressions; lateral lobes broadly rounded and with profile bearing 3–4 short sharp spines. Cell wall with a triangle of three elongate verrucae just above the isthmus, a conspicuous rounded granule at either side of the apical incision, and a few scattered spines within the margins of the lateral lobes. Vertical view oblong-elliptic, with short spines in the polar region and a conspicuous truncate protuberance in the middle of each side. Lateral view of semicells narrowly pyramidate-ovate, with apex rounded and bearing a single erect spine, bases with a conspicuous truncate protuberance. (Tychoplanktont.)

Zygospores unknown.

Cells 26 μ long; 23 μ broad at base of semicells, 16 μ at apex, 6.5 μ at isthmus; depth of apical incision 2.5 μ.

George (rrr).

8. Euastrum elegans (De Brébisson) Kützing. Pl. **56**, Fig. **8**.

Phycol. Germ. 135. 1845; W. & G. S. West, Monogr. Brit. Desm. 2: 48, *pl.* 38, *figs.* 16–21. 1905.

Cells small, length about one and a half times the breadth, deeply constricted, sinus linear and with apex expanded, isthmus narrow; semicells ovate, not incised to form lobes, apex uncinate and with upper margin very strongly inflated and angles apiculate; apical incision deep and narrowly V-shaped; basal angles very broadly rounded. Cell wall smooth, with a triangle of three elongate verrucae just above the isthmus and two small granules on either side of the apical incision and at basal angles of the semicells. Vertical view narrowly elliptic, with a truncate protuberance on either side midway between the poles. Lateral view of semicells ovate and with a protuberance on either side of the base. Chloroplast a massive axial plate with two lobes extending towards each cell face; pyrenoid central and single. (Tychoplanktont.)

Zygospores spherical, with fairly short straight spines that have blunt apices (W. & G. S. West).

Cells 36 μ long; 25 μ broad at base of semicells, 11.3 μ at apex, 5 μ at isthmus; apical incisions about 5 μ deep.

Zygospores (without spines) 20–29 μ in diameter; spines 7.5–9.5 μ long (W. & G. S. West).

George (rrr), Jones (rrr).

9. Euastrum divaricatum Lundell. Pl. **56,** Fig. **9.**

Nova Acta Reg. Soc. Sci. Upsaliae 3 Ser., 8: 21, *pl.* 2, *fig.* 5. 1871.

Cells small, length about one and a quarter times the breadth, deeply constricted, sinus linear, with apical portion slightly inflated, isthmus narrow; semicells truncate-pyramidate, rather distinctly three-lobed, the polar lobe somewhat smaller than the lateral lobes; polar lobe transversely oblong, with apex tumid, angles obtuse and each furnished with a divergent spine, polar incision fairly deep and open; lateral lobes broadly triangular and with rounded angles, with two or three marginal spines of variable length. Cell wall smooth, with a triangle of short linear verrucae above the isthmus and a large rounded granule on either side of the polar incision, about four smaller granules within the margins of the lateral lobes. Vertical view elliptic, with a short spine at each pole and with sides crenulate, median portion of each side with a short truncate protuberance. Lateral view of semicells sublanceolate, with a protuberance on either side of the base. (Tychoplanktont.)

Zygospores unknown.

Cells 43 μ long; 38 μ broad at base of semicells, 20 μ at apex, 7.5 μ at isthmus; apical incisions about 5 μ deep.

Muskallonge (rrr).

10. Euastrum gemmatum De Brébisson. Pl. **57,** Figs. **1–2.**

in Meneghini, Linnaea **14:** 221. 1840; Ralfs, Brit. Desm. 87, *pl.* 14, *figs.* 4A–4E. 1848.

Cells fairly small, length about one and half times the breadth, deeply constricted, sinus linear, with apex inflated and outer margin widely open, isthmus narrow; semicells distinctly three-lobed, polar incision reduced to a broad emargination; lateral incisions deep, broadly open and with rounded apices; lateral lobes subquadrate, with angles broadly rounded and lateral margins deeply retuse. Cell wall finely granulate, with granulation prominent at angles of polar and lateral lobes and on the three protuberances at the bases of the semicells, one above the isthmus and one on each lateral lobe. Vertical view elliptic, with poles rounded and three very prominent protuberances on each side. Lateral view of semicells subrectangular, with bases inflated, polar angles rounded, and apices concave. (Tychoplanktont.)

Zygospores unknown.

Cells 54–56 μ long; 41–43 μ broad at base of the semicells, 20 μ at polar lobe, 12.5 μ at isthmus.

Muskallonge (rrr), Tank (rrr).

11. EUASTRUM VERRUCOSUM Ehrenberg.

Abh. d. k. Akad. d. Wiss. zu Berlin 1833: 247. 1835; Ralfs, Brit. Desm. 79, *pl.* 11, *figs.* 2A–2D. 1848; Carter, Ann. of Bot. 33: 245, *pl.* 17, *figs.* 90–92. 1919.

Cells of medium size, length slightly greater than the breadth, very deeply constricted, sinus with inner half sublinear and apex slightly inflated, outer half gradually opening outwards, isthmus very narrow; semicells truncate-pyramidate, distinctly three-lobed, the polar lobe smaller than the lateral lobes, polar lobe broadly cuneate, with angles rounded and polar incision reduced to an emargination of the polar lobe; incisions forming the lobes deep, sublinear, with rounded apices and widely open outer extremities; lateral lobes broadly cuneate, divided into two lobules by a very deep emargination, superior lobule suberect, inferior lobule horizontally disposed, lobelets broadly rounded. Cell wall distinctly granulate, with granules more pronounced at angles of lobes, bases of semicells with a very large protuberance just above the isthmus and flanked by a smaller protuberance in each of the lateral lobes, protuberances, especially the central one, furnished with somewhat elongate verrucae arranged in concentric circles. Vertical view compressed-hexagonal, with the angles rounded and a conspicuous protuberance in the center of the longer sides, cell outline granulate. Lateral view of semicells urn-shaped, with a deep emargination at the apex. Semicells with two axial chloroplasts, each chloroplast with 4–5 irregular ridges radiating to front and sides of cell wall; pyrenoids single and central in each chloroplast.

Zygospores unknown.

Cells 107 μ long; 108 μ broad (Ralfs).

Not found in the plankton of Wisconsin lakes.

var. REDUCTUM Nordstedt. Pl. 56, Fig. 10.

Lunds Univ. Årsskr. 16: 9, *pl.* 1, *fig.* 14. 1880 (sep.).

Cells smaller, sinus linear, with apex inflated and outer margin open; polar lobes subrectangular, incisions forming lobes more widely open; lateral lobes only slightly emarginate. (Tychoplanktont.)

Cells 58 μ long; 52 μ broad at base of semicells, 23 μ at apical lobe; 12 μ at isthmus.

Devils (rrr), Little Doctor (rrr).

The sinus, lateral incision, shape of the polar lobe and the degree of indentation of the lateral lobes differentiate this variety.

COSMARIUM Corda 1834.

Cells never forming permanent colonies; of variable size though in the majority of cases fairly small, length never more than twice the breadth, generally only slightly greater, in a few species with breadth greater than the length; median constriction usually conspicuous and frequently linear; cells compressed and with front view of semicells quite variable in outline (circular, semicircular, elliptic, ovate, trapezoidal, reniform or subquadrate), at times with the apex elevated but never with the apical elevation incised, central area of semicells at times protuberant. Vertical view generally elliptic, frequently with an elevation in the median portion of the sides. Lateral view of semicells most commonly circular. Cell wall smooth or variously ornamented with granules of varying size or minute verrucae, the ornamentation usually following a definite pattern; never ornamented with spines. Chloroplasts usually single, axial, and with four radiating plates and central pyrenoids; semicells at times with two chloroplasts that have several radiating plates and each chloroplast with a single pyrenoid; a few larger species with four parietal chloroplasts each containing several pyrenoids.

Zygospores generally spherical or ellipsoidal, rarely angular; zygospore wall smooth or papillate to spinate.

Certain of the species in this very large genus can scarcely be distinguished from *Euastrum;* other species resemble closely *Penium* while the rare triradiate varieties might be considered *Staurastra. Euastrum* is separated from *Cosmarium* by the apical incision. *Cosmarium* differs from *Penium* in the compressed cells and from *Staurastrum* in the lack of spines.

Since the drawings in this work have been made from preparations in which in many cases but a single cell of the species was available the side and top views are lacking. These views are essential for the critical interpretation of the various species and certain records of the genus have been omitted since vertical and lateral views were not available. In Wisconsin lakes *Cosmarium* is a comparatively rare organism whereas in certain European lakes it may be present in quantity.

KEY TO THE SPECIES.

Cell wall smooth or punctate, outline never granulate.
  Semicell outline truncate-pyramidate........................(5) *C. granatum*
  Semicell outline not truncate-pyramidate.
    Sinus linear.
      Semicells semicircular..............................(1) *C. circulare*
      Semicells elliptic..................................(2) *C. depressum*
    Sinus open.
      Semicells subcircular...............................(4) *C. contractum*
      Semicells broadly elliptic..........................(3) *C. bioculatum*
Cell wall ornamented and outline granular.
  Apices of semicells not elevated.
    Cell apex flattened.

Lateral margins crenulate .... (6) *C. cyclicum*
Lateral margins granulate.
Granules evenly distributed .... (8) *C. Botrytis*
Granules restricted to margin and center.
Central area not granulate .... (16) *C. ovale*
Central area granulate.
Many granules within margin .... (15) *C. punctulatum*
Few granules with margin .... (14) *C. quinarium*
Cell apex rounded.
Semicells not reniform.
Sinus open, isthmus elongate .... (10) *C. Portianum*
Sinus linear .... (7) *C. amoenum*
Semicells reniform .... (9) *C. reniforme*
Apices of semicells elevated.
Semicells distinctly three-lobed .... (13) *C. protractum*
Semicalls not three-lobed.
Apex of sinus widely inflated .... (12) *C. commissurale*
Apex of sinus not widely inflated .... (11) *C. ornatum*

1. Cosmarium circulare Reinsch. Pl. **57**, Fig. **3**.

Acta Soc. Senckenberg **6**: 5, *pl.* 3, *figs.* C³–C⁴. 1867 (sep.)

Cells fairly large, cell outline circular, deeply constricted, sinus linear and with outer extremity somewhat inflated; semicells semicircular, with basal angles rounded. Vertical view narrowly elliptic. Lateral view of semicells narrowly subcircular. Cell wall minutely punctate. Chloroplasts axial and with two pyrenoids. (Tychoplanktont.)

Zygospores unknown.

Cells 47.5 μ long; 39 μ broad; breadth of isthmus 15 μ.

Cranberry (rrr).

The cells observed were at the minimal dimensions usually given for this species.

2. Cosmarium depressum (Nägeli) Lundell. Pl. **57**, Fig. **4**.

Nova Acta Reg. Soc. Upsaliae 3 Ser., **8**: 38. 1871.

Cosmarium Scenedesmus Delponte, Mem. R. Accad. d. Sci. di Torino 2 Ser., **28**: 101, *pl.* 7, *figs.* 28–34. 1877 (sep.).

Cells slightly under medium size, breadth somewhat greater than the length, very deeply constricted, sinus linear and with outer extremity expanded, isthmus very narrow; semicells transversely elliptic and with dorsal margin more broadly rounded than the ventral, lateral margins broadly rounded. Vertical view elliptic. Lateral view of semicells circular. Cell wall smooth. Chloroplasts with a single pyrenoid. (Tychoplanktont.)

Zygospores spherical, with a smooth wall (Delponte).

Cells 38–51 μ long; 43–52 μ broad; breadth of isthmus 7.5–9 μ.

Fish (rrr), Fowler (rrr), Mercer (rrr), Nicaboyne (rrr), Oconomowoc (rrr), Okauchee (rrr), Sand (rrr).

var. ACHONDRUM (Boldt) W. & G. S. West. Pl. **57**, Figs. **5–6**.

Trans. Roy. Soc. Edinburgh 41: 484. 1905; Monogr. Brit. Desm. 2: 177, *pl.* 62, *figs.* 6–9. 1905.

Semicell shape subhexagonal, apices truncate and dorsal margins with a distinct angularity instead of a smooth curve. (Tychoplanktont.)

Cells 28 μ long; 27 μ broad; breadth of isthmus 7.5 μ; thickness 11 μ.

Found (rr), Sand (rrr), Sishebogema (rr).

The alga as I have found it in Wisconsin lakes is somewhat smaller than the cells of this variety that W. & G. S. West describe from the Scottish plankton. The size and general shape of the Wisconsin specimens is that of *C. abbreviatum* var. *planctonicum* W. & G. S. West but the broader semicells show that it should be referred to *C. depressum*.

3. COSMARIUM BIOCULATUM De Brébisson. Pl. **57**, Figs. **7–8**.

in Ralfs, Brit. Desm. 95, *pl.* 15, *figs.* 5A–5F. 1848.

Cells very small, length and breadth about equal, deeply constricted, sinus linear but open, with outer margin expanded and apex rounded, isthmus narrow; semicells transversely elliptic, with median portion of dorsal and ventral margins flattened, lateral margins very broadly rounded. Vertical view oblong-elliptic. Lateral view of semicells circular. Cell wall smooth. Chloroplasts axial and with one pyrenoid. (Tychoplanktont.)

Zygospores spherical, armed with short blunt conical projections whose length is about twice the breadth (Ralfs).

Cells 18 μ long; 16–17 μ thick; breadth of isthmus 5.5 μ; thickness 7.5 μ.

Dummy (rrr).

4. COSMARIUM CONTRACTUM Kirchner. Pl. **57**, Figs. **9–11**.

Kryptogamenfl. von Schlesiens 2[1]: 147. 1878; Wolle, Desm. U. S. 63, *pl.* 16, *fig.* 1; *pl.* 50, *fig.* 24. 1884; W. & G. S. West, Monogr. Brit. Desm. 2: 170, *pl.* 61, *figs.* 23–25, 34. 1905.

Cells fairly small, length about one and a half times the breadth, very deeply constricted, sinus open but with apex acuminate; semicells elliptic-circular, with ventral margin more rounded than the dorsal. Vertical view very broadly elliptic. Lateral view of semicells circular. Cell wall smooth or very finely punctate. Chloroplasts axial and with a single pyrenoid. (Facultative planktont.)

Zygospores spherical to angular-spherical, smooth-walled (W. & G. S. West).

Cells 29–37 μ long; 22–24 μ broad; 5–6.5 μ broad at isthmus; 15 μ thick.

Diameter of zygospores 31.5–35 μ (W. & G. S. West).

Beaverdam (s), Carroll (r), Horseshoe (rr), Shell (s).

var. PAPILLATUM W. & G. S. West. Pl. **57**, Figs. **12–13.**

Trans. Linn. Soc. 2nd. Ser. Bot. 5: 252, *pl.* 15, *fig.* 21. 1896.

Cells much larger than the type and with sinus more open; lateral angles of semicells with a single long papilla, the papillae being depressed towards the sinus. (Facultative planktont.)

Cells 45–49 μ long; 33–37 μ broad; 10–14 μ broad at isthmus; 25 μ in thickness. Length of papillae 1.5–2 μ.

Devils (s).

The cells of this variety that I have found in the plankton of Devils Lake have the characteristic shape and inward pointing papillae that W. & G. S. West show in their figure but their size is somewhat smaller.

In the plankton of Big Bass, Horseshoe and Shell lakes there was an even smaller form with shorter papillae that were not depressed. This has been given the form name *minor* (Trans. Wis. Acad. Sci., Arts & Lett. **20**: 345, *pl.* 10, *figs.* 2–5. 1922) but might well be considered a distinct variety. The front view of this form is indistinguishable from certain forms of *Staurastrum brevispinum* De Brébisson so that the alga could be considered a compressed form of this *Staurastrum* species. Since no triradiate forms of the alga were found it has been relegated to *Cosmarium* instead of *Staurastrum.*

The dimensions of the form *minor* are: length 31–38 μ; breadth 27–30 μ; breadth of isthmus 8.75–10 μ; thickness 17–25 μ. (Pl. **57,** Figs. **14–17.**)

5. COSMARIUM GRANATUM De Brébisson. Pl. **57,** Fig. **18.**

in Ralfs, Brit. Desm. 96, *pl.* 32, *figs.* 6A–6B. 1848.

Cells fairly small, length about one and a third times the breadth, deeply constricted, sinus linear, with apex slightly inflated and outer margin open, isthmus narrow; semicells pyramidate and with apices truncate, bases of lateral margins subparallel for a short distance and then converging or broadly rounded, apical angles obtuse. Vertical view elliptic. Lateral view ovate. Cell wall smooth to minutely punctate. Chloroplasts axial and with a single pyrenoid. (Tychoplanktont.)

Zygospores unknown.

Cells 41 μ long; greatest breadth of semicell 30 μ, breadth at apex 9 μ; breadth of isthmus 7.5 μ.

Trout (rrr).

6. COSMARIUM CYCLICUM Lundell.

Nova Acta Reg. Soc. Upsaliae 3 Ser., 8: 35, *pl.* 3, *fig.* 6. 1871.

Cells of medium size, subcircular in outline, length and breadth about equal, deeply constricted, sinus narrowly linear, with apex slightly dilated and outer extremity somewhat open, isthmus narrow;

semicells semicircular and with lateral angles somewhat rounded, cell outline crenulate. Vertical view narrowly elliptic but outlines of sides smooth. Lateral view of semicells ovate. Cell wall with 3–5 rows of undulations within the crenae, the undulations forming concentric series; central portion of wall smooth. Chloroplasts axial and with two pyrenoids.

Zygospores unknown.

Cells 49–52 μ long; 52–55 μ broad; 17–19 μ broad at isthmus; thickness 26 μ (Lundell).

Not found in the plankton of Wisconsin lakes.

var. NORDSTEDTIANUM (Reinsch) W. & G. S. West. Pl. **57**, Figs. **19–20.**

Monogr. Brit. Desm. 2: 146, *pl.* 58, *fig.* 12. 1905.

Cosmarium Nordstedtianum Reinsch, Contrib. ad Alg. et Fung. 89, *pl.* 10, *figs.* 11A–11B. 1875.

Semicells with apices truncate and undulation of cell wall within the crenae reduced in number. (Tychoplanktont.)

Cells 52 μ long, 47–51 μ broad; breadth of isthmus 17 μ.

Meta (rrr).

The ornamentation of the cells as I have found it is similar to that given by W. & G. S. West and not the scattered granules covering the entire wall that Reinsch figures.

7. COSMARIUM AMOENUM De Brébisson.

in Ralfs, Brit. Desm. 102, *pl.* 17, *figs.* 3A–3C. 1848.

Cells fairly small, length about twice the breadth, slightly constricted, sinus open to sublinear, isthmus broad; semicells vertically rectangular and with rounded apices. Vertical view broadly elliptic. Lateral view of semicells narrowly oblong-elliptic. Cell wall granulate in outline and with face of cell bearing scattered granules in the apical region and vertical rows of granules extending from apical region to the semicell base, frequently with vertical pairs of granules forming a transverse belt across the base of the semicell. Chloroplasts axial and with two pyrenoids.

Zygospores unknown.

Cells 49 μ long; 24 μ broad; 18 μ broad at isthmus (Ralfs).

Not found in the plankton of Wisconsin lakes.

var. MEDIOLAEVE Nordstedt. Pl. **57**, Fig. **21.**

Bot. Not. 1887: 160. 1887; Kgl. Sv. Vet.-Ak. Handl. 22, No. 8: 50, *pl.* 5, *fig.* 12. 1888.

Semicells with granules lacking in median area of front view and with double row of granules at base of semicells prominent. (Tychoplanktont.)

Cells 39 μ long; 23 μ broad; 15 μ broad at isthmus.

Speese (rr).

These cells are somewhat smaller than other investigators find them but the characteristic ornamentation shows that it is *C. amoenum* var. *mediolaeve.*

8. COSMARIUM BOTRYTIS (Bory) Meneghini. Pl. **57**, Fig. **22.**

Linnaea 14: 220. 1840; Ralfs, Brit. Desm. 99, *pl.* 16, *figs.* 1A–1D. 1848; Carter, Ann. of Bot. 34: 277. 1920.

Cells of medium size, length about a quarter greater than the breadth, deeply constricted, sinus linear, with apex slightly inflated and outer margin open, isthmus narrow; semicells truncate-pyramidate, with basal angles broadly rounded and apical angles more acute; semicell apices distinctly flattened. Vertical view elliptic and with sides inflated in median portion. Lateral view of semicells broadly elliptic. Cell wall coarsely granulate, with granules distributed over the entire wall but not forming a definite pattern. Semicells containing two chloroplasts, each with a central pyrenoid, the axis of each chloroplast giving rise to 4–6 plates that radiate towards the periphery and have their peripheral edges flattened against the cell wall. (Tychoplanktont.)

Zygospores spherical, with numerous short radiating processes that have conical bases and trispinate apices (Ralfs.)

Cells 60–108 μ long; 52–74 μ broad; breadth of isthmus 14–26 μ.

Zygospores 51 μ in diameter, with spines 9.7 μ long (Ralfs).

Beaverdam (rrr), Crab (rrr), Grindstone (rrr), Horseshoe (rr), Muskallonge (rrr), Number One (rrr), Sand (rrr), Whitney (rrr).

9. COSMARIUM RENIFORME (Ralfs) Archer. Pl. **57**, Fig. **23.**

Jour. of Bot. 12: 92. 1874; W. & G. S. West, Monogr. Brit. Desm. 3: 157, *pl.* 79, *figs.* 1–2; *pl.* 82, *fig.* 15. 1908.

Cosmarium margaritiferum var. ***reniformis*** Ralfs, Brit. Desm. 100, ***pl.*** 16, ***fig.*** 2A. 1848.

Cells of medium size, length and breadth about equal, deeply constricted, sinus narrow, with apex greatly inflated and pyriform, outer portion dilated, isthmus narrow and somewhat elongate; semicells reniform, with outline coarsely granulate. Vertical view elliptic. Lateral view of semicells circular. Cell wall with coarse granules covering the entire surface and arranged in a criss-cross series. Chloroplasts axial and with two pyrenoids. (Tychoplanktont.)

Zygospores spherical and with smooth walls (W. & G. S. West).

Cells 54 μ long; 50 μ broad; 16.5 μ broad at isthmus.

Diameter of zygospores 54–60 μ (W. & G. S. West).

Eagle (rrr).

10. COSMARIUM PORTIANUM Archer. Pl. **57**, Fig. **24.**

Quart. Jour. Micr. Sci. 8: 235, *pl.* 11, *figs.* 8–9. 1860.

Cosmarium orbiculatum De Bary non Ralfs, Unters. u. d. Fam. d. Conj. 41, ***pl.* 6, *figs.* 49–50. 1858.**

Cells fairly small, length about one and a half times the breadth, fairly deeply constricted, sinus a semicircular depression with a rounded apex, isthmus elongate; semicells transversely elliptic and with poles quite blunt, semicell outline granulate. Vertical view broadly elliptic. Lateral view of semicells circular. Cell wall granulate, with conspicuous granules arranged in vertical series, at times decussate. Chloroplasts axial and with one pyrenoid. (Tychoplanktont.)

Zygospores spherical, furnished with numerous short conical protuberances with rounded apices (De Bary).

Cells 40 μ long; 28 μ broad; 10 μ broad at isthmus.

Beaverdam (rrr).

11. Cosmarium ornatum Ralfs. Pl. **57**, Fig. **25**.

Ann. & Mag. of Nat. Hist. 14: 392, *pl.* 11, *figs.* 3A–3C. 1844; W. & G. S. West, Monogr. Brit. Desm. 3: 151, *pl.* 78, *figs.* 1–10. 1908.

Cells small, length and breadth about equal, deeply constricted, sinus linear, with apex slightly inflated and outer margin open, isthmus fairly narrow; semicells reniform, with a broad truncate elevation in the apical region, lateral margins broadly rounded, outline of cells granulate. Vertical view elliptic, with poles broadly rounded and central portion of sides somewhat inflated. Lateral view of semicells subcircular and with a somewhat truncate apex. Cell wall with granules within the margin (at times with granules in vertical series) and with a transverse double row of granules below the truncate portion of apex; central area with a circular area of conspicuous granules, face of semicell surrounding central area without granules. Chloroplasts axial and with two pyrenoids. (Tychoplanktont.)

Zygospores spherical, with numerous fairly long stout spines that have conical bases and emarginate apices (W. & G. S. West).

Cells 31 μ long; 33 μ broad; 12.5 μ broad at isthmus.

Zygospores 46–50 μ broad without spines, 70–95 μ with spines (W. & G. S. West).

Crooked (rrr), Muskallonge (rrr).

12. Cosmarium commissurale De Brébisson.

in Meneghini, Linnaea 14: 220. 1840; Ralfs, Brit. Desm. 105, *pl.* 16, *figs.* 8A–8D. 1848.

Cells fairly small, breadth one and a half times the length, very deeply constricted, sinus with inner portion rhomboidal, median portion linear, and outer portion dilated, isthmus very narrow; semicells reniform, with lateral angles broadly rounded and apex truncate and slightly elevated in the median portion. Vertical view elliptic and with a conspicuous semicircular inflation in the median portion. Lateral view of semicells broadly elliptic and with apex flattened. Cell wall (in front view) with outline of truncate apex and lateral angles granulate; median portion of cell face with a ring of granules enclosing scattered granules or with a circular area in which all

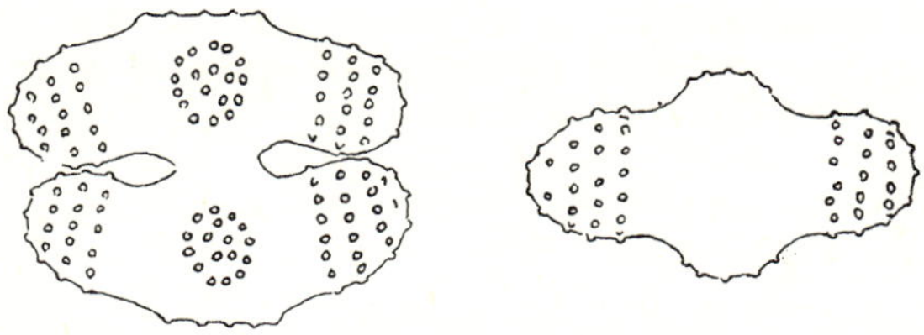

Fig. 3. *Cosmarium commissurale* De Brébisson (x 800). From a herbarium specimen in the British Museum of Natural History collected by De Brébisson at Falaise, France.

granules are irregularly distributed. Vertical view of wall with transverse rows of granules at the poles and with central protuberance granulate. Chloroplast axial and with one pyrenoid.

Zygospores spherical, clothed with numerous long straight spines with emarginate apices and conical bases (Ralfs).

Cells 31 μ long; 39 μ broad; 8 μ broad at isthmus; 21 μ in thickness.

Not found in the plankton of Wisconsin lakes.

var. CRASSUM Nordstedt. Pl. **57**, Fig. **26**.

Vidensk. Medd. f. d. naturh. Foren. i Kiobenhavn 1869: 213, *pl.* 3, *fig.* 19. 1870.

Breadth of cells one to one and a quarter times the length, inner portion of sinus linear and not rhomboidal; semicells elliptic in outline and with granulations more irregular. (Tychoplanktont.)

Cells 32 μ long; 34 μ broad; 9 μ broad at isthmus.

Travis (rrr).

The relative proportions of the cells and the difference in sinus and semicell shape distinguish the variety although there is, in general, the same ornamentation as in the type.

13. COSMARIUM PROTRACTUM (Nägeli) De Bary. Pl. **57**, Fig. **27**.

Unters. u. d. Fam. d Conj. 72. 1858; W. & G. S. West, Monogr. Brit. Desm. 3: 181, *pl.* 82, *fig.* 8; *pl.* 94, *figs.* 4–5. 1908.

Cells fairly small, length and breadth about equal, very deeply constricted, sinus linear, with apex slightly inflated and outer portion somewhat open, isthmus narrow; semicells reniform, with apices considerably elevated and giving the semicells a three-lobed appearance; lateral lobes broadly rounded; apical lobe truncate, with sides subparallel and apex flattened, angles of lobes somewhat rounded, cell outline granulate. Vertical view · elliptic, with poles broadly rounded and sides with a conspicuous median protuberance. Lateral view of semicells ovate and with apices flattened. Cell wall with coarse granules near the margins of lobes, at times with these granules regularly arranged on the lateral lobes; central area with similar large granules arranged in concentric rings, wall between marginal granules and central area granulate but granules considerably smaller. Chloroplasts axial and with two pyrenoids. (Tychoplanktont.)

Zygospores unknown.

Cells 40 μ long; maximal breadth 42.5 μ; breadth at apex 17 μ; breadth at isthmus 12 μ.

Okauchee (rr).

14. Cosmarium quinarium Lundell. Pl. 57, Fig. 28.

Nova Acta Reg. Soc. Upsaliae 3 Ser., 8: 28, *pl.* 2, *fig.* 14. 1871; W. & G. S. West, Monogr. Brit. Desm. 3: 216, *pl.* 85, *figs.* 9–10. 1908.

Cells fairly small, length slightly greater than the breadth, deeply constricted, sinus linear, with apex slightly inflated and outer margin open, isthmus narrow; semicells truncate-pyramidate, with basal angles broadly rounded and apical angles not acute, lateral margins slightly convex and truncate apex flat; outline of semicells granulate. Vertical view elliptic. Lateral view of semicells circular. Cell wall with five conspicuous granules in two rows at center of semicell, the upper row with three and the lower with two granules, with a transverse subapical row of four similar granules; also with several smaller granules near the lateral margins and two or three small granules between the large ones at the center of the cell face. Chloroplast axial and with two pyrenoids. (Tychoplanktont.)

Zygospores unknown.

Cells 42 μ long; 36 μ broad; 10 μ broad at isthmus.

Tank (rrr).

15. Cosmarium punctulatum De Brébisson.

Mem. de la Soc. imp. d. Sc. nat. de Cherbourg 4: 129, *pl.* 1, *fig.* 16. 1856; W. & G. S. West, Monogr. Brit. Desm. 3: 206, *pl.* 84, *figs.* 13–14. 1908; *ibid.* 4: *pl.* 102, *fig.* 22. 1912.

Cells small, length slightly greater than the breadth, deeply constricted, sinus linear, with apex slightly inflated and outer extremity open, isthmus narrow; semicells broadly subtrapeziform, with basal and apical angles broadly rounded, lateral margins convex and apex straight to slightly convex; outline of semicells very finely granular. Vertical view elliptic. Lateral view of semicells circular. Cell wall minutely granulate near the margins and central area, if granulate, with granules approximately the same size as those of the margins. Chloroplasts axial, with one pyrenoid.

Zygospores spherical, with numerous long spines with conical bases and deeply emarginate to bifurcate apices, if bifurcate with the bifurcations emarginate (W. & G. S. West).

Cells 34–36.5 μ long; 31–34 μ broad; 8.5–11 μ broad at isthmus; 17–18.5 μ thick (W. & G. S. West).

Zygospores 42–45 μ in diameter without spines, 65–68 μ with spines (W. & G. S. West).

Not found in the plankton of Wisconsin lakes.

var. SUBPUNCTULATUM (Nordstedt) Børgesen. Pl. **57**, Fig. **29**.

Medd. om Grönland 18: 12. 1894; W. & G. S. West, Monogr. Brit. Desm. 3: 209, *pl.* 84, *figs.* 15–20; *pl.* 85, *figs.* 1–3. 1908.

Semicells with granules of central area conspicuously larger than those near the margin, arrangement of granules at center generally not following a definite pattern. (Tychoplanktont.)

Cells 33 μ long; 30 μ broad; 10 μ broad at isthmus.

George (rrr).

W. & G. S. West recognize the forms Alpha and Beta of the variety, the distinction being based upon smooth or granulate semicell apices. The Wisconsin specimens belong to the variety Beta with granulate apices.

16. COSMARIUM OVALE Ralfs. Pl. **58**, Fig. **1**.

Ann. & Mag. of Nat. Hist. 14: 394 (p. p.), *pl.* 11, *figs.* 7D–7E. 1844; Brit. Desm. 98, *pl.* 15, *figs.* 9A–9B. 1848; Carter, Ann. of Bot. 34: 282, *pl.* 13, *figs.* 82–83. 1920.

Cells very large, length one and a half times to nearly twice the breadth, outline oval, deeply constricted, sinus linear, with apical portion somewhat inflated and outer margin widely open, isthmus narrow; semicells truncate-pyramidate, with basal angles broadly rounded and sides gradually converging to a truncate apex, apex flattened to convex, cell outline with large conical granules with rounded apices, cells frequently without granules at the poles. Vertical view elliptic, with outline smooth except at poles. Lateral view of semicells ovate, with granulation of cell outline confined to apical region. Cell wall with large granules similar to those of the cell outline within the lateral margins but with apical area smooth, entire surface of wall minutely punctate; vertical view with a broad band of coarse granules extending from pole to pole and lateral view of semicells with the band extending from isthmus to apex. Semicells with four parietal chloroplasts, two generally opposite each front face, each chloroplast with numerous vertical laminate projections extending to the cell wall; main body of chloroplast with 6–12 pyrenoids.

Zygospores unknown.

Cells 180 μ long; maximum breadth 90 μ, breadth at apex 11 μ; breadth at isthmus 39 μ.

Devils (rrr).

COSMOCLADIUM De Brébisson 1856.

Cells united by bands of gelatinous material to form spherical to ellipsoidal, sessile or free-floating colonies. Colonies enclosed by hyaline, homogeneous gelatinous envelopes. Cells small, quite similar in shape to those of *Cosmarium;* length somewhat greater than the breadth, median constriction usually deep and sinus linear; semicells usually compressed and front view elliptic. Vertical view elliptic and lateral view of semicells subcircular. Cell wall smooth, without ornamentation aside from pores. Chloroplasts axial, with one pyrenoid.

Zygospores spherical, with irregular, broad, angular projections.

The cells of this genus have the same shape as those of *Cosmarium* but the organization into a definite colony differentiates *Cosmocladium* from *Cosmarium*. With one exception this is the only colonial genus of the family in which cell division takes place in three planes and so gives rise to a colony that is not a filament. The gelatinous envelope enclosing the colony can only be demonstrated by special methods but the gelatinous strands connecting the individual cells are readily discernible. These strands have been the subject of special investigations by Schröder and Lütkemüller and they have shown that the two sister cells are connected by strands that arise near the isthmus of the mother cell at the time of division. These investigators have also demonstrated the presence of definitely arranged pores in the cell walls and Schröder has figured radially arranged denser strands in the colonial envelope, these denser strands apparently originating at the pores of the cell wall.

KEY TO THE SPECIES.

Cells strongly compressed.................................(1) *C. saxonicum*
Cells not compressed.....................................(2) *C. Hitchcockii*

1. Cosmocladium saxonicum De Bary. Pl. **58,** Fig. **2.**

Flora **48:** 329, *pl.* 4, *figs.* 1–3. 1865; Schröder, Ber. d. D. bot. Ges. **18:** 15, *pl.* 1, *figs.* 1–9. 1900.

Cells fairly small, length somewhat greater than the breadth, median constriction relatively deep, sinus sublinear, with apex rounded and outer margin widely open, isthmus fairly narrow; semicells transversely elliptic to subhemispherical, with dorsal margin more rounded than the ventral margin, lateral angles broadly rounded. Vertical view oblong-elliptic, with sides subparallel and poles somewhat flattened. Lateral view of semicells subcircular. Cell wall smooth. Chloroplast axial, with a single central pyrenoid. Colonial envelope a hyaline gelatinous matrix in which there are fairly broad denser strands connecting the cells. (Facultative planktont.)

Zygospores unknown.

Cells 22–30 μ long; 21–26 μ broad; 7–7.5 μ broad at isthmus; 12–14 μ thick. Colonies up to 275 μ in diameter.

Beaverdam (rr), Devils (ccc), Squirrel (rr), North Turtle (sss).

The cells of the species as I have found them in Wisconsin have more rounded semicells than De Bary or Schröder figure. The connecting threads also appear to be the same diameter throughout and do not have the expansion midway between the two cells that these investigators have noted.

2. **Cosmocladium Hitchcockii** (Wolle) comb. nov. Pl. **58,** Figs. **3–9.**

Dictyosphaerium Hitchcockii Wolle, Bull. Torr. Bot. Cl. **12:** 126. 1885; Freshw. Algae of U. S. 186, *pl.* 157, *fig.* 12. 1887.

Dictyocystis Hitchcockii (Wolle) Lagerheim, Nuova Notarisia 5: 226. 1890; W. & G. S. West, Jour. Linn. Soc. Bot. 35: 531. 1903; Brunnthaler in Pascher, Süsswasserfl. Deutschl., Österr. u. d. Schw. 5, Chlorophyceae 2: 185, *fig.* 279. 1915.

Cells small, length about one and three quarters times the breadth, very slightly constricted, sinus a broad very shallow depression, breadth of isthmus nearly equal to that of semicell; semicells broadly ovate and with semicircular apices. Vertical view circular. Lateral view similar to that of front view. Chloroplast axial, usually but one in a cell, with four vertical parietal lobes and a single central pyrenoid. Cells connected by a single delicate strand that usually arises near the isthmus. Inbedded in a homogeneous gelatinous envelope. Gelatinous envelope at times with denser acicular gelatinous rays radiating from the walls of the cells. (Facultative planktont.) Zygospores unknown.

Cells 30–35 μ long with spines, 15–19 μ without spines; 20–30 μ broad with spines, 10–12 μ without spines; isthmus 8–10 μ broad. Colonies up to 120 μ in diameter.

Bear (r), Beaverdam (r), Clare (r), Fish (r), Horseshoe (rrr), Kitten (rr), Lac Court Oreilles (ss), Pokegama (rrr), Reserve (s), Rice (rr), Round (rr), Whitefish (rrr), Wildcat (rr).

In the cell shape and the tendency towards an arrangement in series radiating from a common center this alga shows a great similarity to the *Dictyosphaerium Hitchcockii* of Wolle and I believe that the organism described above is the one which he referred to *Dictyosphaerium.*

On the basis of Wolle's description of the chloroplast, "each cell having the chlorophyll radiately gathered around a large central granule", Lagerheim created the genus *Dictyocystis* since this alga evidently does not belong to *Dictyosphaerium.* In my opinion the cell shape is sufficiently like certain other *Cosmocladium* species [particularly *C. constrictum* (Archer) Joshua] to justify its removal from *Dictyocystis* to *Cosmocladium.* The organization of the colony with strands connecting the cells is a distinct *Cosmocladium* feature, although in most *Cosmocladium* species these strands are broad and appear as a double line. The radiating strands found in certain colonies seems to be good evidence that there are scattered pores in the wall of this alga as in other *Cosmocladium* species, but, unfortunately, empty cells were not available for the accurate determination of this point.

The greatest objection to considering the alga a Desmid is the chloroplast. In all members of the subfamily Placodermae, to which, *Cosmocladium* belongs, there is a chloroplast in each semicell. In *C. Hitchcockii* there is but a single chloroplast that extends from pole to pole. In its axial position and four radiation lobes, however, it is similar to the type of chloroplast found in a majority of the species in the closely related genus *Cosmarium.*

If the alga is to be excluded from *Cosmocladium* I feel that *Dictyocystis* should be retained as a separate genus, that, on account of its axial chloroplast, is closely related to *Asterococcus* of the Palmellaceae.

MICRASTERIAS Agardh 1827.

Cells usually solitary, rarely in filaments, frequently of large size and with the length somewhat greater than the breadth, median constriction always well defined and usually linear to sublinear, isthmus never broad; bilaterally symmetrical and greatly compressed (in one variety radially symmetrical); semicells semicircular to hexagonal, with two incisions from periphery to about half way to center dividing the semicell into three lobes—the central or polar lobe and two lateral lobes; or with four incisions that divide the semicell into one polar and four lateral lobes; depth of incision variable and shape linear to cuneate; polar lobe generally with apex expanded and somewhat emarginate, at times with accessory asymmetrical processes; lateral lobes rarely entire, usually with one, two, three or four series of secondary lobelets, the ultimate lobelets generally with emarginate apices or spines. Cell wall usually smooth, at times with rows of spines near the incisions or marginal denticulations; entire body of cell rarely covered with spines. Chloroplast a flattened massive axial plate having the same general contour as the semicell, with numerous small ridges at right angles to the chloroplast body, lobe of chloroplast in polar lobe of cell frequently deeply incised; pyrenoids numerous, from thirty to a hundred, and fairly evenly distributed throughout the chloroplast. Vertical view usually narrowly fusiform-elliptic and lateral view of semicells narrowly pyramidate.

Zygospores spherical, with long stout radially disposed spines that terminate in simple to quadrifid apices.

The genus comprises a heterogeneous range of cell shapes but the incisions forming the lobes are a very constant feature in all species. The cells are usually compressed and disc-shaped but the discovery of a triradiate variety of *M. pinnatifida* by William West (Jour. of Bot. **27**: 206, *pl.* 291, *fig.* 15. 1889) shows that even in this genus certain qualifying statements must be made about the shape of the cell. Another feature of *Micrasterias* species is the large cell size and, although certain other genera have a few species that approach them in size, large cells are the rule and not the exception in *Micrasterias*.

KEY TO THE SPECIES.

Cells not united in filaments.
  Polar lobe without accessory processes.
    Semicells with two lateral lobes.
      Cells small .... (1) ***M. pinnatifida***
      Cells large.
        Polar incision acute-angled .... (2) ***M. laticeps***
        Polar incision broadly rounded .... (3) ***M. depauperata***

Semicells with four lateral lobes.
Incisions of lateral lobes shallow.
Polar incisions deep .... (5) *M. conferta*
Polar incisions shallow .... (4) *M. truncata*
Incisions of lateral lobes deep.
Cell outline oval.
Lobes with more than one series of incisions.
Sides of incisions not undulate.
Lobelets not deeply emarginate .... (6) *M. denticulata*
Lobelets deeply emarginate.
With a conical projection above the isthmus (9) M. *Thomasiana*
Without a conical projection above the isthmus (8) *M. apiculata*
Sides of incisions undulate .... (7) *M. radiosa*
Lobes with a single incision .... (11) *M. radiata*
Cell outline quadrangular .... (10) *M. Torreyi*
Polar lobe with accessory processes.
Polar angles continued in long processes.
Lateral processes in the same plane .... (13) *M. Nordstedtiana*
Lateral processes not in the same plane .... (14) *M. muricata*
Polar angles with short processes .... (12) *M. americana*
Cells united in filaments .... (15) *M. foliacea*

1. MICRASTERIAS PINNATIFIDA (Kützing) Ralfs. Pl. **59**, Figs. **1–2**.

Brit. Desm. 77, *pl.* 10, *figs.* 3A–3B. 1848; Johnson, Bot. Gaz. 19: 58, *pl.* 6, *figs.* 5–6. 1894; W. & G. S. West, Monogr. Brit. Desm. 2: 80, *pl.* 41, *figs.* 7–13. 1905.

Cells small, breadth slightly greater than the length, deeply constricted, sinus with apical portion linear and outer portion somewhat open, isthmus narrow; semicells three-lobed, polar lobe with basal portion subrectangular and with apical portion laterally expanded and resembling the lateral lobes but smaller, ends of lateral expansions bifid, polar lobe flattened to slightly convex; incisions between polar lobes and lateral lobes very broad and with rounded apices; lateral lobes horizontally disposed, lateral margins somewhat convergent and apex bifid (rarely trifid). Cell wall smooth. Vertical view narrowly elliptic, with poles drawn out and acuminate. Lateral view of semicells ovate-pyramidate, with apices rounded. Chloroplast extending into lateral lobes and lateral extensions of polar lobe, with two prominent vertical ridges; pyrenoids usually three, one in each lobe. (Tychoplanktont.)

Zygospores spherical, with stout tapering spines that have rounded apices (W. & G. S. West).

Cells 54–60 μ long; 61–66 μ broad at the basal lobes, 45–50 μ broad at apex of polar lobes; isthmus 10–12.5 μ broad.

Kawaguesaga (rrr), Jenny (rrr), Lindy (rrr).

Readily distinguished by the small size and the broad incisions between the lobes.

2. MICRASTERIAS LATICEPS Nordstedt. Pl. **59**, Fig. **3**.

Vidensk. Medd. f. d. naturhist. Foreń. i Kiöbenhavn 1869: 220, *pl.* 2, *fig.* 14. 1870; Wolle, Desm. U. S. 115, *pl.* 37, *figs.* 4-5. 1884; Borge, Ark. för Botanik 1: 116, *pl.* 5, *figs.* 11–15. 1903.

Cells large, quadrangular, breadth somewhat greater than the length, very deeply constricted, sinus with apical portion linear and outer portion acute-angled, isthmus narrow; semicells three-lobed, polar lobe transversely fusiform with apex rounded and lateral angles very acute and slightly bent towards the base of the semicell; lateral lobes entire, horizontally disposed, lanceolate and with deeply bifid apices. Cell wall minutely punctate. Vertical view fusiform, with poles acuminate. Chloroplast completely filling the semicell, portion in polar lobe entire, with two conspicuous vertical ridges extending from the nucleus to the apex of the cell, with other ridges following the outline of the cell; pyrenoids 12–15. (Tychoplanktont.)

Zygospores unknown.

Cells 140 μ long; 196 μ broad at basal lobes, 185 μ broad at polar lobe; isthmus 26 μ broad.

Crooked (rrr), Muskallonge (rrr).

Borge has shown that this alga is subject to considerable variation in the height of the lateral lobes and the degree of curvature of the apex of the polar lobe. In view of this wide range of variation it seems unwise to recognize the forms *major* and *minor* of Nordstedt.

The alga should be compared with *M. pinnatifida* (Kützing) Ralfs and *M. oscitans* Ralfs. It differs from the former in the much greater size and from the latter in the greater breadth of the polar lobe and the narrower lateral lobes.

3. Micrasterias depauperata Nordstedt.

Vidensk. Medd. f. d. naturhist. Foren i Kiöbenhavn 1869: 222 (with text fig.). 1870; Cushman, Bull. Torr. Bot. Cl. 31: 395, *text fig.* 1. 1904; Borge, Arkiv för Botanik 15, No. 13: 64, *pl.* 5, *figs.* 16–17. 1919.

Cells of medium size, length about a third greater than the breadth, fairly deeply constricted, sinus sublinear and with outer portion somewhat open, isthmus narrow; semicells three-lobed, incision forming lobes deep and very broadly rounded, polar lobe with a very broad basal portion and upper portion somewhat expanded, polar apex convex and slightly retuse in the median portion, angles of the polar lobe deeply emarginate; basal lobes horizontally disposed, subrectangular, with a shallow semicircular incision and the lobelets deeply emarginate or the incision lacking and basal lobes with one or two teeth. Cell wall punctate. Vertical view broadly fusiform. Lateral view of semicells ovate.

Zygospores unknown.

Cells 81–104 μ long; 64–72 μ broad at basal lobes, 44–63 μ at apex of polar lobes; isthmus 14–16 μ broad; thickness up to 36 μ (Borge).

Not found in the plankton of Wisconsin lakes.

var. Kitchelii (Wolle) W. & G. S. West. Pl. 59, Fig. 4.

Trans. Linn. Soc. 2nd. Ser. Bot. 5: 239. 1896; Cushman, Bull. Torr. Bot. Cl. 31: 396, *text fig.* 2. 1904.

Micrasterias Kitchelii Wolle, Desm. U. S. 116, *pl.* 37, *figs.* 1–3. 1884; Saunders, Proc. Washington Acad. Sci. 3: 408, *pl.* 44, *fig.* 23. 1901.

Cells larger, sinus with extremities more open; polar lobe much shorter; lateral lobes extending beyond the polar lobes and with incisions deeper. (Tychoplanktont.)

Cells 142 μ long; 150 μ broad at basal lobes, 94 μ broad at apex of polar lobe; isthmus 20 μ broad.

Tank (rrr).

4. Micrasterias truncata (Corda) De Brébisson. Pl. **60**, Figs. **1–2**.

in Ralfs, Brit. Desm. 75, *pl.* 8, *fig.* 4; *pl.* 10, *figs.* 5A–5D. 1848; G. S. West, Jour. Linn. Soc. Bot. 34: 383, *pl.* 9, *figs.* 9–16. 1899; W. & G. S. West, Monogr. Brit. Desm. 2: 82, *pl.* 42, *figs.* 1–8; *pl.* 45, *figs.* 5–6. 1905.

Cells relatively small, circular in outline and with poles flattened, deeply constricted, sinus linear and with outer portion slightly open, isthmus narrow; semicells five-lobed, polar incisions somewhat open, comparatively shallow and with apex blunt; lateral incisions about half the depth of the polar incisions and acute-angled to sublinear; polar lobe broadly cuneate, with lateral margins widely divergent, polar apex convex and without a median depression, polar angles usually emarginate-bifid, rarely acuminate; lateral lobes asymmetrical to symmetrical in shape, with a very shallow rounded incision and the lobelets generally strongly emarginate. Cell wall smooth to punctate. Vertical view elliptic, with poles acuminate. Lateral view of semicells ovate-pyramidate. Chloroplasts with two prominent vertical ridges and irregularly disposed accessory ridges; pyrenoids 5–10. (Facultative planktont.)

Zygospores unknown.

Cells 70–87 μ long; 60–98 μ broad at the base of semicell, 55–68 μ broad at apex of polar lobe; isthmus 15–19 μ broad.

Beaverdam (rr), Chetek (rrr), Fish (rrr), Horseshoe (rrr), Sand (rrr), Soft (sss), Speese (rrr), Tank (rrr).

5. Micrasterias conferta Lundell.

Nova Acta Reg. Sci. Upsaliae 3 Ser. 8: 14, *pl.* 1, *fig.* 5. 1871; W. & G. S. West, Monogr. Brit. Desm. 2: 88, *pl.* 43, *figs.* 4–8. 1905.

Cells fairly small, with an elliptic outline, median constriction very deep, sinus linear and with outer extremities overlapping to slightly dilated, isthmus narrow; semicells five-lobed, with incisions forming the polar lobes much deeper than those forming the lateral lobes, polar incisions linear and outwardly curved, lateral incisions linear and straight; polar lobes obversely campanulate with bases narrow and apices flaring, apical margin convex but with median portion emarginate, apical margins with two erect spines near the angles and apical emargination flanked on either side by a small mamillate projection terminating in a short spine; lateral lobules with two series of shallow incisions, the ultimate lobules having their ends emarginate to slightly bispinate. Cell wall smooth to finely punctate, without spines or other ornamentation. Vertical view with sides parallel and ends rounded but terminating in an acute point. Lateral view of semicells oblong, with vertical sides slightly concave, apex rounded and minutely trispinate.

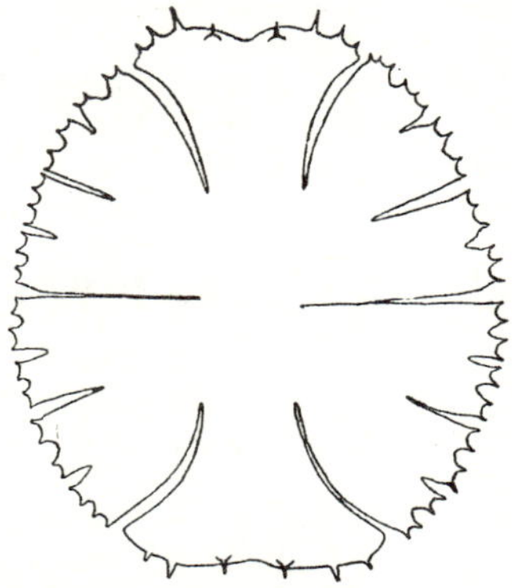

Fig. 4. *Micrasterias conferta* Lundell (x 500). From a plankton specimen collected at Figgenldin, Norway, by H. Huitfeldt–Kaas.

Zygospores unknown.

Cells 100 μ long; 87.5 μ broad at the basal lobes, 40 μ at apex of apical lobe; 17.5 μ broad at isthmus.

Not found in the plankton of Wisconsin lakes.

var. HAMATA Wolle. Pl. **59**, Figs. **6–7.**

Bull. Torr. Bot. Cl. 10: 19, *pl.* 27, *fig.* 1. 1883; W. & G. S. West, Trans. Linn. Soc. 2nd. Ser. Bot. 5: 241, *pl.* 14, *figs.* 8–9. 1896; Monogr. Brit. Desm. 2: 90, *pl.* 43, *figs.* 10–11. 1905.

Incision forming polar lobes more widely open, lateral margin of polar lobe with a semicircular excavation near the apex that causes the lateral margins to be subparallel. Chloroplast extending to margins of lobes, incision in polar lobe of chloroplast shallow, with two longitudinal ridges extending from polar apex to nucleus; pyrenoids 8–10. (Tychoplanktont.)

Cells 75–92 μ long; 73–82 μ broad at base of semicells, 34–37 μ at apex of polar lobe; isthmus 12–13 μ broad.

Des Moines (rrr).

The variety and the type are distinguished by the degree of expansion in the upper part of the polar lobe. In size, general contour, and depth of incision the species resembles *M. papillifera* De Brébisson but the two differ in the shape of the polar lobe. The variety differs from the type in the curvature of the upper part of the lateral margins of the polar lobe, a curvature which causes the incision to be broader and which gives the polar lobe an exserted appearance. W. & G. S. West have shown the close relationship between the type and the variety and have figured specimens with one semicell typical and the other with the polar features of the variety.

6. Micrasterias denticulata De Brébisson. Pl. 59, Fig. 5.

Mem. de la Soc. Acad. Falaise 1835: 54, *pl.* 8. 1835; Ralfs, Brit. Desm. 70, *pl. figs.* 1A–1G. 1848; W. & G. S. West, Monogr. Brit. Desm. 2: 105, *pl.* 49, *figs.* 1–7; *pl.* 50, *figs.* 1–2. 1905; Carter, Ann. of Bot. 33: 299, *pl.* 19, *figs.* 2–8. 1919.

Cells large, broadly elliptic in outline, length somewhat greater than the breadth, very deeply constricted, sinus linear from apex to extremity, isthmus narrow; semicells five-lobed, incisions forming polar and lateral lobes deep and linear; polar lobe with lateral margins of basal portion subparallel and sides of upper portion somewhat divergent, apical margin rounded and not exserted, with a broad emargination in the median portion, angles of polar lobe blunt and rounded; lateral lobes equal in size, cuneate in shape and redivided by two series of linear incisions, the depth of each succeeding incision being about half that of the preceding series, ultimate lobules with broad truncate ends that have rounded angles and a median emargination of variable depth. Cell wall punctate but without ornamentation. Vertical view narrowly elliptic, with poles forming sharp points and median portion of lateral margins with three small rounded elevations. Lateral view of semicells narrowly pyramidate, with the bases bearing an acutely pointed swelling and apex rectangular. Chloroplasts completely filling the cells, portion extending into the polar lobe with a conspicuous excavation, ridges of chloroplasts following the major incisions and not extending across the body of the semicell; pyrenoids numerous, 20–40 in each semicell. (Tychoplanktont.)

Zygospores spherical, with a thick lamellated wall bearing stout spines that are the same diameter throughout and have their apices bifid to trifid, at times with the incised portion recurved (Ralfs).

Cells 236–275 μ long; 195–235 μ broad at base of semicells, 54–68 μ at apex of polar lobes; 28 μ broad at isthmus.

Zygospores 80–116 μ in diameter without spines, 170–195 μ with spines (W. & G. S. West).

Beaverdam (rrr), Bass (rrr), Speese (rrr).

7. Micrasterias radiosa Ralfs non (Lyngbye) Agardh.

Brit. Desm. 72, *pl.* 8, *fig.* 3. 1848.

Micrasterias Sol (Ehrenberg) Kützing, Species Algarum 171. 1849; W. & G. S. West, Monogr. Brit. Desm. 2: 95, *pl.* 46, *figs.* 1–2. 1905.

Cells large, breadth equal to the length, circular in outline, very deeply constricted, sinus linear, with sides undulate and apex acute, isthmus very narrow; semicells five-lobed, incisions between lobes linear and deep; polar lobe arising near the isthmus, with lateral margins subparallel and apical portion somewhat expanded, apical margin deeply emarginate and with lateral margins deeply bifid, emargination flanked on either side by a subapical mamillate projection that bears a single conspicuous erect spine; lateral margins deeply incised, with inner portion of incisions more open and giving the lobelets an undulate appearance; superior lateral lobe usually somewhat larger than the inferior, superior of lateral lobes with three series of incisions the depth of each successive incision a third less

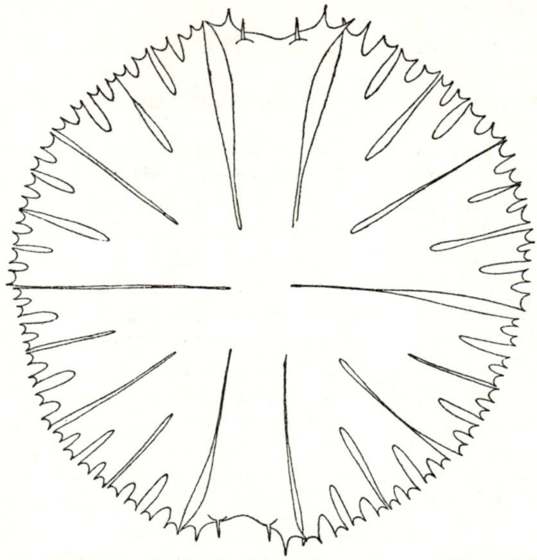

Fig. 5. ***Micrasterias radiosa*** **Ralfs (x 400). A plankton specimen from Grasmere, England, in the G. S. West collection.**

than that of the preceding series, ultimate lobelets deeply emarginate and bifid, inferior lateral lobes sometimes with only two series of incisions. Cell wall smooth. Vertical view fusiform, with central portion slightly elevated and truncate. Lateral view of semicells linear-oblong, with apex tridenticulate. Chloroplast extending to ultimate lobelets, with apex of portion in polar lobe deeply incised, with two conspicuous vertical ridges extending from polar apex to the nucleus and secondary ridges bordering the major incisions; pyrenoids numerous, 50–65, in all parts of the chloroplast.

Zygospores unknown.

Cells 172 μ long; 174 μ broad at the base of the semicells, 50 μ at the apex of the polar lobe; isthmus 20 μ broad.

Not found in the plankton of Wisconsin lakes.

W. & G. S. West have discussed the nomenclature of this alga and shown that what Ralfs calls *M. radiosa* is not *M. radiosa* (Lyngbye) Agardh but *Micrasterias Sol* (Ehrenberg) Kützing (*Euastrum Sol* Ehrenberg). In spite of the fact that the name of *M. Sol* has priority, the ruling that Ralfs' British Desmidieae is to be taken as the starting point for the nomenclature of the Desmids requires the retention of the specific name *radiosa.*

It is distinguished from other deeply incised circular species by the deeper incisions, greater subdivision, and undulate margins of the lobes.

var. ORNATA Nordstedt. Pl. **60**, Figs. **3–4**.

Vidensk. Medd. f. d. naturhist. Foren. i Kiöbenhavn 1869: 223, *pl.* 2, *fig.* 11. 1870.

Micrasterias Sol var. *ornata* Nordstedt. W. & G. S. West, Monogr. Brit. Desm. 2: 97, *pl.* 46, *figs.* 3–4. 1905.

Cells with the same proportions, degree of incision and undulate nature of incisions as in the type; portions of lobes adjacent to sinus, polar incisions, and sometimes lateral incisions with a single row of small spines. (Tychoplanktont.)

Cells 240–248 μ long; 240–258 μ broad at base of semicells, 32–35 μ broad at apex of polar lobe; 25–27.5 μ broad at isthmus.

Beaverdam (rr), Found (rrr), Lindy (rrr), Nicaboyne (rrr), Sand (rrr).

The breadth of the incision is frequently much wider than that which Nordstedt gives in his original figure. The form name *elegantior* has been given to those cells with more open incisions and narrower lobelets by G. S. West (Mém. de la Soc. neuchâteloise d. sc. nat. **5**: 1035, *pl.* 22, *fig.* 44. 1914) and Figure **4** shows a Wisconsin specimen which agrees well with his form.

8. MICRASTERIAS APICULATA (Ehrenberg) Meneghini. Pl. **61**, Fig. **1**.

Linnaea 14: 216. 1840; W. & G. S. West, Monogr. Brit. Desm. 2: 97, *pl.* 47, *figs.* 1–2. 1905; Carter, Ann. of Bot. 33: 300 (with text fig.). 1919.

Cells large, length somewhat greater than the breadth, deeply constricted, sinus with apex linear but outer portion somewhat open, isthmus narrow; semicells five-lobed, with incisions forming lobes extending about half way from the periphery to the cell center, polar incision open and acute-angled, lateral incisions narrow and with slightly divergent to convergent sides; polar lobe with lateral margins subparallel at base and then outwardly divergent at apex, polar apex deeply emarginate and polar angles with two fairly long divergently curved spines, polar apex with another conspicuous erect inwardly curved spine just within the lateral margin, apex frequently with a more delicate spine on each side of the median notch; lateral lobes usually equal in size and with two series of rather wide incisions, the ultimate lobelets terminating in 1–3 conspicuous outwardly curved spines. Cell wall covered with short spines that point towards the periphery of the cell and which are regularly arranged near the incisions but irregularly arranged in the median portion of the lobes; region just above the isthmus with 3–5 (usually four and quadrately arranged) more conspicuous spines. Vertical view rhomboid-elliptic, with acute poles. Lateral view of semicells narrowly truncate-pyramidate. Chloroplasts completely filling the semicells, with apex of portion in the polar lobe deeply incised, with two prominent ridges extending from the nucleus to the apex of the polar lobe and with secondary ridges near each incision of the lateral lobes; about 30 pyrenoids in each chloroplast. (Tychoplanktont.)

Zygospores unknown.

Cells 260 μ long; 240 μ broad at base of semicells, 96 μ broad at apex of polar lobe; 32.5 μ broad at isthmus.
Hill (rrr).

var. FIMBRIATA (Ralfs) Nordstedt. Pl. **62**, Fig. **1**.

Vidensk. Medd. f. d. naturhist. Foren. i Kiöbenhavn **1888**: 190. 1888. W. & G. S. West, Monogr. Brit. Desm. **2**: 99, *pl.* 46, *fig.* 6; *pl.* 47, *figs.* 3–4. 1905.

Micrasterias fimbriata Ralfs, Brit. Desm. 71, *pl.* 8, *fig.* 2. 1848; Wolle, Desm. U. S. 109, *pl.* 36, *figs.* 1–6. 1884; Freshw. Algae of U. S. 338. 1887.

Variety with polar lobes less sharply exserted on account of the narrower polar incisions; polar lobe with the curved spine within the lateral margin lacking but with the delicate spine on either side of the apical notch. Cell wall smooth. (Tychoplanktont.)

Zygospores spherical, with scattered rather slender spines that have furcate apices and with the furcate ends recurved; at times with the notch very deep and extending below the middle of the spine (Wolle).

Cells 293 μ long; 252 μ broad at base of semicells, 67.5 μ broad at apex of the polar lobe; 40 μ broad at the isthmus.

Clear (rrr), Horseshoe (rrr).

Ralfs describes both *M. apiculata* and *M. fimbriata* in his British Desmidieae. The two are not considered distinct species since both have the same recurving of polar spines of different length and the same recurving of the spines at the ends of the lateral lobules. The chief difference in cell shape of the two is the exsertion of the polar lobe and the greater inflation of this lobe in *M. apiculata.* Since *M. apicultata* was described first it is taken as the type and *M. fimbriata* is placed as a variety. The variety *spinosa* of *M. fimbriata* does not seem worthy of varietal recognition so is considered a form which differs from *M. apiculata* var. *fimbriata* in the small spines near the lateral and polar incisions. All transitions from this form to the true variety can be found.

Forma *spinosa* (Bissett) W. & G. S. West (Monogr. Brit. Desm. **2**: 100, *pl.* 47, *fig.* 5. 1905). Cells with rows of spines along the median construction, polar and lateral incisions. Pl. **61**, Fig. **2**. (Tychoplanktont.)

Cells 257 μ long; 220 μ broad at base of the semicells, 70 μ broad at apex of the polar lobe, 39 μ broad at the isthmus.

Crooked (rrr).

9. MICRASTERIAS THOMASIANA Archer. Pl. **60**, Fig. **5**.

Quart. Jour. Micr. Sci. N. S. **2**: 239, *pl.* 12, *figs.* 1–10. 1862; G. S. West, Jour. Linn. Soc. Bot. **34**: 384, *pl.* 9, *figs.* 6–8. 1899; W. & G. S. West, Monogr. Brit. Desm. **2**: 110, *pl.* 51, *figs.* 3–6. 1905.

Cells large, oval in outline, deeply constricted, sinus linear and with outer portion closed, isthmus narrow; semicells five-lobed, lobular incisions narrow and extending slightly more than half the distance from the periphery to the cell center; polar lobe with lateral margins subparallel in basal portion and somewhat outwardly divergent in the apical portion, polar apex not projecting beyond the lateral lobes and with a fairly deep, acute or rounded emargination at the center; lateral angles emarginate to bifid, central emargination at times flanked on either side by an erect spine; lateral lobes of approximately the same size, with three series of incisions, the depth of each successive incision approximately half that of the preceding series, ultimate lobelets with outer face slightly emarginate; face of cell with a transverse row of three large projections just above the isthmus, the central conical projection being flanked on either side by an outwardly curved projection with a mamillate to a truncate apex and with two spines; primary lateral lobules with a conical tooth at the base, secondary lobules sometimes each with a similar but smaller tooth; polar lobe with a basal tooth and sometimes two subapical teeth; teeth always directed outwards and lying in a semicircle that follows the general contour of the semicell. Vertical view narrowly fusiform, with a central conical projection and on either side of this an emarginate projection that faces the poles. Lateral view of semicells narrowly pyramidate, with apex truncate and bases rounded. Chloroplast extending into the ultimate lobelets, portion in polar lobe deeply emarginate, with longitudinal ridges following the major incisions; pyrenoids 20–40. (Tychoplanktont.)

Zygospores spherical, with stout furcate spines (W. & G. S. West).

Cells 225 μ long; 192 μ broad at the base of the semicells; 47.5 μ broad at the apex of the polar lobe; 42.5 μ broad at the isthmus.

Meta (rrr).

The species is readily distinguished from the other deeply incised ones by the basal projections of the semicells and the arc of teeth at the base of the lobelets. These characters are, however, subject to great variation and individuals may be found where the prominence of these features is greatly reduced and where the central basal protuberance and certain of the subperipheral teeth are lacking.

10. Micrasterias Torreyi Bailey. Pl. **62**, Fig. **2**.

in Ralfs, Brit. Desm. 210, *pl.* 35, *fig.* 5. 1848; Wolle, Desm. U. S. 108, *pl.* 30, *figs.* 1–8. 1884.

Cells very large, rectangular in outline, length and breadth about equal, very deeply constricted, sinus linear and with outer portion somewhat open, isthmus very narrow; semicells five-lobed, incisions deep, with apices sublinear and outer portion acute-angled; polar lobe arising near the isthmus, lateral margins of the polar lobe subparallel but widely divergent near the apex, apical margin with a conspicuous median emargination that gives it a concave appearance, median emargination flanked on either side by a simple mamillate swelling; polar angles fairly acute and simple, very rarely emarginate; lateral lobes of unequal size, the superior lateral lobe very deeply incised and one of the lobelets, usually the one adjacent to the polar lobe, in turn deeply incised, at times with both lobelets incised; inferior lateral lobes usually with but a single deep incision; all ultimate lobelets obversely spatulate and with their apices acuminate, truncate, or deeply emarginate. (Tychoplanktont.)

Zygospores unknown.

Cells 410 μ long; 450 μ broad in widest part, 74 μ broad at apex of polar lobe; 27 μ broad at the isthmus.

Lake of the Woods (rrr).

This Desmid stands out from all others in its large size and in the palmate nature of the semicells. Wolle's figures show that it is exceedingly variable in the amount of incision of the lateral lobes and in the nature of the ends of the lobelets.

11. Micrasterias radiata Hassall. Pl. 63, Figs. 1–3.

Brit. Freshw. Algae 386, *pl.* 90, *fig.* 2. 1845; W. & G. S. West, Monogr. Brit. Desm. 2: 113, *pl.* 52, *figs.* 1–4, 6–9. 1905; Carter, Ann. of Bot. 33: 300, *pl.* 20, *figs.* 15–18. 1919.

Micrasterias furcata Ralfs non Agardh, Brit. Desm. 73, *pl.* 9, *figs.* 2A–2B. 1848; Wolle, Desm. U. S. 111, *pl.* 35, *figs.* 5–6. 1884; Johnson, Bot. Gaz. 19: 58, *pl.* 6, *figs.* 7–8. 1894.

Cells fairly large, length usually slightly greater than the breadth, very deeply constricted, sinus diamond-shaped, with apex acute and outer portion somewhat convergent, isthmus narrow; semicells five-lobed, polar lobe with basal portion oblong and upper portion somewhat dilated, angles of polar lobe continued in long slightly attenuated divergent processes with deeply furcate apices, apex of polar lobe somewhat concave; lateral lobes formed by a deep, rather bluntly pointed, lanceolate incision that extends half way from the periphery to the cell nucleus, lobes divided by an acute-angled incision with a rounded apex to form two lobules with subparallel sides and deeply furcate apices. Cell wall smooth. Vertical view narrowly fusiform, with poles acute. Lateral view of semicells narrowly cordate, with apex rounded. Chloroplast extending into the lobules, portion in polar lobe with apex entire, with two vertical ridges extending from the polar apex to the nucleus, ridges of lateral lobes somewhat irregular; pyrenoids 12–15, rather evenly distributed. (Facultative planktont.)

Zygospores unknown.

Cells 172–185 μ long; breadth 142–165 μ, breadth at apex of polar lobe 85–115 μ; breadth at isthmus 18–21 μ.

Beaverdam (rrr), Diamond (rrr), Horseshoe (rr), Kawaguesaga (rrr), Kimball (rrr), Sunday (rr).

W. & G. S. West have given a full nomenclatorial discussion of the species and have shown conclusively why Ralfs' name is untenable.

The deep furcation of the lateral lobes and the wide polar and lateral incisions produce much narrower lobules than those of other species which have the same type of incision. The species is subject to considerable variation in the size and delicacy of the apical and lateral lobes, the short stocky form shown in Figure **2** being much more common than the delicate one of Figure **1**.

The incision of the lateral lobes is also subject to variation and all transition stages may be found from those in which only one lateral lobe is entire to those in which there is no incision of the lateral lobes of either semicell. W. & G. S. West follow Johnson in considering this the normal range for the type and think that no varietal distinction can be made on this reduction of the incision in the lateral lobes. The point at which varietal distinctions shall be drawn is a perennial question for the taxonomist and the treatment of the range of variation rests in a large degree upon the personal view-point. My observation of the alga in Canadian lakes, where in certain stations cells with reduced lobes are found to the exclusion of those with incised lobes, leads me to the conclusion that Wolle was justified in considering the forms with reduced lobes a distinct variety. This view is strengthened by the statement of W. & G. S. West that: "this variation is of far commoner occurrence in the United States than in the British Islands . . . in Britain we have only met this notable variation of the lateral lobes in the Scottish plankton."

A most unusual variation from the normal is shown in Figure 3 where there are three simple processes on the polar lobe of one of the semicells as well as a lack of bifurcation of the lateral lobules on one half of the semicell.

var. SIMPLEX (Wolle) G. M. Smith. Pl. **63**, Fig. **4**.

Trans. Wis. Acad. Sci., Arts & Lett. **20**: 344, *pl.* 9, *figs.* 1–5. 1922.

Micrasterias furcata var. *simplex* Wolle, Bull. Torr. Bot. Cl. **12**: 128, *pl.* 51, *figs.* 6–7. 1885.

Micrasterias radiata Hassall. W. &. G. S. West, Monogr. Brit. Desm. **2**: 113, *pl.* 52, *fig.* 5. 1905.

Micrasterias furcata Ralfs non Agardh. Johnson, Bot. Gaz. **19**: 58, *pl.* 6, *figs.* 9–14. 1894.

Lateral lobes of processes of one or both semicells simple and terminating in a deeply furcate-emarginate apex. (Tychoplanktont.)

Cells 135 μ long; greatest breadth 130 μ, breadth at apex of polar lobe 65 μ; breadth at isthmus 17.5 μ.

Hill (rrr).

In the majority of cases all of the lobes are not reduced and one or more has the deep incision of the type. When this is the case the incision occurs in the lateral lobes adjacent to the polar lobe rather than in the lobes adjacent to the sinus.

var. GRACILLIMUM G. M. Smith. Pl. 64, Fig. 1.

Trans. Wis. Acad. Sci., Arts & Lett. 20: 344, *pl.* 9, *figs.* 6–11. 1922.

Cells slightly larger than the type; base of the polar lobe much narrower; lateral lobes not incised, with parallel sides and apices furcate; horizontally disposed bases of the semicells only slightly tapering. (Facultative planktont.)

Cells 160 μ long; greatest breadth 130 μ, polar lobe 65 μ broad at apex; isthmus 17.5 μ broad.

Lindy (rrr), Little Doctor (rrr).

Distinguished from the foregoing by the much narrower base of the polar lobe and by the nature of the lateral margins. The basal portion of the semicell is likewise much narrower than the basal portion of the semicell in the type, this narrowness giving the cell a more delicate appearance. In this variety no case has been observed where there is an incision of the lateral lobes.

12. MICRASTERIAS AMERICANA (Ehrenberg) Ralfs. Pl. 63, Fig. 5; Pl. 64, Fig. 2.

Brit. Desm. p. XIX. 1848; Wolle, Desm. U. S. 112, *pl.* 32, *fig.* 2. 1884; W. & G. S. West, Monogr. Brit. Desm. 2: 117, *pl.* 53, *figs.* 4–5; *pl.* 54, *figs.* 1–3. 1905.
Micrasterias morsa Ralfs, Brit. Desm. 74, *pl.* 10, *figs.* 1A–1E. 1848.

Cells fairly large, deeply constricted, sinus acute-angled and with apex acuminate, isthmus narrow; semicells five-lobed, incisions between polar and lateral lobes broad and with blunt apices; incisions between lateral lobes narrowly fusiform and with apices rounded; polar lobe with lateral margins parallel at base and then outwardly divergent to form a cuneate apex; apical margin of polar lobe broadly retuse and with lateral margins forming upwardly divergent processes with slightly convergent sides and truncate-emarginate apices; apex of polar lobe with a second similar but smaller process arising near the base of each of the lateral processes, secondary processes less outwardly divergent than the primary processes and asymmetrically disposed, one on each side of the polar lobe; lateral lobes with a shallow broad incision dividing them asymmetrically, the lobules adjacent to the polar incision or to the sinus being longer than those adjacent to the median lateral incision; all lobules with ends truncate-retuse and with 3–4 acute or rounded denticulations; portion of lobules adjacent to the secondary incisions, apical polar incisions and sometimes the portion of the cell bordering the polar incision with marginal denticulations; semicell body with two fairly conspicuous blunt teeth at the base of the polar incisions, sometimes with a central supraisthmal protuberance bearing scattered granules; polar and lateral lobes with a few scattered granules. Vertical view rhomboidal, with poles truncate and polar lobe showing the asymmetrical position of the processes. Lateral view of the semicells narrowly pyramidate, with bases rounded and apices continued in four short processes. Chloroplast extending into all five lobes and with the portion in the polar lobe fairly deeply incised, with two vertical ridges extending into the polar lobe and one or two ridges into each lateral lobe; pyrenoids more numerous in body of the cell than in lobes, 10–20 in each chloroplast. (Tychoplanktont.)

Zygospores unknown.

Cells 125–140 μ long; greatest breadth 104–140 μ, breadth at apex of polar lobe 54–72 μ; breadth at isthmus 27–29 μ.

Found (rr), Horseshoe (rrr), Howey (rrr), Thousand Islands (rrr).

M. americana is subject to considerable variation in the length of the polar processes and the depth of incision of the lateral lobes. The granules of the cell body are always scattered, though they have a tendency to be more numerous and more regularly arranged on the portions bordering the incisions. The two prominent spines at the bases of the polar incisions have not been pointed out by previous investigators of the alga and my attention was first called to them by Mr. Charles Bullard. I find that they are quite constant in all American specimens that I have examined, as well as in European specimens in my herbarium.

M. mahabuleshwarensis Hobson is quite like *M. americana* in the asymmetrical processes of the polar lobes and the general cell shape but differs in the lack of a lateral incision that divides the semicells into five lobes.

13. Micrasterias Nordstedtiana Wolle. Pl. **64**, Fig. **3**.

Bull. Torr. Bot. Cl. **11**: 15. 1884; Desm. U. S. 113, *pl.* 52, *figs.* 3–5. 1884; Johnson, Bull. Torr. Bot. Cl. **22**: 292 *pl.* 239, *fig.* 14. 1895; W. & G. S. West, Trans. Linn. Soc. 2nd. Ser. Bot. **5**: 239, *pl.* 14, *fig.* 4. 1896.

Cells large, length and breadth about equal, deeply constricted, sinus somewhat open and with an acute apex, isthmus narrow; semicells five-lobed, polar incision widely open and with a rounded apex that gives the polar lobe an exserted appearance; polar lobe elongate, with erect subparallel sides that bear on each side of the median portion a simple mamillate projection or a conical projection with truncate ends that have short blunt teeth; upper angles of polar lobe continued in very long, slightly divergent, tapering, cylindrical processes that terminate in three minute blunt spines, polar apex flattened and with a shallow emargination in the median portion, apex with a second, considerably shorter, asymmetrically disposed, pair of similar processes; lateral processes separated by a blunt broad incision, subhorizontally disposed, gradually tapering to blunt apices that have three short divergent spines. Cell wall smooth or finely punctate. Vertical view diamond-shaped, with poles continued in straight cylindrical processes. Lateral view of semicells oblong, bases with a round inflation and apices rectangular. (Tychoplanktont.)

Zygospores unknown.

Cells 203 μ long; greatest breadth 180 μ, breadth at apex of the polar lobe 138 μ; breadth at isthmus 28 μ.

Diamond (rrr).

Having a superficial resemblance to *M. muricata* (Bailey) Ralfs but differing in that the lateral lobes all lie in the same plane, and in the lateral projections of the polar lobes.

14. MICRASTERIAS MURICATA (Bailey) Ralfs. Pl. **64**, Figs. **4–5**.

Brit. Desm. 210. 1848; Wolle, Desm. U. S. 118, *pl.* 31, *figs.* 4–7. 1884; W. & G. S. West, Trans. Linn. Soc. 2nd. Ser. Bot. 5: 239, *pl.* 14, *figs.* 5–6. 1896.

Cells large, length about one and a half times the breadth, fairly deeply constricted, sinus acute-angled at apex and then greatly dilated, outer portion with widely open parallel sides, isthmus relatively broad; semicells five-lobed, polar and lateral incisions very broad and with rounded apices; polar lobe with basal portion broad and rectangular, with upper portion greatly dilated and with the angles continued in long slightly tapering cylindrical processes that are upwardly curved and terminate in three or four minute blunt conical spines; polar apex convex, central portion with a conspicuous narrow emargination and the lateral margins with a second pair of asymmetrically disposed processes that are similar to those at the polar angles; superior lateral lobes divided by an incision to form two long horizontally disposed processes that lie above each other in a front view of the semicell, processes arising from a conical base; inferior lobes submamillate, with the apex continued in a long cylindrical horizontally disposed process, each process of inferior lateral lobe with two similar but shorter processes, the three lying in the same transverse plane at right angles to the long axis of the semicell, all processes of lateral lobes with truncate ends bearing three or four short blunt teeth. Cell wall smooth or finely punctate. Vertical view oblong, sides somewhat tumid and angles continued in long, slightly divergent processes, middle of each end with another similar process. Chloroplast confined to the cell body but with fine laminate projections running into the processes, portion in the polar lobe entire, with two conspicuous longitudinal ridges extending from the polar apex to the nucleus and with curving ridges at the bases of the lateral lobes; pyrenoids 12–15, confined to the body of the cell. (Tychoplanktont.)

Zygospores unknown.

Cells 160–198 μ long; 110–130 μ broad at the inferior lateral lobes, 88–106 μ broad at the superior lateral lobes, 86–96 μ broad at the apex of the polar lobes; isthmus 22–23 μ broad.

Bass (rrr), Bettine (rrr), Carson (rrr), Jenny (rrr), Lindy (rrr), Little Doctor (rrr), Nicaboyne (rrr).

The species has much in common with *Euastrum* where it was first placed by Bailey but it lacks the incision of the polar lobe and the ornamentation of the cell body that characterizes *Euastrum*. There is little danger of confusing it with other *Micrasterias* species.

15. MICRASTERIAS FOLIACEA Bailey. Pl. **65**, Figs. **1–2**.

in Ralfs, Brit. Desm. 210, *pl.* 35, *fig.* 3. 1848; Wallich, Ann. & Mag. of Nat. Hist. 3 Ser., **5**: 280, *pl.* 14, *figs.* 1–4. 1860; Wolle, Desm. U. S. 118, *pl.* 38, *figs.* 10–11. 1884; Johnson, Bot. Gaz. **19**: 56, *pl.* 6, *figs.* 1–4. 1894.

Cells of medium size, united in filaments by the interlocking of the polar lobe, rectangular in outline, very deeply constricted, sinus sublinear, isthmus very narrow; semicells five-lobed; polar lobe exserted, basal fourth of the polar lobe narrow, with erect parallel sides, upper three quarters of the lobe greatly expanded, anvil-shaped, with a very broad deep subrectangular excavation in the median portion of the apex, lateral angles of the polar lobe acuminate, lateral margins of the base of the apical excavation with two erect asymmetrically disposed spines of unequal length; lateral lobes asymmetrical, the superior lobes divergent, the inferior lobes horizontally disposed; superior lobe incised, with the lobelet bordering the polar incision simple and acuminate, with the other lobelet having an incision about half the depth of the primary lobular incision; all ultimate lobelets, except those bordering the polar incisions with truncate-emarginate apices. Cell wall smooth. Vertical view narrowly fusiform, with apices acuminate and median portion somewhat constricted. Chloroplast completely filling the cell, with the portion in the polar lobe entire, without conspicuous ridges; pyrenoids 10–15. (Tychoplanktont.)

Zygospores spherical, furnished with rather numerous, fairly long spines with truncate-emarginate apices (Wallich).

Cells 78–90 μ long; 82–85 μ broad, 45–58 μ broad at the apex of the polar lobe; 15–20 μ broad at the isthmus.

Zygospores 45 μ in diameter (Wallich).

Des Moines (rrr).

XANTHIDIUM Ehrenberg 1837.

Cells of variable, though generally of medium size; length frequently somewhat greater than the breadth, median constriction deep and sinus linear to acute-angled; semicells almost universally compressed, very rarely radially symmetrical, of variable shape, though usually polygonal and with apex flattened but not incised; lateral margins with angles that bear simple, rarely furcate, spines; spines asymmetrically disposed with reference to the vertical plane. Vertical view elliptic. Lateral view of semicells circular to polygonal. Cell wall smooth to punctate, with a thickened protuberant area in the middle of the cell face that frequently is scrobiculate. Chloroplasts at times axial and two in each semicell, at other times parietal and four in each semicell; pyrenoids usually single in each chloroplast.

Zygospores spherical, with scrobiculate walls or with walls bearing simple to furcate spines.

The angular semicells with spines at the angles and the differentiated central area of the cell are the most striking characters of the genus. Except for certain rare triradiate varieties the cells are always compressed. These varieties show a close affinity with *Staurastrum,* while the forms with reduced spines may resemble *Arthrodesmus.*

KEY TO THE SPECIES.

Spines broad furcate processes................................(1) *X. armatum*
Spines simple.
  Semicell with two spines on lateral margin..............(4) *X. subhastiferum*
  Semicell with more than two spines on lateral margins.
    Each margin with two pairs of spines...................(2) *X. antilopaeum*
    Each margin with two pairs and a single spine............(3) *X. cristatum*

1. Xanthidium armatum (De Brébisson) Rabenhorst.

Deutschl. Kryptogamenfl. 2, Abt. 2: 55. 1847 (Ref. Nordst. Index Desm.); Ralfs, Brit. Desm. 112, *pl.* 18, *figs.* A–G. 1848.

Cells large, length one and a quarter to one and a half times the breadth, deeply constricted, sinus acute-angled, with apex acuminate and outer margin widely open, isthmus narrow; semicells sub-octagonal, with margins facing the isthmus slightly convex, lateral margins concave, and apex plane; lateral and superior angles bearing very stout, short, solid processes that terminate in 2–4 tuberculate furcations, lateral margins with a single similar process within each angle; apex also with two similar subapical processes a short distance

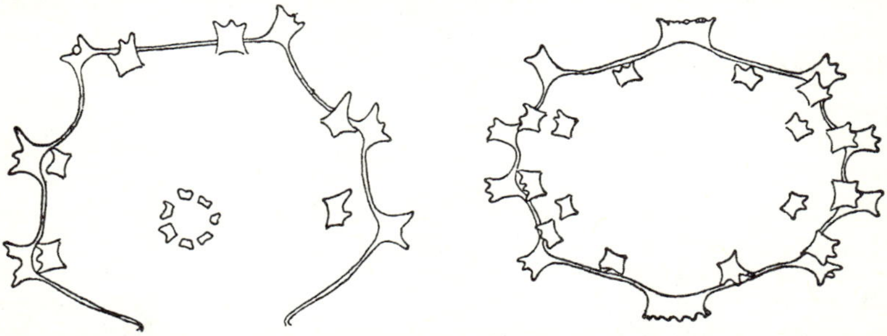

Fig. 6. *Xanthidium armatum* (De Brébisson) Rabenhorst (x 600). From specimens collected at Loch Cormisa, Skye, Scotland, by W. & G. S. West.

inward from the angle; face of semicell with a single prominent solid tuberculate process with a ring of marginal teeth. Vertical view oblong-elliptic, with poles bearing four marginal and 3–4 intra-marginal processes similar to those of the front view, middle of lateral margins with a solid, broad, truncate process with several tubercules at the distal end. Lateral view of semicells ovate-rectangular, with a broad solid tuberculate process at the base, and with four apical processes in addition to a median and basal transverse row of processes. Cell wall punctate. Semicells with four parietal chloroplasts each containing several pyrenoids.

Zygospores spherical, with thick walls containing numerous hemispherical depressions (Ralfs).

Cells 138 μ long with processes, 125 μ long without processes; greatest breadth with processes 91 μ, without processes 80 μ; apex 57 μ broad with processes, 47 μ without processes; isthmus 34 μ broad; thickness 70 μ with processes, 58 μ without processes.

Not found in the plankton of Wisconsin lakes.

var. **mediolaeve** var. nov. Pl. **65**, Figs. **3–4**.

Xanthidium armatum var. *cervicorne* G. M. Smith non W. & G. S. West. Trans. Wis. Acad. Sci., Arts & Lett. 20: 346. 1922.

Cells larger than the type, length (without spines) about a quarter greater than the breadth; concavity of lateral margins much more pronounced; processes at angle solid, much longer, with 3–5 deep furcations, the ultimate furcations being quite sharp; intramarginal processes of superior lateral angles lacking; center of cell smooth, without any sign of a process. (Tychoplanktont.)

Cells 168–175 μ long with processes, 135–138 μ long without processes; greatest breadth 120–123 μ with processes, 92–98 μ without processes; isthmus 31–33 μ broad; processes 15–20 μ long.

Sunday (rrr).

In the elegant antler-like processes of the angles this variety resembles the variety *cervicorne* W. & G. S. West, to which Canadian specimens were referred. Study of European specimens of the varieties *cervicorne* and *fissum* Nordstedt shows that in these two varieties the central protuberance on the face of the semicell is readily observed. Since this central protuberance is not found on the cells that I have collected (both from Wisconsin and Ontario) the American alga is placed as a distinct variety. The intramarginal processes of the superior lateral angles were also lacking in all specimens collected but this feature is not of so great importance as the lack of a central protuberance since the varieties *fissum* and *cervicorne* may at times also lack the superior intramarginal processes.

2. Xanthidium antilopaeum (De Brébisson) Kützing. Pl. **65**, Figs. **5–6**.

Species Algarum 177. 1849; W. & G. S. West, Monogr. Brit. Desm. 4: 63, *pl.* 108, *figs.* 7–18. 1912.

Cells of medium size, length (without spines) about equal to the breadth, deeply constricted, sinus with inner portion linear and apex very slightly dilated, outer portion widely open, isthmus fairly narrow; semicells with ventral margin semielliptic and dorsal margin semihexagonal, lateral angles and angles of apex slightly rounded, basal angles very broadly rounded, apex flattened and bearing two, straight to slightly curved, suberect to widely divergent, fairly long, acutely pointed spines at each lateral margin; lateral angles each with a pair of similar, horizontally disposed spines. Vertical view elliptic, with broadly rounded poles that bear a pair of long divergent spines and a second pair of similar intrapolar spines, lateral margins with a slight median protuberance. Lateral view of semicells subspherical, with a pair of divergent spines at the apex and a second pair in the middle of the cell face. Cell wall finely punctate and with a median thickened area that is variously scrobiculate and frequently brownish in color. Chloroplasts generally four thin parietal plates in each semicell; typically with one pyrenoid in each chloroplast. (Tychoplanktont.)

Zygospores spherical, bearing numerous long delicate spines with conical bases and bifurcate apices (W. & G. S. West.)

Cells 79–102 μ long with spines, 50–65 μ without spines; breadth of cells 80–96 μ with spines, 50–56 μ without spines; isthmus 14–19 μ broad; spines 15–22.5 μ long.

Zygospores 45–58 μ in diameter without spines, 80–98 μ with spines (W. & G. S. West).

Carroll (rrr), Des Moines (rrr), Jag (rrr), Rice (rrr).

var. POLYMAZUM Nordstedt. Pl. **65**, Figs. **7–9**.

Lunds Univ. Årsskr. 9: 38, *pl.* 1, *fig.* 20. 1873.

Cells with angles more rounded than the type, polar spines suberect (very rarely divergent.) Central area smooth but with an arc of emarginate to arcuate verrucae just below the semicell apex. Vertical view oblong-elliptic, with poles strongly emarginate. (Facultative planktont.)

Cells 70–88 μ long with spines, 52–58 μ long without spines; 73–86 μ broad with spines, 52–56 μ broad without spines; isthmus 15–17 μ broad; spines 10–15 μ long.

Beaverdam (ss), Carson (rrr), Cranberry (rrr), Crooked (rrr), Fish (rrr), Found (rr), George (rr), Hill (rr), Horseshoe (rr), Lake of the Woods (rrr), Lindy (rr), Lost (rrr), Lynx (rrr), Meta (rrr), Middle McKenzie (rrr), Nicaboyne (rr), Sand (rrr), Sanford (rrr), Speese (rrr).

The cells of this variety have more rounded angles than the type and as a result a much plumper appearance. The vertical view is also different since the poles are emarginate instead of broadly rounded. The chief feature of the variety, however, is the subapical arc of verrucae. The number and shape of these verrucae varies greatly and in certain specimens there is a double arc of verrucae.

The variety is widely distributed in Wisconsin lakes but never occurs in large quantities.

var. MINNEAPOLIENSE Wolle. Pl. **65**, Figs. **10–11**.

Bull. Torr. Bot. Cl. 11: 16. 1884; Desm. U. S. 94, *pl.* 52, *fig.* 16. 1884; W. & G. S. West, Trans. Linn. Soc. 2nd. Ser. Bot. 5: 252, *text fig.* 2. 1896.

Cell shape and arrangement of spines as in the variety *polymazum;* subapical verrucae generally arranged in an arc, rarely in an elliptical ring, with a small erect spine above or below the verrucae. (Tychoplanktont.)

Cells 78–82 μ long with spines, 50–58 μ long without spines; 79–83 μ broad with spines, 52–55 μ broad without spines; isthmus 16–17.5 μ broad; spines 12.5–15 μ long.

Bass (rr), Diamond (rrr), Horseshoe (rrr), Jenny (rr), Little Doctor (rrr), Meta (rrr), Rice (rr).

The only difference between this variety and the foregoing is the small erect spine associated with the verrucae below the apices of the semicells.

var. DEPAUPERATUM W. & G. S. West. Pl. **66**, Fig. **1**.

Trans. Bot. Soc. Edinburgh 23: 23, *pl.* 1, *figs.* 15–16. 1905.

Semicells with angles more rounded than in the type; spines on lateral margins much smaller and not regularly arranged in pairs, number of spines on a semicell frequently reduced. Central area with scrobiculations in an irregular ring. (Facultative planktont.)

Cells 68 μ long with spines, 65 μ long without spines; 84 μ broad with spines, 66 μ broad without spines; isthmus 27.5 μ broad; spines 7.5–10 μ long.

Otter (rr).

3. XANTHIDIUM CRISTATUM De Brébisson. Pl. **66**, Figs. **2–3**.

in Ralfs, Brit. Desm. 115, *pl.* 19, *figs.* 3A–3F. 1848; Reinsch, Cont. ad Alg. et Fung. 92, *pl.* 16, *fig.* 13. 1875.

Cells of medium size, length (without spines) somewhat greater than the breadth, deeply constricted, sinus sublinear to slightly open, with apex somewhat inflated, isthmus narrow; semicell elliptic-hexagonal, with sides facing the isthmus somewhat convex, lateral margins flattened, and apex plane; apical angles each with a pair of slightly divergent, medium sized, acutely pointed spines; superior lateral angles with a similar pair of somewhat divergent spines; inferior lateral angles with a single spine, basal spines of semicells convergent; face of semicell with a small, circular to elliptic, thickened area that is without ornamentation. Vertical view elliptic, with poles subtruncate and bearing three marginal and two intramarginal slightly divergent spines; lateral margins with a slight protuberance in the median portion. Lateral view of semicells subspherical, with a protuberance on each side and divergent apical and subapical spines. Cell wall finely punctate. Semicells with four parietal laminate chloroplasts, each containing a single pyrenoid. (Tychoplanktont.)

Zygospores spherical, with long spines that have parallel sides and truncate apices (Reinsch).

Cells 65 μ long with spines, 45–48 μ long without spines; 48 μ broad with spines, 35 μ broad without spines; breadth of isthmus 10–12.5 μ; length of spines 5–10 μ.

Zygospores 50.4 μ in diameter; spines 16.8–19.6 μ long (Reinsch).

Chetek (rrr), Found (rrr), Island (rrr), Lac Court Oreilles (rrr).

The single convergent spines at the basal angles of the semicells differentiate this species from all others. The central protuberance is easily overlooked since it is without ornamentation.

var. UNCINATUM De Brébisson. Pl. **66**, Fig. **4**.

in Ralfs, Brit. Desm. 115, *pl.* 19, *figs.* 3D–3F. 1848.

Cells larger, semicells broadly subpyramidate, lateral margins concave between the angles, apex broadly truncate and plane; spines at polar and superior lateral angles suberect, generally slightly curved, spines of basal angle curved and subparallel; center of face of semicell with a broad granulate area, the granules generally in an irregular ring or ellipse. (Tychoplanktont.)

Cells 75 μ long with spines, 60 μ without spines; 66 μ broad with spines, 48 μ broad without spines; isthmus 17.5 μ broad; spines 8–10 μ long.

Tank. (rrr).

The erect curved spines, granulate central area and semicell shape mark the points of difference between the variety and the type. The granulation of the central area is exceedingly variable in both size and number as well as in the arrangement of the granules.

4. XANTHIDIUM SUBHASTIFERUM W. West. Pl. **66**, Figs. **5–12**.

Jour. Linn. Soc. Bot. **29**: 166, *pl.* 22, *fig.* 4. 1892; W. & G. S. West, *ibid.* **35**: 540, *pl.* 16, *figs.* 4–5. 1903; W. & G. S. West, Trans. Roy. Irish Acad. **33**: 102, *pl.* 11, *figs.* 13–15. 1906.

Cells of medium size, length and breadth (without spines) equal, very deeply constricted, sinus acute-angled, with apex acuminate and outer margin widely open, isthmus narrow; semicells transversely elliptic, curvature of ventral and dorsal margins about the same; lateral margins truncate-retuse, with two, long, fairly stout, somewhat divergent, straight or slightly curved spines that lie in the same vertical plane; semicell apices at times with a faint indication of a truncate elevation in the median portion. Vertical view elliptic, poles with two straight spines that lie in the same vertical plane, lateral margins with wall thickened in the median portion. Lateral view of semicells circular, with portion of wall above the isthmus thickened. Cell wall finely punctate, without ornamentation in the thickened portion of the cell face. Chloroplasts parietal, four in each semicell and each with a single pyrenoid. (Euplanktont.)

Zygospores unknown.

Cells 41–73 μ long with spines, 35–52 μ long without spines; 60–101 μ broad with spines, 35–52 μ broad without spines; isthmus 9–14 μ broad; thickness 23–25 μ; spines 10–28 μ long.

Adelaide (rrr), Bear (rr), Beaverdam (c), Birch (ss), Catherine (sss), Catfish (sss), Cranberry (sss), Devils (rrr), Diamond (ss), Duck (rr), Dummy (rr), Found (rrr), Half Moon (rrr), Horseshoe (s), Kawageusaga (rr), Lac du Flambeau (rrr), Mercer (ss), No Mans (ss), Rock (ss), Sanford (rrr), Shell (rr), Sishebogema (ss), Squirrel (rr), Tamarack (rr), North Turtle (rr).

There is considerable variation from the typical structure in this species. One type of variation includes cells with supplementary spines on the apices of the semicells. These may range from a single extra apical spine on one of the two semicells (Figs. **10**) to an extra spine at the lateral margin of both semicell apices. The highest development of this variation observed is shown in Figure **12** where there are two accessory spines arising at the same point. This suggests very strongly a transition to *X. hastiferum* Turner, but the great preponderance of typical *X. subhastiferum* in the collection shows that these specimens are not reduced forms of *X. hastiferum.*

The other type of variation is a reduction of the two spines at the lateral margin to a single spine. In most cases but one lateral margin of the semicell shows this reduction. In the rarer cases where the spines on both lateral margins of a semicell are reduced to a single spine the semicells have much the appearance of *Arthrodesmus convergens* Ehrenberg.

var. TOWERI (Cushman) G. M. Smith. Pl. **66,** Figs. **13–18;** Pl. **67,** Figs. **1–3.**

Trans. Wis. Acad. Sci., Arts & Lett. 20: 346, *pl.* 10, *figs.* 9–13. 1922.

Semicells obversely subtriangular, sinus more open than in the type; ventral margin straight or curved, when curved much more tumid than the dorsal margin; spines much longer in proportion to the breadth of the semicell. (Euplanktont.)

Cells 45–75 μ long with spines, 40–47 μ long without spines; 75–90 μ broad with spines, 39–45 μ broad without spines; isthmus 10–12 μ broad; thickness 24–26 μ; spines 17.5–25 μ.

Carroll (s), Horseshoe (rrr), Shell (cc), Two Sisters (rrr).

This variety differs from the type in the greater proportionate length of the spines and in the difference between the dorsal and ventral margins of the semicells. There are the same variations in the presence of accessory apical spines or in the reduction of spines that are found in the type. The collections from Carroll lake were particularly interesting since they exhibited an almost complete series of transitions from the normal to a form that looks very much like *Arthrodesmus* or certain biradiate *Staurastrum* species.

var. **Johnsonii** (W. & G. S. West) comb. nov. Pl. **66,** Figs. **4–7.**

Xanthidium hastiferum var. *Johnsonii* W. & G. S. West, Jour. Linn. Soc. Bot. 33: 299. 1898.

Xanthidium antilopaeum var. *javanicum* Johnson non Nordstedt, Bull. Torr. Bot. Cl. 22: 295, *pl.* 239, *fig.* 10. 1895.

Semicells subsemicircular to subhexagonal, ventral margin with straight to tumid sides, apical margin strongly tumid, at times with median portion somewhat elevated; lateral margins with two, fairly stout, straight to strongly curved spines that lie in the same vertical plane. Cell wall with a transverse to arcuate row of irregularly disposed subapical pores. (Facultative planktont.)

Cells 52–63 μ long with spines, 48–57 μ long without spines; 75–95 μ broad with spines, 48–53 μ broad without spines; isthmus 12.5–15 μ broad; spines 10–20 μ long.

Beaverdam (s), Lindy (rrr), Sand (rr), Silver (rrr).

The chief character of this variety, which Johnson erroneously referred to *X. antilopaeum* var. *javanicum* Nordstedt, is the rather irregular row of subapical pores. W. & G. S. West have considered this a variety of *X. hastiferum* Turner but as the accessory apical processes are of quite rare occurrence I feel that the variety belongs rather to *X. subhastiferum.* The peculiar cell shown in Figure **7** was found in the plankton of Beaverdam lake.

STAURASTRUM Meyen 1829.

Cells variable in size, with the length (without processes, spines or other appendages) generally somewhat greater than the breadth, median constriction usually deep though at times scarcely discernible, sinus generally acute-angled and isthmus narrow; semicells cylindrical, spherical, ellipsoidal, triangular, campanulate, cyathiform, hexagonal or variously shaped; with the cell wall smooth or punctate, without ornamentation or bearing granules, mucros, denticulations, simple to emarginate verrucae, spines, or combinations of these ornamentations; angles of a majority of the species continued in processes of variable length, though usually long, that in turn almost always have some sort of ornamentation and usually terminate in a truncate end with short divergent spines, at times with bifurcate ends that resemble spines; processes ordinarily arising in one whorl though sometimes in two transverse series of whorls. Vertical view at times with a bilateral symmetry though commonly with a radial symmetry; radially symmetrical species with triangular forms predominating though the 4–8 angled or radiate species not uncommon and with the number of angles as high as eleven in one whorl. Chloroplast generally with a simple axial mass containing one pyrenoid and with deeply incised lobes of variable size running to each angle of the cell, at times, especially in the larger species, with several pyrenoids in each lobe and with various irregularities; in very rare instances with parietal chloroplasts. Cells, particularly the plankton species, at times enclosed in a spherical gelatinous envelope.

Zygospores infrequently angular, usually spherical and with numerous conical elevations that are continued in long spines that have simple or furcate apices.

The genus comprises so many species of such variable shape that it is exceedingly difficult to describe it in anything but the most general terms. In spite of the groups of species which at first glance appear to have little in common with one another there are so many transition forms that it is impossible to break it up into a number of subgenera. The simplest members of the genus resemble triangular *Cosmaria* that are without ornamentation, in fact it is impossible in certain cases to distinguish between varieties of *Cosmarium* species that are triangular in vertical view and *Staurastrum* species. These simple species lead up to those which have long spines at the angles and then to those which have spines or denticulations on both the angles and the cell body. The majority of the species have the cell angles continued in hollow processes, the triangular ones being the most numerous and probably the most typical for the genus. On one hand some of the triangular species have become compressed and form a group with a bilateral symmetry. On the other hand the radial symmetry of the species with more than three processes is very striking. The climax of development of processes is reached in those species in which there are two distinct whorls of processes on each semicell.

In addition to the close connection noted above between the simple *Staurastrum* species and *Cosmarium* there is also a close relationship between other *Staurastrum* species, particularly the biradiate simple-spined species, and *Arthrodesmus*. The triradiate varieties of one or two *Xanthidium* species also show a striking resemblance to *Staurastrum* species.

The genus is by far the most important one of the family in the plankton community. The total number of species and varieties found compares favorably with the total number of species of all the other genera of the Desmids, while the number of individuals observed is greatly in excess of all the other Desmids. It is in this genus that we find the largest number of forms which are true plankton organisms while many more that are known from the benthos seem to thrive even better under plankton conditions. In general the species with long processes, probably because of the greater buoyancy, seem best fitted for living in midlake, although some of the species without processes but with a conspicuous gelatinous envelope are also well equipped for the plankton.

KEY TO THE SPECIES.

(A) *Angles of cells not continued in hollow processes.*

(B) *Without spines at angles or on cell body.*

Cell wall smooth.
- Isthmus not elongate.
  - Angles without mucros.
    - Ventral margin less rounded than the dorsal.
      - Cells small....................................(3) *S. muticum*
      - Cells large....................................(1) *S. grande*
    - Ventral margin more rounded than the dorsal...........(2) *S. subgrande*
  - Angles with mucros...................................(4) *S. brevispinum*
- Isthmus elongate..............................(5) *S. pseudopachyrhynchum*

Cell wall with granules.
  Lateral angles rounded....................................(6) *S. alternans*
  Lateral angles acute or truncate.
    Vertical view with angles truncate....................(7) *S. denticulatum*
    Vertical view with angles acute.
      Angles with a single curved spine.......................(9) *S. lunatum*
      Angles with two divergent spines.........................(8) *S. avicula*

(B) *Cells with long spines at angles or on cell body.*

Spines confined to angles.
  Vertical view triangular.
    Isthmus elongate.
      Apex of semicells concave.
        Vertical view with sides indented.................(10) *S. aristiferum*
        Vertical view with sides not indented...............(11) *S. curvatum*
      Apex of semicells convex.
        Spines short and upwardly divergent................(12) *S. dejectum*
        Spines long and subparallel.......................(13) *S. cuspidatum*
    Isthmus not elongate.
      Semicells transversely elliptic............................(15) *S. Dickiei*
      Semicells obversely triangular.
        Cells small and angles submamillate.................(16) *S. connatum*
        Cells large and angles acute......................(14) *S. megacanthum*
  Vertical view quadrangular.
    Angles with a single spine.............................(10) *S. aristiferum*
    Angles with more than one curved spine.............(17) *S. quadrangulare*
Spines on angles and on body of semicells.
  Spines all the same size...........................(18) *S. breviaculeatum*
  Spines of two sizes.
    Median portion of cell apex without spines.
      Semicells elliptic...............................(20) *S. minnesotense*
      Semicells hexagonal...............................(21) *S. spiculiferum*
    Median portion of cell apex with spines..................(19) *S. setigerum*

(A) *Angles of cells continued in hollow processes.*

(C) *Processes in one whorl.*

(D) *Vertical view triangular.*

Processes smooth ....(22) *S. brachiatum*
Processes variously ornamented.
  Apex of semicells without verrucae.
    Ends of processes with long divergent teeth........(23) *S. pseudopelagicum*
    Ends of processes with short teeth.
      Semicell bases with a transverse ring of granules........(24) *S. cingulum*
      Semicell bases without transverse ring................(25) *S. paradoxum*
  Apex of semicells with verrucae.
    Two verrucae on each side of cell.
      Processes short ....(26) *S. protectum*
      Processes long.
        Ends with two long divergent spines.................(34) *S. lacustre*
        Ends with short divergent teeth.
          Processes subparallel ....(29) *S. floriferum*
          Processes upwardly curved.......................(30) *S. Bullardii*
    More than two verrucae on each side of the cell.

With apical and subapical verrucae.
Two subapical verrucae on each cell side..............(31) *S. vestitum*
Four to seven subapical verrucae on each side.........(32) *S. anatinum*
With apical verrucae only.
Semicell base inflated and triangular.............(28) *S. longiradiatum*
Semicell base not triangular..........................(27) *S. gracile*

(D) *Vertical view not triangular.*

(E) *Semicells biradiate.*

Semicells twisted at isthmus.
Cells very small....................................(33) *S. tetracerum*
Cells of medium size................................(35) *S. contortum*
Semicells not twisted at isthmus.
Ends of processes all not in the same place..(40) *S. leptocladum* var. *sinautum*
Ends of processes all in same plane.
Apex without verrucae or spines.
Cell wall thicker in apical portion..................(37) *S. americanum*
Cell wall of uniform thickness.
Processes relatively short.............(33) *S. tetracerum* var. *validum*
Processes long..................................(36) *S. chaetoceras*
Apex with verrucae or spines.
One spine or verruca on apex......................(40) *S. leptocladum*
More than one spine or verruca on apex.
Terminal spines of processes unequal..............(38) *S. grallatorium* var. *forcipigerum*
Terminal spines of processes of same size.
Vertical view without a truncate projection.
Without ornamentation below apical verrucae.
3 apical verrucae........................(38) *S. grallatorium*
7–10 apical verrucae.........................(39) *S. anchora*
With ornamentation below apical verrucae.......(41) *S. Johnsonii*
Vertical view with a truncate projection.
With a basal row of granules.
Processes straight............................(42) *S. natator*
Processes upwardly curved.....(40) *S. leptocladum* var. *insigne*
Without a basal row of granules.................(43) *S. urinator*

(E) *Semicells with four or more processes.*

Four processes on each semicell.
Outline of processes smooth.
Processes straight.
Cell body gradually fusing with processes............(22) *S. brachiatum*
Cell body distinct at point of juncture.........(44) *S. subnudibrachiatum*
Processes with a knee-like bend.....................(45) *S. inconspicuum*
Outline of processes not smooth.
Vertical view without emarginate verrucae.
Processes with concentric rings of granules...........(25) *S. paradoxum*
Processes without rings of granules..(47) *S. arachne* var. *curvatum* (p. p.)
Vertical view with emarginate verrucae.
Processes convergent.
Vertical view with four verrucae on cell body........(49) *S. ankyroides*
Vertical view with several verrucae...................(46) *S. cerastes*
Processes subparallel to outwardly divergent.
Upper margin only of processes crenulate..........(48) *S. pentacerum* var. *tetracerum*
Both margins crenulate............................(47) *S. arachne*

Five or more processes on each semicell.
Short processes with long terminal spines.....(55) *S. brasiliense* var. *Lundellii*
Long processes with short terminal teeth.
Processes upwardly divergent.
Cells of medium size....................(47) *S. arachne* var. *curvatum*
Cells large....................................(52) *S. limneticum*
Processes subparallel to convergent.
Convergent processes.
Verrucae on cell body conspicuous....(49) *S. ankyroides* var. *pentacerum*
Verrucae on cell body inconspicuous..................(47) *S. arachne*
Subparallel processes.
Cell body without verrucae...........................(53) *S. radians*
Cell body with verrucae.
Only upper margin of processes crenulate.... ....(48) *S. pentacerum*
Both margins of processes crenulate.
Processes with transverse rings of granules..........(51) *S. rotula*
Processes with verrucae at base.................(50) *S. Ophiura*

C. *Processes in two whorls.*

Vertical view triangular.
Outline of processes smooth.
Processes short......................................(54) *S. furcatum*
Processes long..............................(56) *S. tohopekaligense*
Outline of processes not smooth.........................(57) *S. furcigerum*
Vertical view circular.
Superior whorl with four bifurcate processes...........(58) *S. leptacanthum*
Superior whorl with six trifurcate processes................(59) *S. Arctiscon*

1. Staurastrum grande Bulnheim. Pl. **67**, Figs. **8–9**.

Hedwigia **1861:** 51, *pl.* 9, *fig.* 14. 1861; W. & G. S. West, Monogr. Brit. Desm. **4:** 140, *pl.* 119, *figs.* 11–13, *pl.* 120, *fig.* 1. 1912; Carter, Ann. of Bot. **34:** 308, *pl.* 15, *figs.* 39–42. 1920.

Cells fairly large, length equal to or greater than the breadth, deeply constricted, sinus acute-angled, isthmus narrow; semicells transversely elliptic, with dorsal margin more rounded than the ventral, dorsal margin at times flattened at the apex, lateral angles rounded. Vertical view triangular, sides of cell body slightly concave, angles rounded to acutely pointed. Cell wall finely punctate. Chloroplast variable; with a small axial central mass and two large laminate blades to each angle of the cell; or without the central mass and a parietal arrangement; pyrenoids normally seven, one in the central mass and one in each blade, at times more numerous. (Tychoplanktont.)

Zygospores polygonal in profile, furnished with numerous stout, curved to straight spines that taper to an acute point (W. & G. S. West).

Cells 68 μ long; 60 μ broad at the base of the semicell, 28 μ broad at the apex; isthmus 14–15 μ broad.

Diameter of zygospores without spines 60 μ, with spines 100 μ (W. & G. S. West).

Bass (rrr), Devils (rrr).

The few specimens found in our lakes have more rounded angles than those from British lakes and they approach the variety *parvum* W. West in shape but not in size.

2. Staurastrum subgrande Borge.

Arkiv för Botanik 15, No. 13; 46, *pl.* 4, *fig.* 9. 1918.

Cells large, length 1.3–1.5 times the breadth, deeply constricted, sinus widely open and acute-angled; semicells transversely ovoid, with the ventral margin more rounded than the dorsal; lateral angles broadly rounded and without ornamentation. Vertical view triangular, with sides very slightly retuse and angles broadly rounded. Cell wall smooth.

Zygospores unknown.

Cells 76–83 μ long; 57–59 μ broad; isthmus 18.5–20 μ broad (Borge).

Not found in the plankton of Wisconsin lakes.

var. **minor** var. nov. Pl. **67,** Figs. **10–12.**

Cells about half the size of the type and somewhat broader in proportion to the length; dorsal margin of semicells much less rounded than the ventral. (Facultative planktont.)

Cells 36–40 μ long; 28–32 μ broad; isthmus 8–9 μ broad.

Devils (r).

3. Staurastrum muticum De Brébisson. Pl. **67,** Figs. **13–15.**

in Meneghini, Linnaea **14**: 228. 1840; Ralfs, Brit. Desm. 125, *pl.* 21, *figs.* 4A–4D; *pl.* 34, *fig.* 13. 1848; W. & G. S. West, Monogr. Brit. Desm. **4**: 133, *pl.* 118, *figs.* 16–20. 1912.

Cells small, length and breadth about equal, deeply constricted, sinus open to sublinear, with an acute apex, isthmus narrow; semicells transversely elliptic, with the curvature of both margins the same or with the dorsal margin more rounded than the ventral; angles of semicells broadly rounded. Vertical view triangular, with sides markedly concave and angles broadly rounded. Cell wall smooth. (Tychoplanktont.)

Zygospores spherical, armed with stout, comparatively short spines that are bifurcate at the distal end (De Brébisson).

Cells 27 μ long; 24–27.5 μ broad; isthmus 8.5 μ broad.

Diamond (rrr).

W. & G. S. West state that "both triangular and quadrangular vertical views are common" but I have observed only the triangular individuals. It may be distinguished from all other smooth-walled species without ornamentation by the size, the ratio between length and breadth, and by the shape of the semicells.

4. Staurastrum brevispinum De Brébisson.

in Ralfs, Brit. Desm. 124, *pl.* 34, *figs.* 7A–7D. 1848; W. & G. S. West, Monogr. Brit. Desm. 4: 145, *pl.* 123, *figs.* 1–3. 1912.

Cells fairly small, length and breadth equal, deeply constricted, sinus open and with apex sublinear to acute, isthmus fairly narrow; semicells broadly elliptic, with ventral side more rounded than the dorsal, median portion of the strongly tumid ventral side truncate

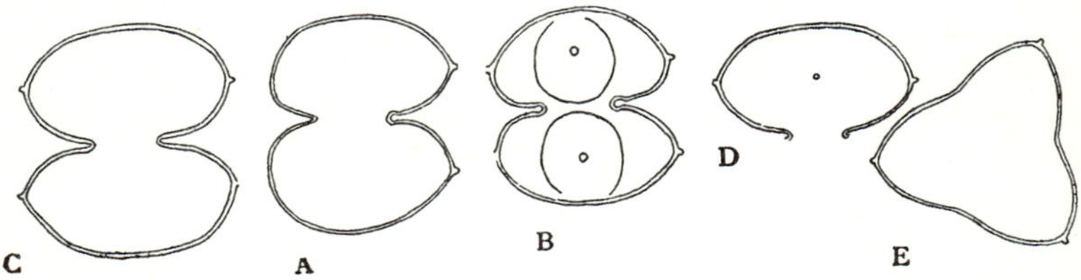

Fig. 7. *Staurastrum brevispinum* De Brébisson (x 600). Plankton specimens from Irish lakes. Figs. A–B from Lough Corrib, Fig. C from Lough Leane, Figs. D–E from Lough Neagh. All from the G. S. West collection.

to very slightly retuse, lateral angles rounded and terminating in a small mucro. Vertical view triangular, sides of the cell body somewhat retuse and angles subacute to rounded; angles with a single rounded mucro. Cell wall smooth. Chloroplast axial, with a small central mass and two laminate arcuate arms running to each cell angle; pyrenoid single and central.

Zygospores unknown.

Cells 35–42 μ long; 35–42 μ broad; isthmus 10–13 μ broad.

Not found in the plankton of Wisconsin lakes.

The type of the species varies in the amount of curvature of the upper and lower halves of the semicells and the incurving of the sides of the cells when viewed from above. The mucro at the angles distinguishes the species from all others found in the state.

var. Boldtii Lagerheim. Pl. **67**, Figs. **16–17.**

Nuova Notarisia **8**: 163. 1893.

Staurastrum brevispinum forma Boldt, Öfvers. kgl. Sv. Vet.-Ak. Förh. 42, No. 2: 113, *pl.* 5, *fig.* 30. 1885.

Length of cells one and a quarter times the breadth, apical portion of sinus more linear than the type; apex of semicells more tumid. Vertical view triangular, with sides retuse. (Facultative planktont.)

Cells 40 μ long; 36–38 μ broad; isthmus 9 μ broad.

Shell (r).

Approaching the variety *alatum* W. & G. S. West, but with the curvature of the two sides of the semicells about the same.

var. RETUSUM Borge. Pl. **67**, Figs. **18–19.**

Bih. Kgl. Sv. Vet.-Ak. Handl. **19**, Afd. **3**, No. **5**: 36, *pl.* 3, *fig.* 42. 1894.
Staurastrum brevispinum var. *Boldtii* forma *retusum* (Borge) W. & G. S. West, Monogr. Brit. Desm. **4**: 148, *pl.* 123, *fig.* 6. 1912.

Cell shape as the foregoing but with the apices of semicells retuse in the median portion. (Facultative planktont.)
Cells 35–50 μ long; 38–40 μ broad; isthmus 11–12 μ broad.

Beaverdam (r), Fish (rrr), Sishebogema (rrr).

W. & G. S. West prefer placing the alga as a form of the variety *Boldtii* but I think it more logical to follow Borge in considering it a distinct variety.

var. **tumidum** var. nov. Pl. **67**, Figs. **20–25.**

Length of semicells about one and a half times the breadth, sinus widely open and acute-angled; semicells obsemicircular, with apices tumid and subtruncate in the median portion. Vertical view of cells triangular, with sides somewhat tumid (not retuse) and angles acute. (Euplanktont?).
Cells 42–52 μ long; 32–41 μ broad; breadth of isthmus 11–13 μ.

Shell (ss).

When the cells are viewed from above the variety is sharply differentiated from the others of the species by the sharp angles and the outward swelling of the sides of the cells. It is the only variety in which the sides of the cells are not retuse. In the front view the proportions are different, no other variety having a length of one and a half times the breadth.

5. STAURASTRUM PSEUDOPACHYRHYNCHUM Wolle. Pl. **68**, Figs. **1–3.**

Desm. U. S. 125, *pl.* 51, *figs.* 32–35. 1884.

Cells small, length about equal to the breadth, deeply constricted, sinus open and with a broad rounded apex, isthmus elongate and fairly narrow; semicells rhomboidal-elliptic, with the dorsal margin convex and more tumid in the median region, lateral angles very broadly rounded. Vertical view triangular, with the sides strongly retuse and the angles broadly rounded. Cell wall smooth and of uniform thickness throughout. (Tychoplanktont.)
Zygospores unknown.
Cells 32–34 μ long; 28–35 μ broad; isthmus 9–10.5 μ broad.

Dummy (rr).

The general shape of the cell is quite similar to *S. pachyrhynchum* Nordstedt but the cell wall does not have the thickening at the angles so characteristically present in *S. pachyrhynchum*. All of the specimens observed were triangular in vertical view.

6. STAURASTRUM ALTERNANS De Brébisson. Pl. **68**, Fig. **4**.

in Ralfs, Brit. Desm. 132, *pl.* 21, *figs.* 7A–7C. 1848; W. & G. S. West, Monogr. Brit. Desm. 4: 170, *pl.* 126, *figs.* 8–9. 1912.

Cells small, length about equal to the breadth, deeply constricted, sinus acute-angled and with apex subacute, isthmus narrow; semicells transversely elliptic, with the dorsal margin more flattened than the ventral, lateral angles broadly rounded; cell body with scattered granules that are arranged in concentric rings near the angles. Vertical view triangular, sides concave and angles rounded, the angles of one semicell alternating with those of the other. (Tychoplanktont.)

Zygospores spherical, furnished with spines that are bifurcate at their apices (Ralfs).

Cells 35 μ long; 37 μ broad; isthmus 10 μ broad.

Speese (rrr).

7. STAURASTRUM DENTICULATUM (Nägeli) Archer. Pl. **68**, Figs. **5–7**.

in Pritchard, A History of Infusoria, 4th. Ed. 738. 1861; W. & G. S. West, Trans. Roy Irish Acad. 33: 103, *pl.* 11, *fig.* 11. 1906.

Phycastrum denticulatum Nägeli non Kützing, Gatt. einz. Algen 128, *pl.* 8, *fig.* C3. 1849.

Cells small, breadth slightly greater than the length, fairly deeply constricted, sinus widely open and with apices fairly acute, isthmus narrow; semicells broadly triangular-ellipsoid, with ventral margin more strongly rounded than the dorsal, angles truncate and with two very short divergent spines, angles of semicells with two to four concentric rings of conical granules or short spines, the profile of the cell being distinctly scalloped between the spines; median portion of cell body smooth. Vertical view triangular, with sides strongly retuse and angles truncate and bearing two short divergent spines; central portion of cell body smooth, angles with two to four transverse rings of denticulations. Chloroplast with a small central mass and two laminate plates running to each angle of the cell; pyrenoid single and central. (Facultative planktont.)

Zygospores unknown.

Cells 26–29 μ long; 32–34 μ broad with spines, 28–32 μ broad without spines; isthmus 10 μ broad; spines 1.5–2 μ long.

Beaverdam (rr), Lac Court Oreilles (sss).

Differing from *S. avicula* De Brébisson in the fewer concentric rings of granules at the angles of the cells and the shape of the semicells. The angles in the vertical view are distinctly truncate and bear two spines (four in an end view), while in the vertical view of *S. avicula* the angles are sharp and bear a single spine (two in an end view).

Irish specimens examined in the G. S. West collection have one or two more rings of granules at the angles of the semicells than do the Wisconsin specimens, but this difference is of little importance.

8. STAURASTRUM AVICULA De Brébisson. Pl. 68, Figs. 8–10.

in Ralfs, Brit. Desm. 140, *pl.* 23, *figs.* 11A–11B. 1848.

Cells small, length about three quarters the breadth, deeply constricted, sinus widely open and apex with acute angles, isthmus fairly narrow; semicells obversely triangular, with the dorsal margin somewhat inflated; superior angles truncate and with two very short divergent denticulate spines; body of semicell with scattered granules in the central portion of the apex and the granules in concentric rings in the region of the angles. Vertical view triangular, with the sides slightly retuse and the angles acute, angles terminating in two short spines that lie in the same vertical plane (usually only one visible). Chloroplast axial, with a small central body and two arcuate lobes running to each angle of the cell; pyrenoid single and central. (Facultative planktont.)

Zygospores unknown.

Cells 31 μ long; 37 μ broad with spines, 31 μ broad without spines; isthmus 10 μ broad; spines 2–3 μ long.

Bear (rrr), Duck (sss), Horseshoe (rrr), Shell (rr).

9. STAURASTRUM LUNATUM Ralfs.

Brit. Desm. 124, *pl.* 34, *figs.* 12A–12B. 1848; W. West, Jour. Linn. Soc. Bot. 29: 173. 1892.

Cells small, breadth (without spines) slightly greater than the length, deeply constricted, sinus acute-angled, isthmus fairly broad; semicells obsemicircular, with apex convex; angles of semicells with a single stout upwardly curved spine. Vertical view triangular, with angles rounded and bearing a straight stout spine at each angle; sides of cell retuse. Cell wall punctate-granulate.

Zygospores unknown.

Cells 25 μ long without spines; 36 μ broad with spines, 30 μ broad without spines; isthmus 12.5 μ broad (W. West).

Not found in the plankton of Wisconsin lakes.

var. PLANCTONICUM W. & G. S. West. Pl. 68, Figs. 11–13.

Jour. Linn. Soc. Bot. 35: 546, *pl.* 16, *figs.* 11–12. 1903.

Cells somewhat larger and broader than the type; angles of semicells more arcuate in front view and with terminal spine much smaller; cell wall finely granulate, with granules in concentric rings near the angles but scattered in the central portion of the cell. Vertical view with lateral margins very slightly retuse. (Euplanktont.)

Cells 35 μ long; breadth with spines 45–54 μ, without spines 40–48 μ; isthmus 10–12.5 μ broad; spines 2.5–4 μ long.

Chain (rr), Clear Crooked (rr), Cranberry (rrr), Found (rrr), Granite (r), Lac Court Oreilles (r).

The upwardly curved short spines at the angles of the semicells combined with the granulate cell wall distinguish this species. The variety has shorter spines and a more regular arrangement of the granules.

10. Staurastrum aristiferum Ralfs.

Brit. Desm. 123, *pl.* 21, *figs.* 2A–2B. 1848.

Cells of medium size, length (without spines) slightly greater than the breadth, deeply constricted, sinus widely open and with a rounded apex, isthmus elongate and narrow; semicells obversely triangular, with dorsal margins somewhat retuse and ventral margins indented midway between the isthmus and the apex; angles mamillate and

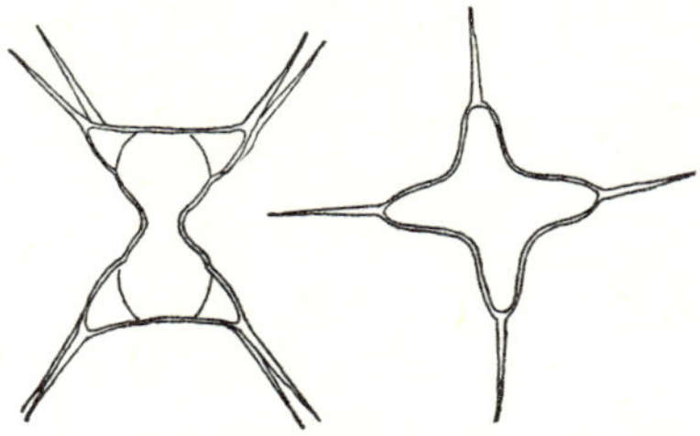

Fig. 8. *Staurastrum aristiferum* Ralfs (x 625). From plankton specimens collected in Capel Curig lakes, Wales, by W. & G. S. West.

terminating in long delicately tapering spines with acute apices, spines outwardly divergent. Vertical view quadrangular (rarely triangular), with sides strongly retuse, angles rounded and bearing a single straight delicately tapering spine. Cell wall smooth.

Zygospores unknown.

Cells 58 μ long with spines, 27.5 μ long without spines; breadth 45 μ with spines, 25 μ without spines; isthmus 6 μ broad; spines 15–18 μ long.

Not found in the plankton of Wisconsin lakes.

var. **indentatum** var. nov. Pl. **68,** Figs. **14–17.**

Cells smaller than the type, isthmus narrower and not elongate; ventral margin of semicells slightly tumid and smooth in outline, angles mamillate. Vertical view triangular, sides slightly concave and with a sharp indentation midway between the angles. (Euplanktont.)

Cells 35–44 μ long with spines, 20–22 μ long without spines; breadth 46–52 μ with spines, 22–25 μ without spines; isthmus 5–6.5 μ broad; spines 15–18 μ long.

Devils (r).

The typical *S. aristiferum* differs from other *Staurastra* both in the indentation of the ventral margins of the semicells in the front view and in the indentation of the sides in the vertical view. This variety has the indentation in the vertical but not in the front view. It might well be placed with other species bearing a single spine at each angle of the semicells but the indentation of the cell shows that it belongs to *S. aristiferum*.

11. Staurastrum curvatum W. West. Pl. **69**, Figs. **4–9.**

Jour. Linn. Soc. Bot. 29: 172 *pl.* 22, *fig.* 13. 1892; W. & G. S. West, *ibid.* 35: 543, *pl.* 17, *fig.* 12. 1903.

Cells of medium size, length (with spines) somewhat less than the breadth, deeply constricted, sinus widely open and with apex very blunt and rounded (semicircular), isthmus narrow, elongate and subrectangular; semicells obversely subtriangular to sublunate, ventral margins slightly convex, dorsal margins concave; angles fairly sharp and continued in long stout straight spines that taper to a sharp point, spines strongly divergent. Vertical view triangular, sides of cells retuse and angles acute; angles each bearing a long stout gradually attenuated spine. Cell wall smooth. Chloroplast having a fairly massive triangular central mass with two parallel laminate blades running to each angle of the cell; pyrenoid single and central. (Euplanktont.)

Zygospores unknown.

Cells 52–56 μ long with spines, 25–28 μ long without spines; breadth 64–80 μ with spines, 28–38 μ without spines; isthmus 7.5–8.5 μ broad; spines 21–25 μ long.

Dummy (ss).

Resembling *S. megacanthum* Lundell but differing in the smaller size of the body of the cell, the more divergent spines, and the rounded apex of the sinus.

var. **elongatum** var. nov. Pl. **69**, Figs. **10–15.**

Cells with isthmus elongate and cylindrical; incurving of semicell apices more pronounced. Vertical view with sides of the cell more retuse. (Euplanktont.)

Cells 40–50 μ long with spines, 25–29 μ long without spines; breadth with spines 56–70 μ, without spines 25–33 μ; isthmus 6–7 μ broad; spines 17.5–25 μ long.

Blue Gill (rrr), Carroll (sss), Lindy (sss).

This variety has an elongation of the isthmus that is comparable to the elongation of the isthmus of *S. cuspidatum* De Brébisson. It is readily distinguished from *S. cuspidatum* by the side of the cells and by the arrangement of the spines. The sides of the cells in a vertical view of the variety are more incurved than those of the type and as a result the body of the cell has a more delicate appearance.

12. Staurastrum dejectum De Brébisson. Pl. **68**, Figs. **18–24.**

in Meneghini, Linnaea **14**: 227. 1840; Ralfs, Brit. Desm. 121, *pl.* 20, *figs.* 5A–5M. 1848.

Cells fairly small, length (without spines) about equal to the breadth, deeply constricted, sinus acute-angled and with the apex broadly rounded, isthmus narrowly elongate and subrectangular; semicells obversely subsemicircular, with dorsal margin slightly tumid, angles acute to rectangular and furnished with a single short acute spine, spines strongly outwardly divergent (suberect). Vertical view triangular, sides of cell retuse in median portion, angles acute, furnished with a single short straight acute spine. Cell wall smooth. Chloroplast with a small central mass and two conspicuous laminate blades leading to each angle of the cell; pyrenoid single and central. (Facultative planktont.)

Zygospores spherical, furnished with numerous stout straight spines with broad bases and rounded apices (Ralfs).

Cells 25–32 μ long with spines, 25–28 μ long without spines; breadth 29–41 μ with spines, 24–32 μ without spines; isthmus 5.5–6.5 μ broad; spines 2.5–5 μ long.

Zygospores 36 μ in diameter, spines 12–20 μ long (Ralfs).

Carson (r), Lac du Flambeau (rrr), Big McKenzie (rrr), Middle McKenzie (rrr), Mud (rr), Travis (sss).

The distinctive features of this species are the elongate isthmus and the upwardly slanting, comparatively small spines. In the elongate isthmus it resembles *S. cuspidatum* De Brébisson but the length of the spines and broader semicells readily distinguish the two. The cell shape is somewhat variable and at times the peculiar isthmus does not stand out as prominently as in Ralfs' figures.

var. INFLATUM W. West. Pl. **68**, Figs. **25–26.**

Jour. Linn. Soc. Bot. **29:** 170, *pl.* 22, *fig.* 11. 1892; W. & G. S. West, Trans. Roy. Irish Acad. **32:** 44, *pl.* 2, *fig.* 23. 1902.

Cells somewhat larger, sinus with a fairly acute apex and isthmus not elongate; apex of semicells more tumid. Vertical view with very short spines at angles and with sides more retuse. (Tychoplanktont.)

Cells 31–34 μ long; breadth with spines 36–41 μ, without spines 31–35 μ; isthmus 6–7 μ broad; spines 2–4 μ long.

Fish (rrr).

Differing from the type in the lack of elongation at the isthmus and the rounded dorsal margins of the semicells. The short outwardly divergent spines and ventral sides of the semicells are like those of *S. dejectum* so that the variety is well placed although it does not have the characteristic isthmus.

13. STAURASTRUM CUSPIDATUM De Brébisson. Pl. **68**, Figs. **27–34.**

in Meneghini, Linnaea **14:** 226. 1840; Ralfs, Brit. Desm. 122, *pl.* 21, *figs.* 1A–1E; *pl.* 33, *fig.* 10. 1848.

Cells small, breadth (with spines) about twice the length, deeply constricted, sinus widely open and with apex broadly rectangular, isthmus narrow, rectangular, greatly elongated; semicells obversely triangular, with ventral margins slightly convex and dorsal margins convex to slightly concave; lateral angles acute and furnished with a single, medium-sized, acutely pointed, straight spine; spines usually subparallel, though at times somewhat convergent or divergent. Vertical view triangular (very rarely quadrangular), with sides of cells retuse in the median portion; angles acute, furnished with a simple straight spine. Cell wall smooth. Chloroplast triangular in outline, fairly massive, with two parallel laminate lobes running to each angle of the semicell; pyrenoid single and central. (Facultative planktont.)

Zygospores spherical, with several mamillate projections, each terminating in a short stout spine (Ralfs).

Cells 23–28 μ long; breadth with spines 42–53 μ, without spines 19–26 μ; breadth isthmus 5 μ; length of spines 15–20 μ.

Diameter of zygospores 30 μ (Ralfs).

Beaverdam (rrr), Big Bass (rr), Bird (rrr), Chain (sss), Clare (rrr), Crawling Stone (rr), Devils (rr), Duck (rr), Dutchmans (rrr), Grindstone (rrr), Half Moon (rrr), Horseshoe (r), Jag (rrr), Lac Court Oreilles (rrr), Lac du Flambeau (rrr), Little Bass (rrr), Little Doctor (rrr), Lost (rr), Big McKenzie (rr), Mercer (rrr), Mud (rr), Muskallonge (rr), Nicaboyne (rrr), Owen (ss), Sishebogema (rr), Tank (sss), Travis (rrr), Whitefish (sss), Wolf (rrr).

The species can be distinguished at once from other triradiate species with a single spine at each angle of the semicells by the smaller size and the elongate isthmus.

var. CANADENSE G. M. Smith. Pl. **69**, Figs. **1–3**.

Trans. Wis. Acad. Sci. Arts & Letts. 20: 349, *pl.* 10, *figs.* 23–24. 1922.

Cells with a distinct indentation at the point of union of the two semicells; spines usually divergent. (Euplanktont.)

Cells 37–39 μ long with spines, 26–32 μ long without spines; breadth 53–58 μ with spines, 21–23 μ without spines; isthmus 5 μ broad; spines 18–22 μ long.

Clear Crooked (rr).

14. STAURASTRUM MEGACANTHUM Lundell. Pl. **69**, Figs. **16–21**.

Nova Acta Reg. Soc. Sci. Upsaliae 3 Ser. 8: 61, *pl.* 4, *fig.* 1. 1871.

Cells of medium size, about as long as broad (without spines), deeply constricted, sinus acute-angled to subrectangular, with an acute apex; semicells obversely triangular, with apex generally tumid, rarely straight; lateral margins of cells straight to slightly convex, angles with a single stout straight elongate spine; spines usually subparallel, rarely subdivergent. Cell wall smooth to finely punctate. Vertical view triangular, with sides somewhat retuse in the central portion, angles acute and with a single stout straight spine. (Facultative planktont.)

Zygospores unknown.

Cells 36–46 long with spines, 36–40 μ long without spines; breadth with spines 58–72 μ, without spines 41–49 μ; breadth of isthmus 10–11.5 μ; spines 12.5–17 μ long.

Adelaide (r), Bass (rr), Little Bass (rrr), Bear (rrr), Crooked (sss), Devils (rrr), Diamond (r), Duck (rrr), Dummy (sss), Grindstone (rrr), Jenny (rr), Jones (rrr), Lac Court Oreilles (ss), Lindy (rr), Lynx (rr), Nicaboyne (rrr), Reserve (sss), Sanford (rrr), Tank (rrr), Turtle (rr), Whitefish (cc).

The species should be compared with *S. dejectum* De Brébisson, *S. Dickiei* Ralfs, and *S. curvatum* W. West. It differs from all of them in the stouter spines at the angles, in size and in shape of the cells. Lundell states that there are both triangular and quadrangular members of the species but I have observed the triangular specimens only in the plankton of Wisconsin lakes.

var. SCOTICUM W. & G. S. West. Pl. **69**, Figs. **22–24**; Pl. **70**, Figs. **1–2.**

Jour. Linn. Soc. Bot. **35**: 544, *pl.* 16, *fig.* 8. 1903.
S. megacanthum forma Borge, Bot. Not. 1897: 213, *pl.* 3, *fig.* 7. 1897.

Sinus subrectangular to obtuse-angled, isthmus proportionally narrower; apices of semicells generally concave, rarely straight; spines at angles longer and always divergent. Vertical view with sides of cells more retuse. (Euplanktont.)

Cells 47–54 μ long with spines, 36–38 μ long without spines, breadth 67–82 μ with spines, 40–46 μ without spines; isthmus 8–10 μ broad; spines 20–25 μ long.

Bass (sss), Diamond (r), Horseshoe (rr), Jones (r), Little Doctor (rrr), Meta (rr), Shell (c).

The variety is more graceful than the type and differs in having concave apices of the semicells and longer spines.

15. STAURASTRUM DICKIEI Ralfs.

Brit. Desm. 123, *pl.* 21, *figs.* 3A–3B. 1848; W. West, Jour. Linn. Soc. Bot. **29**: 171. 1892; G. S. West, Brit. Freshw. Algae 141, *figs.* 52A–52C. 1904.

Cells fairly small, breadth (with spines) about one and a quarter times the length, deeply constricted, sinus sublinear, opening outwards and with an acute apex, isthmus narrow; semicells transversely elliptic, the dorsal margin somewhat more tumid than the ventral margin, angles furnished with a single short stout acute spine, spines strongly convergent. Vertical view triangular, with sides of cells somewhat retuse and angles rounded and furnished with a short stout straight spine. Cell wall smooth.

Zygospores spherical, clothed with fairly numerous straight spines with acutely pointed apices (G. S. West).

Cells 35 μ long; breadth with spines 42 μ, without spines 29 μ; isthmus 6 μ broad (W. West).

Not found in the plankton of Wisconsin lakes.

var. MAXIMUM W. & G. S. West. Pl. **70**, Figs. **3–5.**

Trans. Linn. Soc. 2nd. Ser. Bot. **5**: 72, *pl.* 8, *fig.* 19. 1895; *ibid.* 255, *pl.* 18, *fig.* 13. 1896.

Cells about twice the size of the type, sinus acute-angled and with an acute apex; semicells more narrowly elliptic and with the curvature of dorsal and ventral margins the same; spines at angles stout, fairly long, and strongly convergent. (Tychoplanktont.)

Cells 44–47.5 μ long; breadth with spines 62–72 μ, without spines 45–53 μ; breadth of isthmus 12 μ; spines 12–14 μ long.

Beaverdam (rrr), Fish (rr).

The dimensions of the alga as I have found it are those given by W. & G. S. West for the plant as originally discovered in Madagascar and not as they found it in North America. Like their North American specimens, however, the spines are strongly convergent and not recurved. The specimens from Beaverdam Lake (Fig. **5**) were slightly smaller than those from Fish Lake (Figs. **3–4**) but the stoutness of the spines and the open sinus show that they belong to the variety and not to the type.

16. Staurastrum connatum (Lundell) Roy & Bisset.

Jour. of Bot. **24**: 237. 1886.

Staurastrum dejectum var. *connatum* Lundell, Nova Acta Reg. Soc. Upsaliae 3 Ser., **8**: 60, *pl.* 3, *fig.* 28. 1871.

Cells small, length (without spines) about equal to the breadth, deeply constricted, sinus widely open and acute-angled; semicells obsemicircular, ventral margins rounded and apex slightly retuse; superior angles submamillate and bearing a single straight fairly short delicate spine, spines strongly divergent. Vertical view triangular, sides of cells slightly retuse and angles rounded, angles with a single short spine. Cell wall smooth.

Zygospores unknown.

Cells 22.4 μ long without spines; 21 μ broad without spines; isthmus 6.6 μ broad; spines 8–13 μ long (Lundell).

Not found in the plankton of Wisconsin lakes.

var. rectangulum Roy & Bisset. Pl. **70**, Figs. **6–7**.

Jour. of Bot. **24**: 237, *pl.* 268, *fig.* 12. 1886.

Sinus subrectangular and with a subacute apex; ventral margins of semicells only slightly rounded and angles less mamillate. (Facultative planktont.)

Cells 28 μ long with spines, 17.5 μ long without spines; breadth 31–32.5 μ with spines, 17 μ without spines; breadth of isthmus 4.5 μ; spines 6–7.5 μ long.

Pine Tree (r).

The isthmus of the specimens that I have collected is less rectangular than Roy and Bisset show in their figure. The alga shows considerable resemblance to *S. cuspidatum* var. *divergens* Nordstedt but the isthmus is far too short for *S. cuspidatum*.

17. Staurastrum quadrangulare De Brébisson forma. Pl. **70**, Figs. **8–9**.

in Ralfs, Brit. Desm. 128, *pl.* 22, *figs.* 7A–7C; *pl.* 34, *figs.* 11A–11B. 1848; W. & G. S. West, Trans. Linn. Soc. 2nd. Ser. Bot. 5: 257, *pl.* 16, *figs.* 16–17. 1896; Borge, Arkiv för Botanik 1: 108, *pl.* 4, *fig.* 17. 1903.

Cells small, length about equal to the breadth, median constriction fairly deep, sinus subrectangular and with a blunt apex; semicells rectangular, with apex and lateral margins straight, lateral margin with a short curved spine at the superior and inferior angles, spines outwardly divergent; apex with a short perpendicular spine just within each lateral margin. Vertical view quadrangular, with strongly retuse sides and flattened emarginate corners each bearing two curved spines, basal portion of the semicell extending beyond the truncate angle, rounded, and with a single straight spine. (Tychoplanktont.)

Zygospores unknown.

Cells 27.5 μ long with spines, 22 μ long without spines; breadth with spines 25 μ, without spines 20 μ; breadth isthmus 10 μ.

Lindy (rrr).

But two cells of this alga have been seen and although they undoubtedly belong to the species they do not agree precisely with any published figures.

18. **Staurastrum breviaculeatum** sp. nov. Pl. **70**, Figs. **10–13**.

Cells small, length and breadth about equal, deeply constricted, sinus acute-angled and with apex subacuminate, isthmus narrow; semicells transversely elliptic, dorsal margin sometimes flattened in the median portion; lateral angles broadly rounded and bearing four short quadrately arranged divergent spines; cell body with two transverse rows of 4–7 spines, the lower row half way between the isthmus and the cell apex, the upper midway between the lower row and the apex; cell apex with median portion bare and with margins towards the angles with 3–4 outwardly pointing spines. Vertical view triangular, sides of cells emarginate and angles broadly rounded and bearing four quadrately arranged short divergent spines; center of cell body with a triangular ring of spines, the sides of the triangle lying parallel to the margins of the cell and each side composed of three outwardly facing spines, angles of the central triangle connected to the corresponding angle of the cell by a short row of spines; lateral margins of cells with 6–8 erect spines and a second arcuate intramarginal row of spines that are of similar size and arrangement; outline of cell slightly emarginate between adjacent spines. Chloroplast with a small central mass and two laminate blades running to each angle of the cell; pyrenoid single and central. (Euplanktont.)

Zygospores unknown.

Cells 38–46 μ long with spines, 36–42 μ long without spines; breadth with spines 37–55 μ, without spines 34–44 μ; isthmus 9–13 μ broad; spines 2–5 μ long.

Bass (rrr), Devils (rrr), Speese (rrr), Sunday (rrr), Tank (rrr), Travis (r).

This species is related to the *Staurastrum teliferum-setigerum* group of species but differs in having the spines at the angles shorter and the same size as those of the cell body. The regular arrangement of the spines on the cell body in the vertical view is also more pronounced in this species.

The spines in the few specimens from Tank lake were much smaller than those from the other lakes but they were arranged in the same regular pattern (Figs. **16–18**).

19. Staurastrum setigerum Cleve. Pl. **70**, Figs. **19–25.**

Öfvers. Kgl. Sv. Vet.-Ak. Förh. 20: 490, *pl.* 4, *fig.* 4. 1864; Lundell, Nova Acta Reg. Soc. Sci. Upsaliae 3 ser., 8: 64, *pl.* 4, *fig.* 3. 1871; Roy & Bisset, Ann. Scottish Nat. Hist. 1893: 243, *pl.* 3, *fig.* 9. 1893.

Cells of medium size, length (without spines) equal to the breadth, deeply constricted, sinus open but with an acute apex, isthmus narrow; semicells subelliptic, with the central portion of the rounded apex flattened, lateral margins truncate and with two stout divergent spines; dorsal margin with a stout spine near the angles and four more delicate spines on the apex; face of semicells with a transverse row of long delicate spines. Vertical view triangular, with the sides retuse and the angles acute; angles with three stout spines in the same vertical plane, two at the angle and one a short distance back from the angle; sides with four more delicate spines of the same length on each lateral margin, margin retuse between spines; body of semicells with another curved row of similar spines within the margin and one or two spines continuing the line formed by the stout spines on the angles. Chloroplasts axial, with a small central mass and lobes very deeply incised to form two arcuate plates that run to each angle; pyrenoid single. (Facultative planktont.)

Zygospores unknown.

Cells 52–57 μ long with spines, 42–48 μ long without spines; breadth with spines 60–67 μ, without spines 42–55 μ; breadth of isthmus 11.5–12.5 μ; length of stout spines 8–13 μ, of delicate spines 5–11 μ.

Beaverdam (sss), Fish (s), Nicaboyne (rrr).

The vertical view of this species shows more of the distinctive features than does the front view. The three stout spines forming a vertical row at the angle and the four delicate spines on each margin are quite constant features. The arrangement of the spines within the margin is subject to considerable variation and the description of their arrangement given above is found only in symmetrically developed specimens; the number of interior spines is also subject to variation. An atypical specimen is shown in Figure **24** where there are accessory stout spines at the angles and a large number of spines at the center of the cell.

var. OCCIDENTALE W. & G. S. West. Pl. **71**, Figs. **1–4.**

Trans. Linn. Soc. 2nd. Ser. Bot. 5: 260, *pl.* 16, *fig.* 27. 1896.

Cells somewhat smaller; semicells narrowly elliptic and with apices more flattened; stout spines at angles as in the type; delicate spines on the face of the semicells reduced to one near each angle. Vertical view with more rounded angles and margins with but two delicate spines on each side of the cell; central portion of the cell with fewer and more scattered spines. (Facultative planktont.)

Cells 41–45 μ long with spines, 32–33 μ long without spines; breadth with spines 57–63 μ, without spines 41–42 μ; breadth of isthmus 11.5–12.5 μ; stout spines 10–12.5 μ long, slender spines 4–9 μ long.

Lac Court Oreilles (ss), Reserve (rrr).

The semicells of this variety are more compressed in the front view than are those of the type. In a vertical view, aside from the more rounded angles, the occurrence of only two delicate marginal spines on each side is the most striking feature. In this respect the cells show a great resemblance to *S. subteliferum* Roy & Bisset, but the semicells of the two are quite different in front view.

var. PECTINATUM W. & G. S. West. Pl. **71**, Figs. **5–8.**

Trans. Linn. Soc. 2nd. Ser. Bot. 5: 260, *pl.* 16, *fig.* 28. 1896.

Cells with approximately the same shape and size as the type but with longer spines. Vertical view having delicate marginal spines confined to the angles and spines at the center of the cell more numerous. (Facultative planktont.)

Cells 56–60 μ long with spines, 45–49 μ long without spines; breadth with spines 58–66 μ, without spines 44–49 μ, breadth of isthmus 12.5–17 μ; length of stouter spines 6.5–12.5 μ, length slender spines 5–9 μ.

Dummy (rrr), Owen (rrr), Horseshoe (rrr), Big St. Germain (rrr), Sand (ss).

W. & G. S. West give only a front view of this variety so that determination of the variety is somewhat doubtful.

var. BREVISPINUM G. M. Smith. Pl. **71**, Figs. **9–13.**

Trans. Wis. Acad. Sci. Arts & Lett. 20: 350, *pl.* 11, *figs.* 6–8. 1922.

Cells smaller, length and breadth about equal, sinus acute-angled and with an acute apex; semicells narrowly elliptic and with lateral margins flattened; lateral margins with four stout spines that are quadrately arranged and with one or two stout spines on the dorsal margin just back of the lateral spines; body of semicell with two transverse rows of smaller spines. Vertical view triangular, with angles rounded and sides very sightly concave; margin of cell bearing 5–6 more delicate and shorter spines and with portion between each pair of spines concave, body of cell with a second arcuate row of spines within each margin, center with a ring of spines and a row of spines leading to each angle or with the rows only and the ring missing. (Euplanktont.)

Cells 39–41 μ long; breadth with spines 42–53 μ, without spines 35–40 μ; isthmus 12–14 μ broad; stout spines 5–7.5 μ long, slender spines 3–7 μ long.

Carroll (r).

This might be considered a distinct species but the arrangement of the spines and the two sizes of spines is that of *S. setigerum.* The variety also differs in the quadrispinate endings of the angles of the semicells.

20. Staurastrum minnesotense Wolle. Pl. 71, Figs. 14–15.

Bull. Torr. Bot. Cl. 12: 6, *pl.* 47, *figs.* 7–8. 1885; Freshw. Algae of U. S. 43, *pl.* 57, *figs.* 7–8. 1887; W. & G. S. West, Trans. Linn. Soc. Bot. 2nd. Ser. 5: 260, *pl.* 18, *fig.* 15. 1896.

Cells large, length and breadth about equal, deeply constricted, sinus widely open and with a more or less rounded apex, isthmus relatively narrow; semicells broadly elliptic, generally with the dorsal margin more tumid than the ventral, lateral angles each furnished with two long stout divergent spines that lie in the same vertical plane; face of the semicell with a second series of more delicate long spines, one just inside of each angle and two just below the apex. Vertical view triangular, with median portion of sides of the cells slightly retuse to slightly tumid and the portion near the angles straight; angles subrectangular to obtuse, terminating in two long stout spines that lie in the same vertical plane; each side of the cell with a single long delicate spine perpendicular to the cell wall that lies on the margin a short distance from the angle and with another similar intramarginal spine within each spine at the margin. Cell wall coarsely punctate. (Tychoplanktont.)

Zygospores unknown.

Cells 145 μ long with spines, 125 μ long without spines; breadth of cells with spines 172 μ, without spines 112 μ; isthmus 42.5 μ broad; stout spines 25–32 μ long, slender spines 25–28 μ long.

Muskallonge (rr).

This handsome Desmid cannot be confused with any other species. It is the only one with spines of two sizes which has punctate walls and anywhere near this size. The arrangement of the spines, particularly in the vertical view, is also unique. W. & G. S. West have called attention to certain inaccuracies in the drawings of Wolle and my observations on Wisconsin specimens confirm their statement that the angles in a vertical view of the cell are not nearly so sharp as Wolle shows them. There is some variation in the size and length of the spines but the two terminal spines at the angles are always stouter than the other four. Certain of the slender spines may also be lacking so that the symmetrical arrangement of three pairs of spines at each angle is not found in every case.

21. **Staurastrum spiculiferum** sp. nov. Pl. **71,** Figs. **16–20.**

Cells fairly small, length (without spines) slightly greater than the breadth, deeply constricted, sinus acute-angled and with an acute apex, isthmus narrow; semicells hexagonal-rectangular, with the dorsal margin subconvex to subconcave; lateral angles broadly truncate and with the truncate faces somewhat convergent towards the isthmus; semicells ornamented with delicate spines of two different lengths, superior and inferior angles of lateral margin with a single long spine, the two spines diverging, cell apex with a shorter spine within each lateral margin and face of semicells with a similar spine near each lateral margin. Vertical view triangular, median portion of the sides with a slight truncate elevation; angles subrectangular and bearing two long straight spines that lie in the same vertical plane, cell margins with a single short vertical spine near each angle and two similar short spines within each angle. Cell wall finely punctate. (Euplanktont.)
Zygospores unknown.

Cells 29–45 μ long with spines, 23–25 μ long without spines; breadth with spines 30–40 μ, without spines 20–24 μ; breadth at isthmus 6.5–8.5 μ; length of spines at angles 5–12.5 μ, length interior spines 3.5–5 μ.

Pine Tree (r). Long (rr).

In the shape of the cell the species has a certain resemblance to *S. quadrangulare* De Brébisson but the spines are much too fine for the alga to be considered a variety of *S. quadrangulare*. In the spines of two sizes, as well as in the arrangement of the spines it resembles *S. minnesotense* Wolle, but the shape of the cells and the sizes of the two algae are quite different.

The spines of the alga in the specimens from Long lake were shorter than those from Pine Tree lake and there were at times three long spines at the angles of the cells, two inferior and one superior spine.

22. Staurastrum brachiatum Ralfs. Pl. **72,** Figs. **1–4.**

Brit. Desm. 131, *pl.* 23, *figs.* 9A–9G. 1848; Eichler, Pamietnik Fizyograficzny **12:** 164, *pl.* 10, *fig.* 37. 1893; W. & G. S. West, Jour. Roy. Micr. Soc. **1896:** 159, *pl.* 4, *figs.* 53–56. 1896; G. S. West, Jour. Linn. Soc. Bot. **34:** 391, *pl.* 11, *figs.* 5–15. 1899.

Cells small, length (with processes) about equal to the breadth, deeply constricted, sinus widely open, with apex rounded to acute, isthmus relatively broad; semicells obversely triangular, with the ventral margins straight to slightly convex and the dorsal margins somewhat concave; angles continued in short hollow processes, the line of demarcation between cell body and processes very indistinct, processes terminating in three (rarely two) blunt teeth with rounded ends.

Vertical view triangular (rarely quadrangular), with sides retuse and the angles continued in broad processes, processes of one semicell usually alternating with those of the other. Cell wall smooth and without ornamentation. Chloroplast axial, with two lobes running to the end of each process; pyrenoid single and central. Cells frequently imbedded in a copious spherical gelatinous envelope with a radially fibrillar structure. (Facultative planktont.)

Zygospores pentahedral to octohedral, with sides straight; angles rectangular and continued in blunt spines (Ralfs, Eichler) or angles rounded and spines scarcely visible (W. & G. S. West).

Cells 28–35 μ long with processes, 15–17.5 μ long without processes; breadth with processes 27.5–34 μ; breadth at isthmus 8–9 μ.

Carroll (rrr), Dummy (s), Little Bass (rrr).

G. S. West has discussed the variation of this species and states that the processes may be long and somewhat tapering or short and blunt. The apices may be either bi- or trifurcate and with the furcations ranging from a rather long incision (usually in the individuals with bifurcate processes) to small rounded knobs that show but little furcation.

23. Staurastrum pseudopelagicum W. & G. S. West. Pl. **72**, Figs. **5–7**.

Jour. Linn. Soc. Bot. 35: 547, *pl.* 18, *figs.* 1–3. 1903; Trans. Roy. Soc. Edinburgh 41: 504. 1905.

Cells fairly small, length (without processes) about twice the breadth, deeply constricted, sinus widely open to acute-angled and with an acuminate apex, isthmus narrow; semicells obversely semicircular to subtriangular, with apices flattened to convex; angles of semicells prolonged into long divergent processes that have their apices terminating in two or three long outwardly curved and divergent spines; subapical portion of processes with concentric rings of granules, basal portion with scattered granules; apex of semicells smooth or irregularly granulate. Vertical view triangular, with sides concave to convex and with the angles continued in long processes that terminate in three to two spines, when bispinate with the spines in the same vertical plane; body of the cell smooth to granulate; processes ornamented as in the front view. (Euplanktont.)

Zygospores unknown.

Cells 40–55 μ long with processes, 24–30 μ long without processes; breadth with processes 54–71 μ, without processes 18–31 μ; breadth at isthmus 6.5–8.5 μ; length of terminal spines 5–8 μ.

Devils (rrr), Shell (sss).

The figures which W. & G. S. West give of the alga have a stouter cell body than is shown in my drawing. The more delicate form that I have figured is almost identical with a form of *S. psuedopelagicum* in the plankton from Loch Morar, Scotland, in the G. S. West collection. Since no form name was given to this I have not named the Wisconsin form. The original description states that the processes are terminated by two large divergent spines but in the local collections trispinate specimens are more abundant than bispinate. This point is of no taxonomic significance since both may occur on different processes of the same semicell. Although W. & G. S. West make no mention of the fact, the same variation occurs sparingly in the specimens from Loch Morar.

The distinguishing features of the species are the shape of the semicells, the nature of the sinus, the granulation of the cell wall, and the terminal spines of the processes. It should be compared with *S. pelagicum* W. & G. S. West, with *S. lacustre* G. M. Smith, and with certain forms of *S. paradoxum* Meyen (especially those figured by W. & G. S. West, Jour. Linn. Soc. Bot. **35**: 548, *pl.* 18, *fig.* 5. 1903).

var. **tumidum** var. nov. **Pl.** 72, **Figs.** 8–11.

Sinus always acute-angled; divergence of the processes not so marked and the transition from the body of the cell to the processes more gradual, ends of processes generally with three straight divergent fairly long spines, rarely with two or four spines; concentric granulation of processes much more pronounced than in the type; curvature of the apex of the semicells much greater and with one or two short spines near the point of origin of each process. Vertical view triangular, body of the cell granular to smooth. (Euplanktont.)

Cells 40–55 μ long with processes, 24–31 μ long without processes; breadth with processes 54–80 μ, without processes 19–25 μ; breadth at isthmus 6.5–8 μ; length of terminal spines 5–7 μ.

Devils (rrr), Grindstone (sss), Lac Court Oreilles (ss), Lac du Flambeau (rrr), Whitefish (ss).

The variety differs from the type in the greater curvature of the cell apex, different shape of the sinus, and greater regularity of granulation. The terminal spines are conspicuous, as in the type, but are straight, not curved, and more variable in number.

24. Staurastrum cingulum (W. & G. S. West) G. M. Smith. Pl. **72**, Figs. **12–14**.

Trans. Wis. Acad. Sci. Arts & Lett. 20: 353. 1922.

Staurastrum paradoxum var. *cingulum* W. & G. S. West, Jour. Linn. Soc. Bot. 35: 548, *pl.* 18, *figs.* 6–7. 1903.

Cells of medium size, length (without processes) about one and three quarters that of the breadth, slightly constricted, sinus a small subrectangular depression, isthmus broad; semicells subcylindrical, without a basal inflation and with a ring of 12–15 minute spines (7–9 visible in a front view) just above the isthmus; apex of semicells flattened or slightly tumid, with processes arising just below the apex, processes slightly attenuated and upwardly curved, terminating in three minute divergent spines; apical portion of semicells and processes covered with minute spines that are generally scattered but at times arranged in concentric rings near the point of juncture of cell body and process. Vertical view triangular, with the angles continued in long straight processes whose ornamentation appears the same as in a front view; central portion of body of the cell without spines. (Euplanktont.)

Zygospores unknown.

Cells 60–100 μ long with processes, 35–45 μ long without processes; breadth with processes 65–105 μ, without processes 18–30 μ; breadth at isthmus 9–11.5 μ.

Balsam (rrr), Bear (r), Beaverdam (ss), Bloom (rr), Crawling Stone (rrr), Deer (rrr), Dummy (ss), Found (r), Granite (ss), Half Moon (r), Little Bear (rrr), Lost (r), Mercer (rrr), Meta (rr), Mud (rr), Pokegama (rr), Sand (sss), Sishebogema (ss), Two Sisters (rr), Wapagossett (rr), Whitefish (r).

Since this differs from *S. paradoxum* Meyen in the cylindrical body of the semicells, the nature of the sinus, and the slender upwardly curved processes as well as in the basal ring of spines it seems better to consider this a distinct species and not a variety of *S. paradoxum*. It should also be compared with *S. longiradiatum* W. & G. S. West.

25. Staurastrum paradoxum Meyen. Pl. **72**, Figs. **15–22**; Pl. **73**, Figs. **1–2**.

Nova Acta Phys. Med. Ac. Caes.-Leop. Nat. Cur. **14**[2]: 777, *pl.* 43, figs. 37–38. 1829; Ralfs, Brit. Desm. 138, *pl.* 23, *figs.* 8A–8F. 1848; Roy & Bisset, Ann. Scottish Nat. Hist. **1893**: 241. 1893; W. & G. S. West, Jour. Linn. Soc. Bot. **35**: 548, *pl.* 18, *figs.* 4–5. 1903; Trans. Bot. Soc. Edinburgh **23**: 27, *pl.* 2, *figs.* 33–35. 1905.

Cells of medium size, length somewhat greater than the breadth (without processes), median constriction fairly deep, sinus widely open and with an acute apex, isthmus fairly narrow; semicells obversely triangular to subcampanulate, with the dorsal margin flattened to slightly convex; angles continued in long somewhat attenuated processes that terminate in truncate ends that are tri- or quadrispinate; processes always divergent but the degree of divergence exceedingly variable; body of semicell with scattered granules and processes with concentric rings of granules. Vertical view triangular, with the angles continued in processes whose length is at least equal to the diameter of the cell body; central area of body of the cell with scattered granules or with the granules confined to the marginal region; processes with concentric rings of granules. Chloroplast axial, with a bilobed process running to each angle of the semicell; pyrenoid single and central. (Facultative planktont.)

Zygospores spherical, with a few long spines that terminate in bifurcate or trifurcate apices (Roy and Bisset.)

Cells 37–65 μ long with processes, 25–35 μ long without processes; breadth with processes 57–88 μ; without processes 18–25 μ; breadth at isthmus 7.5–11 μ.

Big Bass (sss), Beaverdam (rrr), Bettine (rrr), Carroll (sss), Clare (rr), Clear Crooked (rr), Crab (ss), Cranberry (r), Crawling Stone (ss), Devils (rrr), George (rrr), Granite (rrr), Horseshoe (rrr), Jenny (rrr), Lac du Flambeau (rrr), Little Rice (sss), Long (rrr), Lost (rrr), Middle McKenzie (rrr), Pine (rrr), Rice (rr), Red Bass (rrr), Sand (rrr), Shell (sss), South Crab (rrr), Tenderfoot (rrr), Tied Canoe (rrr), Travis (ss).

Although the genus *Staurastrum* was based upon this species and many observers have recorded it, the specific limits are poorly defined, even at the present day. After examining plankton material from many European and American lakes I have reached the conclusion that the essential characters are the relatively long and divergent processes and an absence of a row of verrucae just within and parallel to the margin of the cell body in a vertical view of the cell.

Staurastrum gracile Ralfs is an alga that is exceedingly difficult to separate from *S. paradoxum* but *S. gracile* usually has shorter subparallel processes and always a row of emarginate verrucae just within the lateral margins in a vertical view of the cell. *S. paradoxum* is subject to considerable variation in size and degree of divergence of the processes. Under plankton conditions the granulation becomes more pronounced, the granules at times becoming very short spines, while the terminal spines of the processes are also much larger. The present practice among phycologists is not to attempt a naming of the various forms of the species as it is found in the plankton.

var. LONGIPES Nordstedt. Pl. **73**, Figs. **3–6**.

Lunds Univ. Arsskr. **9**: 35, *pl.* 1, *fig.* 17. 1873; W. & G. S. West, Trans. Roy. Soc. Edinburgh **41**: 504, *pl.* 7, *fig.* 13. 1905.

Semicells cyathiform, with apices flattened; angles continued in very long graceful upwardly turned processes with finely crenulate margins and terminating in four divergent spines; cell body smooth and without ornamentation. Vertical view triangular, with the sides of the body of the cell concave and the processes with the same appearance as in the front view. (Euplanktont.)

Cells 67–108 μ long with processes, 22–28 μ long without processes; breadth with processes 67–114 μ, without processes 18–22 μ; breadth at isthmus 7.5–9 μ.

Little Bass (rr), Bear Trap (rr), Camp (rr), Center (sss), Chetek (rr), Devils (sss), Dummy (ss), Fowler (rr), Granite (ss), Hill (rrr), Hooker (rr), Lac la Belle (rrr), Mud (r), Nancy (s), Red Bass (rrr), Round (rr), Big St. Germain (rrr), Shell (rrr), Sishebogema (rrr), Speese (rrr), Two Sisters (rr), Trout (rr), Vermilion (r), Whitefish (ss).

The great length of the processes in proportion to the length of the cell body is the most striking feature of this variety. There is considerable variation in the dimensions of the cells and I have found all gradations from the typical variety to the form *permagna* of W. & G. S. West (*l. c.* 504, *pl.* 7, *fig.* 14).

var. PARVUM W. West. Pl. **73,** Figs. **7–15.**

Jour. Linn. Soc. Bot. 29: 182, *pl.* 23, *fig.* 12. 1892.

Cells much smaller; processes more delicate as in the variety *longipes* but much shorter. (Facultative planktont.)

Cells 20–50 μ long with processes, 14–27 μ long without processes; breadth with processes 38–60 μ; without processes 12–20 μ; breadth at isthmus 5–6 μ.

Adelaide (ss), Big Bass (ss), Bloom (rrr), Bone (ss), Big Butternut (rrr), Little Butternut (ss), Duck (ss), Dunham (rrr), East (rr), Granite (r), Hooker (rrr), Long (rrr), Loveless (rr).

26. STAURASTRUM PROTECTUM W. & G. S. West.

Ann. Roy. Bot. Garden Calcutta 6[2]: 217, *pl.* 16, *fig.* 12. 1907.

Cells fairly small, length about half the breadth (with processes), deeply constricted, sinus widely open and with apex acute, isthmus narrow; semicells broadly obversely triangular, with lower margins straight and apex slightly convex, angles continued in divergent processes of medium length; processes irregularly and finely granulate, terminating in two long divergent spines; apex of semicells with four bidenticulate verrucae; median part of body of the semicells with two similar subapical verrucae. Vertical view triangular, with sides slightly convex to slightly concave, angles with stout processes terminating in two spines that usually do not lie in the same vertical plane; median portion of each lateral margin with two bidenticulate verrucae and central portion of body of the cell with a ring of similar verrucae, two verrucae lying opposite each angle of the cell.

Zygospores unknown.

Cells 46 μ long with processes, 25 μ long without processes; breadth with processes 66 μ; breadth at isthmus 8.5 μ (W. & G. S. West).

Not found in the plankton of Wisconsin lakes.

var. PLANCTONICUM G. M. Smith. Pl. **74,** Figs. **1–4.**

Trans. Wis. Acad. Sci., Arts & Lett. 20: 355, *pl.* 12, *figs.* 8–12. 1922.

Cells with stouter processes, granulation of processes concentric and granules larger; verrucae on sides of semicells lacking, apical verrucae tridenticulate and forming a much smaller ring. Vertical view with sides of cell strongly retuse. (Euplanktont.)

Cells 36–39 μ long with processes, 25–28 μ long without processes; breadth with processes 46–52 μ, without processes 30–34 μ; breadth at isthmus 6–7 μ; terminal spines 5–6.5 μ long.

Carroll (rr), Rudolph (rrr).

This organism shows such a strong resemblance to *S. protectum* in shape of the semicells, divergent spines on the processes, and apical verrucae that I feel it should be considered a variety. It differs primarily in the rings of granules on the processes and a lack of subapical verrucae.

27. Staurastrum gracile Ralfs. Pl. 73, Figs. 16–18.

Ann. & Mag. of Nat. Hist. **15**: 155, *pl.* 11, *figs.* 3A–3B. 1845; Brit. Desm. 136, *pl.* 22, *figs.* 12A–12D. 1848; W. & G. S. West, Jour. Roy. Micr. Soc. **1897**: 495, *pl.* 6, *fig.* 27. 1897; W. & G. S. West, Jour. Linn. Soc. Bot. 35: 529. 1903.

Cells of medium size, length generally greater than the breadth (without processes), deeply constricted, sinus widely open and with an acute apex, isthmus fairly narrow; semicells broadly triangular to cyathiform; dorsal margin slightly concave to somewhat convex, with a row of small emarginate verrucae, apex of semicell body sometimes with a short vertical row of granules below each apical verruca; angles continued in hollow processes of variable length, processes slightly attenuated, with transverse rings of granules and terminating in three or four small spines, processes parallel to widely divergent. Vertical view triangular, with sides of body of the cell slightly concave and angles continued in straight processes; cell body with an intramar-

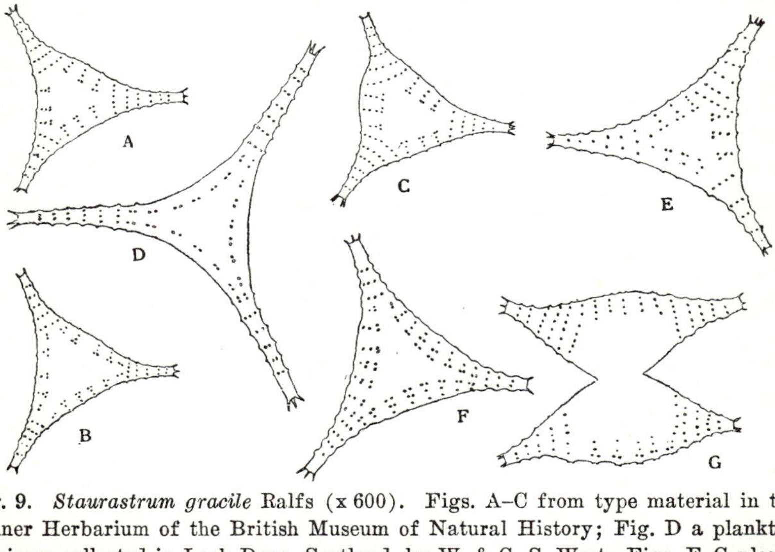

Fig. 9. *Staurastrum gracile* Ralfs (x 600). Figs. A–C from type material in the Jenner Herbarium of the British Museum of Natural History; Fig. D a plankton specimen collected in Loch Doon, Scotland, by W. & G. S. West; Figs. E–G plankton specimens from Norwegian lakes, Fig. E from Orrvandet, Figs. F, G from Lemonsjøen.

ginal row of pairs of granules (the verrucae of the front view) and at times short rows of granules at right angles to the sides of the cell, center of the cell without ornamentation; processes with transverse rings of granules and trispinate or quadrispinate. Chloroplast with a fairly small axial mass and two laminate blades running to each angle of the cell and sometimes continuing a short distance into the processes; pyrenoid single and central. (Facultative planktont.)

Zygospores spherical, with numerous straight long spines that have inflated bases and apices with widely spreading bifurcations (W. & G. S. West).

Cells 95–107 μ long with processes, 46–49 μ long without processes; breadth with processes 93–118 μ, without processes 35–39 μ breadth at isthmus 10 μ.

Diameter of zygospores without spines 32 μ, with spines 60 μ (W. & G. S. West).

Des Moines (rr), Horseshoe (rr), Kawaguesaga (c).

Staurastrum gracile and *S. paradoxum* Meyen are closely related species and ones which are inextricably confused in phycological literature. The description and figure of the older species *(S. paradoxum)* might apply to many of the numerous triangular species with processes at the angles. Ralfs, the discoverer of *S. gracile* distinguished between the two on the basis of smaller size and divergent processes of *S. paradoxum* in his British Desmidieae; a work which has been taken as the starting point for the nomenclature of the family. Ralfs' description says nothing about the ornamentation of the body of the cell of *S. gracile* but the fortunate discovery of type material in the Jenner herbarium at the British Museum shows that the vertical view always has an intramarginal row of granules that lie in pairs and have a vertical row of granules running to the cell margin from each pair of granules. In this type material the processes are very short (Text Figs. **9A–9C**). No specimens exactly like this were found in the plankton catches of the G. S. West collections, although specimens from Norwegian lakes in the Huitfeld–Kaas collection were quite like them (Text Figs. **9F–9G**). After examining the series found in the British plankton, as contained in the G. S. West collection, I feel that shape of the semicells and length or divergence of the processes cannot be used in distinguishing between *S. gracile* and *S. paradoxum* and that on these bases the two species should be united. The intramarginal row of granules in pairs (the apical verrucae of the front view) is a feature which separates the two and all specimens with these verrucae should be referred to *S. gracile.*

The species is undoubtedly of much more limited distribution than has been supposed and many references to it are unquestionably dealing with *S. paradoxum.* The Wisconsin specimens have long divergent processes and look, in vertical view, like the specimens found in Loch Doon, Scotland, by W. & G. S. West which they called a "very large form" (Text Fig. **9D**).

28. STAURASTRUM LONGIRADIATUM W. & G. S. West. Pl. 74, Figs. 5–11.

Trans. Linn. Soc. 2nd. Ser. Bot. 5: 267, *pl.* 17, *fig.* 23. 1896; G. S. West, Jour. Linn. Soc. Bot. 39: 28. 1909.

Staurastrum gracile var. *bulbosum* G. M. Smith *non* W. West, Trans. Wis. Acad. Sci., Arts & Lett. 19: 649. 1918.

Cells of medium size, breadth (with processes) 1.5–2 times the length, slightly constricted, sinus widely open, isthmus relatively broad; semicells campanulate and with cyathiform bases, apex flattened and ornamented with emarginate verrucae; angles of semicells continued in long divergent, parallel or slightly convergent processes with gradually attenuated margins, upper margin of processes deeply crenulate, crenulations on lower surface of processes shallow, ends truncate and with four short divergent spines. Vertical view triangular; body of the cell with straight to slightly concave sides, with an intramarginal row of two or three bidenticulate verrucae on each side of the cell; angles continued in long straight processes with undulate to subcrenulate margins and a single row of simple verrucae down the center. Vertical view of basal portion of the semicell triangular with the apices flattened and bearing one or two conical verrucae. Chloroplasts with a small axial mass and two laminate lobes running part way into each process; pyrenoid single and central. (Euplanktont.)

Zygospores unknown.

Cells 46–96 μ long with processes, 41–49 μ long without processes; breadth with processes 73–108 μ, without processes 20–28 μ; breadth at isthmus 8.5–10 μ.

Amicoy (sss), Bear (ss), Beartrap (rrr), Beaverdam (rr), Birch (aaa), Bird (rrr), Blake (r), Big Butternut (ss), Little Butternut (rr), Butternut (aa), Chetac (cc), Chetek (cc), Chetek (ss), Deer (ss), Des Moines (rrr), Duck (s), Dummy (ss), Hooker (rrr), Horse (cc), Horseshoe (ss), Found (r), Granite (r), Grassy (rrr), Lac Court Oreilles (rrr), Lake of the Woods (rr), Little Bass (rr), Long (aa), Lost (rr), Loveless (rr), Big McKenzie (r), Middle McKenzie (rrr), Nicaboyne (rr), Pokegama (ss), Poplar (rrr), Prairie (rr), Reserve (r), Rice (rrr), Round (r), St. Croix (rr), Big. St. Germain (r), Sand (r), Shell (rr), Silver (rr), Squirrel (rrr), Trout (rrr), Vermilion (sss).

This alga has considerable resemblance to *S. gracile* Ralfs in the shape of the semicells and especially to the variety *cyathiforme* in the swollen bases of the semicells. In *S. longiradiatum,* however, this swollen base is triangular in vertical view while in *S. gracile* var. *cyathiforme* W. & G. S. West it is circular.

The determination of this species is based upon a comparison of the Wisconsin material with specimens from the plankton of Yan Yean (Australia) in the G. S. West collection. In the figures of the alga published by W. & G. S. West the ratio between length of the semicell and breadth of the swollen basal portion is approximately 1:1 but in both the Australian and Wisconsin material this ratio is roughly 2:1. There is also a greater variation in the divergence of the processes than is given in the original description.

The dimensions of the alga as I have found it agree well with the variety *major* W. & G. S. West, which was described at the same time as the type, but since the dimensions of Australian specimens which G. S. West referred to the type are those of the variety *major* I feel that the variety should not be recognized.

29. STAURASTRUM FLORIFERUM W. & G. S. West. Pl. **74,** Figs. **12–18.**

Trans. Linn. Soc. 2nd. Ser. Bot. 5: 267, *pl.* 18, *fig.* 1. 1896.

Cells of medium size, breadth (with processes) two to three times the length, fairly deeply constricted, sinus widely open and with an acute apex, isthmus fairly narrow; semicells obversely trapeziform, with the ventral margins slightly convex and the dorsal margin flattened, cell apex with a row of emarginate verrucae; angles continued in long horizontal processes with subparallel sides, superior margin of processes with short acute spines, inferior margin of processes generally with shorter similar spines, ends of processes with three fairly stout short divergent spines, processes parallel to divergent. Vertical view triangular, cell body with straight sides and an intramarginal hexagonal ring of emarginate verrucae (two within each lateral margin); angles continued in long hollow slightly tapering processes with faintly undulate lateral margins, basal portion of processes with a double row of spines, apical portion with a single row, ends of processes with three stout divergent spines. Chloroplast with a small central mass and two long parallel plates running to each angle and continuing for a short distance into the processes; pyrenoid single and central. (Facultative planktont.)

Zygospores unknown.

Cells 23–44 μ long with processes, 22–26 μ long without processes; breadth with processes 52–86 μ, without processes 19–26 μ; breadth at isthmus 6.5–8 μ.

Lake of the Woods (rrr), Paddock (rrr), Pokegama (r).

This alga has not been reported since it was described twenty-five years ago. I have found it at three stations, though never in any great quantity. There is more variation in both the amount of divergence of the processes and the dimensions of the cells than W. & G. S. West describe.

The length of the processes in the individuals from Pokegama lake was somewhat greater than that found in the other examples (Figs. **16–18**) and the front view of these individuals suggests *S. anatinum* var. *longibrachiatum* but the ornamentation of the top view is quite distinctive and the two need never be confused.

30. **Staurastrum Bullardii** sp. nov. Pl. **74,** Figs. **19–23;** Pl. **75,** Figs. **1–3.**

Cells fairly large, breadth (with processes) equal to or slightly greater than the length, deeply constricted, sinus widely open and with a sublinear apex, isthmus narrow; semicells obsemicircular, with apices truncate; angles continued in very long slightly attenuated quadrispinate processes, superior and inferior margins of processes crenulate, median portion at times with a row of granules; processes at first outwardly divergent and then upwardly curved; cell apex with two emarginate verrucae in the median portion, cell body without other ornamentation. Vertical view triangular, with the sides of the body of the cell straight and the angles continued in long straight processes with faintly undulate margins; body of the cell with two intramarginal emarginate verrucae in the middle of each side. Chloroplast with a small central mass and two short arcuate lobes extending to each angle; pyrenoid single and central. (Euplanktont.)

Zygospores unknown.

Cells 70–88 μ long with processes, 23–30 μ long without processes; breadth with processes 81–100 μ, without processes 20–24 μ; breadth at isthmus 8–10 μ.

Balsam (ss), Big Bass (r), Big McKenzie (rr), Birch Island (rrr), Blake (ss), Carroll (c), Clear Crooked (rr), Cochrane (rr), Crooked (rrr), Deer (ss), Des Moines (s), Dutchmans (rrr), Found (rrr), Grindstone (ss), Half Moon (s), High (rrr), Lac Court Oreilles (sss), Lac du Flambeau (rrr), Long (s), Lost (r), Loveless (rr), Mercer (ss), Middle McKenzie (ss), Mud (rr), Owen (cc), Reserve (sss), Round (s), Rudolph (rrr), Sand (rr), Squirrel (rr), Whitefish (ss), Wolf (rrr).

The species is dedicated to Mr. Charles Bullard in grateful appreciation of many courtesies extended during the preparation of the section of this work dealing with the Desmidiaceae. There is a certain suggestion of *S. paradoxum* var. *longipes* Nordstedt in the length and delicate nature of the processes, but in the shape of the cell body, apical verrucae, and curvature of the processes it is quite unlike it.

31. Staurastrum vestitum Ralfs. Pl. **75**, Figs. **4–5.**

Brit. Desm. 143, *pl.* 23, *figs.* 1A–1E. 1848; Delponte, Mem. d. R. Acad. d Sc. di Torino 2 Ser., 28: 158, *pl.* 12, *figs.* 46–49. 1877 (sep.); Wolle, Desm. U. S. 138, *pl.* 45, *figs.* 28–30. 1884.

Cells of medium size, length about half the breadth (with processes), deeply constricted, sinus widely open and with apex acuminate, isthmus narrow; semicells broadly obtriangular, with lateral margins and apex straight, angles gradually attenuated into quite stout processes; processes parallel, with four to five concentric rings of conical spines that give the upper and lower margins a coarsely crenulate appearance, ends of processes with three short divergent spines; apex of semicells with four emarginate verrucae, center of face of semicell with two verrucae lying below the central verrucae of the apical series. Vertical view triangular, with the sides straight and the angles with a process that has the same appearance as in the front view; median portion of each side of the cell with two elongate bifid to emarginate verrucae projecting at right angles to the cell wall, marginal verrucae sometime flanked by similar but smaller verrucae; interior of cell with an

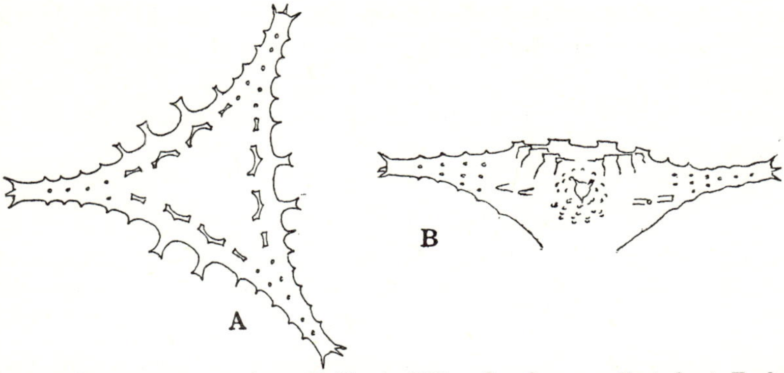

Fig. 10. *Staurastrum vestitum* Ralfs (x 600). Specimens collected at Tarbert, Lewis, Scotland, by W. & G. S. West.

intramarginal row of four verrucae on each side, the two central verrucae arcuate and tridenticulate, the two terminal verrucae of each row broad and emarginate. (Tychoplanktont.)

Zygospores spherical, clothed with elongate processes with inflated bases and bifurcate apices (Wolle).

Cells 40 μ long; breadth with processes 58 μ, without processes 25 μ; breadth at isthmus 10 μ.

Found (rrr), Lost (rrr).

The few examples seen in Wisconsin lakes have a much more abrupt transition from the body of the semicells to the processes (as seen in vertical view) than is ordinarily the case. Since our specimens are not quite typical a figure is given of an English plankton specimen.

The determination of this species is an easy matter when the cells are viewed from above, the long verrucae projecting at right angles from the cell immediately separating it from the other species.

var. SUBANATINUM W. & G. S. West. Pl. **75**, Figs. **6–10.**

Trans. Roy. Irish Acad. 32: 54, *pl.* 1, *fig.* 28. 1902.

Cells with smaller bodies but longer and less robust processes; length of processes greater than the diameter of the cells; ornamentation as in the type but with the verrucae smaller and the terminal spines of the processes larger; processes slightly divergent. Chloroplast axial, with the central mass small and with a broad deeply furcate lobe running to each angle of the cell; pyrenoid single and central. (Euplanktont.)

Cells 30–42 μ long with processes, 22–30 μ long without processes; breadth with processes 70–81 μ, without processes 23–32 μ; breadth at isthmus 10.5–11.5 μ.

Diamond (sss), Jag (rr), Red Bass (rrr).

This variety is very happily named since it bears a striking resemblance to some of the pelagic varieties of *S. anatinum* Cooke & Wills.

The presence of the characteristic verrucae on the side of the semicells shows, however, that the variety belongs with *S. vestitum*. In the Wisconsin plankton these lateral verrucae are not so long as those figured by W. & G. S. West, but in all other particulars the two are identical.

32. Staurastrum anatinum Cooke & Wills.

Grevillea **9**: 92, *pl.* 139, *fig.* 6. 1881; W. & G. S. West, Trans. Roy. Irish Acad. **32**: 54, *pl.* 1, *figs.* 24–25. 1902; Carter, Ann. of Bot. **34**: 307, *pl.* 14, *figs.* 19–20. 1920.

Cells fairly large, breadth (with processes) one and a half times to twice the length, deeply constricted, sinus acute and with apex acuminate, isthmus narrow; semicells broadly cyathiform, with the dorsal margin slightly convex and the ventral margin rounded, angles continued in fairly long hollow somewhat attenuated processes that usually terminate in three stout divergent spines, superior and inferior margins of processes coarsely serrate and body of processes with a median row of granules, processes subparallel to divergent; body of semicells with an apical row of 5–7 emarginate verrucae and another lower transverse subapical row of verrucae, lower part of body of semicells smooth. Vertical view triangular, with sides of cell body slightly concave to somewhat convex; angles continued in long straight processes with serrate margins and a single median row of granules, ends trispinate; sides of body of cells with emarginate verrucae and central area with similar verrucae (that appear as pairs of granules) just within the lateral margins and connecting with the axial rows of granules of the processes. Chloroplast with a very small central mass and a large broadly lanceolate lobe with a deep incision running to each angle of the cell; pyrenoids usually three, one in each lobe.

Zygospores unknown.

Cells 65 μ long with processes, 46 μ long without processes; breadth with processes 101–113 μ; breadth at isthmus 15.5 μ (W. & G. S. West).

Not found in the plankton of Wisconsin lakes.

var. longibrachiatum W. & G. S. West. Pl. **75**, Figs. **11–16.**

Trans. Roy. Soc. Edinburgh **41**: 504, *pl.* 7, *figs.* 8–9. 1905.

Semicells obversely trapeziform; with the apex flattened or slightly concave; body of semicells without the subapical verrucae; processes long, gradually attenuated and with concentric rings of granules at the bases, ends tri- or quadrispinate, processes only slightly divergent. (Euplanktont.)

Cells 43–68 μ long with processes, 30–35 μ long without processes; breadth with processes 80–92 μ, without processes 29–37 μ; breadth at isthmus 10.5–12 μ.

Adelaide (rr), Crooked (ss), Meta (ss), Nancy (c), Sanford (rrr), Speese (s).

The primary distinction of the variety is the great length of the processes, the length being greater than the breadth of the body of the semicells. It also differs from the type in that the processes are less divergent and have concentric rings of granules at the base. The breadth with processes of the Wisconsin specimens is not as great as that of specimens from Scotland.

var. CURTUM G. M. Smith. Pl. **75**, Figs. **17–20.**

Trans. Wis. Acad. Sci., Arts & Lett. **20**: 354, *pl.* 12, *figs.* 6–7. 1922.

Cells smaller, semicells broadly cyathiform and with apex flattened, processes short and only subdivergent, cell body without the subapical row of verrucae. (Euplanktont.)

Cells 35–48 μ long with processes, 25–28 μ long without processes; breadth with processes 55–74 μ, without processes 24–33 μ; breadth at isthmus 9–11 μ.

Bass (rrr), Beaverdam (rrr), Cranberry (rr), Crawling Stone (rr), Found (rrr), Hill (r), Horseshoe (rrr), Lac Court Oreilles (rrr), Lindy (ss), Long (rrr), Jones (rrr), Nicaboyne (rrr), Sand (rrr).

The variety is distinguished from another small variety (*pelagicum* W. & G. S. West), found in British lakes, by the different proportions between length and breadth of the semicells, smaller degree of divergence of the processes, cell size, and the lack of the subapical verrucae. The cell shape and ornamentation is quite similar to the variety *longibrachiatum* but the processes are very much shorter.

var. **denticulatum** var. nov. Pl. **75**, Figs. **21–25.**

Shape of the semicells as in the variety *curtum;* processes with very long denticulations on the upper and the lower margins; cell apex with small straight linear verrucae and small linear subapical verrucae. (Euplanktont.)

Cells 36–50 μ long with processes, 26–34 μ long without processes; breadth with processes 60–86 μ; without processes 26–34 μ; breadth at isthmus 10.5–12 μ.

Carroll (sss), Jag (rr), Mud (sss), Sishebogema (rrr), Tied Canoe (rrr).

The variety is distinguished by the great development of the teeth that ornament the processes and the reduction in size of the apical and subapical verrucae of the body of the cell. These smaller verrucae have the same position as those of the type but are not nearly so conspicuous, especially in a vertical view. In one case they were not regularly arranged but were scattered coarse granules (Fig. **21**).

The spiny nature of the processes is quite similar to that of an alga from the plankton of Canadian lakes that I have considered a variety *(verrucosum)* of *S. aspinosum* Wolle. The relative length of the processes and the breadth of the cell body as well as the ornamentation of the cell body are quite distinct in the two algae.

33. Staurastrum tetracerum (Kützing) Ralfs.

Ann. & Mag. of Nat. Hist. **15**: 150, *pl.* 10, *figs.* 1A–1C. 1845; Brit. Desm. 137, *pl.* 23, *figs.* 7A–7F. 1848; Lundell, Nova Acta Reg. Soc. Sci. Upsaliae, 3 Ser. **8**: 68. 1871.

Cells very small, length (with processes) about equal to the breadth, deeply constricted, sinus obtuse-angled and with apices subacute, isthmus narrow; semicells obversely triangular, with apices very slightly retuse; angles continued in delicate slightly tapering processes with sharply crenulate margins; ends of processes without teeth (?). Vertical view compressed and body of semicells not much broader than the processes. Semicells almost invariably twisted at the isthmus.

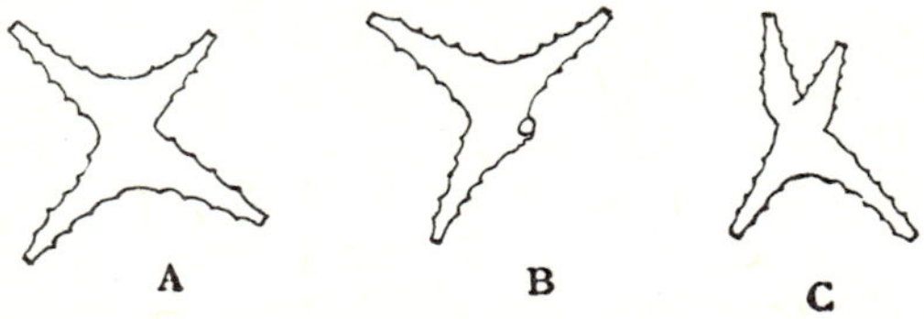

Fig. 11. *Staurastrum tetracerum* Ralfs (x 800). Specimens collected at Moidart, Scotland, by W. & G. S. West.

Zygospores spherical, with about sixteen processes that are dilated at the base and furcate at the apex (Lundell).

Cells 20–23 μ long with processes, 10 μ long without processes; breadth with processes 22.5 μ; breadth at isthmus 5 μ.

Zygospores 30 μ in diameter with processes, 16 μ without processes (Lundell).

Not found in the plankton of Wisconsin lakes.

var. validum W. & G. S. West. Pl. 76, Figs. 9–10.

Jour. Roy. Micr. Soc. **1897**: 495, *pl.* 6, *fig.* 25. 1897.

Cells somewhat larger, deeply constricted and with sinus rectangular; semicells obversely triangular and with point of origin sharply defined; length of processes somewhat greater than the breadth of the semicells; processes with margins subparallel and finely crenulate, terminating in four minute spines; body of semicell smooth but with a finely crenulate outline; semicells not twisted at the isthmus. (Facultative planktont.)

Cells 36–43 μ long with processes, 16–20 μ long without processes; breadth with processes 46–52 μ; without processes 14–17 μ; breadth at isthmus 6–7 μ; thickness 7.5 μ.

Duck (r).

The variety as observed in our plankton has the processes more finely crenulate than W. & G. S. West show and does not have the smooth outline of the body of the semicell. It might possibly be a small variety of *S. chaetoceras* (Schröder) G. M. Smith but has been excluded since no granules were observed within the margins of the body or the processes of the cells.

var. EVOLUTUM W. & G. S. West. Pl. 76, Figs. 11–14.

Trans. Bot. Soc. Edinburgh 23: 25, *pl.* 2, *fig.* 31. 1905.

Cells larger and with much longer and more delicate processes. Vertical view triangular, with the sides slightly retuse and the angles continued in straight processes with finely crenulate margins. Processes of one semicell alternating with those of the other. (Facultative planktont.)

Cells 40–51 μ long with processes, 23–25 μ long without processes; breadth with processes 51–63 μ, without processes 16–18 μ; breadth at isthmus 4.5–6 μ.

Middle McKenzie (ss).

var. TRIGONUM Lundell. Pl. 76, Figs. 15–16.

Nova Acta Reg. Soc. Sci. Upsaliae 3 Ser., 8: 69, 1871; W. West, Jour. Linn. Soc. Bot. 29: 187. 1892.

Cells small, fairly deeply constricted; processes stout in proportion to the size of the cells, divergent and with three or four deep crenulations on both the upper and the lower margins, median portion of processes with small spines. Vertical view triangular, with sides slightly retuse and angles with stout straight processes, lateral margins of processes deeply crenulate and ends quadrispinate. (Tychoplanktont.)

Cells 32 μ long with processes, 22 μ long without processes; breadth with processes 42 μ; without processes 15 μ; breadth at isthmus 7 μ.

Little Bass (rr), Crab (rrr), Horseshoe (rrr), Jones (rrr), Red Bass (rrr), Whitefish (rr).

Lundell mentions triangular forms of *S. tetracerum* and William West gives their dimensions. I have found a minute triradiate alga which I believe to be the one described by Lundell.

34. STAURASTRUM LACUSTRE G. M. Smith. Pl. 76, Figs. 1–8.

Trans. Wis. Acad. Sci., Arts & Lett. 20: 355, *pl.* 12, *figs.* 13–15. 1922.

Cells of medium size, length (with processes) about equal to the breadth, fairly deeply constricted, sinus widely open and with apex acute, isthmus fairly narrow; semicells obversely triangular, with ventral margins straight or slightly tumid, dorsal margin flat, usually with the median portion truncate and elevated; angles continued in long slightly tapering processes that terminate in two very long divergent spines, proximal portion of processes with upper and lower surfaces coarsely crenulate and at times a median row of conical granules, distal portion always smooth, processes always strongly divergent; cell body without ornamentation except for two apical emarginate verrucae. Vertical view triangular, sides straight or slightly retuse; angles continued in long processes with subparallel sides and terminating in two long spines that lie in the same vertical plane, processes generally smooth in outline, sometimes with 2–3 transverse rings of coarse conical granules in the basal portion; median portion of base of processes with an axial or a double row of granules; body of semicells with two emarginate verrucae just within each lateral margin. Vertical view of chloroplast with a small central mass and two long arcuate lobes running to the base of each process; pyrenoid single and central. (Euplanktont.)

Zygospores unknown.

Cells 70–86 μ long with processes, 25–34 μ long without processes; breadth with processes 77–105 μ, without processes 17–20 μ; breadth at isthmus 8–9 μ; length of terminal spines 9–11 μ.

Bear (rrr), Devils (rrr), Dummy (r).

A very distinctive species in which the long straight divergent processes termining in two divergent spines are quite unlike those of other plankton species of the genus. The ornamentation of the body of the cell and the base of the processes is somewhat variable. In some cases the bases of the processes have three or four rings of conical granules, in others the granules are found only on the upper and lower sides, while in still others the granules are lacking. The emarginate verrucae at the apex of the cell are at times absent. This lack of an ornamentation of the body of the cell is not always correlated with a lack of ornamentation of the processes since specimens with the best development of ornamentation of processes may have a lack of body ornamentation or vice versa.

35. **Staurastrum contortum** sp. nov. Pl. **76,** Figs. **17–20.**

Cells of medium size, breadth (with processes) generally about twice the length, deeply constricted, sinus obtuse-angled and widely open, isthmus narrow; semicells broadly obtriangular, with ventral and dorsal margins straight; body of semicells small and not sharply differentiated from the long attenuated processes; processes divergent, with margins sharply crenulate and ends tridenticulate. Semicells almost invariably twisted, usually at right angles to each other. Vertical view narrowly elliptic, with the poles prolonged into straight processes with undulate sides and trispinate ends; processes of the two semicells generally cruciform. Chloroplast, laminate, with one pyrenoid.

Zygospores unknown.

Cells 36–67 μ long with processes, 12 μ long without processes; breadth with processes 44–119 μ, without processes 10.5–12 μ; breadth at isthmus 6 μ; thickness 6 μ.

Big Bass (rrr), Bone (ss), Found (rrr), Grindstone (rr).

Twisting of the semicells is well known in some of the other small biradiate species of the genus but semicells lying at right angles to each other are quite unusual in the other species. This species shows a great resemblance to the twisted forms of *S. excavatum* W. & G. S. West that G. S. West found in the Australian plankton. (Jour. Linn. Soc. Bot. **39**: 71, *pl.* 6, *figs.* 19–20. 1909) but cannot be placed in *S. excavatum* since it lacks the characteristic apical excavation. The body of the cell of *S. contortum* is much smaller and less sharply defined than those of other biradiate species.

36. **Staurastrum Chaetoceras** (Schröder) comb. nov. Pl. **76,** Figs. **21–24;** Pl. **77,** Fig. **1.**

Staurastrum polymorphum Meyen (?) var. *chaetoceras* Schröder in Zacharias, Forschungsbr. a. d. Biol. Stat. su Plön **6**[2]: 131, *text figs.* A–C. 1898.

Cells fairly large, length (with processes) about equal to the breadth, deeply constricted, sinus obtuse-angled and with a blunt apex, isthmus narrow; semicells obversely triangular, with the ventral margins sublinear and the apex flattened; angles continued in long divergent straight processes, the angle of divergence being a continuation of the angle of the sinus, processes slightly attenuated, delicate, with transverse concentric rings of minute granules and terminating in four minute teeth; body of semicells with outline finely crenulate and with a few scattered subapical granules, at times with a transverse row of granules just above the isthmus. Vertical view narrowly elliptic (very rarely triangular) with the poles continued in straight processes whose ornamentation is as in the front view. (Euplanktont.)

Zygospores unknown.

Cells 46–89 μ long with processes, 19–22 μ long without processes; breadth with processes 65–89 μ, without processes 16–22 μ; breadth at isthmus 5.5–6.5 μ; thickness 10 μ.

Bear (rrr), Big Butternut (rrr), Blake (ss), Bone (r), Camp (r), Center (r), Chain (rrr), East (rr), Found (rr), Half Moon (r), Little Bass (s), Little Butternut (rr), Lost (r), Big McKenzie (rrr), Middle McKenzie (rr), Pokegama (ss), Whitefish (rrr), Wolf (rrr).

In describing the alga Schröder placed it as a variety of *S. paradoxum* Meyen (by mistake written *S. polymorphum* Meyen) but I feel that if it is to be given varietal rank it should be a variety of *S. tetracerum* (Kützing) Ralfs. In spite of occasional triradiate semicells it should be grouped with the biradiate species and separated from the other species of this type by the lack of verrucae at the apex of the semicells. Most of the biradiate species also lack the concentric rings of granules on the processes that are found in *S. Chaetoceras.* These rings are easily overlooked, especially when the cell contents are present. In certain stations forms with exceptionally long processes were observed (Fig. **1**).

37. STAURASTRUM AMERICANUM (W. & G. S. West) G. M. Smith. Pl. 77, Fig. 2.

Trans. Wis. Acad. Sci., Arts & Lett. **10**: 351. 1922.

Staurastrum grallatorium var. *americanum* W. & G. S. West, Trans. Linn. Soc. 2nd. Ser. Bot. **5**: 265, *pl.* 17, *fig.* 15. 1896.

Cells of medium size, length (with processes) about two-thirds the breadth, deeply constricted, sinus widely open and with an acute apex, isthmus very narrow; semicells obsemicircular, with the apex somewhat convex, dorsal margin without ornamentation but with the wall thicker on the dorsal margin than elsewhere; angles continued in long processes that have subparallel sides, upper and lower margins of processes crenulate, processes subparallel to divergent and with ends minutely trispinate. Vertical view elliptic, with poles continued in hollow processes with undulate sides, body of cell without ornamentation. (Tychoplanktont.)

Zygospores unknown.

Cells 61 μ long with processes, 29 μ long without processes; breadth with processes 90 μ, without processes 26 μ; breadth at isthmus 6.5 μ.

Clare (rrr).

Although but two specimens were seen both agreed well with W. & G. S. West's description. The lack of ornamentation on the apex of the semicells and the thickened wall at this point are quite unlike other biradiate species. The shape of the semicells, lack of apical ornamentation and terminal ornamentation of the processes is so unlike that of *S. grallatorium* Nordstedt that the alga should be considered a distinct species and not a variety of *S. grallatorium.*

38. STAURASTRUM GRALLATORIUM Nordstedt. Pl. **77**, Figs. **3–5**.

Vidensk. Medd. f. d. naturhist. Foren. i Kiöbenhavn **1869**: 228, *pl.* 4, *fig.* 52. 1870.

Cells somewhat under medium size, breadth (with processes) about twice the length, slightly constricted, sinus very widely open and with a rounded indentation of the relatively broad isthmus; semicells narrowly campanulate, apex truncate and with a median portion bearing three emarginate verrucae and somewhat elevated above the rest of the apex; body of semicells, except for the verrucae, without ornamentation; angles continued in long slightly attenuated hollow processes with sharply crenulate upper and lower margins, ends of processes deeply bifurcate, the bifurcations (not spines) being a continuation of the processes and hollow, processes subparallel. Vertical view elliptic, body of semicells relatively small and with the lateral margins greatly rounded, poles continued in long processes with straight or slightly undulate margins, processes terminating in two arms that lie in the same vertical plane. (Tychoplanktont.)

Zygospores unknown.

Cells 42–48 μ long with processes, 32–34 μ long without processes; breadth with processes 69–86 μ, without processes 20 μ; breadth at isthmus 8.5–9 μ; thickness 14 μ.

Tank (rrr).

Superficially the species resembles *S. Johnsonii* W. & G. S. West but differs in the parallel processes, lack of subapical and isthmial ornamentation, termination of the processes, and shape of the semicells. Nordstedt's figure shows five elevated verrucae in the front view but in all of the specimens that I have seen there are only three. The shape of the semicells, especially the small body of the cell in a vertical view, and the branching of the ends of the processes show that the organism which I have found is *S. grallatorium.* The processes do not end in spines as do the other biradiate species of the genus, instead the process ending is bifurcate and distinctly hollowed.

var. FORCIPIGERUM Lagerheim. Pl. **77**, Figs. **6–7**.

Öfvers. kgl. Sv. Vet.-Ak. Förh. **42**, No. **7**: 249, *pl.* 27, *fig.* 27. 1886.
Staurastrum saltans var. *forcipigerum* Lagerheim, *ibid.* **44**, No. **8**: 539. 1887.

Semicells trapeziform, with apex truncate and bearing two elongate subapical verrucae, apex with two fairly long divergent spines asymmetrically disposed on the sides of the elongated median portion; processes with margins more coarsely crenulate and with fewer crenulations, furcation of ends of processes unequal, the fork on the side towards the isthmus always the longer. (Tychoplanktont.)

Cells 45 μ long with processes, 38 μ long without processes; breadth with processes 65–72 μ; without processes 21–25 μ; breadth at isthmus 7.5–8.5 μ.

Hillman (rrr).

After describing the alga as a variety of *S. grallatorium* Lagerheim transferred it to *S. saltans* Joshua, an alga which was not correctly figured until some years later (W. & G. S. West, Trans. Linn. Soc. **2**nd. Ser. Bot. **6**: 188, *pl.* 22, *figs.* 13–14. 1902). Mr. Charles Bullard has shown me a preparation in which one semicell is typical *S. grallatorium* and the other the variety *forcipigerum.* From this it is evident that if the variety is to be recognized it belongs to *S. grallatorium.*

39. STAURASTRUM ANCHORA W. & G. S. West. Pl. **77**, Figs. **8–9**.

Trans. Linn. Soc. 2nd. Ser. Bot. **5**: 266, *pl.* 17, *figs.* 21–22. 1896.

Cells large, breadth (with processes) twice the length, slightly constricted, sinus sublinear, isthmus relatively broad; semicells campanulate, with bases slightly inflated, apices flattened and smooth; angles of semicells prolonged in long somewhat tapering processes that end in three divergent spines of medium size, upper and lower margins of processes deeply crenulate, processes parallel to somewhat convergent; median area of apex of cell body with a transverse subapical row of 8–10 emarginate verrucae. Vertical view elliptic, with a row of 8–10 emarginate verrucae just within each lateral margin; poles of cells continued in long tapering processes that have the same appearance as in the front view. (Tychoplanktont.)

Zygospores unknown.

Cells 68 μ long; breadth with processes 143–162 μ, without processes 45–58 μ; breadth at isthmus 13.5 μ; thickness 24 μ.

Des Moines (rrr), Fish (rrr), Owen (rrr).

The largest of the biradiate members of the genus and one with a distinctive shape of the semicells and subapical ornamentation.

40. Staurastrum leptocladum Nordstedt. Pl. **78,** Figs. **1–7.**

Vidensk. Medd. f. d. naturhist Foren. i Kiöbenhavn **1869:** 228, *pl.* 4, *fig.* 57. 1870; W. & G. S. West, Jour. Linn. Soc. Bot. 33: 317. 1898.

Staurastrum leptocladum var. *cornutum* Wille, Bih. Sv. Vet.-Ak. Handl. **8,** No. **18:** 19, *pl.* 1, *fig.* 39. 1884; Johnson, Bull. Torr. Bot. Cl. 21: 289, *pl.* 211, *fig.* 3. 1894; W. & G. S. West, Trans. Linn. Soc. 2nd. Ser. Bot. 5: 79, *pl.* 9, *figs.* 12–13. 1895.

Cells fairly large, breadth (without processes) about half the length, slightly constricted, sinus a small semicircular depression, isthmus relatively broad; semicells subcampanulate, with bases inflated and apex strongly tumid; angles continued in long slightly attenuated processes that are at first outwardly divergent and then curve gracefully away from the isthmus, upper and lower margins of processes finely crenulate and ends bispinate; basal inflation of body of the semicells usually with a single transverse row of conical or vertically elongate granules, rarely smooth, median portion of semicell body with a tumid protuberance and dorsal margin with a single vertical subapical spine. Vertical view narrowly elliptic, with poles continued in straight slightly tapering processes with undulate margins, ends with two small spines that lie in the same vertical plane; median part of lateral margin slightly truncate. (Facultative planktont.)

Zygospores unknown.

Cells 71–95 μ long with processes, 35–41 μ long without processes; breadth with processes 67–105 μ, without processes 18–21 μ; breadth at isthmus 6–7 μ; thickness 10–12 μ.

Bear (sss), Bone (r), Carrol (s), Chain (sss), Clare (ss), Cranberry (rrr), Deer (s), Devils (r), Diamond (s), Grindstone (rrr), Half Moon (rrr), Horseshoe (c), Jag (rr), Little Butternut (rrr), Loveless (rr), Nancy (rrr), Sand (rrr), Sishebogema (rr), Trapper (sss), Whitney (rr).

The species and its varieties are distinguished by the curvature of the processes, at first divergent and then upwardly curved. Nordstedt in his description does not mention the subapical spine, a feature upon which Wille based his variety *cornutum.* My experience with the alga in Ontario, New York and Wisconsin has been so similar to that of W. & G. S. West that I agree fully with them when they say: "there can be no doubt that *S. leptocladum* always possesses the two spines at the apex of the semicells, this was made the character for a variety *(cornutum)* by Wille but we think it highly probable that these spines had been broken off in the specimens originally described by Nordstedt. All of the examples we have ever seen, from many diverse parts of the world, possessed these spines, and it is worthy of note that other observers only record the variety *cornutum,* a fact which confirms our suspicions as to the invariable presence of the apical spines in uninjured specimens."

In some of the specimens that I have seen the transverse row of granules on the basal swelling of the semicell was lacking, though it usually is a conspicuous feature.

var. DENTICULATUM G. M. Smith. Pl. **77**, Figs. **10–14.**

Trans. Wis. Acad. Sci., Arts & Lett. **20**: 351, *pl.* 11, *fig.* 14. 1922.

Variety with the subapical spines replaced by a broad emarginate elevation. Vertical view with the median portion of the lateral margins with a truncate elevation. (Facultative planktont.)

Cells 79–95 μ long with processes, 38–43 μ long without processes; breadth with processes 71–98 μ, without processes 18–21 μ; breadth at isthmus 6–7 μ; thickness 10–12 μ.

Balsam (s), Little Bass (rrr), Big McKenzie (c), Middle McKenzie (s), Oak (rrr), Pokegama (c).

var. INSIGNE W. & G. S. West. Pl. **77**, Fig. **15**; Pl. **78**, Figs. **8–11.**

Trans. Linn. Soc. 2nd. Ser. Bot. **5**: 266, *pl.* 17, *fig.* 17. 1896.

Cells with the same general outline as the type, though with the upward curvature of the processes more pronounced; ventral margin of the body of the semicells with crenulations or emarginate verrucae extending from the base of the processes to the vicinity of the basal swelling; dorsal margin with 4–6 emarginate verrucae that lie just within the apex of the semicells and form a transverse row; center of semicells with a circular protuberance. Vertical view rhomboidal, with the median protuberance of the lateral margins truncate and with the poles continued in long straight processes with undulate margins. (Facultative planktont.)

Cells 55–100 μ long with processes, 36–42 μ long without processes; breadth with processes 86–110 μ, without processes 20–26 μ; breadth at isthmus 6–8 μ; thickness 15–18 μ.

Blake (s), George (r), Lake of the Woods (rrr), Nicaboyne (rrr), Big St. Germain (rrr).

A variety which is unlike the others of the species in the subapical ornamentation but which should be kept a variety of *S. leptocladum* because of the characteristic curvature of the processes.

var. SINUATUM Woll. Pl. **78**, Figs. **12–14**.

Bull. Torr. Bot. Cl. **10:** 20, *pl.* 27, *fig.* 24. 1883.

Cells with the body much narrower and with the bases of the semicells not inflated; without ornamentation except for an apical notch; processes arising at the same point and with the same curvature as those of the type but not lying in the same plane. (Euplanktont.)

Cells 65–88 μ long with processes, 25–29 μ long without processes; breadth with processes 80–93 μ, without processes 12–15 μ; breadth at isthmus 7–8 μ; thickness 10 μ.

Bass (sss), Grassy (rrr), Lindy (c), Little Doctor (ss), Lynx (rrr), Speese (rrr).

This variety is unique in the peculiar twisting of the processes and in the nature of the upper portion of the semicells, features which are seen to the best advantage in lateral views of the cells. The identification of the Wisconsin material with Wolle's variety *sinuatum* is open to question since he based his variety on the divergence of the processes and the emargination of the semicell apices. Wolle does not mention the variety in his general work on the Desmids of the United States so, perhaps, decided that it was not worthy of recognition. The front view of our alga is so like that which Wolle gives that it seems better to revive his old name than to try to consider it new.

The specimens from Blue Gill Lake were not twisted, the four ends of the processes lying in the same plane. The crenulation of the processes is also more marked and the ends are quadrispinate. This form has been given the name **planum.** Its length with processes is 88 μ, without processes 28 μ; breadth with processes 110 μ, without processes 15 μ, breadth at isthmus 9 μ. (Pl. **78**, Fig. **15**).

41. STAURASTRUM JOHNSONII W. & G. S. West. Pl. **79**, Figs. **1–6**.

Trans. Linn. Soc. 2nd. Ser. Bot. **5:** 266, *pl.* 17, fig. 16. 1896.

Staurastrum leptocladum Johnson non Nordstedt, Bull. Torr. Bot. Cl. **21:** 288, *pl.* 211, *fig.* 2. 1894.

Cells of medium size, breadth (with processes) about one and a half times the length, slightly constricted, sinus a small semicircular depression, isthmus relatively broad; semicells subcampanulate, with inflated bases and slightly tumid apices, angles continued in long gradually tapering processes that terminate in three divergent spines, superior margin of processes crenulate, inferior margin also crenulate but with the crenulations less pronounced than those of the superior margin, median area of processes with a row of minute denticulations, processes somewhat divergent; body of semicells with two transverse rows of granules on the basal inflation, the row farthest from the isthmus frequently poorly developed and irregular, dorsal portion of body of semicells with a transverse subapical row of 5–7 emarginate verrucae and below this a transverse row of rings or masses of small granules the row forming a continuation of the median row of granules of the processes. Vertical view elliptic, with poles continued in straight slightly tapering processes terminating in three divergent spines, lateral margins of processes crenulate and with a single median row of granules; sides of body of cells with an intramarginal row of emarginate verrucae and a marginal row of verrucae. (Facultative planktont.)

Zygospores unknown.

Cells 59–85 μ long with processes, 35–41 μ long without processes; breadth with processes 79–110 μ, without processes 19–24 μ; breadth at isthmus 9–12 μ; thickness 12–13 μ.

Bass (ss), Bear (rrr), Carroll (sss), Carson (rr), Crooked (sss), Des Moines (rrr), Devils (r), Diamond (s), Hill (rrr), Horseshoe (sss), Howey (rr), Lindy (ss), Meronk (rr), Pine (rr), Shell (sss), Silver (rrr), South Crab (rrr), Sunday (rr).

The alga sometimes occurs in quantity in certain of the lakes. It is distinguished from others of the same general type by the size, divergence of the processes, and ornamentation. I have found that the divergence of the processes is generally greater than W. & G. S. West or Johnson figure but that this greater degree of divergence is not caused by a curvature of the processes. The subapical ornamentation of the body of the semicells is characteristic but subject to considerable fluctuation. In some specimens the granules below the emarginate verrucae are arranged in distinct rings, in others there are two opposed arcs of granules, in still other examples these arcs of granules are fused to form small arcuate verrucae.

In some of the lakes the cells were consistently smaller than the type and I have considered this a separate form. This form, **parvum** (Pl. **79,** Fig. **6**), has the same general shape and ornamentation as the type and differs only in the smaller size. Its length with processes is 42–68 μ, length without processes 25–31 μ; breadth with processes 62–68 μ, without processes 18–21 μ; breadth at isthmus 9 μ. Its distribution is as follows:

Found (rrr), Horseshoe (cc), Lynx (r), Pine Tree (ss), Travis (ss).

var. **depauperatum** var. nov. Pl. **79,** Figs. **7–11.**

Staurastrum leptocladum var. *divergens* G. M. Smith non Wolle; Trans. Wis. Acad. Sci., Arts & Lett. 19: 649. 1918.

Semicells campanulate and with basal swelling less pronounced; processes with the same kind of divergence as the type or with a slight curvature; apex of semicells with the subapical row of verrucae and second subapical row greatly reduced; bases of semicells without transverse rows of granules. (Euplanktont?)

Cells 55–86 μ long with processes, 47–52 μ long without processes; breadth with processes 98–125 μ, without processes 22–30 μ; breadth at isthmus 9–10.5 μ.

Bass (rrr), Beaverdam (rrr), Crab (rr), Cranberry (rrr), Crawling Stone (sss), Duck (rrr), George (r), Grindstone (rr), Lac Court Oreilles (r), Long (rr), Reserve (r), Whitefish (sss).

The general outline of the semicells in this variety is that of *S. Johnsonii* except for the basal swelling. The ornamentation of the base is lacking and the apical verrucation is much simpler. In some of the specimens there is a median row of granules on the processes, a feature found in *S. Johnsonii,* but not in the closely related *S. leptocladum* Nordstedt.

42. Staurastrum natator W. West. Pl. **79**, Fig. **12.**

Jour. Linn. Soc. Bot. **29:** 183, *pl.* 23, *fig.* 14. 1892.

Cells of medium size, length (without processes) about twice the breadth, slightly constricted, sinus a small semicircular depression, isthmus fairly broad; semicells subrectangular to subtrapeziform, with the lateral margins slightly convex and the apex straight; central portion of the apex with three tridenticulate verrucae in which the central spine is more prominent than the other two; angles of cells continued in long slightly curved processes that are ornamented with concentric rings of granules, ends of processes with three small

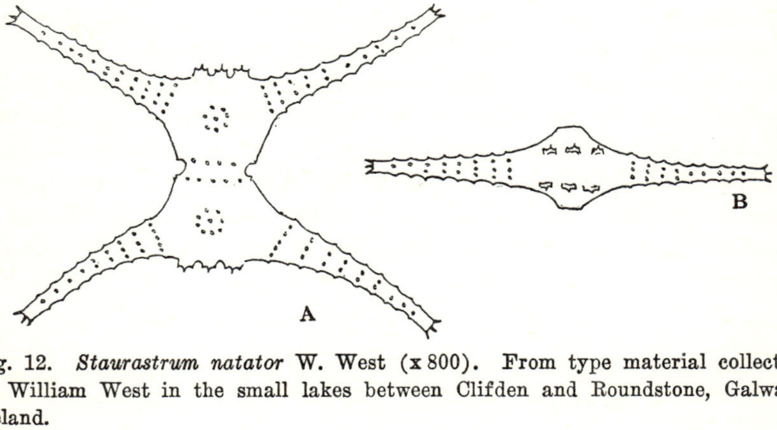

Fig. 12. ***Staurastrum natator*** W. West (x 800). From type material collected by William West in the small lakes between Clifden and Roundstone, Galway, Ireland.

subdivergent spines, processes divergent; central area of body of semicells with a protuberance bearing a ring of 6–8 granules that enclose one or two similar granules, basal portion of semicells with a transverse row of minute granules, 6–8 of which are visible in a front view of the cell. Vertical view compressed and subelliptic, with the transition from the body of the cell to the processes gradual, central portion of the lateral margins with a truncate projection and margins also with an intramarginal row of tridenticulate verrucae; processes with concentric rings of granules. (Tychoplanktont.)

Zygospores unknown.

Cells 60 μ long with processes, 34 μ long without processes; breadth with processes 88 μ, without processes 20 μ; breadth at isthmus 11.5 μ.

Little Doctor (rrr).

The species has not been observed since it was described by William West. In reëxamining type material in the G. S. West collection I have noted a ring of granules at the base of the semicells, a feature which was overlooked in the original description. A figure of an Irish specimen is also given for comparison with the Wisconsin material.

var. CRASSUM W. & G. S. West. Pl. **79,** Figs. **13–15.**

Trans. Linn. Soc. 2nd. Ser. Bot. 5: 265, *pl.* 17, *fig.* 14. 1896.

Semicells with a much larger ring of granules on the central protuberance; apical verrucae with denticulations of the same size. Vertical view with truncate projection much larger than that of the type. (Facultative planktont.)

Cells 42–50 μ long with processes, 30–33 μ long without processes; breadth with processes 58–70 μ, without processes 18–21 μ; breadth at isthmus 7.5–8.5 μ; thickness 20 μ.

Horseshoe (r).

S. natator can be separated from other biradiate species by the central protuberance, a feature which is much more pronounced in the variety *crassum.* When due allowance is made for the defects of draftmanship frequently found in the drawings of Wolle there is a possibility that *S. natator* var. *crassum* is identical with his *S. minneapoliense* (Bull. Torr. Bot. Cl. **12**: 5, *pl.* 47, *figs.* 11–13. 1885). The front view of *S. minneapoliense* is, as W. & G. S. West have pointed out, quite different from that of *S. natator* var. *crassum.*

43. **Staurastrum urinator** sp. nov Pl. **79,** Figs. **16–18.**

Cells fairly large, breadth (with processes) about one and a quarter times the length, slightly constricted, sinus a small acute-angled indentation with a blunt apex, isthmus relatively broad; semicells campanulate with the lateral margins of the bases semicircular and generally bearing an erect spine in the median portion; apices flattened and with three somewhat emarginate verrucae; angles continued in long slightly attenuated processes that terminate in three short divergent spines, superior and inferior margins of processes with conspicuous teeth that point away from the body of the semicell, median area of processes smooth, processes outwardly divergent. Vertical view narrowly elliptic, lateral margin with three emarginate verrucae and a truncate projection in the median portion, poles continued in straight processes with smooth margins and a median row of spines. (Facultative planktont.)

Zygospores unknown.

Cells 68–79 μ long with processes, 26–32 μ long without processes; breadth with processes 81–89 μ, without processes 19–21 μ; breadth at isthmus 9–10.5 μ; thickness 15 μ.

Little Doctor (r).

The species resembles *S. natator* W. West in the truncate projection on the face of the middle of the semicell but differs in the angle of divergence of the processes as well as the shape and ornamentation of the semicells. It is given the name *urinator* because of the resemblance of a side view to a diver.

44. Staurastrum subnudibrachiatum W. & G. S. West.

Trans. Roy. Soc. Edinburgh 41: 502, *pl.* 7, *figs.* 18–19. 1905.

Cells small, breadth (with processes) about one and a half times the length, slightly constricted, sinus widely open and with a blunt apex, isthmus fairly broad; semicells subspherical, with the upper portion continued in smooth gradually attenuated processes that terminate in two blunt teeth, ends of processes at times obtuse and entire, processes slightly divergent. Vertical view four or five radiate, with the processes of one semicell alternating with those of the other semicell, processes slightly attenuated and with the length about equal to the diameter of the cell; body of the cells gradually fusing with the processes, margin of cell retuse between the bases of the processes. Cell wall without ornamentation. Cells generally enclosed in a wide gelatinous envelope.

Zygospores unknown.

Cells 40–44 μ long with processes, 31–37 μ long without processes; breadth with processes 53–60 μ, without processes 20–22 μ; breadth at isthmus 15–15.5 μ (W. & G. S. West).

Not found in the plankton of Wisconsin lakes.

var. **incisum** var. nov. Pl. **80**, Figs. **1–4**.

Cells smaller than the type, length (without processes) about one and a half times the breadth, median constriction very slight; semicells subspherical-quadrangular, with the processes arising abruptly from the body of the semicell; processes of one semicell opposite those of the other semicell, processes with parallel sides and deeply bifurcate apices, the bifurcations tending to lie in the vertical plane of the cell, furcations with rounded ends. Vertical view always tetragonal, with the body of the cell circular and the transition from process to body abrupt. Chloroplast axial, massive, with two long plate-like extensions running into each process, frequently with a plate entering each branch of the end of the process; pyrenoids single and central. Cells imbedded in a copious spherical gelatinous envelope. (Euplanktont.)

Cells 38–41 μ long with processes, 20–21 μ long without processes; breadth with processes 43–52 μ, without processes 17–19 μ; breadth at isthmus 10–13.5 μ.

Carson (rrr), Devils (rrr), Whitney (r).

This variety has considerable resemblance to *S. subnudibrachiatum* in the smooth processes and their general relation to the body of the cell. The variety differs, however, in the much sharper transition from body to processes, more rounded semicells, smaller median constriction, parallel sides of processes, and stronger forking of the ends of the processes. The general relation of process to body of semicell is more like that of *S. digitatum* G. S. West (Jour. Linn. Soc. Bot. **39**: 66, *pl.* 5, *figs.* 11–12. 1909) but the processes are totally different. The forking of the processes is quite like that of *S. brachiatum* Ralfs but the bodies of the semicells are different. The extent to which the chloroplast extends into the processes is quite unique, since it can be traced to a point beyond the forking of the processes.

45. Staurastrum inconspicuum Nordstedt.

Lunds Univ. Årsskr. 9: 26, *pl.* 1, *fig.* 11. 1873.

Cells small, length (with processes) about equal to the breadth, median constriction slight, sinus semicircular, isthmus elongate and broad; semicells subquadrate, with the lateral margins somewhat inflated, apex truncate and with median portion inflated, angles with short hollow processes that are at first parallel to the apex of the semicell and then bend upwards sharply with a knee-like joint at the point of bending, ends of processes truncate; body of cell and processes smooth. Vertical view quadrangular, with the sides emarginate and the angles continued in straight processes that are inflated and truncate at the ends.

Zygospores unknown.

Cells 17–26 μ long with processes, 14–15.5 μ long without processes; breadth with processes 17–26 μ; breadth at isthmus 7–9 μ (Nordstedt.)

Not found in the plankton of Wisconsin lakes.

var. **planctonicum** var. nov. Pl. **80**, Figs. **5–7**.

Cells with the isthmus narrower and greatly elongate; processes longer and upward bending more pronounced; ends of processes with four minute spines. (Euplanktont.)

Cells 26–30 μ long with processes, 17–19 μ long without processes; breadth with processes 22–30 μ, without processes 12.5 μ; breadth at isthmus 5–6 μ.

Jag (r).

This variety approaches the variety *gracilior* Maskell but has much slenderer processes and cell body.

46. STAURASTRUM CERASTES Lundell. Pl. 80, Figs. 8–10.

Nova Acta Reg. Soc. Sci. Upsaliae 3 Ser., 8: 69, *pl.* 4, *fig.* 6. 1871; W. & G. S. West, Trans. Linn. Soc. 2nd. Ser. Bot. 5: 268, *pl.* 18, *fig.* 4. 1896.

Cells of medium size, breadth and length about equal, slightly constricted, sinus a slight wedge-shaped constriction, isthmus relatively broad; semicells trapeziform-rectangular, tapering only slightly towards the isthmus, dorsal margin rounded, angles continued in stout tapering hollow processes with truncate ends that terminate in three or four minute teeth, processes strongly convergent; base of semicells with conical papillae or mucros just above the isthmus, one usually lying midway between each adjacent pair of processes, rarely with the mucros directly below the processes; median portion of semicell with two transverse rows of bifid emarginate verrucae that are continued in the processes as a single row of conical verrucae; apex of semicells with emarginate verrucae that continue for a ways along the dorsal margin of the processes and then give place to simple conical verrucae; ventral margin of processes with simple verrucae. Vertical view quadrangular (in Wisconsin specimens) with the sides retuse and the angles continued in stout straight processes that gradually taper to a truncate end; central area of the cell body with a rectangle of twelve emarginate verrucae (three on each side) whose sides parallel those of the cell, angles of this design with two parallel rows of emarginate verrucae leading to the bases of the processes and then giving way to a double or to a single row of conical verrucae than continues to the end of the processes; lateral margins of cells with emarginate verrucae; lateral margins of processes with short outwardly divergent spines. (Tychoplanktont.)

Zygospores unknown.

Cells 50 μ long; breadth with processes 45–52 μ, without processes 25 μ; breadth at isthmus 11 μ.

Found (rrr), Rudolph (rrr), Speese (rrr).

The alga as found in Wisconsin is always quadrangular although triangular specimens are well known in Europe. There were also but four supraisthmal mucros in all individuals examined. These resemble, in this respect, the type figured by Lundell and not the type with a transverse row of mucros above the isthmus that W. & G. S. West show. Welsh plankton material examined in the G. S. West collection also had the transverse row of conical papillae at the base of the semicells. It is possible that a separation of forms can be made on this feature of basal ornamentation but for the present the two are considered identical.

47. STAURASTRUM ARACHNE Ralfs.

Ann. & Mag. of Nat. Hist. 15: 157, *pl.* 11, *figs.* 6A–6B. 1845; Brit. Desm. 136, *pl.* 23, *figs.* 6A–6B. 1848.

Cells fairly small, length about half the breadth (with processes), deeply constricted, sinus widely open from a subacuminate apex, isthmus narrow; semicells broadly campanulate, with apices slightly tumid and with pairs of minute denticulations between the processes, body of semicells smooth; angles continued in long slightly tapering processes that terminate in three minute denticulations, dorsal margin of processes finely crenulate, ventral margin smooth or slightly crenulate, processes strongly convergent. Vertical view five-radiate (rarely six-radiate), the processes of one semicell alternating slightly with those of the other semicell; body of cell pentagonal (or hexagonal) with the sides somewhat emarginate, ornamented with a

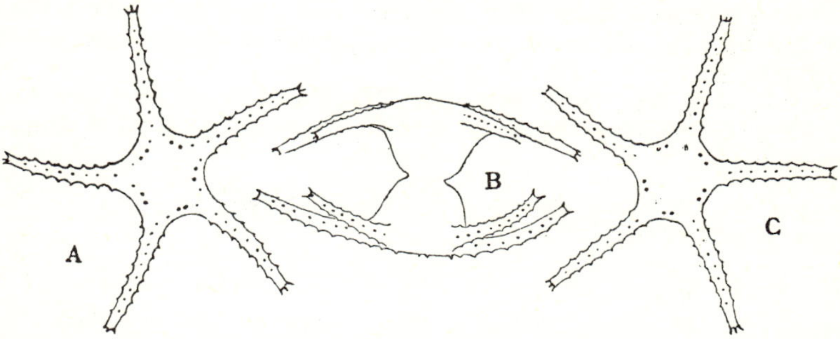

Fig. 13. *Staurastrum arachne* Ralfs (x 625). From plankton material collected in the Capel Curig lakes, Wales, by W. & G. S. West.

pair of intramarginal denticulations between each pair of processes, sometimes with two more delicate denticulations, one on each side towards the base of the processes; central area without ornamentation; processes tapering slightly, straight or curved in the same direction, lateral margins finely crenulate and with a single median row of minute denticulations. Chloroplast axial, fairly massive, with two laminate lobes running into each process; pyrenoid single and central.

Zygospores unknown.

Cells 32 μ long; breadth with processes 60–65 μ, without processes 15–18 μ; breadth at isthmus 7.5 μ.

Not found in the plankton of Wisconsin lakes.

Resembling *S. pentacerum* (Wolle) G. M. Smith in the five-radiate character of the semicells but much smaller and with a totally different shape of the semicells in front view. In this species the convergence of the processes is a constant feature (in the type) whereas in *S. pentacerum* the processes are parallel. The ornamentation of the cell body, as seen in the vertical view, is subject to some variation; the two denticulations lying midway between the processes are always present but the accessory denticulations flanking them are frequently lacking. The processes may be straight or all may be curved in the same direction as in *S. cyrtocerum* De Brébisson.

var. CURVATUM W. & G. S. West. Pl. **80**, Figs. **11–14.**

Jour. Linn. Soc. Bot. 35: 549, *pl.* 18, *fig.* 9. 1903.

Cells somewhat larger than the type, less deeply constricted; semicells obversely subtriangular and with apices bearing small truncate verrucae; processes outwardly divergent. Vertical view four- or five-radiate, with central area of cell smooth or with a truncate verruca midway between each pair of adjacent processes. Chloroplast axial, with a small central body and a lobe arising midway between bases of processes, lobes with laminate ends, a single lobe running into the base of adjacent processes; pyrenoid single and central. (Euplanktont.)

Cells 36–40 μ long with processes, 18–25 μ long without processes; breadth with processes 62–74 μ, without processes 18–20 μ breadth at isthmus 8.5 μ.

Carson (rr), Crab (sss), Dummy (rr), Dutchmans (rrr), Jag (rr), Lac Court Oreilles (rr), Shell (rrr).

The chief feature of the variety is the outward curving of the processes in the front view of a cell. There is also a difference in the ornamentation of the top of the cell and the cell shape. In the Wisconsin plankton the four-radiate specimens predominate.

48. STAURASTRUM PENTACERUM (Wolle) G. M. Smith. Pl. **80**, Figs. **15–18.**

Trans. Wis. Acad. Sci., Arts & Lett. 20: 356, *pl.* 12, *figs.* 19–22. 1922.

Staurastrum Ophiura var. *pentacerum* Wolle, Bull. Torr. Bot. Cl. 9: 28, *pl.* 13, *fig.* 5. 1992; Desm. U. S. 134, *pl.* 43, *fig.* 13; *pl.* 44, *fig.* 3. 1884.

Cells of medium size, length about one and a half times the breadth (without processes), slightly constricted, sinus a small semicircular indentation, isthmus broad; semicells trapeziform-quadrate, with the apex slightly tumid and broader than the base, body of the semicell without ornamentation and apex with a row of subapical verrucae, processes arising just below the apex, subparallel, somewhat attenuated and terminating in three short divergent spines, dorsal margin finely crenulate, ventral margin smooth. Vertical view five-radiate (rarely six-radiate), with the processes of one semicell alternating with those of the other semicell; body of cell pentagonal or hexagonal, with sides retuse, and with a single broad emarginate verruca between each pair of processes, at times with two smaller verrucae one on each side of this larger verruca, central area without ornamentation; processes slightly tapering and with lateral margins crenulate or with two rows of spines. Choloroplast axial, massive, with reniform lobes arising between the bases of adjacent processes and with a single plate-like process running a short distance into each adjacent process. (Facultative planktont.)

Zygospores unknown.

Cells 36–41 μ long; breadth with processes 85–105 μ, without processes 18–20 μ; breadth at isthmus 10 μ.

Adelaide (rrr), Bass (sss), Little Bass (r), Bear (rrr), Blue Gill (ss), Little Butternut (rr), Canteen (rrr), Carson (rrr), Cochrane (rrr), Crooked (ss), Devils (ss), Diamond (ss), Dummy (s), Dunham (rrr), George (rr), Half Moon (rrr), Jag (sss), Jenny (rr), Lake of the Woods (rrr), Lindy (r), Little Doctor (rrr), Long (rrr), Lynx (rr), Meta (ss), Nicaboyne (rr), Oak (r), Pine (rr), Rudolph (r), Sand (rrr), Sanford (rrr), Silver (rrr), Sishebogema (rrr), Speese (rr), Streitor (rrr), Sunday (rr), Tank (ss), Trapper (rrr), Whitney (rrr).

As I have stated elsewhere *(l. c.)* this alga seems sufficiently distinct from *S. Ophiura* Lundell to warrant recognition as a species. It differs from *S. Ophiura* in the relative proportions of the cells, smaller size, smaller number of processes, and different ornamentation, both in the apical verrucae and lack of spines on the bases of the semicells. The processes do not have the prolonged lateral verrucae of *S. Ophiura* and lack both the median row of spines and ornamentation of the ventral side.

var. TETRACERUM (Wolle) G. M. Smith. Pl. **80**, Figs. **19–21**.

Trans. Wis. Acad. Sci., Arts & Lett. 20: 357, *pl.* 13, *figs.* 3–5. 1922.
Staurastrum Ophiura var. *tetracerum* Wolle, Bull. Torr. Bot. Cl. 9: 28, *pl.* 13, *fig.* 4. 1882; Desm. U. S. 134, *pl.* 44, *figs.* 1–2. 1884; W. & G. S. West, Trans. Linn. Soc. 2nd. Ser. Bot. 5: 268, *pl.* 18, *fig.* 16. 1896.

Cells with the same relative proportions and shape as the type; processes subparallel. Vertical view tetragonal and processes of one semicell alternating with those of the other semicell; processes with spines more prominent, sometimes with spines in the portion near the base bifid and with a median row of spines. (Facultative planktont.)

Cells 32–35 μ long; breadth with processes 85–120 μ, without processes 18–20 μ; breadth at isthmus 10 μ.

Bass (rrr), Bird (sss), Lindy (cc), Nicaboyne (rr), Sand (rrr), Speese (rr).

A superficial examination of *S. ankyroides* Wolle might lead to the conclusion that it is an extremely well developed specimen of this variety. *S. ankyroides* differs in the nature of the lateral ornamentation of the processes, convergence of the processes and structure of the verrucae on the apices of the semicells.

49. STAURASTRUM ANKYROIDES Wolle.

Bull. Torr. Bot. Cl. **11**: 14, *pl.* 44, *fig.* 4. 1884; Desm. U. S. 137, *pl.* 51, *fig.* 4. 1884; G. M. Smith, Trans. Wis. Acad. Sci., Arts & Lett. **20**: 358, *pl.* 13, *figs.* 13–15. 1922.

Cells large, length about three times the breadth (without processes), slightly constricted, sinus a small semicircular depression, isthmus broad; semicells cylindrical, with bases swollen and apex tumid, upper portion of semicell prolonged into four processes, the processes of one semicell alternating with those of the other semicell,

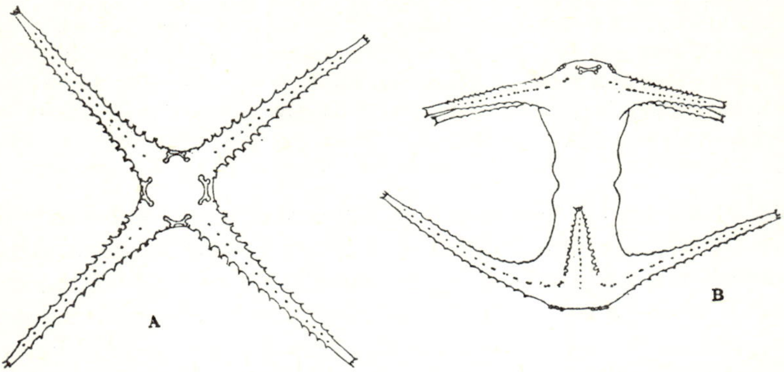

Fig. 14. *Staurastrum ankyroides* Wolle (x 500). From a plankton collection made in Burned Rock lake, Ontario, Canada.

processes strongly convergent, with the upper margin more sharply crenulate than the lower margin, median area of processes with a row of bifid verrucae at the base and simple spines at the apex, ends of processes with three minute spines; apex of semicells bearing subapical emarginate verrucae. Vertical view quadrangular, angles with long gracefully attenuated processes that terminate in three spines, bases of processes with elongate bifid lateral verrucae and apices with simple lateral spines, median axis of processes with a single row of spines; body of cell with a conspicuous arcuate quadrigranulate verruca midway between every pair of processes.

Zygospores unknown.

Cells 60–68 μ long; breadth with processes 92–132 μ, without processes 20–24 μ; breadth at isthmus 11–13.5 μ.

Not found in the plankton of Wisconsin lakes.

var. **pentacladum** var. nov. Pl. **81,** Figs. **1–2.**

Semicells with five processes; processes of one semicell alternating with those of the other semicell; apical verrucae on body of cell more nearly semicircular and closer together. (Euplanktont?)

Cells 120–150 μ broad with processes, 25–28 μ broad without processes; breadth at isthmus 14 μ.

Crooked (rrr), Little Doctor (rrr).

The essential difference between the variety and the type is the five-radiate character of the cell. I have collected the alga in considerable quantity both in New York and Ontario (Canada) and find it constantly four-radiate. On the other hand the few individuals found in the two stations in Wisconsin were always five-radiate.

The alga should be compared with *S. pentacerum* (Wolle) G. M. Smith from which it differs in the more elegant body of the cell, much stouter lateral verrucae on the processes, ornamentation of under side of the processes, arcuate verrucae on the apex of the semicells, and in the size of the cells. Although the processes resemble those of tetragonal and pentagonal forms of *S. Ophiura* Lundell it differs in the more delicate body of the cell, inflation of the base of the semicells, lack of a basal ring of conical spines, and most of all in the totally different ornamentation of the cell apices.

50. Staurastrum Ophiura Lundell. Pl. **81**, Figs. **3–4.**

Nova Acta Reg. Soc. Sci. Upsaliae 3 Ser., **8**: 69, *pl.* 4, *fig.* 7. 1871.

Cells very large, length two and half to three and a half times the breadth (without processes), slightly constricted, sinus small and semicircular, isthmus broad; semicells cyathiform, with strongly convex apices, base of semicells with a ring of stout conical verrucae with rounded apices and with four or five of these verrucae visible in a front view; superior portion of semicells prolonged into gracefully attenuated subparallel to convergent processes, upper surfaces of processes more sharply crenulate than the lower, ends of processes trispinate, median area of processes with bifid verrucae on the portion near the body of the cell and simple spines on the apical portion; rounded portion of semicell above the processes with a bifid verruca midway between apex and the insertion of the processes. Vertical view seven-rayed (rarely six- or eight-rayed), with rays gradually attenuated and trispinate, with lateral margins at base of processes bearing elongate emarginate to bifid verrucae and apical portion with simple spines, processes with a median row of simple spines; body of semicell with two broad emarginate verrucae at the margin between the insertion of each pair of processes; central portion of body with a ring of elongate bifid verrucae, the number of verrucae corresponding to the number of processes, verrucae alternating with, not opposite, the processes. (Tychoplanktont.)

Zygospores unknown.

Cells 75–82 μ long; breadth with processes 115–160 μ, without processes 32–35 μ; breadth at isthmus 23–25 μ.

Bass (rrr), Horseshoe (rrr), Lake of the Woods (rrr), Muskallonge (rrr).

This conspicuous and unique alga varies somewhat in the number, length, and convergence of the processes. In the Wisconsin plankton seven-rayed cells are the rule whereas in the English plankton eight- and nine-rayed specimens predominate. There is likewise variation in the ornamentation of the cell and the verrucae forming the central ring at the apex may be near the margin or the center. Lundell states that these verrucae are quadrifid but I have only noted bifid apical verrucae. Likewise the two emarginate verrucae between each pair of rays may be within the margin of the cell or projecting beyond it. The lateral verrucae of the processes are generally bifid, though cases have been found in which they are reduced to simple spines.

51. Saurastrum rotula Nordstedt. Pl. **81**, Figs. **5–8.**

Vidensk. Medd. f. d. naturh. Foren. i Kiöbenhavn **1869**: 227, *pl.* 4, *fig.* 38. 1870; Wolle, Desm. U. S. 135, *pl.* 44, *figs.* 13–14. 1884; W. & G. S. West, Trans. Linn. Soc. 2nd. Ser. Bot. **5**: 268, *pl.* 18, *fig.* 11. 1896.

Cells fairly large, length about twice the breadth (without processes), deeply constricted, sinus widely open and with a subacuminate apex, isthmus narrow; semicells subhexagonal, with portion below the processes straight-sided and that above truncate; body of semicells with a transverse row of subapical bidenticulate verrucae; processes arising midway between the apex and isthmus of semicells, subparallel, slightly attenuated and terminating in three short divergent spines, body of processes with three or four rings of sharp denticulations. Vertical view nine-rayed (very rarely eight-rayed), with the processes straight and slightly attenuated, ends of processes trispinate; body of cells with a single bidenticulate verruca at the base of each process, the verrucae forming an intramarginal ring. Cell wall smooth or finely punctate. (Tychoplanktont.)

Zygospores unknown.

Cells 40–45 μ long; breadth with processes 92–102 μ, without processes 22–30 μ; breadth at isthmus 10.5–12 μ.

Little Butternut (rrr), Des Moines (rrr), George (rrr), Round (rrr).

W. & G. S. West's criticism of Wolle's drawing which shows rectangular instead of conical apical papillae is, at least as far as Wisconsin specimens are concerned, without justification; the apical verrucae are distinctly bifid and not conical as their drawing shows. The species should be compared with *S. limneticum* var. *cornutum* G. M. Smith which appears closely related but is not, for reasons stated below, placed as a variety of *S. rotula.*

52. Staurastrum limneticum Schmidle.

Engl. Bot. Jahrb. **26**: 52, *pl.* 4, *fig.* 5. 1899; G. S. West, Jour. Linn. Soc. Bot. **38**: 129. 1907.

Cells large, length (without processes) about twice the breadth, deeply constricted, sinus widely open and with an acute apex, isthmus fairly narrow; semicells elliptic-obsemicircular, with the ventral margin more rounded than the dorsal (almost semicircular); processes arising at a point about a third of the distance from the apex to the isthmus, fairly delicate and very gracefully upwardly curved; lower margin of processes always crenulate and upper surface smooth or very shallowly crenulate, ends with three or four short divergent spines; body of semicells without ornamentation. Vertical view pentagonal or hexagonal, with sides slightly retuse and angles continued in long slightly tapering processes, lateral margins smooth, ends of processes trispinate to quadrispinate, processes of one semicell opposite those of the other semicell; central portion of cell body without ornamentation.

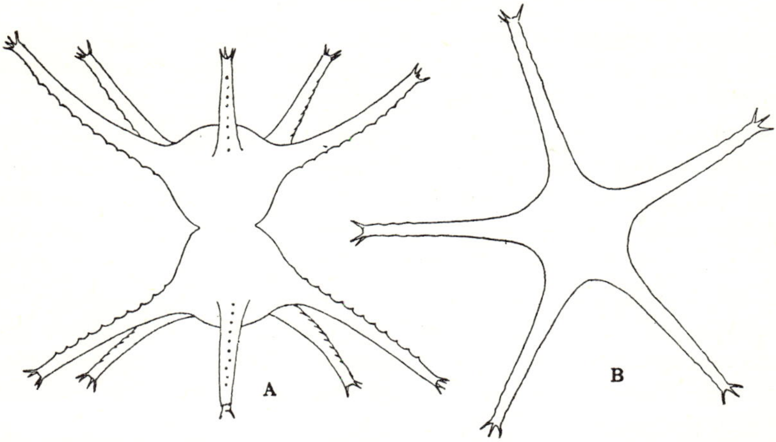

Fig. 15. *Staurastrum limneticum* Schmidle (x 600). From plankton material collected in Lake Victoria Nyanza, Africa.

Cells 46–51 μ long; breadth with processes 100–130 μ, without processes 28–36 μ; breadth at isthmus 14 μ (G. S. West).

Zygospores of the type unknown.

Not found in the plankton of Wisconsin lakes.

var. **cornutum** var. nov. Pl. **82,** Figs. **1–2.**

Staurastrum limneticum var. *burmense* G. M. Smith non W. & G. S. West, Trans. Wis. Acad. Sci., Arts & Lett. 20: 359, *pl.* 13, *figs.* 6–11. 1922.

Cells shorter and narrower than the type; length about two and a half times the breadth (without processes), sinus widely open and with apex rounded, isthmus narrow; semicells with sides of bases sublinear and processes upwardly divergent at apices but with bases subparallel, ornamented with three or four rings of conical teeth and ends of processes with three fairly long divergent spines, top of semicell with subapical elongate bidenticulate verrucae. Vertical view with length of processes one and a half times the diameter of the cell, five- to seven-rayed, with the processes straight and ornamented as in the front view; body of the cell with a bidenticulate verruca lying near the base of each process, verrucae forming a ring within the margin of the body of the cell; processes of one semicell alternating with those of the other semicell. (Euplanktont.)

Zygospores spherical, with long slender spines that have their apices bifurcate and the bifurcations recurved.

Cells 48–52 μ long; breadth with processes 78–88 μ, without processes 20 μ; breadth at isthmus 9 μ.

Diameter of zygospores with spines 85–92 μ, without spines 42–44 μ; length of spines 20–24 μ.

Diamond (sss).

This is a clearly defined variety in which there are strong arguments for referring it to *S. rotula* Nordstedt or to *S. limneticum.*

The ring of bifurcate apical verrucae and ornamentation of the processes is more like that of *S. rotula.* The curved apices of the cells, smaller number of processes, and most of all the graceful upward curvature of the processes shows a greater resemblance to *S. limneticum.* The argument for referring this alga to *S. limneticum* is much stronger when the variety *burmense* W. & G. S. West (Ann. Roy. Bot. Garden Calcutta **6**: 222, *pl.* 16, *fig.* 13. 1907) is taken into consideration. I have found that five- and six-radiate individuals are widely distributed in Ontario (Canada) and that these show a striking resemblance to the variety *burmense,* both in shape of the semicells and the curvature of the processes. A comparison of Canadian specimens with type material of the variety *burmense* from the G. S. West collection shows that the two are identical except for the apical verrucae.

The alga has been found in but one lake in Wisconsin and the cells are almost invariably with seven processes. The apical verrucae are also slightly stouter than those of cells found in the Ontario plankton but it does not seem advisable to separate the two on this account.

53. Staurastrum radians W. & G. S. West. Pl. **82,** Figs. **6–7.**

Jour. Linn. Soc. Bot. 33: 318, *pl.* 18, *fig.* 18. 1898.

Cells fairly large, length (with processes) about half the breadth, fairly deeply constricted, sinus obtuse-angled and with the apex subacute, isthmus broad; semicells trapeziform in outline, with the apex slightly convex and the angles continued in straight short processes with coarsely crenulate margins, ends of processes with three or four stout divergent spines, processes parallel; body of semicell without ornamentation. Vertical view circular, with eight radiating processes, processes straight, coarsely crenulate in outline, bases of lateral margins with an elongate emarginate outgrowth on each side; apex of processes with three or four small stout divergent spines; body of cell without ornamentation. (Tychoplanktont.)

Zygospores unknown.

Cells 40 μ long; breadth with processes 70 μ, without processes 30 μ; breadth at isthmus 17.5 μ.

Des Moines (rrr), Big McKenzie (rrr).

A distinctive alga which is easily recognized in a vertical view.

54. Staurastrum furcatum (Ehrenberg) De Brébisson. Pl. **83,** Figs. **1–3.**

Mem. de la soc. imp. d. Sc. nat. de Cherbourg 4: 136. 1856; Rabenhorst, Flora Eur. Algarum 3: 218. 1868; Wolle, Desm. U. S. 150, *pl.* 40, *figs.* 40–41; *pl.* 48, *figs.* 15–16; *pl.* 52, *fig.* 34. 1884; W. West, Jour. Roy. Micr. Soc. 1890: 292, *pl.* 5, *fig.* 11. 1890.

Xanthidium furcatum Ehrenberg, Abh. Ak. d. Wiss. zu Berlin 1833: 318. 1834; Ralfs, Brit. Desm. 213. 1848.

Cells small, length (with processes) slightly greater than the breadth, deeply constricted, sinus acute-angled and with the apex acuminate, isthmus fairly narrow; semicells transversely elliptic-hexagonal, with ventral margin tumid and dorsal margin truncate; cell body with two whorls of processes, the three lower processes arising midway between the isthmus and the apex of the semicell; the six processes of the upper whorl subapical in insertion; processes of the lower whorl at right angles to the long axis of the cell, with subparallel lateral margins and ends deeply bifurcate; processes of the superior whorl similar to those of the lower whorl but lying at an angle of 45 degrees with the long axis of the cell; cell wall smooth. Vertical view triangular, sides of cell body straight and angles continued in very short processes; sides of cell each with two short intramarginal processes, one near each angle of the cell. (Tychoplanktont.)

Zygospores spherical, with numerous elongate spines that are bifurcate at their apices (Rabenhorst).

Cells 27–30 μ long with processes, 18–20 μ long without processes; breadth with processes 22–27 μ, without processes 14 μ; breadth at isthmus 7–8 μ.

Lindy (rrr).

The species resembles *Staurastrum tohopekaligense* Wolle in the double whorl of three inferior and six superior processes. The two are distinguished, however, by the smaller size, relatively shorter processes, different cell shape, and merging of the lower whorl of procsses with the cell body in the vertical view of *S. furcatum. S. senarium* (Ehrenberg) Ralfs is a closely related species of the same cell size but with accessory processes in the lower whorl.

The use of Ralfs' British Desmidieae as the starting point for nomenclature involves certain difficulties in determining the correct name of the species. Ralfs described the species under the name *S. spinosum* and then in the appendix recognizes the *Xanthidium furcatum* of Ehrenberg. He suggests that this *Xanthidium* may belong to *Staurastrum.* Since Ralfs recognizes both names the older one is taken.

55. Staurastrum brasiliense Nordstedt.

Vidensk. Medd. f. d. naturhist. Foren. i Kiöbenhavn **1869:** 227, *pl.* 4, *fig.* 39. 1870. (*non* Lundell, Nova Acta Reg. Soc. Sci. Upsaliae 3 Ser., **8:** 73, *pl.* 5, *fig.* 2. 1871.).

Cells large, breadth (with spines) one and a third times the length, deeply constricted, sinus widely open and with apex acute; isthmus broad, about half the diameter of the cell; semicells broadly cuneate, with the ventral margins slightly convex and the apex somewhat concave; angles truncate and with three very long stout divergent spines, the two inferior ones lying in the same horizontal plane, the superior one inserted at an angle. Vertical view four- (very rarely five-) radiate, the angles bearing three (very rarely four) outwardly divergent stout spines; sides of cells deeply emarginate.

Zygospores unknown.

Cells 56–77 μ long with spines 43–51 μ long without spines; breadth with spines 65–99 μ, without spines 25–37 μ (Nordstedt).

Not found in the plankton of Wisconsin lakes.

var. LUNDELLII W. & G. S. West. Pl. **82**, Figs. **3–5.**

Trans. Linn. Soc. 2nd. Ser. Bot. 5: 259. 1896.

Staurastrum brasiliense Lundell (non Nordstedt), Nova Acta Reg. Soc. Sci. Upsaliae 3 Ser., 8: 73, *pl.* 5, *fig.* 2. 1871.

Cells (without spines) about twice the size of the type; apex of semicells straight or only slightly retuse. Vertical view five-sided (rarely six-sided); emargination of sides subsemicircular. Cell wall smooth or finely punctate. Chloroplast with a central mass and massive rays towards each angle, rays each with several smaller laminate rays; pyrenoids numerous, in both the central portion and the rays. (Facultative planktont.)

Cells 145 μ long with spines, 74 μ long without spines; breadth with spines 148 μ, without spines 92 μ; breadth at isthmus 38 μ; spines 27–37 μ long.

Bass (rrr), Diamond (r).

56. STAURASTRUM TOHOPEKALIGENSE Wolle.

Bull. Torr. Bot. Cl. 12: 128, *pl.* 51, *figs.* 4–5. 1885; Freshw. Algae of U. S. 45, *pl.* 59, *figs.* 4–5. 1887; W. & G. S. West, Trans. Linn. Soc. 2nd. Ser. Bot. 6: 180, *pl.* 21, *fig.* 26. 1902.

Cells of medium size, breadth (without processes) somewhat less than the length, fairly deeply constricted, sinus widely open and with apex acute, isthmus narrow; semicells with bases obsemicircular and apices flattened; with three lateral processes arising about half way from apex to the isthmus, lateral processes parallel; upper part of semicells with six subapical processes arising just back of the lateral margins, subapical processes markedly divergent; all processes hollow, with subparallel sides and length about equal to the diameter

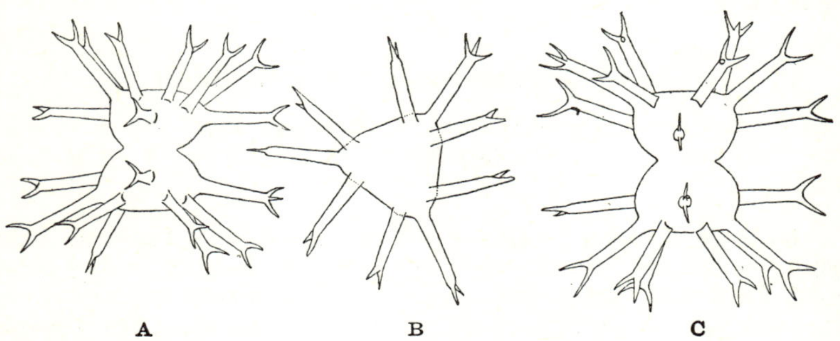

Fig. 16. *Staurastrum tohopekaligense* Wolle (x 500). Plankton specimens from lakes in Ontario, Canada. Figs. A, C from Silver Lake, Fig. B from Gull Lake.

of the cell, ends of processes with two (seldom three) long divergent hollow spines. Cell wall smooth. Vertical view triangular, with sides straight but frequently bending in the vicinity of the angles to form a sharply pointed angle; angles prolonged into processes, each lateral margin of cell with two perpendicular processes arising one near each angle but inside the margin of the cell.

Zygospores unknown.

Cells 70–83 μ long with processes 32–39 μ long without processes; breadth with processes 67–75 μ, without processes 28–32 μ; breadth at isthmus 15 μ; length of terminal spines 4.5–6.5 μ.

Not found in the plankton of Wisconsin lakes.

This is a well defined species but one which is subject to considerable variation and one where, owing to the incomplete original description and poor figures, there is some doubt as to the type. The varieties *trifurcatum* and *quadrangulare* were described by W. & G. S. West in 1895 (Trans. Linn. Soc. 2nd. Ser. Bot. **5**: 80, *pl.* 9, *figs.* 8–9) and the variety has been observed by them again in Ceylon in 1902 *(l. c.)* and in the Scottish plankton in 1905 (Trans. Roy. Soc. Edinburhg **41**: 503, *pl.* 7, *fig.* 7). Two years later in discussing the plankton of the Central African lakes G. S. West (Jour. Linn. Soc. Bot. **38**: 130, 1907) mentions quadrangular and triangular forms that "were typical as regards proportions and length of processes."

After a careful comparison of specimens from Burma and from the plankton of Victoria Nyanza (Africa), Scotland, New York, Ontario and Wisconsin I have reached the conclusion that text figure **16** is that of the type. It is characterized by processes which are about as long as the diameter of the cell, a flat cell apex, triangular body and long divergent spines at the ends of the processes. The figure of an American specimen from Wolle given by W. & G. S. West in 1902 *(l. c.)*, which shows two supplementary processes on each face of the cell, has not been observed. This shows that the number of processes is no sure criterion in determining varieties in this species. In my opinion the ends of the processes in the West drawing are too rounded. The number of spines at the ends of the processes is generally two but I have found that occasionally the processes on either the superior or the inferior whorls may be trispinate.

The varieties *trifurcatum* and *quadrangulare* which have been found in Madagascar, Burma, and the African plankton should be considered synonyms of *S. tohopekaligense* var. *nonanum* (Turner) Schmidle.

var. **brevispinum** nom. nov. Pl. **82**, Figs. **8–11**.

Staurastrum tohopekaligense var. *trifurcatum* W. & G. S. West, Trans. Roy. Soc. Edinburgh **41**: 503, *pl.* 7, *fig.* 7. 1905 (*non* Trans. Linn. Soc. 2nd. Ser. Bot. **5**: 80, *pl.* 9, *fig.* 8. 1895).

Cells somewhat smaller and with diameter of the cell greater than the length of the processes. Processes trifurcate or bifurcate but with much shorter spines. Chloroplast massive, with a broad lobe arising opposite each side and sending a single ray into the two lateral processes; processes at angles with a single ray arising from the base of the chloroplast. (Facultative planktont.)

Cells 48 μ long with processes, 26 μ long without processes; breadth with processes 50 μ, without processes 24 μ; breadth at isthmus 11 μ.

Carroll (rrr), Horseshoe (rrr), Lake of the Woods (rrr).

This alga as found in Wisconsin is identical with specimens from the Scotch plankton. It is distinguished from the type by the shorter processes and much shorter terminal spines of the processes. I have found specimens in which all processes are bispinate and others in which all are trispinate as well as intermediates between these two extremes.

57. Staurastrum furcigerum De Brébisson. Pl. **83**, Figs. **4–7.**

in Meneghini, Linnaea 14: 226. 1840; G. S. West, Brit. Freshw. Algae 172, *fig.* 65G. 1904.

Didymocladon furcigerus Ralfs, Brit. Desm. 144 *pl.* 33, *figs.* 12A–12D. 1848.

Cells of medium size, length (with processes) slightly greater than the breadth, deeply constricted, sinus acute-angled and with the apex acuminate, isthmus narrow; semicells broadly hexagonal, with sides bordering the sinus slightly convex and the apex flattened or faintly concave; angles with two whorls of hollow processes, those of the inferior whorl stouter and attenuated to a blunt end furnished with three short divergent spines, body of processes with four to six concentric rings of small conical granules; processes of superior whorl more delicate, not so strongly attenuated and ending in three small divergent spines, ornamented with concentric rings of granules; processes of inferior whorl parallel, processes of superior whorl suberect to strongly divergent. Body of cell without ornamentation. Vertical view triangular; sides somewhat concave and angles continued in gently tapering processes with truncate trispinate ends, processes with concentric rings of granules; body of cell with three smaller radially arranged processes, one at each angle of the cell; central portion of cell body without ornamentation. (Facultative planktont.)

Zygospores spherical, furnished with numerous long spines that have conical bases and bifurcate apices (G. S. West).

Cells 60–66 μ long with processes, 31–38 μ long without processes; breadth with processes 52–60 μ, without processes 28 μ; breadth at isthmus 12.5–13.5 μ.

Beaverdam (rrr), Des Moines (rrr), Duck (c), Granite (rr), Half Moon (rrr), Hill (rrr), Lac Court Oreilles (rrr), Silver (rrr).

var. ARMIGERUM forma **gracillimum** n. forma. Pl. **83**, Figs. **8–11**.

Staurastrum furcigerum var. *armigerum* (De Brébisson) Nordstedt Vidensk. Medd. f. d. naturh. Foren. i Kiöbenhavn **1888**: 207. 1888.

Staurastrum armigerum De Brébisson, Mem. de la soc. imp. d. Sc. nat. de Cherbourg **4**: 136, *pl.* 1, *fig.* 22. 1856.

Superior whorls containing six (not three) processes arising in pairs at the base of the processes of the inferior whorl and lying (in the vertical view) at right angles to the sides of the cell. Processes long and delicate and with their apices bifurcate. (Facultative planktont.)

Cells 63–68 μ long with processes, 30–32 μ long without processes; breadth with processes 67–73 μ, without processes 27–30 μ; breadth at isthmus 12–15 μ.

Shell (r).

Three very similar species have been described in which there are three processes in the lower whorl and six in the upper. Nordstedt *(l. c.)* has shown that two of these become the varieties *eustephanum* (Ehrenberg) Nordstedt and *armigerum* (De Brébisson) Nordstedt of *S. furcigerum;* while the third *S. pseudofurcigerum* Reinsch cannot be recognized but belongs in part to the variety *armigerum* and in part to the variety *eustephanum*. The variety *armigerum* is characterized by long delicate processes with undulate margins and bifid ends while the processes of the variety *eustephanum* are short and have the ornamentation of the type.

58. STAURASTRUM LEPTACANTHUM Nordstedt. Pl. **83**, Figs. **12–14**.

Vidensk. Medd. f. d. naturhist. Foren. i Kiöbenhavn **1869***;* 229, *pl.* 4, *fig.* 46. 1870.

Cells fairly large, length (with processes) about the same as the breadth, fairly deeply constricted, sinus widely open and with the apex subacute, isthmus narrow; semicells subcircular to subhexagonal in outline, with the ventral margins slightly tumid, dorsal margins and apex tumid; cell body punctate and with processes arising in two whorls; lower whorl arising midway between the isthmus and the apex of the semicells, containing six processes (four visible in front view); processes subparallel, long and very slightly attenuated, terminating in a bifurcation that is continued in short spines, processes without ornamentation; superior whorl of processes subapical in insertion, processes four in number and similar to those of the lower whorl but lying at an angle of 45 degrees with the vertical axis of the semicell. Vertical view hexagonal, with sides slightly concave and angles continued in straight processes that terminate in a bifurcation, the two arms of the bifurcation lying in the same vertical plane, processes without ornamentation; central portion of cell body with four similar processes. Chloroplast with a comparatively small central mass and a single ray running into each process of both whorls and continuing nearly to the point of bifurcation. (Facultative planktont.)

Zygospores unknown.

Cells 100–110 μ long with processes, 41–45 μ long without processes; breadth with processes 100–114 μ, without processes 25–28 μ; breadth at isthmus 13 μ; length of terminal spines 11–14 μ.

Bass (rrr), Crooked (rrr), Diamond (ss), Dummy (rr), Horseshoe (rr), Lake of the Woods (ss), Mud Hen (rrr), Shell (r), Tank (rrr).

The species shows a certain resemblance to *S. tohopekaligense* Wolle in the two whorls of deeply bifurcate processes but the two differ in size of the cells, shape of the cells and number of processes. In Wisconsin specimens the lower whorl always has six and the upper whorl four processes but forms with six lower and six upper, nine lower and six upper, as well as eight lower and four upper have been described.

59. Staurastrum Arctiscon (Ehrenberg) Lundell. Pl. **84,** Figs. **1–2.**

Nova Acta Reg. Soc. Sci. Upsaliae 3 Ser., 8: 70, *pl.* 4, *fig.* 8. 1871.

Cells large, length slightly greater than the breadth (with processes), fairly deeply constricted, sinus open and acute-angled, isthmus broad; semicells subspherical, with bases obsemicircular and apices strongly tumid to semicircular; body of semicells without ornamentation, median portion of semicell body with a whorl of nine (five visible in front view) long stout very slightly tapering hollow processes, processes parallel to slightly divergent; apex of semicell with a whorl of six similar processes that are subapical in insertion and lie at an angle of 45 degrees with the lower series; both series of processes with three or four transverse rings of conspicuous conical granules, ends of processes with three stout hollow divergent spines. Vertical view circular, with nine straight marginal processes and an inner whorl of six intramarginal processes, structure of processes as in front view. Cell wall smooth or finely punctate. Chloroplast a central axial mass with a single arm running a short distance into each process; pyrenoid single and central. (Facultative planktont.)

Zygospores unknown.

Cells 120 μ long with processes, 70 μ long without processes; breadth with processes 114–120 μ, without processes 44–53 μ; breadth at isthmus 29 μ.

Bass (rrr), Beaverdam (r), Clear Crooked (rrr), Des Moines (rrr), Diamond (ss), Found (rrr), Horseshoe (rrr), Lake of the Woods (ss), Silver (rrr), South Crab (rrr), Tank (rrr).

A very distinctive species where the number of processes in the two whorls is quite constant. Aside from differences in the length of the processes, which is generally shorter in American than in European individuals, the variation is confined to the number of rings of granules on the processes and the angle of the sinus.

var. GLABRUM W. & G. S. West. Pl. **83**, Fig. **15**; Pl. **84**, Figs. **3–5.**
Trans. Linn. Soc. 2nd. Ser. Bot. 5: 269, *pl.* 18, *fig.* 14. 1896.

Processes frequently shorter and stouter, without the transverse rings of granules, outline straight or somewhat undulate. Cell wall smooth or punctate. (Facultative planktont.)

Cells 105–120 μ long with processes, 62–65 μ long without processes; breadth with processes 106–115 μ, without processes 47–50 μ; breadth at isthmus 25–26 μ.

Black Oak (rrr), Des Moines (rrr), Devils (rrr), Fish (rr), Long (rrr), Middle McKenzie (rrr), Mud Hen (rrr), Muskallonge (r), Poplar (sss), Reserve (rrr), Rooney (rr), Sand (rr), Soft (cc).

Intermediate stages between this variety and the type are sometimes found but generally the two are quite distinct.

ARTHRODESMUS Ehrenberg 1838.

Cells solitary, generally small, with the length usually equal to the breadth, usually compressed and bilaterally symmetrical in front view, rarely triangular in vertical view and with a radial symmetry, median constriction pronounced and with sinus widely open to linear; semicells in front view generally triangular, rarely elliptic, subtrapeziform or subrectangular, with simple straight or curved spines at the angles, spines lying in the same plane. Vertical view usually elliptic and with spines at the poles, rarely triangular. Lateral view of semicells subcircular. Cell wall of uniform thickness and without ornamentation on body of semicells. Cells sometimes embedded in a wide gelatinous sheath with blunt radial strands of denser mucilage. Chloroplasts axial, with one pyrenoid.

Zygospores spherical; with walls smooth or bearing simple spines.

This genus was orginally founded on the compressed nature of the cells in combination with simple spines at the angles and lack of ornamentation on the body wall. We know now, however, that there are bilaterally symmetrical *Staurastrum* species with spines at the angles of the semicells and smooth walls. W. & G. S. West have also described radially symmetrical varieties of certain *Arthrodesmus* species so that there is no sharp distinction between *Arthrodesmus* and *Staurastrum*. The systematic treatment of the group is facilitated by considering it a separate genus and not a subgenus or section of *Staurastrum*. On the other hand, species like *A. octocornis* Ehrenberg and *A. tenuissimus* Archer approach some of the smaller *Xanthidium* species. The central portion of the cell face in *Xanthidium* is generally thickened or protruding, while in *Arthdrodesmus* it is not differentiated.

KEY TO THE SPECIES.

Semicells with two spines.
  Shape of semicells approximately elliptic.
    Apex of semicells a smooth curve.
      Spines short and convergent .... (1) *A. convergens*
      Spines long, parallel to divergent .... (3) *A. subulatus*
    Apex truncate in median portion .... (4) *A. michiganensis*
  Shape of semicells not elliptic.
    Diameter of semicells greater than length of spines.
      Isthmus elongate .... (6) *A. constrictus*
      Isthmus not elongate.
        Spines convergent .... (7) *A. Ralfsii*
        Spines divergent .... (5) *A. phimus*
    Diameter of semicells less than length of spines.
      Apex of semicells retuse .... (8) *A. quiriferus*
      Apex of semicells not retuse.
        Semicells subrectangular .... (2) *A. Bulnheimii*
        Semicells subtriangular.
          Apex flat .... (9) *A. incus*
          Apex convex .... (10) *A. triagularis*
Semicells with four spines .... (11) *A. octocornis*

1. Arthrodesmus convergens Ehrenberg. Pl. 84, Fig. 6.

Die Infusionsthierchen 152, *pl.* 10, *figs.* 18A–18B. 1838; Ralfs, Brit. Desm. 118, *pl.* 20, *figs.* 3A–3D. 1848; G. S. West, Jour. Linn. Soc. Bot. 34: 397, *text fig.* 4. 1899; W. & G. S. West, Monogr. Brit. Desm. 4: 106, *pl.* 116, *figs.* 4–13. 1912.

Cells of medium size, breadth (without spines) slightly greater than the length, very deeply constricted, sinus narrowly linear and with outer portion widely open, isthmus narrow; semicells transversely elliptic, with the apex more rounded than the base of the semicell, lateral margins broadly rounded and furnished with a single stout tapering spine, spines strongly convergent and with the extremities of opposite pairs frequently apposed. Vertical view narrowly elliptic, with a short spine at each pole. Chloroplast axial, with a single conspicuous pyrenoid.

Zygospores spherical, with a smooth wall (W. & G. S. West).

Cells 30–45 μ long; breadth with spines 45–75 μ, without spines 32–48 μ; isthmus 10–11 μ broad; spines 12–16 μ long.

Diameter of zygospores 39–48 μ (W. & G. S. West).

Diamond (rrr), Muskallonge (rrr).

Variation is much more pronounced in this species than in any other of the genus. This is particularly true of the spines and it is not at all unusual to find those of one semicell reduced or even entirely lacking. No particular significance can be attached to this suppression of spines since when new semicells are formed they are generally quite typical.

The species is the largest of the genus and is characterized by the elliptic semicells and the short converging spines. It has a certain resemblance to *A. subulatus* Kützing but the latter has long parallel to divergent spines and a different shape of the semicells.

2. Arthrodesmus Bulnheimii Raciborski. Pl. **84,** Figs. **7–8.**

Pamietnik Wydz. Akad. Umiej. w Krakowie 17: 95, *pl.* 16, *fig.* 17. 1889; W. & G. S. West, Monogr. Brit. Desm. 4: 105, *pl.* 116, *figs.* 1–2. 1912.

Cells of medium size, length (without spines) slightly greater than the breadth, deeply constricted, sinus linear, isthmus narrow; semicells transversely subrectangular, with lateral margins subparallel and apex very slightly tumid, basal angles more rounded than the apical angles, apical angles with long fairly stout gradually attenuated spines, spines strongly divergent to suberect. Vertical view elliptic, with a single straight fairly stout spine at each pole. Cell wall smooth. Cells enclosed in a spherical gelatinous envelope with a radially fibrillar structure. (Tychoplanktont.)

Zygospores unknown.

Cells 47–60 μ long with spines, 20–22 μ long without spines; breadth with spines 42–50 μ, without spines 20–23 μ; breadth at isthmus 5–8 μ; spines 18–25 μ long.

Lindy (rrr).

The transversely subrectangular semicells combined with the conspicuous long spines, and especially the suberect position of these spines, differentiate the alga from the others of the genus.

3. Arthrodesmus subulatus Kützing. Pl. **84,** Figs. **9–10.**

Species algarum 176. 1849; W. & G. S. West, Monogr. Brit. Desm. 4: 109, pl. 116, *fig.* 14; *pl.* 117, *fig.* 1. 1912.

Cells of medium size, length (without spines) slightly greater than the breadth, median constriction deep, sinus widely open and with the apex acute, isthmus narrow; semicells obsemicircular, with bases convex and apices markedly tumid; lateral angles rounded and each bearing a single long gradually attenuated spine, spines parallel to slightly divergent. Vertical view elliptic, with a single spine at each pole. Chloroplast axial, with a single pyrenoid. (Facultative planktont.)

Zygospores spherical, with smooth walls (W. & G. S. West).

Cells 27–34 μ long; breadth with spines 53–80 μ, without spines 18–28 μ; breadth at isthmus 6–7 μ; spines 20–25 μ long.

Zygospores 30–36 μ in diameter (W. & G. S. West).

Diamond (rrr), Dummy (rrr), George (rr), Lindy (rrr), Tank (rrr), Travis (r), Whitney (rrr).

The distinguishing features of *A. subulatus* are the elevated apices and the semicircular bases of the semicells. It shows a certain resemblance to *A. convergens* Ehrenberg but the two are quite distinct.

var. **Nordstedtii** var. nov. Pl. **85,** Figs. **1–3.**

Apices of semicells slightly tumid, not strongly convex; spines always somewhat divergent. Vertical view narrowly elliptic. (Tychoplanktont.)

Cells 35–38 μ long with spines, 27.5–30 μ long without spines; breadth with spines 68–85 μ, without spines 22–27 μ; breadth at isthmus 6–8 μ; thickness 12 μ; spines 22–26 μ long.

Horseshoe (rrr), Meta (rr).

This variety differs from the type in the great reduction in the curvature of the portion of the semicell lying above the spines. In this respect the shape of the semicells shows a close affinity with the *A. subulatus* forma *major* figured by Nordstedt (Vidensk. Medd. f. d. naturh. Foren. i Kiöbenhavn **1869**: 232, *pl.* 4, *fig.* 59. 1870) but the Wisconsin cells being much smaller have been considered distinct from the form *major*.

4. Arthrodesmus michiganensis Johnson. Pl. **85**, Fig. **4**.

Bull. Torr. Bot. Cl. **22**: 296, *pl.* 232B, *fig.* 12. 1895.

Cells somewhat under medium size, length about one and a half times the breadth (without spines), median constriction deep, sinus open, with apex broadly rounded, isthmus narrow and elongate; semicells broadly reniform-pyramidate, with the basal halves obelliptic and the apices truncate-pyramidate, flanks of truncate portion retuse; lateral angles of semicells broadly rounded and each with a long somewhat attenuated curved spine, spines inwardly convergent or outwardly divergent. Vertical view elliptic, with a single straight spine at each pole. (Tychoplanktont.)

Zygospores unknown.

Cells 37 μ long; breadth with spines 65 μ, without spines 27 μ; breadth at isthmus 5.5 μ; spines 21–23 μ long.

Soft (rrr).

This rare alga has not been reported since it was discovered. The specimens that I have seen are proportionally narrower than Johnson figures them and the spines may be convergent.

5. Arthrodesmus phimus Turner. Pl. **85**, Fig. **5**.

Kgl. Sv. Vet.-Ak. Handl. **25**, No. **5**: 136, *pl.* 12, *fig.* 9. 1893; W. & G. S. West, Trans. Linn. Soc. 2nd. Ser. Bot. **6**: 192, *pl.* 22, *fig.* 21. 1902; Monogr. Brit. Desm. **4**: 104, *pl.* 115, *figs.* 15–16. 1912.

Cells small, breadth (without spines) somewhat greater than the length, median constriction deep, sinus acute-angled and with the apex acuminate, isthmus fairly narrow; bases of semicells obsemicircular, apices strongly retuse, lateral angles blunt and each terminating in a single short pointed spine, spines divergent. Vertical view narrowly elliptic, with a short spine at each pole. Chloroplasts axial and with a single pyrenoid. (Tychoplanktont.)

Zygospores unknown.

Cells 36 μ long with spines, 25 μ long without spines; breadth with spines 42 μ, without spines 35 μ; breadth at isthmus 7.5 μ; spines 10 μ long.

Clear Crooked (rrr), Owen (rrr), Sand (rrr).

The few individuals that have been seen agree well with W. & G. S. West's drawings of the alga as it is found in India. More reliance should be put in their figures than on that of Turner which is merely a copy of a sketch by Wallich. The organism belongs in the group of short-spined species and is distinguished by the marked concavity of the apex of the semicell.

6. ARTHRODESMUS CONSTRICTUS G. M. Smith. Pl. **85**, Figs. **6–8.**

Trans. Wis. Acad. Sci., Arts & Lett. **20**: 348, *pl.* 10, *figs.* 17–18. 1922.

Cells fairly small, breadth (without spines) somewhat greater than the length, deeply constricted, sinus a broad excavation with a subrectangular apex and sides opening outwards, isthmus narrow and greatly elongate; semicells broadly obversely triangular to transversely elliptic, apex somewhat tumid, lateral angles of semicells acute and each with a single fairly long gradually attenuated spine, spines divergent. Vertical view narrowly elliptic and with a single straight spine at each pole. (Euplanktont.)

Zygospores unknown.

Cells 27–35 μ long with spines, 20–25 μ long without spines; breadth with spines 52–58 μ, without spines 25–28 μ; breadth at isthmus 4–5 μ; spines 15–18 μ long.

Bass (rrr), Blue Gill (rr), Carson (rrr), Horseshoe (rrr), Jenny (rrr), Lynx (rr), Pine Tree (rrr), Pokegama (rrr).

Showing a close relationship to both *A. incus* (De Brébisson) Hassall and *A. triangularis* Lagerheim, but differing in the much greater breadth of the semicell in proportion to the length. An elongate isthmus similar to that found in this species is known in certain varieties of *A. incus* but the narrowness of the isthmus is much more pronounced in *A. constrictus.*

7. ARTHRODESMUS RALFSII W. West. Pl. **85**, Figs. **9–12.**

Jour. Linn. Soc. Bot. **29**: 168. 1892.

Arthrodesmus incus var. *Beta* Ralfs, Brit. Desm. 118, *pl.* 20, *figs.* 4E–4L. 1848.

Arthrodesmus incus var. *Ralfsii* (W. West) W. & G. S. West, Monogr. Brit. Desm. **4**: 95, *pl.* 114, *figs.* 2–4. 1912.

Arthrodesmus incus var. *Ralfsii* forma *latiuscula* W. & G. S. West, Monogr. Brit. Desm. **4**: 96, *pl.* 114, *fig.* 5. 1912; G. M. Smith, Trans. Wis. Acad. Sci., Arts & Lett. **19**[1]: 647. 1918.

Cells small, length about equal to the breadth (without spines), median constriction fairly deep, sinus widely open and with apex semicircular, isthmus broad but not elongated; semicells obversely trapeziform, with bases straight to slightly convex, apex straight or in rare cases slightly concave; lateral angles of semicells acute and bearing short stout gradually attenuated spines; spines convergent. Vertical view elliptic, with a single spine at each pole. (Facultative planktont.)

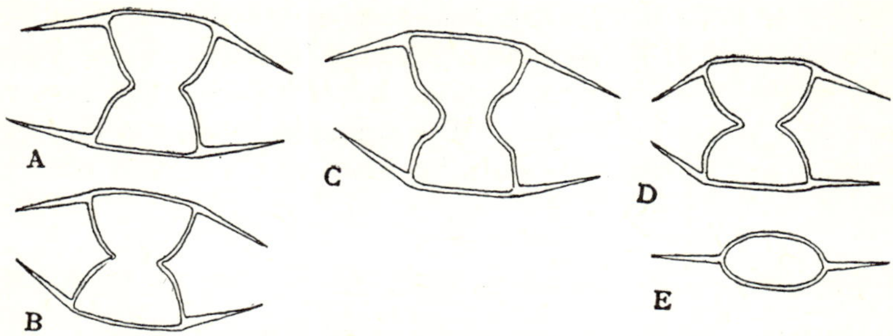

Fig. 17. *Arthrodesmus Ralfsii* W. West (x 700). From material in the Jenner Herbarium of the British Museum of Natural History.

Zygospores spherical, with simple long spines (Ralfs).

Cells 16–18 μ long; breadth with spines 28–31 μ, without spines about 16 μ; breadth at isthmus 7.5–8.5 μ; thickness 8–10 μ; length of spines 7.5–8.5 μ.

Dummy (rrr), George (rr), Horseshoe (rr), Pardee (rr).

This alga was first considered a distinct species by William West but later re-referred to *Arthrodesmus incus* where Ralfs originally described it as an unnamed variety.

I have examined type material collected by Ralfs and preserved in the Jenner herbarium of the British Museum and feel that the alga is sufficiently distinct to warrant the specific recognition that William West first gave it. The most striking character is the convergence of the spines. *Arthrodesmus Ralfsii* also differs from *A. incus* (De Brébisson) Hassall in the ratio between length and breadth as well as in the shape of the semicells.

The specimens from the Jenner herbarium are about one and a half times the sizes of the Wisconsin specimens and are somewhat broader than what W. & G. S. West consider typical *A. Ralfsii* since they also include the form *latiuscula* of W. & G. S. West.

var. **Brebissonii** (Raciborski) comb. nov. Pl. **85**, Figs. **13–17.**

Arthrodesmus incus var. *Brebissonii* Raciborski, Pamietnik Wydz. Akad. Umiej. w Krakowie 17: 96, *pl.* 6, *fig.* 15. 1889; de Toni, Nuova Notarisia 5: 68. 1890.

Sinus widely open and with an acute, not rounded, apex; isthmus narrow; semicells obversely triangular, with the bases straight or slightly tumid; spines much stouter and longer than the type and more markedly convergent. (Facultative planktont.)

Cells 21–27 μ long; breadth with spines 47–60 μ, without spines 21–25 μ; breadth at isthmus 6.5–7 μ; thickness 13 μ; spines 16–18 μ long.

Chain (rrr), Mud (sss).

Differing from the foregoing in the triangular semicells, narrower isthmus, as well as in the greater length and degree of convergence of the spines.

8. Arthrodesmus quiriferus W. & G. S. West. Pl. **85**, Fig. **18**.

Jour. Linn. Soc. Bot. 35: 542, *pl.* 17, *figs.* 9–10. 1903; Monogr. Brit. Desm. 4: 101, *pl.* 115, *figs.* 6–7. 1912.

Cells fairly large, length (without spines) slightly greater than the breadth, median constriction deep, sinus widely open and with a rounded apex, isthmus narrow and slightly elongate; semicells obversely subtriangular, with the basal margins slightly convex and apex markedly retuse; angles of semicells somewhat rounded and each furnished with a single extremely long gradually attenuated spine, spines markedly divergent. Vertical view elliptic, with a long spine at each pole. Chloroplasts single, axial, with one pyrenoid. (Euplanktont.)

Zygospores unknown.

Cells 50 μ long with spines, 35 μ long without spines; breadth with spines 91 μ, without spines 30 μ; breadth at isthmus 7 μ; length of spines 28 μ.

Clear Crooked (rrr), Speese (rr), Squirrel (rr).

The individuals that I have seen in Wisconsin lakes are not exactly typical, but after a comparison of them with type material from Loch Luichart (Scotland) in the G. S. West collection I feel convinced that they are a form of this species midway between the type and the form *compacta* W. & G. S. West. Unlike the British specimens the Wisconsin plants do not show a twisting at the isthmus.

The species should be compared with *Arthrodesmus incus* (De Brébisson) Hassall from which it differs in the longer spines and concave apices of the semicells.

9. Arthrodesmus incus (De Brébisson) Hassall. Pl. **85**, Figs. **19–22**.

Brit. Freshw. Algae 357, *pl.* 85, *fig.* 10. 1845; W. & G. S. West, Monogr. Brit. Desm. 4: 90, *pl.* 113, *figs.* 13–15. 1912.

Cells small, length (without spines) about equal to the breadth, median constriction deep, sinus widely open and with an acute apex, isthmus narrow but not elongate; semicells obversely triangular, with straight or slightly convex sides, apex straight or somewhat concave, angles of semicells acute and each with a single long gradually attenuated spine, spines subparallel to strongly divergent. Vertical view elliptic, with a long spine at each pole. Chloroplasts axial, with a single pyrenoid. (Facultative planktont.)

Zygospores structure not definitely known, probably spherical and with simple spines (W. & G. S. West).

Cells 27–50 μ long with spines, 22–27 μ long without spines; breadth with spines 65–80 μ, without spines 21–26 μ; breadth at isthmus 6–7.5 μ; spines 25–30 μ long.

Red Bass (rrr), Chain (rr), Devils (r), Speese (r).

Ralfs grouped several species under *A. incus* so that the typical *A. incus* of the early investigators is largely a matter of conjecture. I have followed W. & G. S. West in considering the above described alga the typical *A. incus* although it should be noted that their conception of the species is based upon phylogenetic rather than upon taxonomic grounds.

A. incus belongs to the group of species with long spines and straight-sided cells that includes *A. triangularis* Lagerheim and *A. quiriferus* W. & G. S. West. It is distinguished from *A. triangularis* by the shorter isthmus, straight apex of the semicell, and divergent spines.

var. EXTENSUS Andersson. Pl. **85**, Figs. **23–26**.

Bih. Kgl. Sv. Vet.-Ak. Handl. **16**, Afd. **3**, No. **5**: 13, *pl.* 1, *fig.* 7. 1890. Eichler & Raciborski, Rozprawy Akad. Umiej. w Krakowie 2 Ser. **6**; 119, *pl.* 3, *fig.* 23. 1893.

Arthrodesmus incus var. *vulgaris* Eichler & Raciborski, Rozprawy Akad. Umiej. w Krakowie 2 Ser. **6**: 119, *pl.* 3, *figs.* 22, 24. 1893.

Sinus widely open and semicircular to semielliptic, isthmus elongate; semicells trapeziform, with basal angles rounded and superior angles sharper and with fairly short spines, spines parallel to divergent. (Facultative planktont.)

Cells 20–25 μ long; breadth with spines 45–62 μ, breadth without spines 15–17.5 μ; breadth at isthmus 5–6 μ; length of spines 15–22.5 μ.

Bear (rrr), Devils (r), George (rrr), Horseshoe (rrr), Jones (rrr), Lindy (ss), Lynx (rrr), Oak (rrr), Rudolph (rrr), Tank (r), Wolf (rrr).

This variety is characterized by the elongate isthmus and the angle at the base of the semicell, features which make the shape of the semicells trapeziform instead of triangular. In Wisconsin specimens the spines at the angles are usually subparallel instead of divergent.

var. **praelongus** var. nov. Pl. **85**, Figs. **27–29**.

Cells larger, sinus trapeziform; semicells obversely triangular, spines at the angles very long. (Euplanktont?)

Cells 21–22.5 μ long; breadth with spines 61–65 μ, without spines 17.5–20 μ; breadth at isthmus 5 μ; thickness 9 μ; length of spines 22–28 μ.

Crab (rrr), Whitney (rrr).

Very similar to the foregoing in the elongate isthmus but with a sinus that has rounded and not angular sides. The spines are also longer than those of other varieties of the species. The variety might be considered a biradiate variety of *Staurastrum cuspidatum* De Brébisson.

10. ARTHRODESMUS TRIANGULARIS Lagerheim. Pl. **86**, Fig. **1**.
Öfvers. Kgl. Sv. Vet.-Ak. Förh. **42**, No. **7**: 244, *pl.* 27, *fig.* 22. 1885; W. & G. S. West, Ann. Roy. Bot. Garden Calcutta **6**: 211, *pl.* 15, *fig.* 7. 1907; Monogr. Brit. Desm. **4**: 97, *pl.* 114, *figs.* 11–13, 17. 1912.

Cells small, length equal to or slightly greater than the breadth (without spines), median constriction deep, sinus widely open and with a blunt flattened apex, isthmus narrow and elongate; semicells obversely triangular, with slightly convex sides and a distinctly convex apex which is slightly retuse in the central portion; angles of semicells somewhat pointed and each with a fairly long gradually attenuated spine, spines parallel to somewhat but not markedly divergent. Vertical view elliptic, with a single long spine at each pole. Chloroplast axial and with a single pyrenoid. (Tychoplanktont.)

Zygospores unknown in the type.

Cells 15–23 μ long; breadth with spines 52–65 μ, without spines 14–22 μ; breadth at isthmus 5 μ; length of spines 18–22.5 μ.

Dummy (rr), Tied Canoe (rrr).

The type of this species is much scarcer than the varieties *rotundatus* and *subtriangularis*. The species is characterized by the convex apices of the semicells and the length of the spines.

var. **rotundatus** (Raciborski) comb. nov. Pl. **86**, Figs. **2–4**.

Arthrodesmus incus var. *rotundata* Raciborski, Pamietnik Wydz. Akad. Umiej. w Krakowie **17**: 96, *pl.* 6, *fig.* 16. 1889; de Toni, Nuova Notarisia **5**: 69. 1890; Eichler & Raciborski, Rozprawy Akad. Umiej. w Krakowie 2 Ser. **6**: 120, *pl.* 3, *figs.* 16–17. 1893.

Arthrodesmus triangularis var. *inflatus* W. & G. S. West, Jour. Linn. Soc. Bot. **33**: 320. 1898; Monogr. Brit. Desm. **4**: 99 *pl.* 114, *figs.* 14–15. 1912.

Arthrodesmus triangularis forma W. West, Jour. Linn. Soc. Bot. **29**: 168, *pl.* 24, *fig.* 19. 1892.

Cells slightly larger than the type; apices of semicells more convex and indentation at center of apex very much reduced or lacking. Elongation of isthmus not so conspicuous. (Facultative planktont.)

Zygospores spherical, sparsely clothed with long attenuated spines.

Cells 28–30 μ long; breadth with spines 70–76 μ, breadth without spines 21 μ; breadth at isthmus 6–7 μ; length of spines 20–25 μ.

Diameter of zygospores 24–30 μ; with spines, 13–15 μ without spines.

Carroll (rr), Horseshoe (rr), Shell (rr), Streiter (rrr).

W. & G. S. West seem to have overlooked *Arthrodesmus incus* var. *rotundatus* when they described their variety *inflatus* of *A. triangularis*. The two are identical and Raciborski's name for the variety takes precedence.

A few zygospores of this species were noted in one of the collections and this is the first record of zygospore structure in the species.

var. SUBTRIANGULARIS (Borge) W. & G. S. West. Pl. **86**, Figs. **5–7**.

Trans. & Proc. Bot. Soc. Edinburgh **23**: 24, *pl.* 2, *fig.* 36. 1905.

Arthrodesmus incus var. *subtriangularis* Borge, Bot. Not. **1897**: 212, *pl.* 3, *fig.* 4. 1897.

Cells somewhat larger and with stouter spines; basal portion of semicells indented just above the isthmus or with an irregular angular swelling, apex of semicells more elevated than the type and with a pronounced indentation in the central portion, spines generally somewhat divergent. (Euplanktont.)

Cells 35–43 μ long with spines, 30–32 μ long without spines; breadth with spines 75–78 μ, without spines 22–23 μ; breadth at isthmus 7–8.5 μ; length of spines 25 μ.

Blue Gill (sss), Crab (rrr), South Crab (sss), Crooked (rrr), Little Doctor (ss).

The variety differs in the irregularity of the outline of the basal portion of the semicells. This irregularity is usually produced by an indentation (Fig. **5**) but individuals have been observed where there is angular swelling instead of an indentation (Figs. **6–7**) and which, as far as the bases of the semicells are concerned, resemble varieties of *Arthrodesmus Ralfsii* W. West.

11. ARTHRODESMUS OCTOCORNIS Ehrenberg. Pl. **86**, Figs. **8–12**.

Die Infusionsthierchen 152. 1838; Wolle, Desm. U. S. 97, *pl.* 24, *figs.* 21–23. 1884; W. & G. S. West, Monogr. Brit. Desm. **4**: 111, *pl.* 117, *figs.* 6–10. 1912.

Xanthidium (?) *octocorne* (Ehrenberg) Ralfs, Brit. Desm. 116, *pl.* 20, *figs.* 2A–2E. 1848.

Cells small, length (without spines) about one and a quarter times the breadth, median constriction fairly deep, sinus semicircular, isthmus relatively broad; semicells trapeziform, with concave bases and lateral margins markedly incurved, the four rounded angles of semicells bearing a single long gradually attenuated spine, spines on lower angles parallel, those on upper angles divergent. Vertical view elliptic, with two spines inserted one above the other at each pole. Cell wall of uniform thickness and without ornamentation. Chloroplasts axial, with one pyrenoid. (Facultative planktont.)

Zygospores spherical to subspherical, with a few simple somewhat attenuated spines (W. & G. S. West).

Cells 35–44 μ long with spines, 14–20 μ long without spines; breadth with spines 25–38 μ, without spines 10–16 μ; breadth at isthmus 5–6 μ; thickness 6.5 μ; length of horizontal spines 8–19 μ, length of oblique spines 11–17 μ.

Zygospores 28 μ in diameter with spines, 15 μ without spines (W. & G. S. West).

Blue Gill (rrr), Speese (rrr), Travis (c).

At first glance this species appears to be a minute *Xanthidium* but phycologists have excluded it from *Xanthidium* because there is no thickening of the cell wall on the face of the semicells. It is a species that cannot be confused with any other of the genus *Arthrodesmus* being the only one with trapeziform semicells that bears long spines at the angles.

The alga occurred in sufficiently quantity in Travis Lake to warrant the assumption that in this case it was multiplying rapidly under plankton conditions.

ONYCHONEMA Wallich 1860.

Cells generally small, breadth usually somewhat greater than the length. United to form permanent filaments by two apical processes that lie on opposite sides of the apex and overlap adjacent cells of the filament. Cells compressed, deeply constricted and generally with a closed sinus; semicells elliptic, with a smooth or serrate outline, lateral angles smooth or furnished with a single long convergent spine. Cell wall smooth or with transverse rows or granules or pores, at times finely punctate. Vertical view fusiform-elliptic, with the apical processes of the front view alternately arranged and the poles rounded or bearing a single straight spine. Lateral view of semicells usually circular. Chloroplasts axial and with one pyrenoid. Colonies generally enclosed in a fairly broad gelatinous sheath.
Zygospores spherical and ornamented with short spines.

The genus is closely related to *Sphaerozosma* Corda but differs in the nature of the apical processes. In *Onychonema* these processes hold the cells together in a filament by overlapping adjacent cells while in *Sphaerozosma* the processes merely touch corresponding processes of adjacent cells.

KEY TO THE SPECIES.

Lateral angles without spines................................(1) *O. filiforme*
Lateral angles with spines......................................(2) *O. laeve*

1. Onychonema filiforme (Ehrenberg) Roy & Bisset. *Pl.* **86**, Figs. **13–14**.

Jour. of Bot. 24: 242. 1886; Lütkemüller, Beitr. z. Biol. d. Pflanzen 8: 365, *pl.* 18, *fig.* 28. 1902.

Cells small, breadth slightly greater than the length, deeply constricted, sinus sublinear, with the apex acute and the outer margin widely open, isthmus narrow; semicells broadly elliptic, dorsal margin more rounded than the ventral, lateral angles broadly rounded and without spines, apical processes quite long, divergent, and with a very slight swelling at the ends; adjacent cells of the filament not in contact except by the processes. Lateral view of semicells subcircular. Cell outline smooth and wall with two transverse rows of pores. Chloroplasts axial and with one pyrenoid. (Tychoplanktont.)
Zygospores unknown.

Cells 12–18 μ long with apical processes, 8.5–12.5 μ long without processes; breadth 8.5–10 μ; breadth at isthmus 4 μ; length of apical processes 2.5–3.5 μ.

Dutchmans (rrr), Lindy (rr), Tank (rrr).

2. ONYCHONEMA LAEVE Nordstedt.

Vidensk. Medd. f. d. naturhist. Foren. i Kiöbenhavn **1869**: 206, *pl.* 3, *fig.* 34. 1870; W. & G. S. West, Ann. Roy. Bot. Garden Calcutta **6**: 224, *pl.* 12, *figs.* 8–10. 1907.

Cells somewhat under medium size, breadth (without spines) one and a half to two times the length, very deeply constricted, sinus linear, with apex slightly inflated and outer margin widely open, isthmus narrow; semicells transversely elliptic, with the ventral margin more rounded than the dorsal and furnished with two erect spine-like apical processes that terminate in small knobs; lateral angles rounded and each furnished with a single fairly long incurved spine. Cell outline smooth and the wall without granulation. Vertical view narrowly fusiform-elliptic, with poles bearing a single straight spine. Filaments at times enclosed in a gelatinous sheath.

Zygospores (of the variety *micracanthum*) spherical, furnished with numerous small sharp spines (W. & G. S. West).

Cells 16–17 μ long; breadth with spines 31–36.5 μ, without spines 21–25 μ; breadth at isthmus 6 μ (W. & G. S. West).

Not found in the plankton of Wisconsin lakes.

var. LATUM W. & G. S. West. Pl. **86**, Fig. **15**.

Trans. Linn. Soc. 2nd. Ser. Bot. **5**: 232, *pl.* 12, *fig.* 18. 1896.

Cells broader than the type, sinus open and with an acute or slightly inflated apex, apical margin of semicells with a truncate elevation in the median portion, adjacent cells of the filament with their truncate apical portions in contact; apical processes slightly divergent. (Facultative planktont.)

Cells 21–29 μ long with apical processes, 16–20 μ without processes; breadth with spines 27–36 μ, without spines 21–26 μ; breadth at isthmus 5–6 μ; length of apical processes 3–4 μ; length lateral spines 5–6 μ.

Beaverdam (rrr), Devils (sss), Diamond (rrr), Horseshoe (rrr), Little Doctor (rrr), Soft (rrr), Sunday (rrr).

Although the variety differs from the type in the length of the lateral spines and nature of the isthmus the most striking feature is the truncate apical elevation. All specimens found in Wisconsin lakes can be referred to this variety.

SPHAEROZOSMA Corda 1834.

Cells usually small and with the breadth generally somewhat greater than the length. United to form permanent filaments by short apical processes that lie on opposite sides of the apices of the semicells and interlock with corresponding processes from adjacent cells; processes frequently simple granules, never overlapping the adjacent cells. Cells compressed, generally moderately constricted and with an open sinus; semicells elliptic to oblong, lateral margins smooth or granulate, never with spines. Vertical view usually elliptic and lateral view of semicells subspherical. Cell wall smooth, punctate, or with granules arranged in a definite pattern. Chloroplasts axial and with one pyrenoid.

Zygospores generally spherical and with smooth or spinate walls.

The genus stands between *Spondylosium* and *Onychonema*. The short interlocking apical processes or granules are the most prominent diagnostic character. It is difficult to differentiate between many *Spondylosium* and *Sphaerozosma* species since these granules may be greatly reduced and easily overlooked.

KEY TO THE SPECIES.

Apical processes somewhat elongate........................(1) *S. Aubertianum*
Apical processes simple granules.
  Lateral margins of cells granulate........................(3) *S. excavatum*
  Lateral margins of cells smooth............................(2) *S. exiguum*

1. SPHAEROZOSMA AUBERTIANUM W. West.

Jour. of Bot. **27**: 206, *pl.* 291, *fig.* 17. 1889; Jour. Linn. Soc. Bot. **29**: 115, *pl.* 19, *fig.* 1. 1892; W. & G. S. West, Trans. Linn. Soc. 2nd. Ser. Bot. **5**: 230, *pl.* 12, *figs.* 7–8. 1896.

Cells of moderate size, breadth somewhat greater than the length, fairly deeply constricted, sinus with the apex acute and the outer margin widely open, isthmus fairly narrow; semicells elliptic to oblong-elliptic, with lateral margins conspicuously flattened, ventral and dorsal margins with the same degree of curvature; apical processes reduced to two small elongate granules that lie some distance from each other; lateral margins of semicells with two small granules on each side. Lateral view of semicells subspherical, the pairs of granules on side of semicell forming a linear series.

Zygospores spherical to subspherical, with numerous long spines that have broadly conical bases and simple acute apices (W. & G. S. West).

Cells 16.5–19 μ long; 18–23 μ broad; isthmus 5–8 μ broad (W. West).

Diameter of zygospores with spines 37.5–42 μ, without spines 19–24 μ (W. & G. S. West).

Not found in the plankton of Wisconsin lakes.

var. ARCHERII (Gutwinski) W. & G. S. West. Pl. **86**, Fig. **16**.

Trans. Linn. Soc. 2nd. Ser. Bot. **5**: 230. 1896.

Sphaerozosma Archeri Gutwinski, Bot. Cent. **43**: 66. 1890; Sprawozd. kom. fizyjogr. Akad. Umiej. w Krakowie **27**: 29, *pl.* 1, *fig.* 4. 1891.

Cells larger; semicells transversely elliptic and with the lateral margins more rounded. Face of semicells with two transverse rows of granules. Apical processes close together and more elongate than the type. (Facultative planktont.)

Zygospores spherical, with longer spines than the type (Gutwinski).

Cells 22–27 μ long with apical processes, 17.5–20 μ long without processes; breadth 30–35 μ; breadth at isthmus 10–12 μ; length of apical processes 2–3 μ.

Beaverdam (rr), Devils (rr), Horseshoe (rrr), Razorback (rrr), Shell (rrr), Fish (rrr).

2. Sphaerozosma exiguum Turner. Pl. **86**, Fig. **21**.

Kgl. Sv. Vet.-Ak. Handl. 25, No. 5: 141, *pl.* 18, *fig.* 16. 1893.

Cells small, length and breadth about equal, sinus fairly deep, broad and with a rounded apex, isthmus relatively narrow; semicells transversely elliptic, with the ventral margin more rounded than the dorsal, lateral margins very broadly rounded and without granules; apical processes simple granules that lie some distance from each other. Chloroplasts axial and with one pyrenoid. (Tychoplanktont.)

Zygospores unknown.

Cells 14 μ long, 12.5 μ broad; isthmus 6.5 μ broad.

Carson (rr), Jenny (rrr).

Except for the pair of granules between adjacent cells this looks very much like a small specimen of *Spondylosium planum* (Wolle) W. & G. S. West.

3. Sphaerozosma excavata Ralfs. Pl. **86**, Figs. **17–18**.

Ann. & Mag. of Nat. Hist. 16: 15, *pl.* 3, *figs.* 8A–8C. 1845; Brit. Desm. 67, *pl.* 6, *figs.* 2A–2G. 1848.

Cells small, length slightly greater than the breadth, fairly deeply constricted, sinus a broad semicircular depression, isthmus elongate and fairly narrow; semicells oblong-elliptic, with the ventral margins more rounded than the dorsal, apex flattened, with the apical processes reduced to simple granules that lie some distance apart; lateral margins flattened, with 2–3 small granules, at times with one or two granules within the margins. Vertical view oblong-elliptic, with rounded poles. Lateral view of semicells subcircular, with the sides compressed. Chloroplast axial and with one pyrenoid. (Facultative planktont).

Zygospores spherical, with smooth walls (Ralfs).

Cells 14 μ long, 11–12.5 μ broad; breadth at isthmus 6.5 μ; thickness 5 μ.

Diamond (rrr), Horseshoe (r), Lac Court Oreilles (rrr), Little Doctor (rrr), Two Sisters (rr).

var. **Westii** var. nov. Pl. **86**, Figs. **19–20.**

Sphaerozosma excavata var. W. & G. S. West, Trans. Linn. Soc. 2nd. Ser. Bot. 5: 231, *pl.* 12, *fig.* 9. 1896.

Cells with the isthmus less elongate; lateral margins of semicells with a single granule. (Facultative planktont.)

Cells 10.5 μ long; 7.5 μ broad. 5 μ broad at isthmus; 5 μ thick.

Carroll (r).

W. & G. S. West found but a single filament of this variety and so hesitated to describe it as new. I have found it in sufficient quantity in Carroll Lake to determine that the single granule is a constant character and hence constitutes a varietal difference.

SPONDYLOSIUM De Brébisson 1844.

Cells usually of moderate size, ratio between length and breadth variable; united to form permanent filaments. Cells moderately to deeply constricted, sinus linear to somewhat open, with an acute to rounded apex; semicells usually compressed, rarely radially symmetrical, elliptic to oblong or triangular, apices flattened, at times with a truncate elevation or with a slight concavity, never with apical processes. Cell wall smooth or punctate, very rarely with granules arranged in a definite pattern. Chloroplasts axial and usually with but a single pyrenoid. Filaments with or without a gelatinous envelope.

Zygospores usually spherical and with smooth walls.

The older phycologists did not differentiate between *Spondylosium* and *Sphaerozosma* and as a result all of the earlier described species have been referred to both genera. The current conception differentiates *Spondylosium* from *Sphaerozosma* and from *Onychonema* on the absence of apical processes; all filamentous compressed species without apical processes and with a well-defined median constriction being referred to *Spondylosium*, *S. moniliforme* Lundell occupies a rather analamous position since the cells of this species are triangular in vertical view and not compressed. In the classification of the filamentous species the presence or absence of a gridle-like thickening at the point of cell division also plays an important rôle. *Spondylosium* is one of the genera where there is no such thickening.

KEY TO THE SPECIES.

Vertical view compressed.
  Cell apices not elevated..(1) *S. planum*
  Cell apices elevated in median portion........................(2) *S. pulchrum*
Vertical view triangular..(3) *S. moniliforme*

1. SPONDYLOSIUM PLANUM (Wolle) W. & G. S. West. Pl. **86**, Fig. **22**; Pl. **87**, Figs. **1–2**.

Jour. Linn. Soc. Bot. 40: 430, *pl.* 19, *figs.* 5–8. 1912.

Cells under medium size, compressed, breadth slightly greater than the length, moderately constricted, sinus widely open, with the apex broadly rounded and the outer margin open, isthmus fairly narrow; semicells oblong elliptic, with the ventral margin more rounded than the dorsal, lateral margins very broadly rounded, with apices of semicells flattened. Vertical view oblong-elliptic. Lateral view of semicells subcircular. Cell wall smooth. Chloroplasts axial and with a single pyrenoid. Filaments neither twisted nor enveloped in a gelatinous sheath; region of contact between adjacent cells very broad. (Facultative planktont.)

Zygospores unknown.

Cells 12–18 μ long; 12–22 μ broad; 7–12 μ broad at isthmus; 8–10 μ thick.

Beaverdam (rr), Blue Gill (sss), Carroll (rrr), Catfish (rr), Clear Crooked (rrr), Cranberry (ss), Crawling Stone (ss), Deer (rrr), Devils (r), Duck (rrr), Dunham (rrr), Found (s), Horseshoe (rr), Lac Court Oreilles (r), Lake of the Woods (rrr), Mercer (rrr), Middle McKenzie (rrr), Reserve (r), Rock (rrr), Shell (s), Whitefish (rr).

The species is subject to considerable variation in the size of the cells but the shape of the cells is a quite constant feature. It is a species that has become widely distributed as a plankton organism.

2. SPONDYLOSIUM PULCHRUM (Bailey) Archer. Pl. **87**, Figs. **3–4**.

in Pritchard, A History of Infusoria 4th. Ed., 724. 1861; Borge, Ark. för Botanik 15, No. 13: 72. 1918.

Sphraerozosma pulchrum Bailey in Ralfs, Brit. Desm. 209, *pl.* 35, *figs.* 2A–2B. 1848.

Sphaerozosma pulchrum var. *inflatum* Wolle, Desm. U. S. 29, *pl.* 49, *fig.* 8. 1884.

Cells large, compressed, breadth about twice the length, very deeply constricted, sinus sublinear, with the apex acute and the outer margin very slightly open; semicells transversely elliptic, curvature of the dorsal and the ventral margins the same, lateral margins very broadly rounded, semicell apices with a conspicuous truncate elevation in the central region, adjacent cells of filament in mutual contact only at truncate apical elevations. Lateral view of semicells subspherical, with apex truncate and flattened. Cell wall finely punctate. Chloroplasts axial, with 6–10 pyrenoids. (Facultative planktont.)

Zygospores (of the variety *brasiliense*) spherical, with smooth walls (Borge).

Cells 45–54 μ long with apical elevation, 38–45 μ long without elevation; breadth 76–83 μ; breadth at isthmus 18–22 μ.

Bass (rrr), Little Doctor (rr), Nell (s), Speese (rrr).

The species has been bandied back and forth between *Sphaerozosma* and *Spondylosium,* largely because of the interpretation of the apical elevation of the semicells. Since these are elevations of the semicell apex and not apical processes the alga is a *Spondylosium* species.

There has been even more dispute as to the identiy of the organism ever since its description. This confusion has been increased by certain European investigators referring another *Spondylosium* species to it; whereas in reality *S. pulchrum* is not found in Europe. Borge has reviewed with great care all of the accounts of the species and has come to the conclusion that most records of *S. pulchrum* should be referred to the variety *brasiliense* of Nordstedt (Wittrock & Nordstedt, Alg. Exs. No. 369. 1880), a variety that has long been overlooked; and that the variety *inflatum* Wolle is in reality the type. The variety *brasiliense* is distinguished by the reduced apical elevation, more linear sinus, and especially by the compressed median portion of the lateral margins in the vertical view.

In certain cases the Wisconsin specimens do not have the polar elevations in contact but the polar elevations are connected by very short broad gelatinous bands. In one filament also the cells were spirally twisted and not in a flat ribbon.

3. Spondylosium moniliforme Lundell. Pl. 87, Figs. 5–7.

Nova Acta Reg. Soc. Sci. Upsaliae 3 Ser., 8:92, *pl.* 5, *fig.* 16. 1871.

Cells of medium size, radially symmetrical, length about twice to one and a half times the breadth, very deeply constricted, sinus widely open and with the apex rounded, isthmus narrow and somewhat elongate; semicells subtriangular, with the ventral margins rounded and the lateral margins convex; lateral angles broadly rounded and the polar angle rounded or slightly flattened where the cells are in mutual contact. Vertical view triangular, with the angles broadly rounded and the sides strongly concave. Cell wall smooth. Chloroplasts axial, with a small central mass and two broad plates running to each angle of the cell; pyrenoid single and central. Filaments not twisted, imbedded in a copious hyaline gelatinous envelope; area where cells are in apical contact narrow. (Facultative planktont.)

Zygospores unknown.

Cells 30–32.5 μ long; 20–24 μ broad; isthmus 10 μ broad.

Beaverdam (rr).

Differing from other species in the radial symmetry of the vertical view of the cells.

HYALOTHECA Ehrenberg 1841.

Cells united to form long filaments. Filaments not twisted but generally enclosed by a copious gelatinous envelope that at times is radially fibrillar. Cells usually somewhat longer than broad, cylindrical to discoidal, median constriction very slight; ends of cells plane and without projections, polar angles usually rounded. Vertical view circular. Cell wall without ornamentation aside from transverse ridges or in rare cases transverse rows of granules just beneath the apices of the semicells. Chloroplasts axial and with several vertical lobes that extend to the cell walls; pyrenoids single and central. Cells without a girdle-like thickening at the point of division.

Zygospores spherical and smooth-walled.

This is distinguished from the other filamentous genera by the cylindrical cells and lack of apical processes. Examination of mature cells might lead to the conclusion that there is no real distinction between *Hyalotheca* and *Desmidium* but the manner of cell division in the two genera is quite different.

KEY TO THE SPECIES.

Cells without a transverse subapical ridge....................(1) *H. dissiliens*
Cells with a transverse subapical ridge.........................(2) *H. mucosa*

1. HYALOTHECA DISSILIENS (Smith) De Brébisson. Pl. **87**, Figs. **8–9.**

in Ralfs, Brit. Desm. 51, *pl.* 1, *figs.* 1A–1I. 1848.

Cells fairly small, breadth one and a quarter to two times the length, very faintly constricted, sinus a slight annular depression midway between the poles, breadth of isthmus essentially that of the semicells; cells cylindrical-discoidal, with flattened apices and slightly rounded angles. Vertical view circular. Cell wall smooth, without ornamentation aside from an annular ridge just above the isthmus. Chloroplasts axial, with about six longitudinal rays that extend to the cell walls; pyrenoids single and central; cells with but a single chloroplast until nearly mature. Gelatinous colonial envelope broad. (Facultative planktont.)

Zygospores spherical, with smooth walls (Ralfs).

Cells 10–20 μ long; 20–27 μ broad.

Bird (rr), Carroll (rrr), Crab (rrr), Crooked (rrr), Devils (rrr), Howey (r), Lynx (rr).

2. HYALOTHECA MUCOSA (Dillwyn) Ehrenberg. Pl. **87**, Fig. **10.**

Ber. Verh. Preuss. Akad. d. Wiss. Berlin 1840: 212. 1841; Ralfs, Brit. Desm. 53, *pl.* 1, *figs.* 2A–2D. 1848; Lagerheim, Öfvers. Kgl. Sv. Vet.-Ak. Förh. 1883, No. 2: 54. 1883.

Cells fairly small, length slightly greater than the breadth, not constricted; shape of cells cylindrical, with the ends flattened and the angles rounded. Vertical view circular. Lateral walls with a double transverse subapical ridge at either end. Chloroplast axial, with broad vertical rays; pyrenoid single and central. Gelatinous sheath of filament broad, at times with a radial structure. (Facultative planktont.)

Zygospores spherical, with smooth dark yellow walls (Lagerheim).

Cells 14–20 μ long; 17 μ broad.

Diameter of zygospores 30 μ.

Beaverdam (r), Duck (rrr), Fish (rrr), Horseshoe (rrr), Kimball (rrr), Lindy (rrr), Little Doctor (rrr), Meta (rrr), Mud Hen (rrr), Muskallonge (rr), Rooney (rr), Sand (rr).

The most conspicuous difference between this and the foregoing species is the transverse double ridge near the apex of the semicell. In most specimens this ridge appears to be a homogeneous band but in some cases it is a transverse double ring of angular granules.

DESMIDIUM C. A. Agardh 1824.

Cells united to form long filaments. Filaments usually spirally twisted and enclosed in a broad tough gelatinous envelope. Median constriction generally shallow and poorly defined. Cells not marked compressed, usually with a radial symmetry; breadth mostly about twice the length; semicells trapezoidal in front view and with the apices plane or depressed in the median portion and forming an elliptic open space between adjacent cells. Lateral view identical with the front view or subrectangular to octagonal. Vertical view triangular, quadrangular or broadly elliptic. Cell wall not ornamented. Chloroplasts axial, with one or more pyrenoids.

Zygospores formed between the conjugating cells or within the wall of one of the two conjugating cells; spherical to ellipsoidal; with smooth walls.

Although the genus does not contain many species the shape of the cells is somewhat variable and at first glance the species do not appear to belong together. At one time or another attempts have been made to break up *Desmidium* into smaller genera but the current practice is to place in the same genus all filamentous species without apical processes that have smooth walls and develop a girdle-like thickening at the time of division that pushes back into the old semicells.

KEY TO THE SPECIES.

Cell apices plane.
- Vertical view angular.......................................(1) *D. Swartzii*
- Vertical view elliptic.......................................(4) *D. Grevillii*

Cell apices with broad depressions.
- **Median constriction well marked......................(2) *D. Aptogonum***
- **Median constriction poorly defined......................(3) *D. Baileyi***

DESMIDIUM SWARTZII C. A. Agardh. Pl. **88,** Figs. **1–2,**

Systema Algarum 9. 1824; Ralfs, Brit. Desm. 61, *pl.* 4, *figs.* A–F. 1848.

Cells united in permanent filaments; filaments spirally twisted (one complete turn every 12–20 cells), enclosed by a broad firm gelatinous envelope that frequently is radially fibrillar. Cells slightly under medium size, breadth about twice the length, moderately constricted, sinus with inner portion having sides subparallel and apex rounded, outer portion widely open, isthmus very broad; semicells broadly trapezoidal, with lateral margins bearing in the median portion an acute upturned corner, semicells apices plane, without a depression that forms open spaces between adjacent cells. Cells separated by a flattened gelatinous cushion that is thickest at the outer margin. Vertical view triangular, with sides concave but slightly tumid in the median portion, angles broadly rounded (subcircular). Cell wall smooth. Chloroplast with a fairly massive central axial mass and with broadly ovate projections running to the angles of the cells, projections incised about half the distance from the apex to the central mass; pyrenoids in each lobe and at times in the central mass. (Tychoplanktont.)

Zygospores subspherical to ellipsoidal, with broadly rounded poles, smooth-walled (Ralfs).

Cells 13–17.5 μ long; 35–42 μ broad; 27–36 μ broad at isthmus.

Beaverdam (rrr), Diamond (rrr), Devils (rrr), Fish (rrr), Found (rr), George (rrr), Horseshoe (rrr), Howey (rrr), Jag (rrr), Lac Court Oreilles (rrr), Little Doctor (rrr), Meta (rrr), Sand (rrr), Thousand Island (rrr).

This, the first member of the family to be described, is sharply differentiated from other species of the genus except *D. Aptogonum* De Brébisson. It is readily distinguished from *D. Aptogonum* by the lack of open spaces between contiguous cells.

2. DESMIDIUM APTOGONUM De Brébisson. Pl. **88,** Figs. **3–4.**

Mem. soc. acad. d. sc. de Falaise 1835: 268, *pl.* 2. 1835; W. & G. S. West, Trans. Linn. Soc. 2nd. Ser. Bot. 5: 233, *pl.* 12, *fig.* 24. 1896.

Aptogonum Desmidium (Ehrenberg) Ralfs, Brit. Desm. 64, *pl.* 32, *figs.* 1A–1H. 1848.

Cells united in permanent filaments that are spirally twisted (one complete turn in every 15–20 cells); filaments enclosed at times by a firm gelatinous envelope. Cells somewhat under medium size, breadth about twice the length, moderately constricted, sinus with sides subparallel and apex rounded, outer portion widely open, isthmus very broad; semicells trapeziform, with lateral margins bearing at the base an acutely upturned angle; semicell apices with a deep broad semielliptic to subrectangular depression that forms an elliptic to subrectangular open space between adjacent cells. Vertical view triangular, with sides slightly concave and angles broadly rounded. Cell wall smooth. (Tychoplanktont.)

Zygospores spherical to broadly elliptic, with smooth walls (Ralfs).

Cells 18–20 μ long; breadth 31–43 μ; breadth at isthmus 23–35 μ.

Zygospores 22–27 μ long, 20–25 μ broad (W. & G. S. West).

Devils (rrr).

Ralfs thought that the large foramen between the cells of the filaments in this species and *D. Baileyi* (Ralfs) Nordstedt justified placing them in a separate genus *(Aptogonum)* but *Aptogonum* is not recognized by phycologists of the present generation.

3. Desmidium Baileyi (Ralfs) Nordstedt. Pl. **88,** Figs. **5–7.**

Lunds Univ. Arsskr. **16:** 4. 1880 (sep.); Kgl. Sv. Vet.-Ak. Handl. **22,** No. **8:** 25, *pl.* 2, *figs.* 4–5. 1888.

Aptogonum Baileyi Ralfs, Brit. Desm. 208, *pl.* 35, *figs.* 1A–1C. 1848.

Cells united to form filaments that are straight (not spirally twisted) and devoid of a gelatinous sheath. Cells fairly small, with the length and breadth about equal, median constriction reduced to a faint undulation; cell outline rectangular, with the lateral margins parallel and apices with a broad deep semielliptic depression that gives the appearance of short erect truncate projections at either lateral margin of the semicells apices. Adjacent cells of filaments almost completely separated by elliptical to subrectangular openings. Vertical view triangular, with slightly convex sides and rounded angles. Cell wall smooth. (Facultative planktont.)

Zygospores spherical and with smooth walls (Ralfs).

Cells 20–26 μ long; 21–27.5 μ broad.

Beaverdam (rrr), Crooked (rrr), Diamond (rrr), Found (r), Little Doctor (rrr), Muskallonge (rrr), Speese (rrr).

Sharply differentiated from the other species by the rectangular cells in which there is only the faintest indication of a median constriction. The filaments of this species, like those of *D. aptogonum* De Brébisson, have elliptic open spaces separating adjacent cells.

4. Desmidium Grevillii (Kützing) De Bary. Pl. **88,** Fig. **8.**

Unters. u. d. Fam. d. Conj. 42, *pl.* 4, *figs.* 30–31. 1858.

Desmidium cylindricum Greville, Scott. Crypt. Fl. 5: *pl.* 293. 1827.

Didymoprium Grevillii Kützing, Phycol. gen. 166. 1843; Ralfs, Brit. Desm. 57, *pl.* 2, *figs.* A–K. 1848.

Cells united to form filaments that are spirally twisted (about one complete turn in every ten cells); filaments enclosed in a broad firm gelatinous envelope; envelope frequently with radial fibrillae. Cells of medium size, breadth about twice the length, slightly constricted, sinus a small semicircular depression, isthmus very broad; semicells trapezoidal, basal angles very broadly rounded, apical angles obtuse; lateral margins frequently with a sharp projection midway between base and apex, the projections on opposite sides being connected by a faint line; apex flat, without open spaces between adjacent cells. Lateral view of cells octagonal, breadth about one and a half times the length, without a constriction in the median portion. Vertical view broadly elliptic, with a distinct nodule at each pole. Cell wall smooth, except for faint vertical striae just above the isthmus.

Zygospores subspherical, with smooth walls (De Bary).

Cells 20–24 μ long; 45–48 μ broad; breadth at isthmus 40–42 μ.

Found (rr), No Mans (rrr).

The species is widely known under the name *D. cylindricum.* In founding the genus *Didymoprium* Kützing, for some unknown reason, gave the alga a new specific name, a name which Ralfs adopted in his British Desmidieae. The specific name *cylindricum* antedates the name *Grevillii* but the use of Ralfs as the starting point for nomenclature of the Desmidiaceae necessitates the use of the latter name.

In the conjugation of this species zygospore formation takes place within one of the old cells and not between the conjugating cells. This feature, however, has not been considered a justification for separating it from other species of the genus.

GYMNOZYGA Ehrenberg 1841.

Cells united to form long filaments. Filaments rarely twisted, frequently enclosed by a gelatinous sheath. Cells barrel-shaped, not compressed, length about twice the breadth, median constriction slight, at times a small groove running around the cell; apices flattened and without protuberances. Cell wall smooth or with vertical striae near the poles and at the isthmus. Chloroplasts axial and with radiating vertical laminae; pyrenoids single and central.

Zygospores spherical to broadly ellipsoidal, with smooth walls.

In the division of the *Gymnozyga* cell there is the same girdle-like thickening that pushes back into the old semicell at the time of division that is found in *Desmidium.*

GYMNOZYGA MONILIFORMIS Ehrenberg. Pl. 87, Fig. 11.

Berichte Verh. Preuss. Akad. d. Wiss. Berlin **1840**: 212. 1841.

Cells barrel-shaped, united in permanent filaments, length about twice the breadth, sinus a small semielliptic depression, isthmus very broad; semicells with sides converging to a truncate apex, lateral margins distinctly concave and with apical angles only slightly rounded. Cell wall with vertical striae in the polar regions and at times with other vertical striae at the bases of the semicells. Vertical view circular. Chloroplasts axial, with 6–8 delicate laminate vertical processes to the cell wall; pyrenoids single and central. (Facultative planktont.)

Zygospores spherical to ellipsoidal and with broadly rounded poles, zygospore wall smooth (Ralfs).

Cells 26–35 μ long; breadth at base of semicells 15–22 μ, at apex of semicells 10–15 μ; breadth at isthmus 13–19 μ.

Bettine (rrr), Blue Gill (rr), Carson (rr), Crooked (r), Howey (rr), Jag (rrr), Jones (rrr), Little Doctor (sss), Lynx (rrr), Red Bass (rr), Big St. Germain (r), Whitney (r).

PLATE 52.

All figures x 500.

PLATE 52

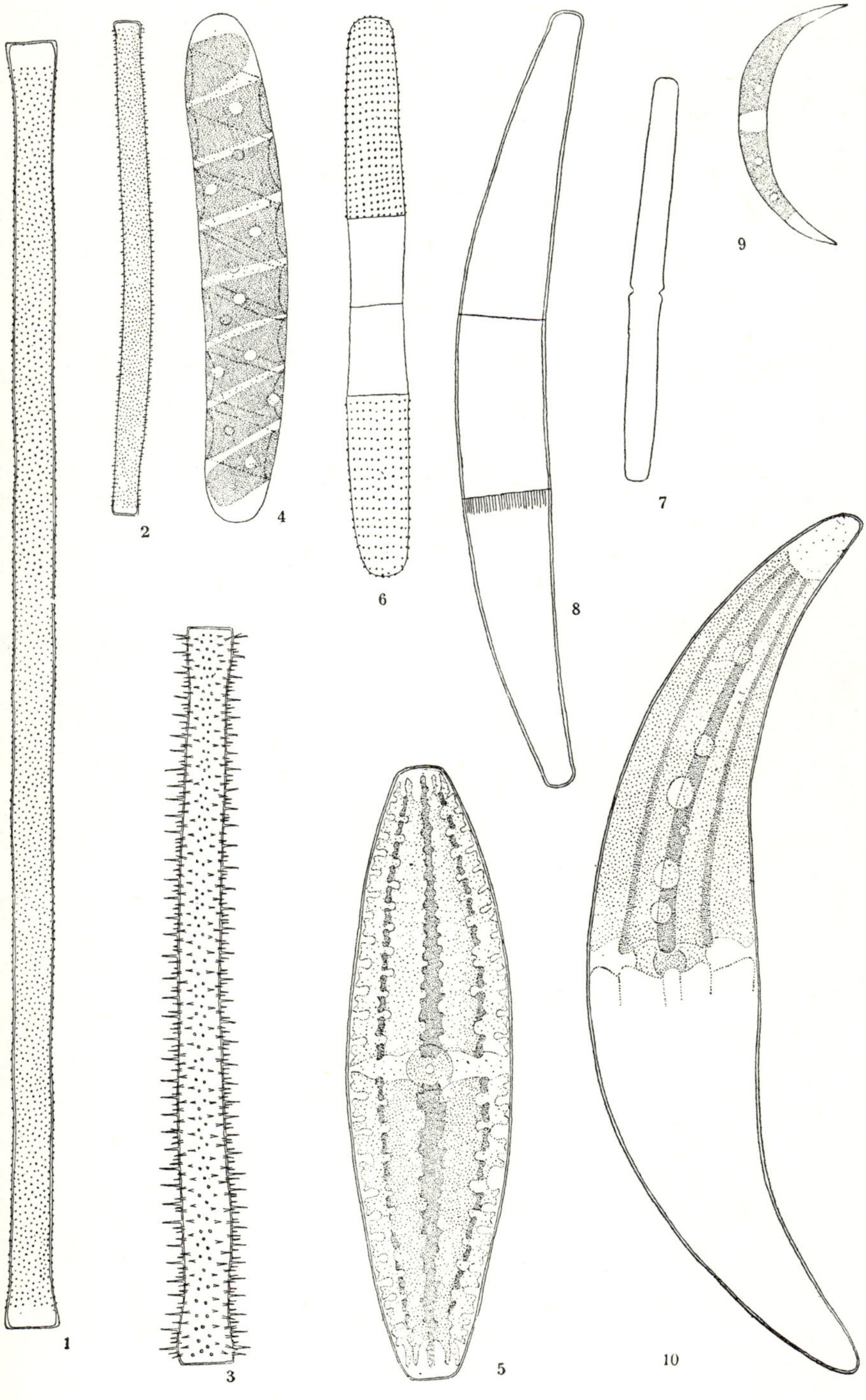

PLATE 53.

All figures x 400.

		Page
Fig. 1.	*Closterium acerosum* (Schrank) Ehr.	10
Fig. 2.	*Closterium gracile* var. *elongatum* W. & G. S. West	11
Fig. 3.	*Closterium aciculare* var. *subpronum* W. & G. S. West	11
Fig. 4.	*Closterium acutum* (Lyngb.) De Bréb.	11
Fig. 5.	*Closterium Ralfsii* var. *hybridum* Rab.	12
Fig. 6.	*Closterium Kuetzingii* De Bréb.	12
Fig. 7.	*Closterium decorum* De Bréb.	13

PLATE 53

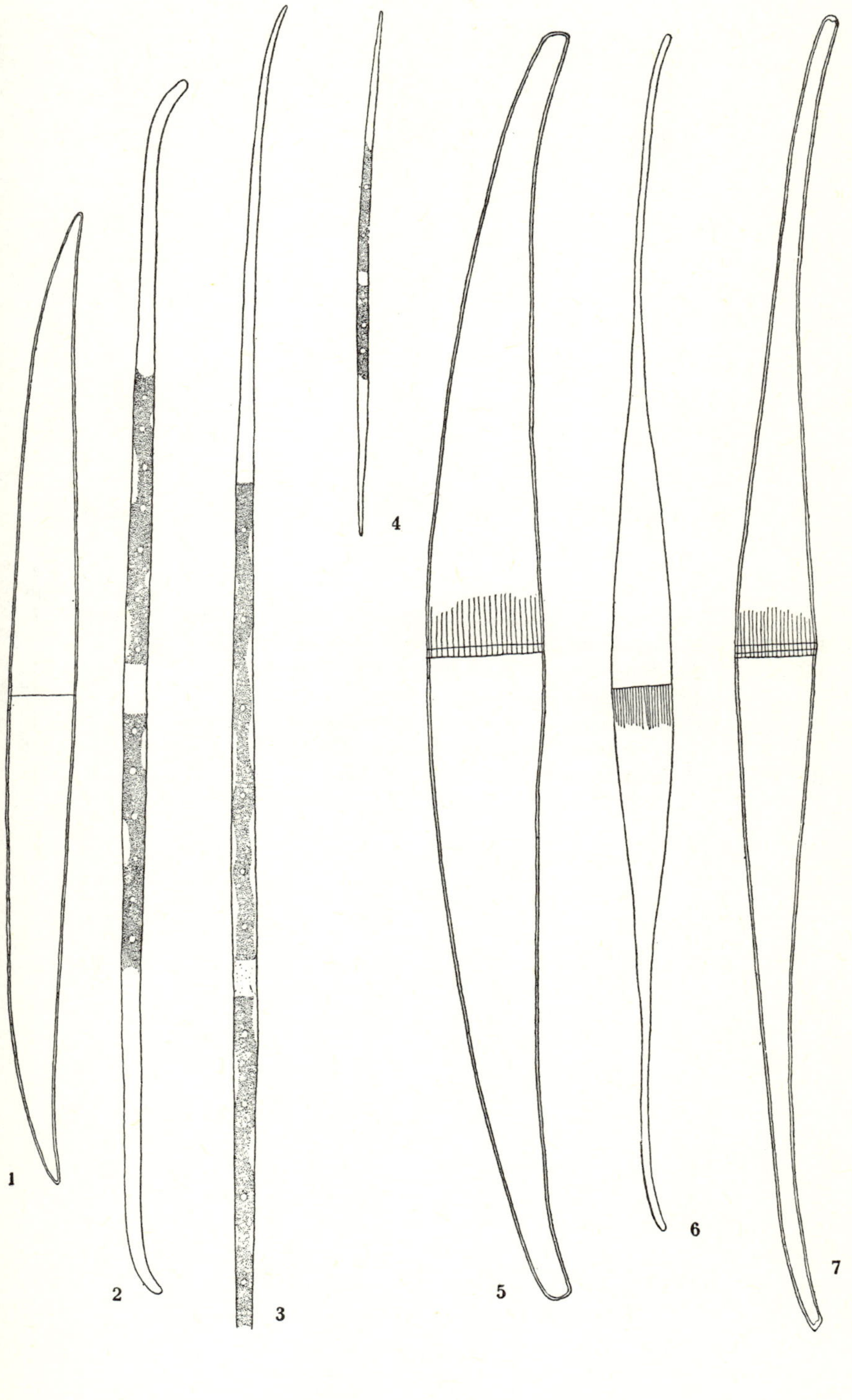

PLATE 54.

All figures x 400.

		Page
Fig. 1.	*Pleurotaenium trabecula* var. *rectum* (Delp.) W. & G. S. West..	14
Figs. 2–4.	*Pleurotaenium subcoronulatum* var. *detum* W. & G. S. West.....	14
Figs. 5–6.	*Pleurotaenium Ehrenbergii* (De Bréb.) D. By.................	15
Figs. 7–8.	*Pleurotaenium Ehrenbergii formae*..........................	15
Fig. 9.	*Pleurotaenium Ehrenbergii* var. *undulatum* Schaarsch..........	15
Fig. 10.	*Pleurotaenium Ehrenbergii* var. *elongatum* W. West..........	16
Fig. 11.	*Pleurotaenium nodosum* (Bail.) Lund.......................	16

PLATE 54

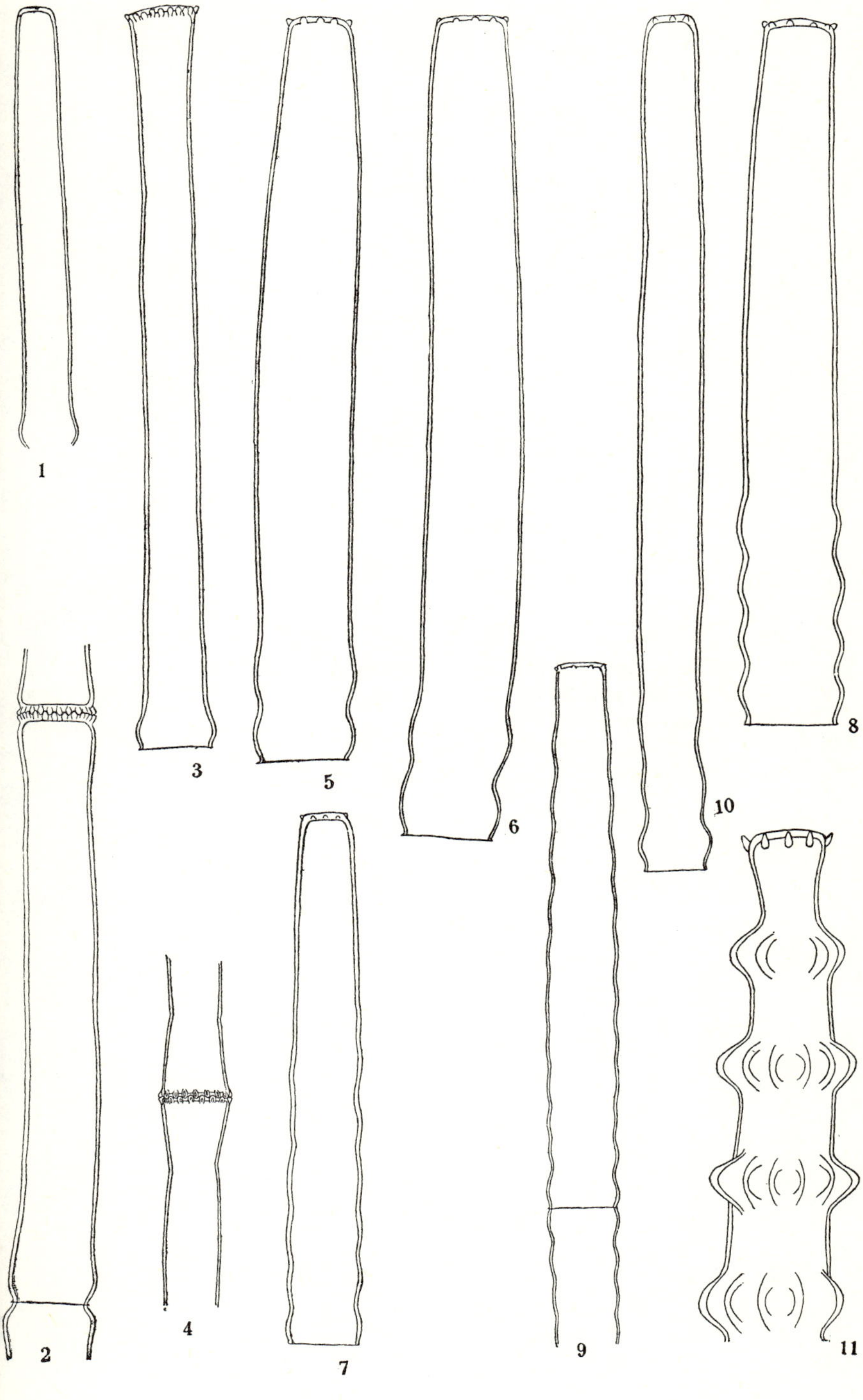

PLATE 55.

All figures x 400.

PLATE 55

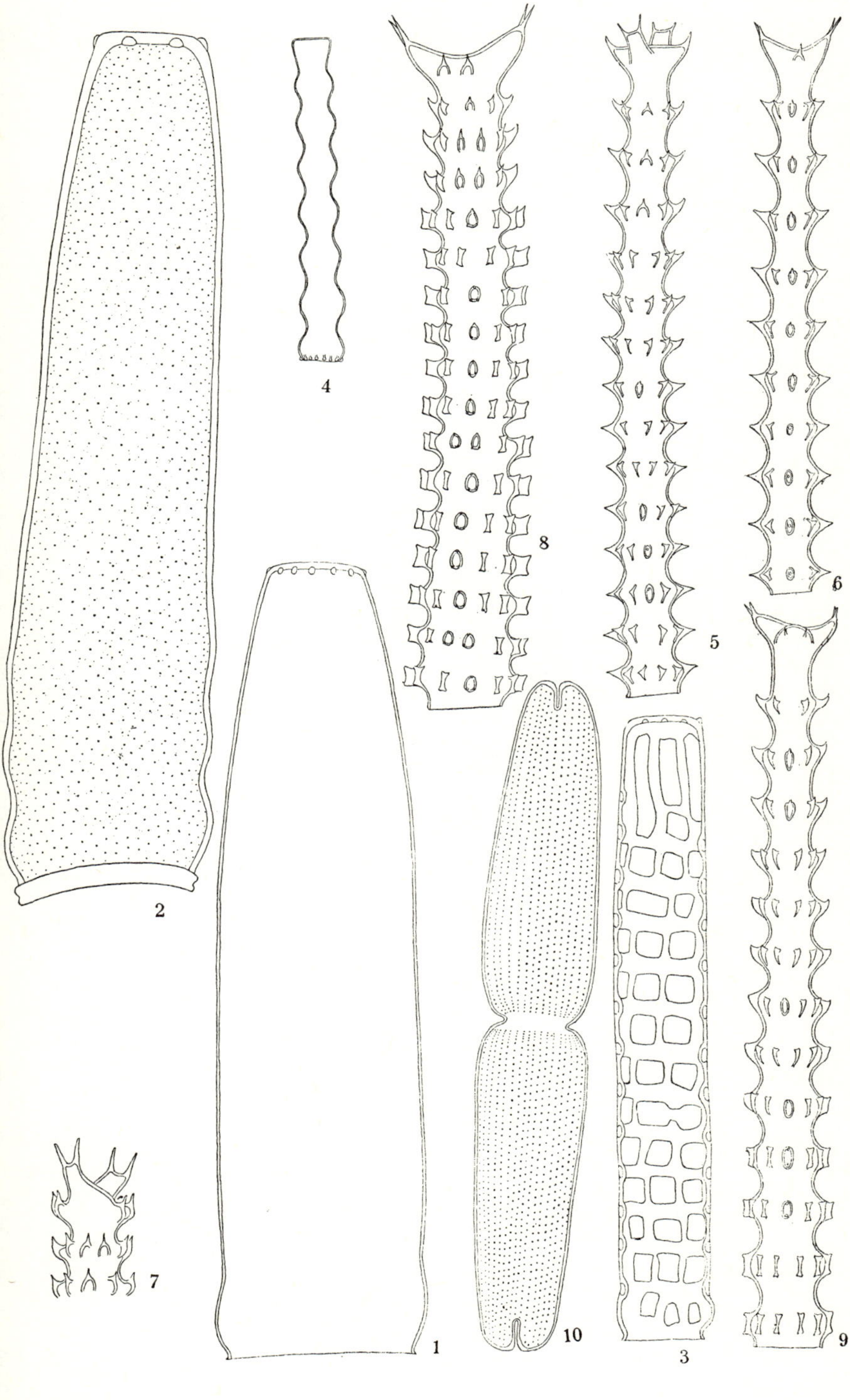

PLATE 56.

All figures x 600.

PLATE 56

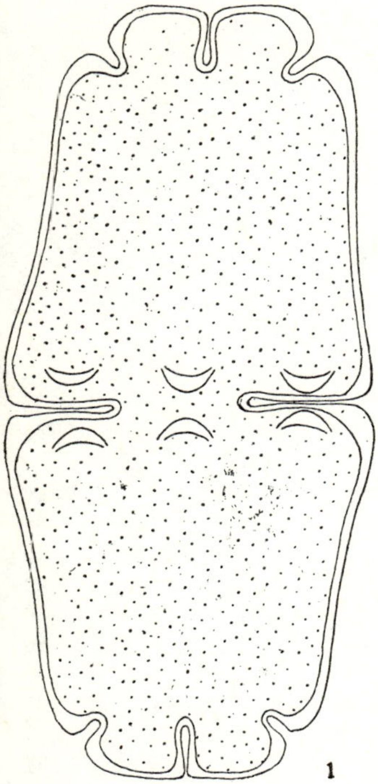

1

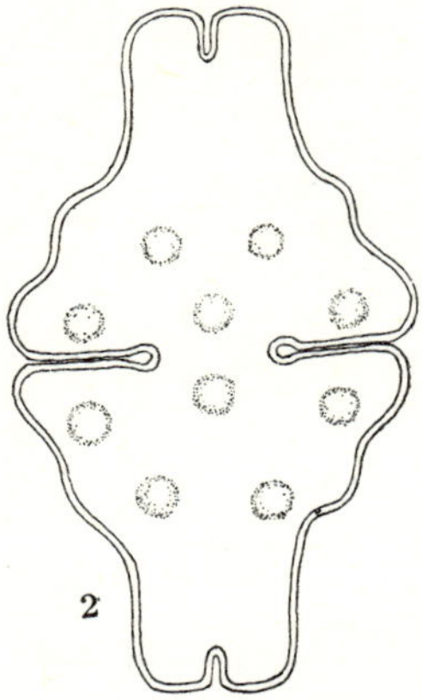

2

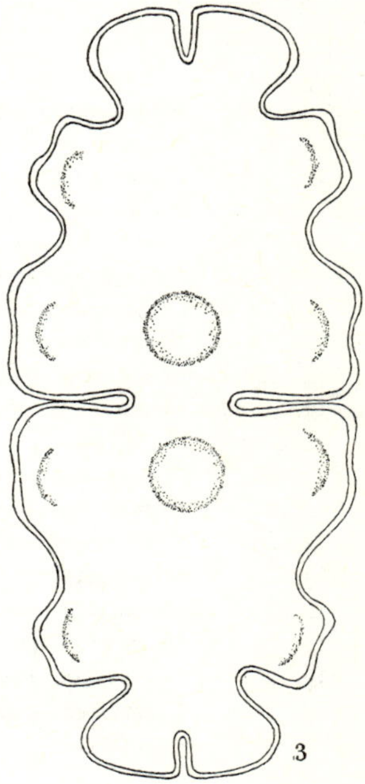

3

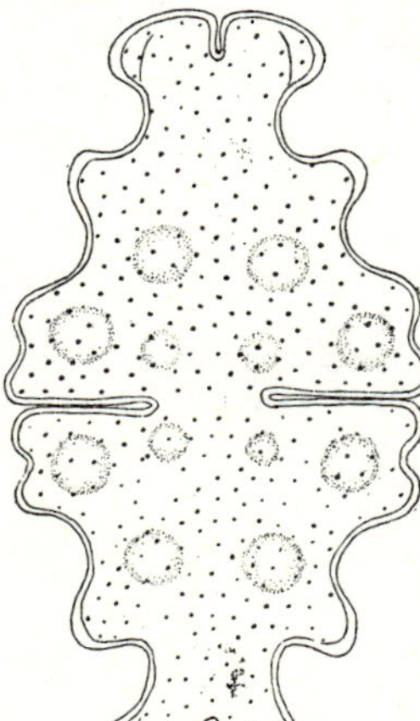

4

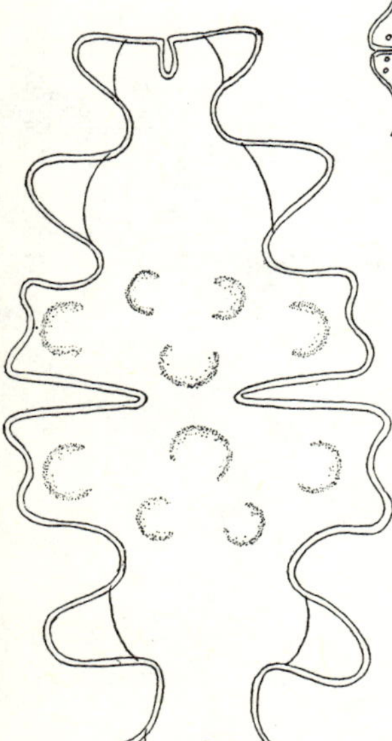

5

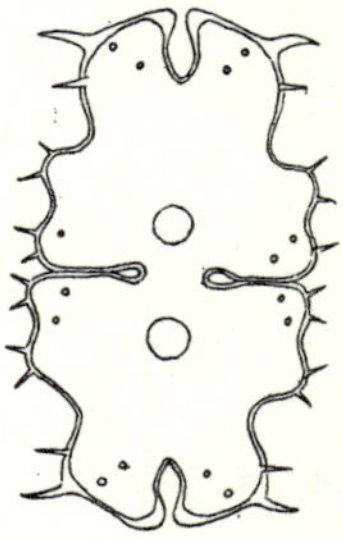

6

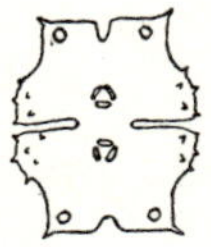

7

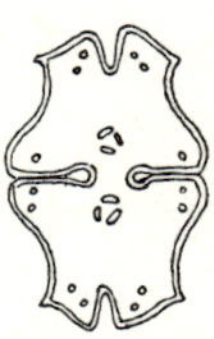

8

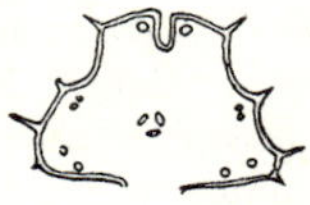

9

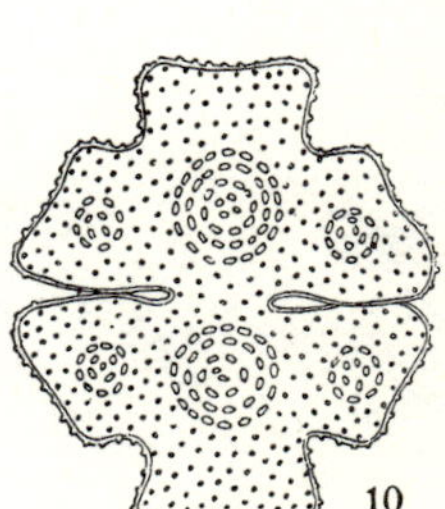

10

PLATE 57.

All figures x 600.

PLATE 57

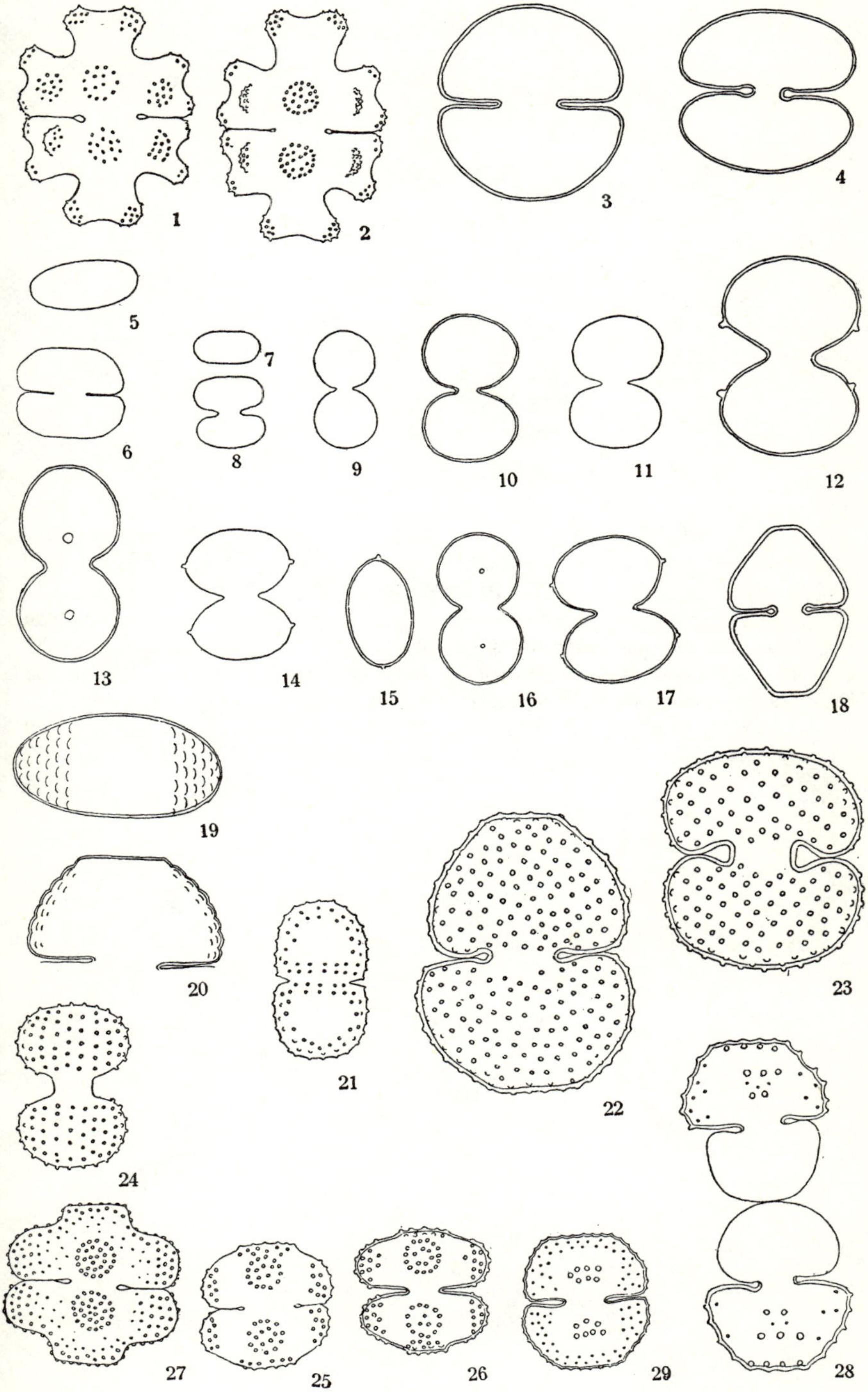

PLATE 58.

		Page
Fig. 1.	*Cosmarium ovale* Ralfs (x 400)	37
Fig. 2.	*Cosmocladium saxonicum* De Bary (x 400)	38
Figs. 3–9.	*Cosmocladium Hitchcockii* (Wolle) comb. nov. (Fig. 3, x 400; 4, x 500; 5-a, x 1000)	38

PLATE 58

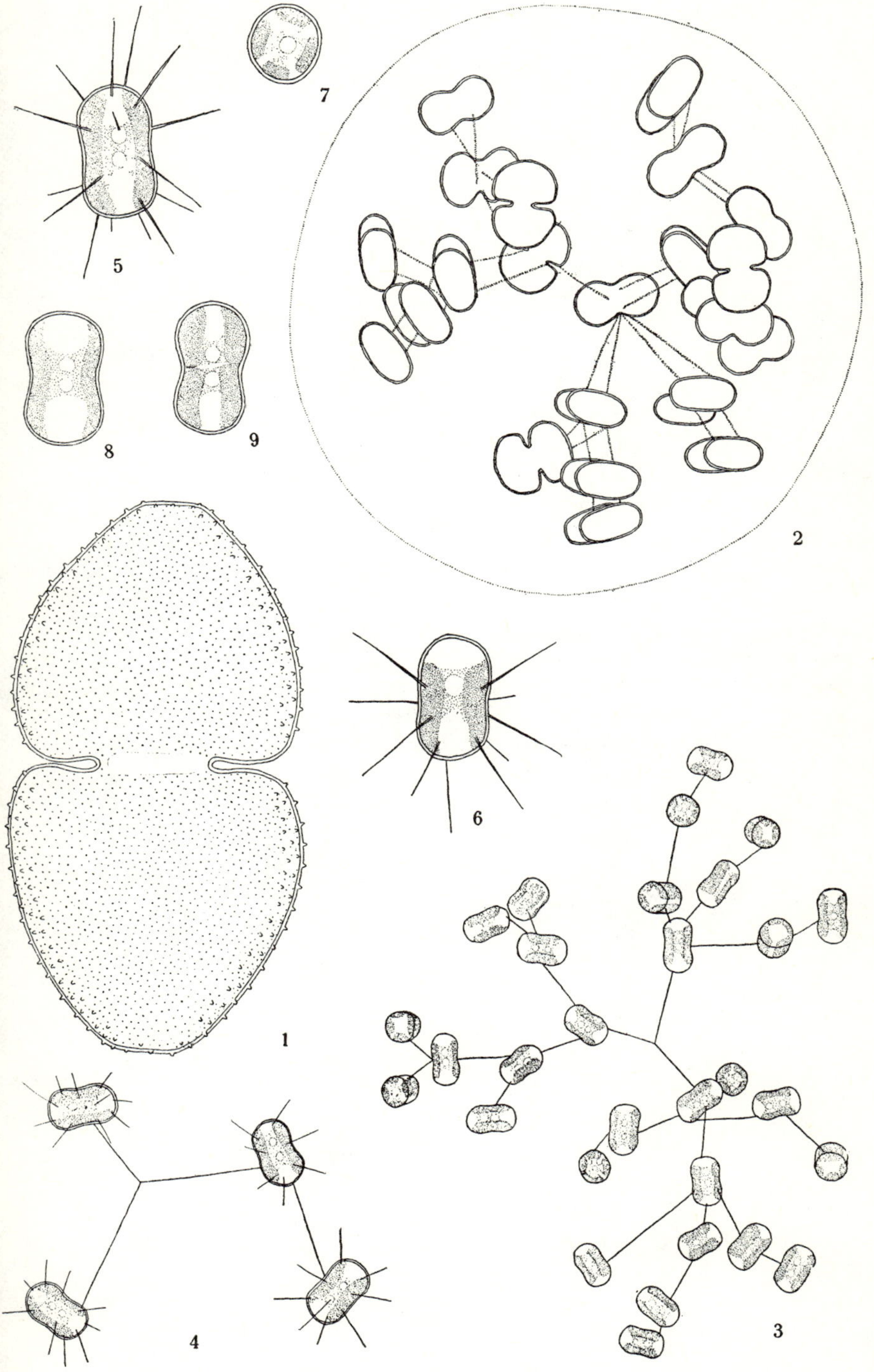

PLATE 59.

All figures x 400.

		Page
Figs. 1–2.	*Micrasterias pinnatifida* (Ktz.) Ralfs........................	41
Fig. 3.	*Micrasterias laticeps* Nordst...............................	41
Fig. 4.	*Micrasterias depauperata* var. *Kitchelii* (Wolle) W. & G. S. West	42
Fig. 5.	*Micrasterias denticulata* De Bréb..........................	45
Figs. 6–7.	*Micrasterias conferta* var. *hamata* Wolle....................	44

PLATE 59

PLATE 60.

Figs. 1–3, 5 x 400; Fig. 4 x 200.

		Page
Figs. 1 2.	*Micrasterias truncata* (Corda) De Bréb.	43
Fig. 3.	*Micrasterias radiosa* var. *ornata* Nordst.	47
Fig. 4.	*Micrasterias radiosa* var. *ornata* forma *elegantior* G. S. West	47
Fig. 5.	*Micrasterias Thomasiana* Arch.	48

PLATE 60

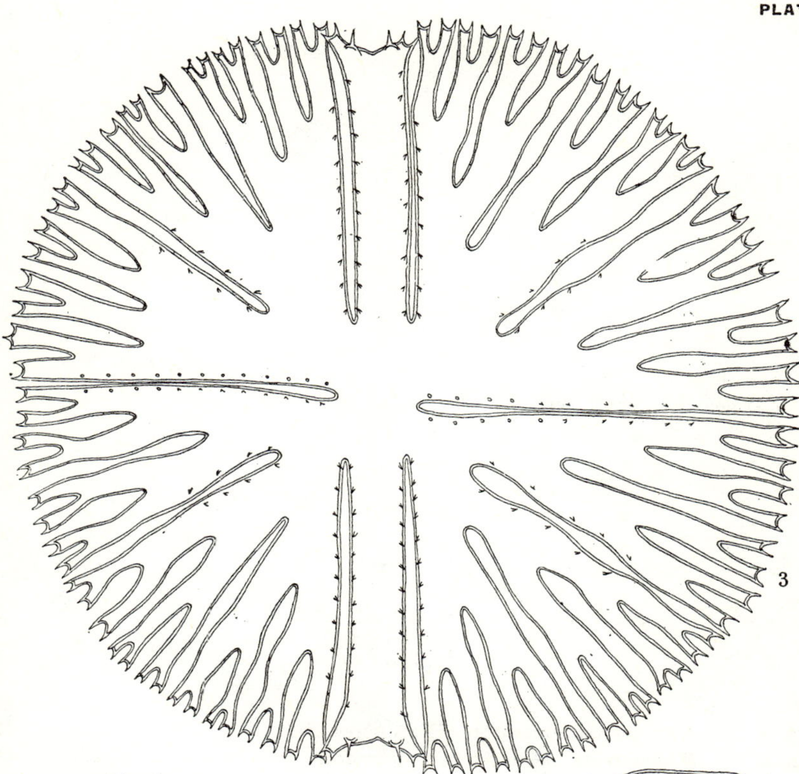

3

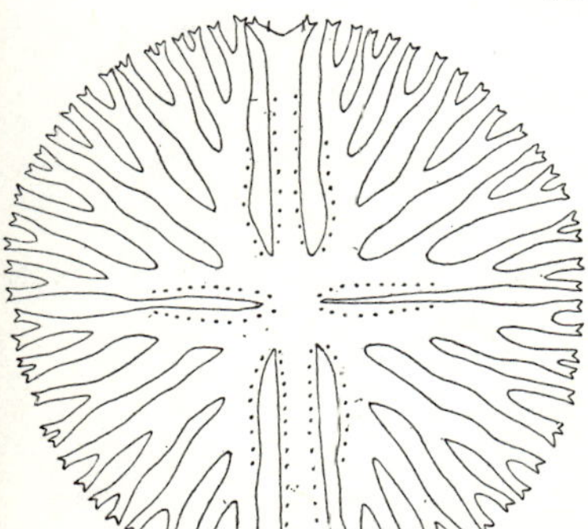

4

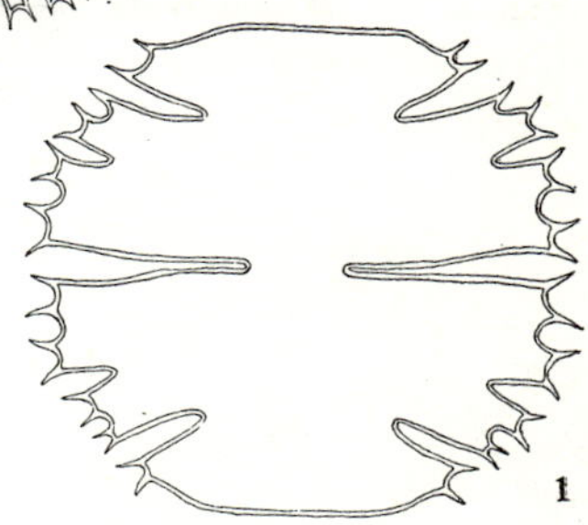

1

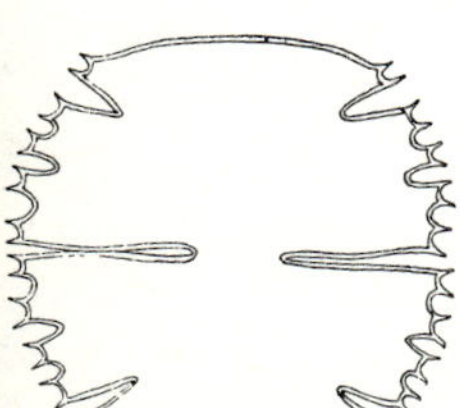

2

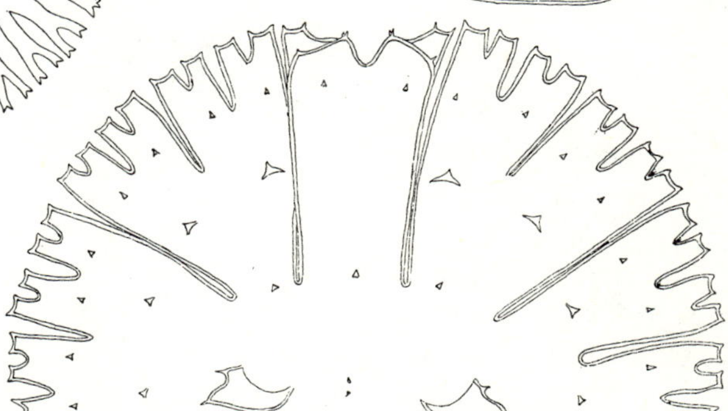

5

PLATE 61.

All figures x 400.

		Page
Fig. 1.	*Micrasterias apiculata* (Ehr.) Menegh.	47
Fig. 2.	*Micrasterias apiculata* var. *fimbriata* forma *spinosa*	48

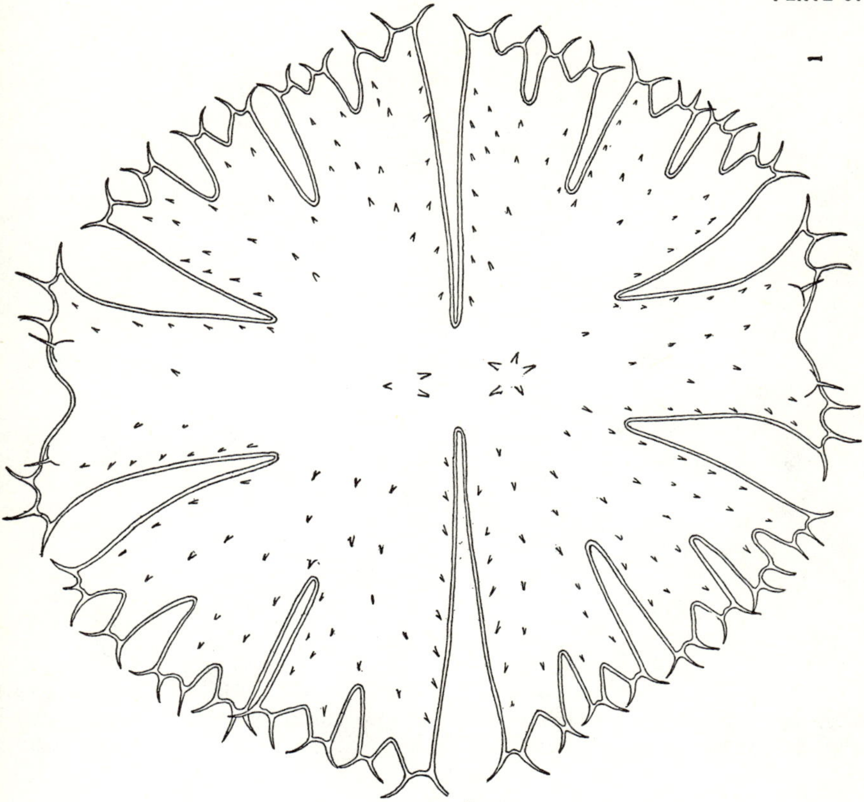

1

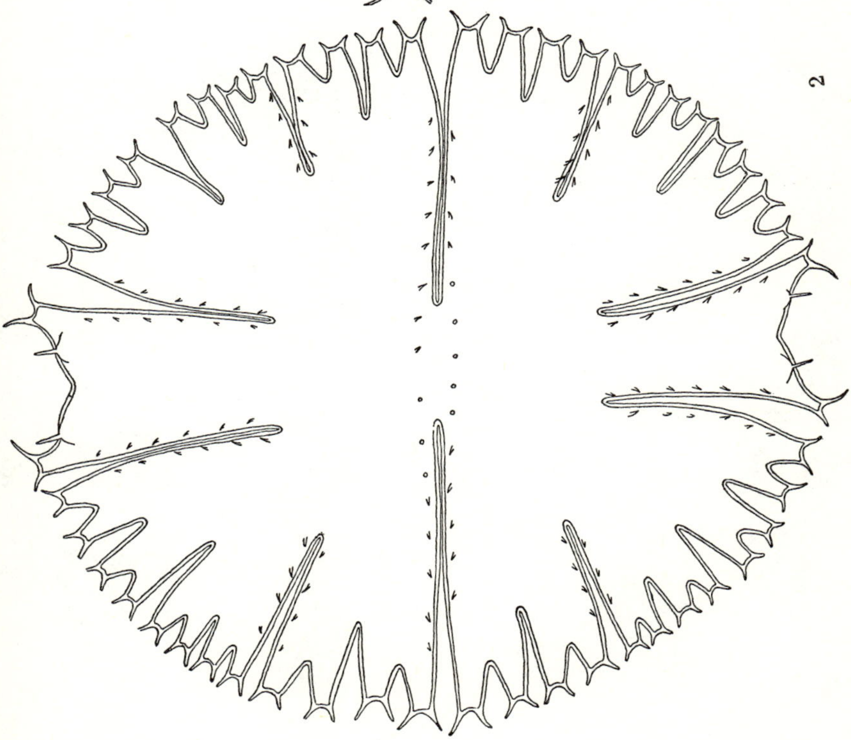

2

PLATE 62.

PLATE 62

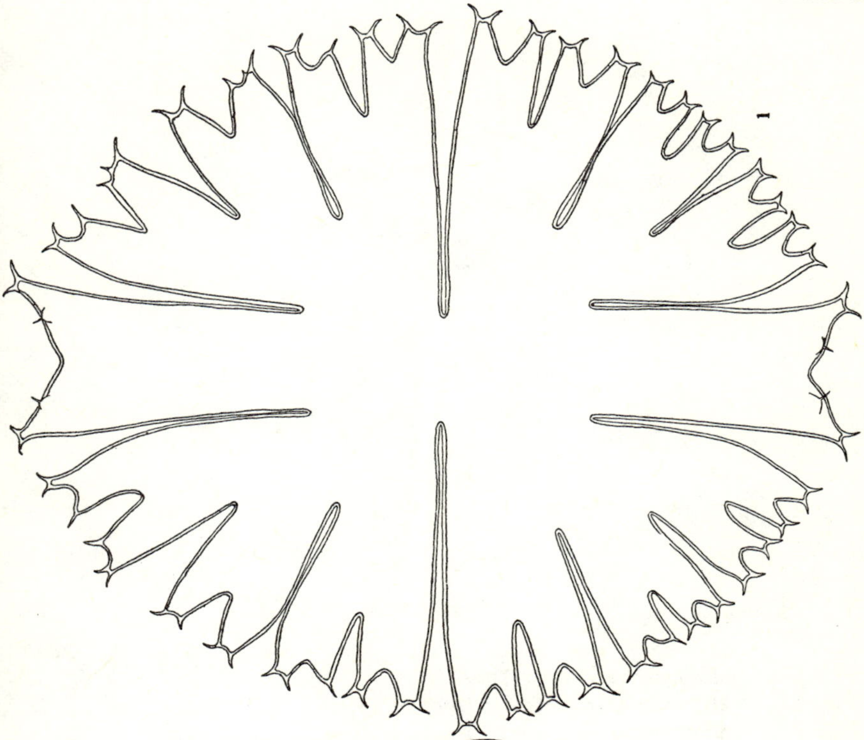

1

2

PLATE 63.

All figures x 400.

		Page
Figs. 1–3.	*Micrasterias radiata* Hass	50
Fig. 4.	*Micrasterias radiata* var. *simplex* (Wolle) G. M. Smith	51
Fig. 5.	*Micrasterias americana* (Ehr.) Ralfs	52

PLATE 63

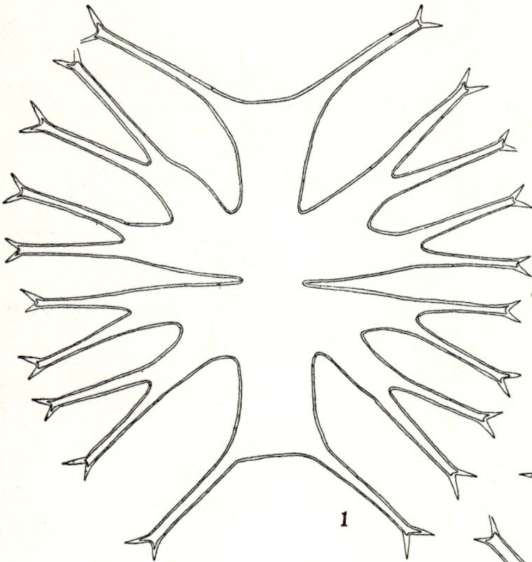

1

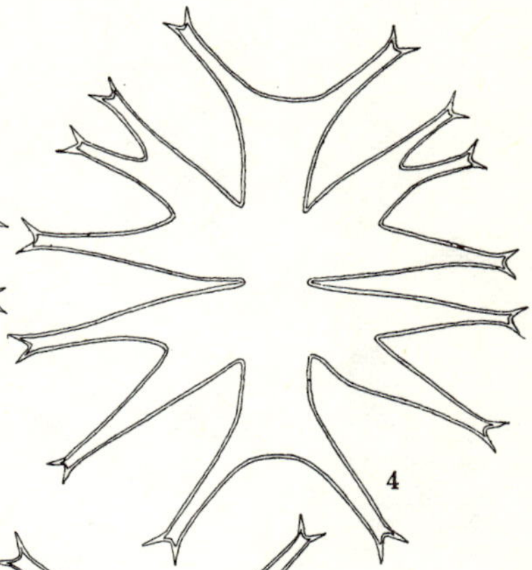

4

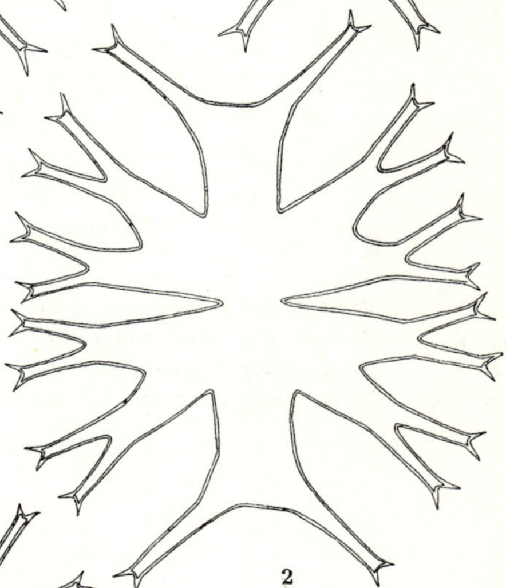

2

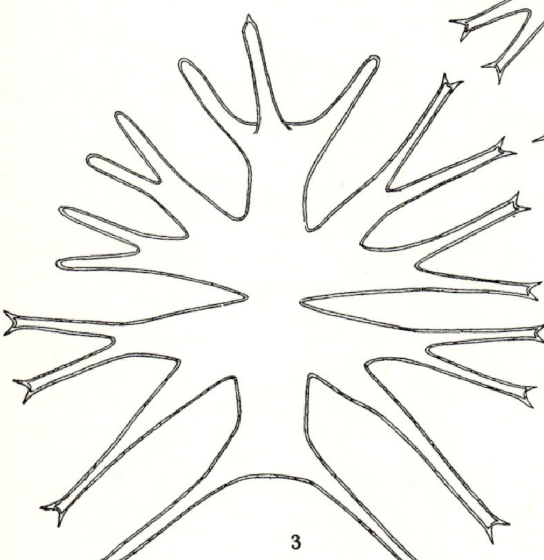

3

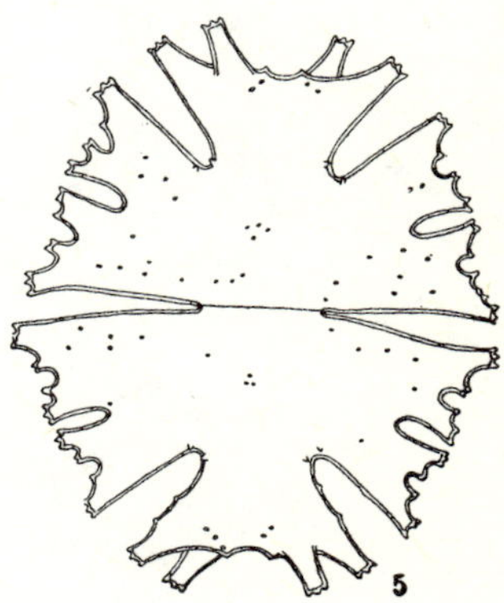

5

PLATE 64.

All figures x 400.

		Page
Fig. 1.	*Micrasterias radiata* var. *gracillimum* G. M. Smith	52
Fig. 2.	*Micrasterias americana* (Ehr.) Ralfs	52
Fig. 3.	*Micrasterias Nordstedtiana* Wolle	53
Figs. 4–5.	*Micrasterias muricata* (Bail.) Ralfs	54

PLATE 64

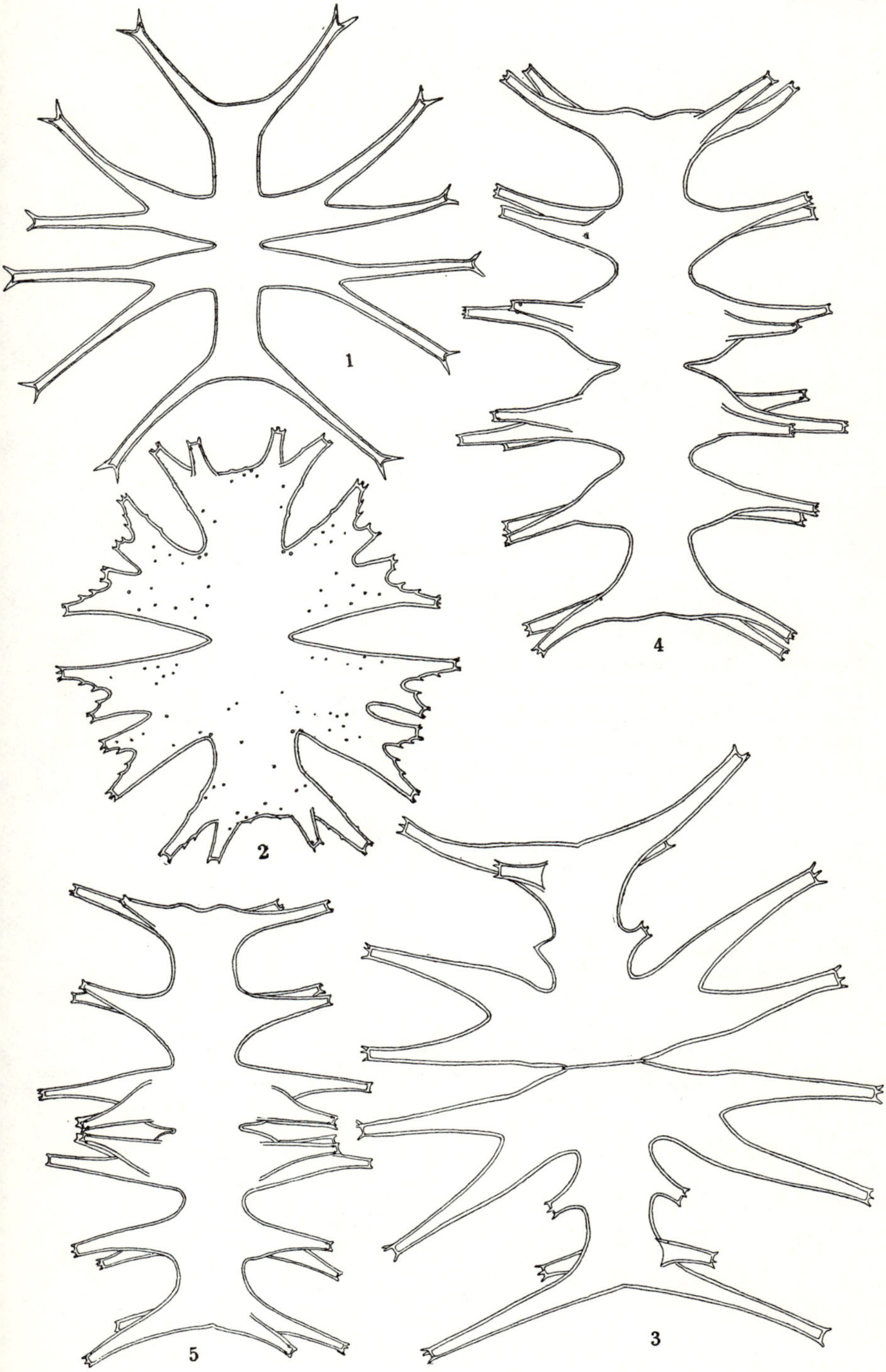

PLATE 65.

Fig. 1, x 250; Figs. 2–11, x 500.

PLATE 65

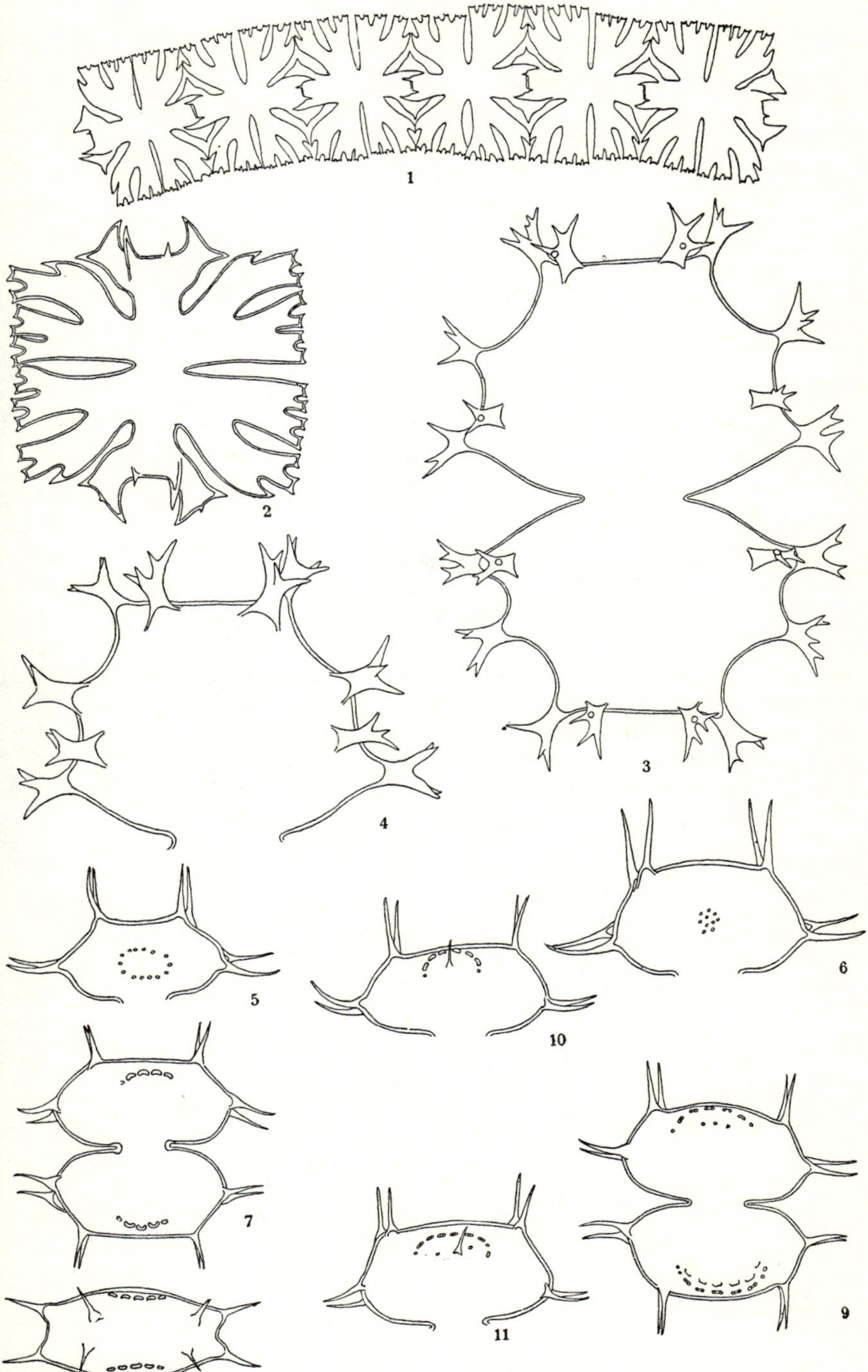

PLATE 66.

All figures x 500.

		Page
Fig. 1.	*Xanthidium antilopaeum* var. *depauperatum* W. & G. S. West	59
Figs. 2–3.	*Xanthidium cristatum* De Bréb.	59
Fig. 4.	*Xanthidium cristatum* var. *uncinatum* De Bréb.	60
Figs. 5–12.	*Xanthidium subhastiferum* W. West	60
Figs. 13–18.	*Xanthidium subhastiferum* var. *Toweri* (Cushm.) G. M. Smith	61

PLATE 66

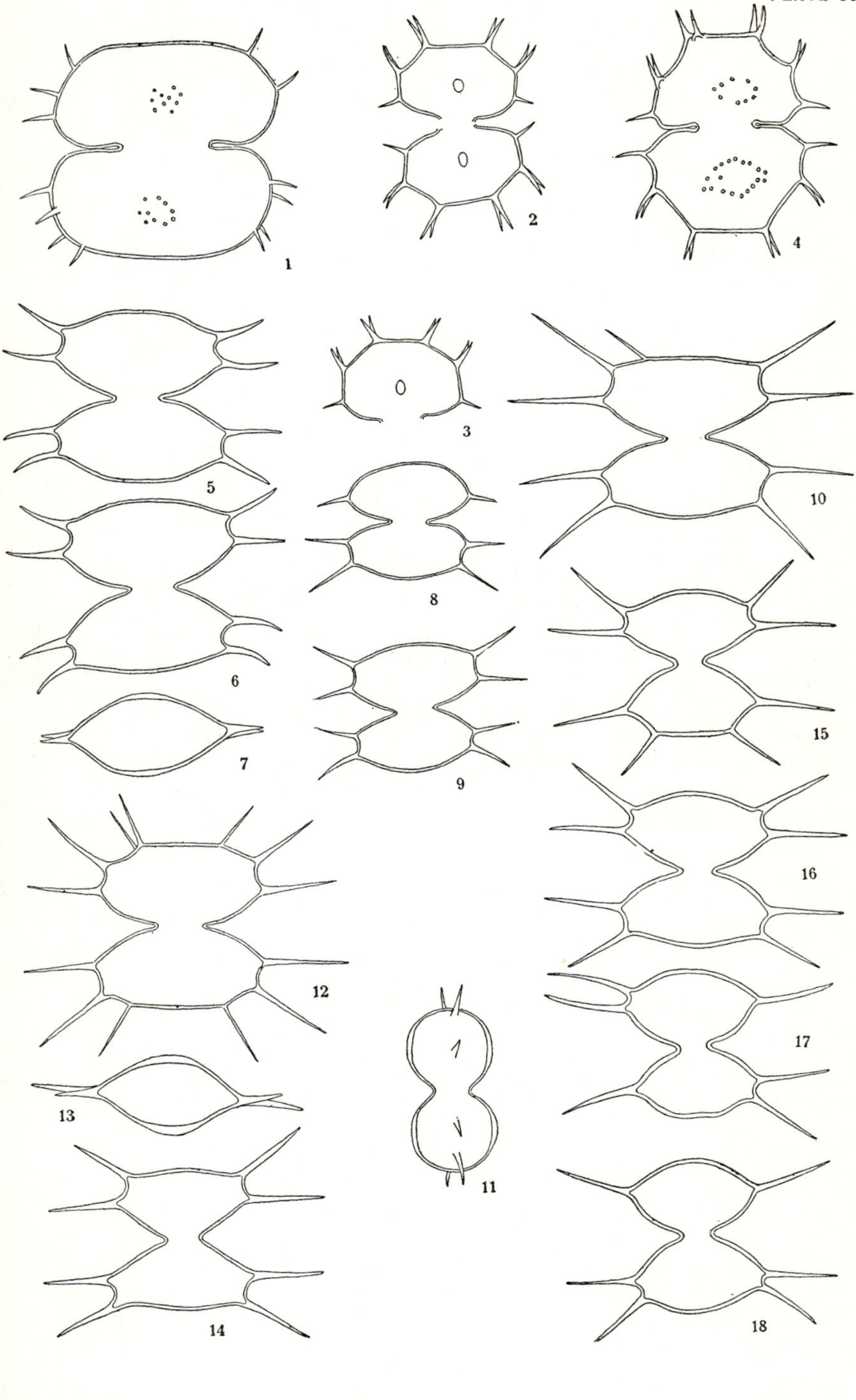

PLATE 67.

All figures x 500.

PLATE 67

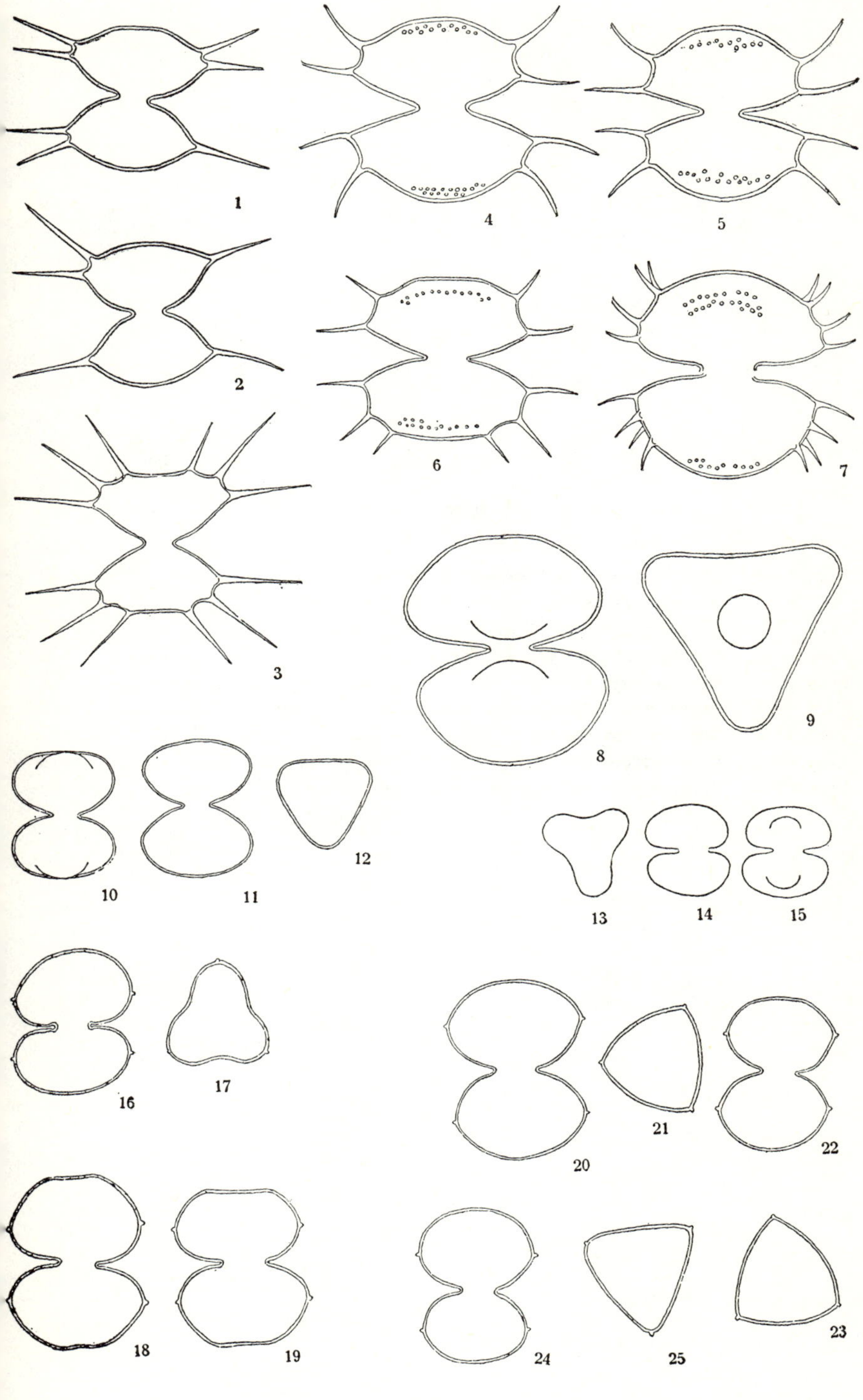

PLATE 68.

All figures x 600.

PLATE 68

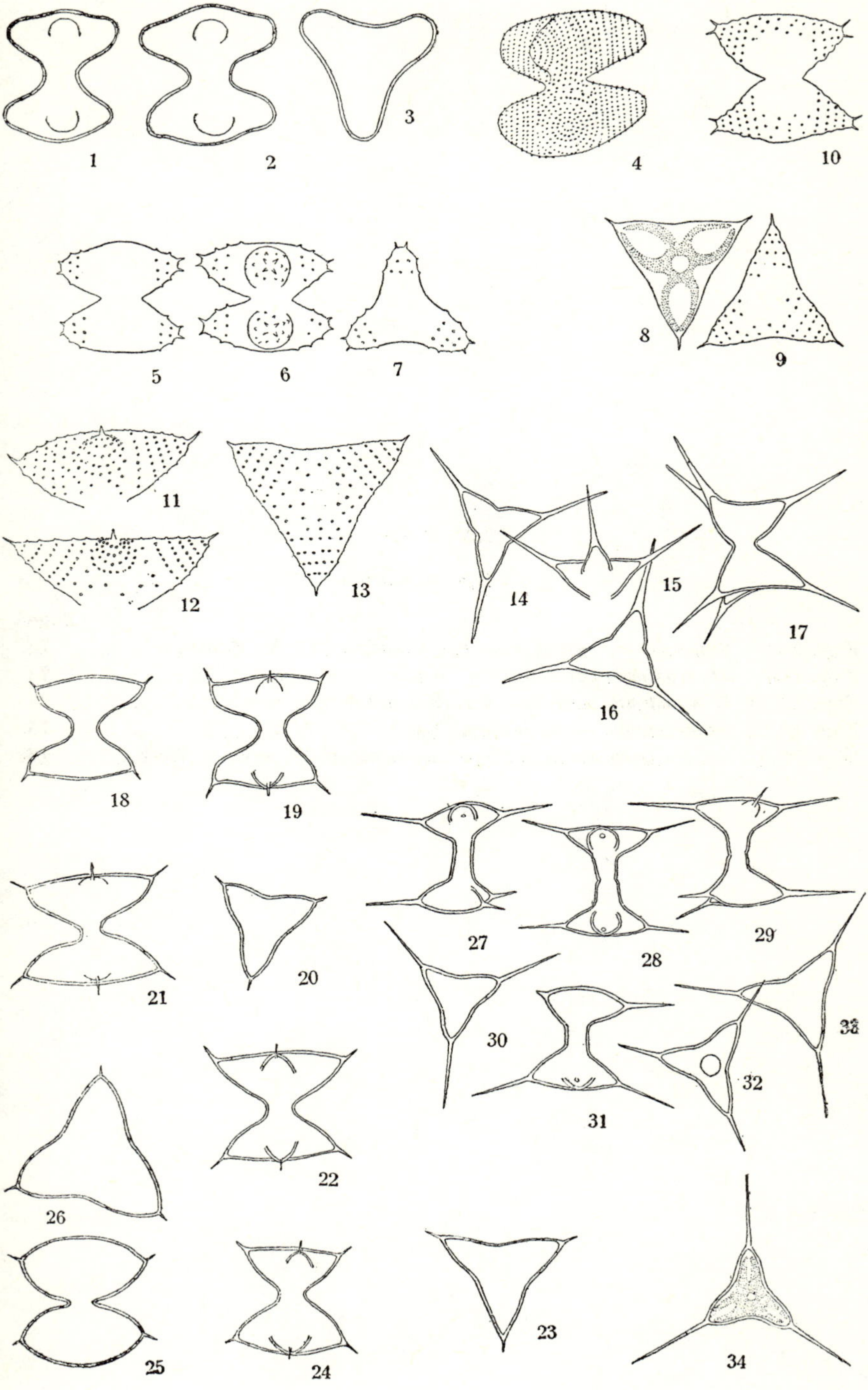
1
2
3
4
10
8
9
5
6
7
11
12
13
14
15
17
16
18
19
21
20
27
28
29
30
31
32
33
26
22
25
24
23
34

PLATE 69.

All figures x 500.

		Page
Figs. 1–3.	*Staurastrum cuspidatum* var. *canadense* G. M. Smith..........	75
Figs. 4–9.	*Staurastrum curvatum* W. West..............................	73
Figs. 10–15.	*Staurastrum curvatum* var. *elongatum* var. nov...............	73
Figs. 16–21.	*Staurastrum megacanthum* Lund..............................	75
Figs. 22–24.	*Staurastrum megacanthum* var. *scoticum* W. & G. S. West......	76

PLATE 69

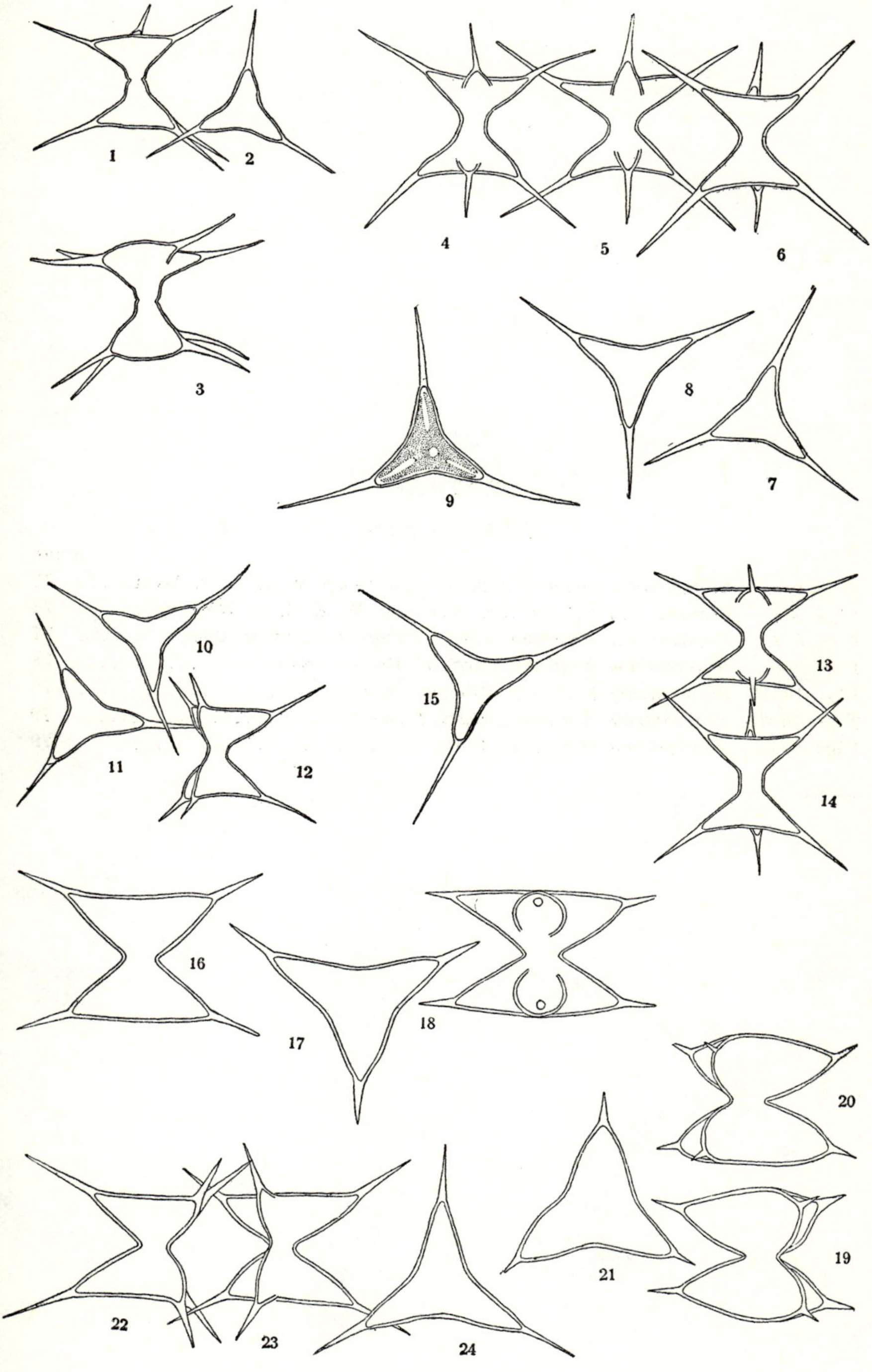

PLATE 70.

All figures x 500.

PLATE 70

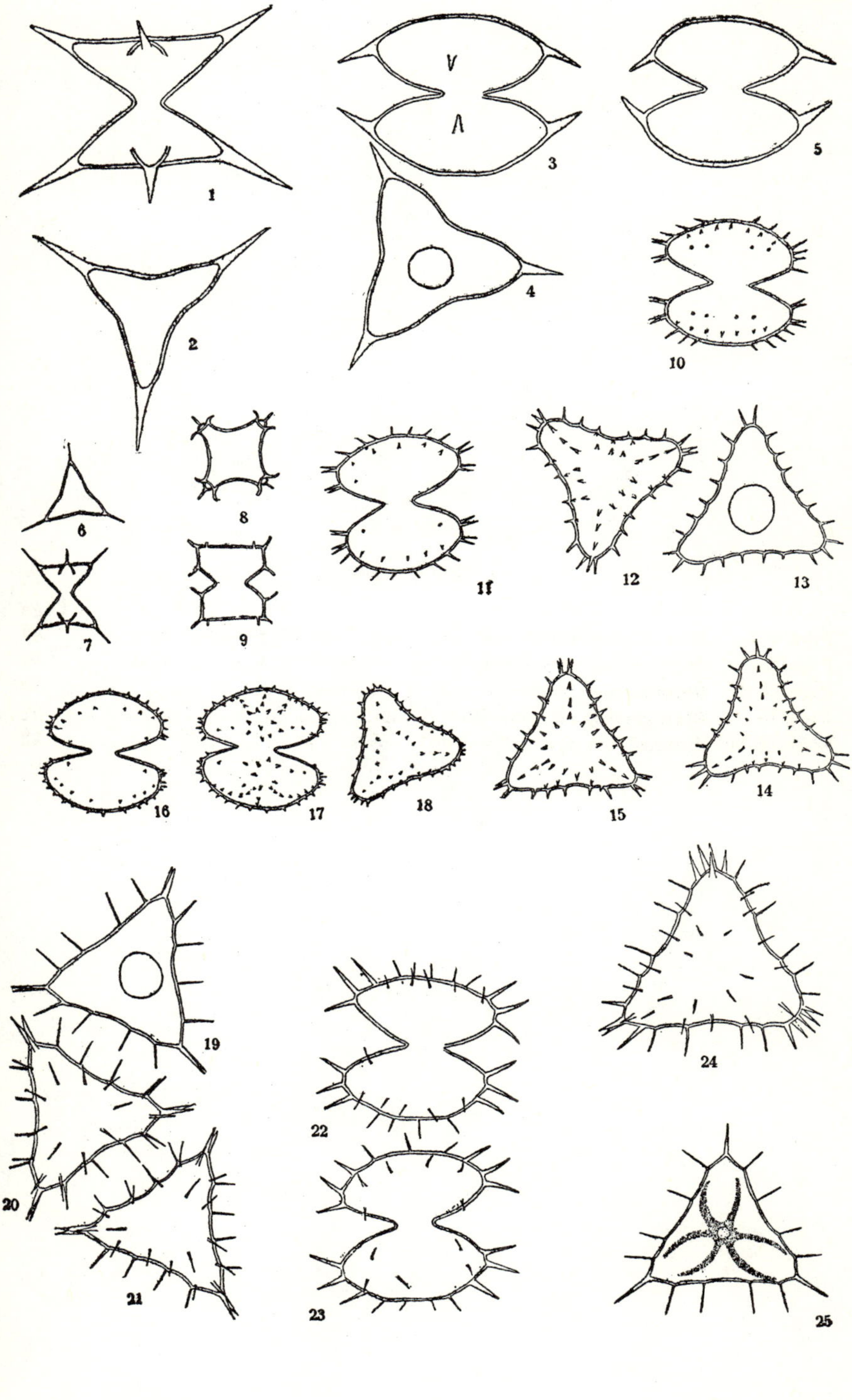

PLATE 71.

All figures x 500.

		Page
Figs. 1–4.	*Staurastrum setigerum* var. *occidentale* W. & G. S. West......	80
Figs. 5–8.	*Staurastrum setigerum* var. *pectinatum* W. & G. S. West......	80
Figs. 9–13.	*Staurastrum setigerum* var. *brevispinum* G. M. Smith.........	80
Figs. 14–15.	*Staurastrum minnesotense* Wolle..........................	81
Figs. 16–20.	*Staurastrum spiculiferum* sp. nov.........................	82

PLATE 71

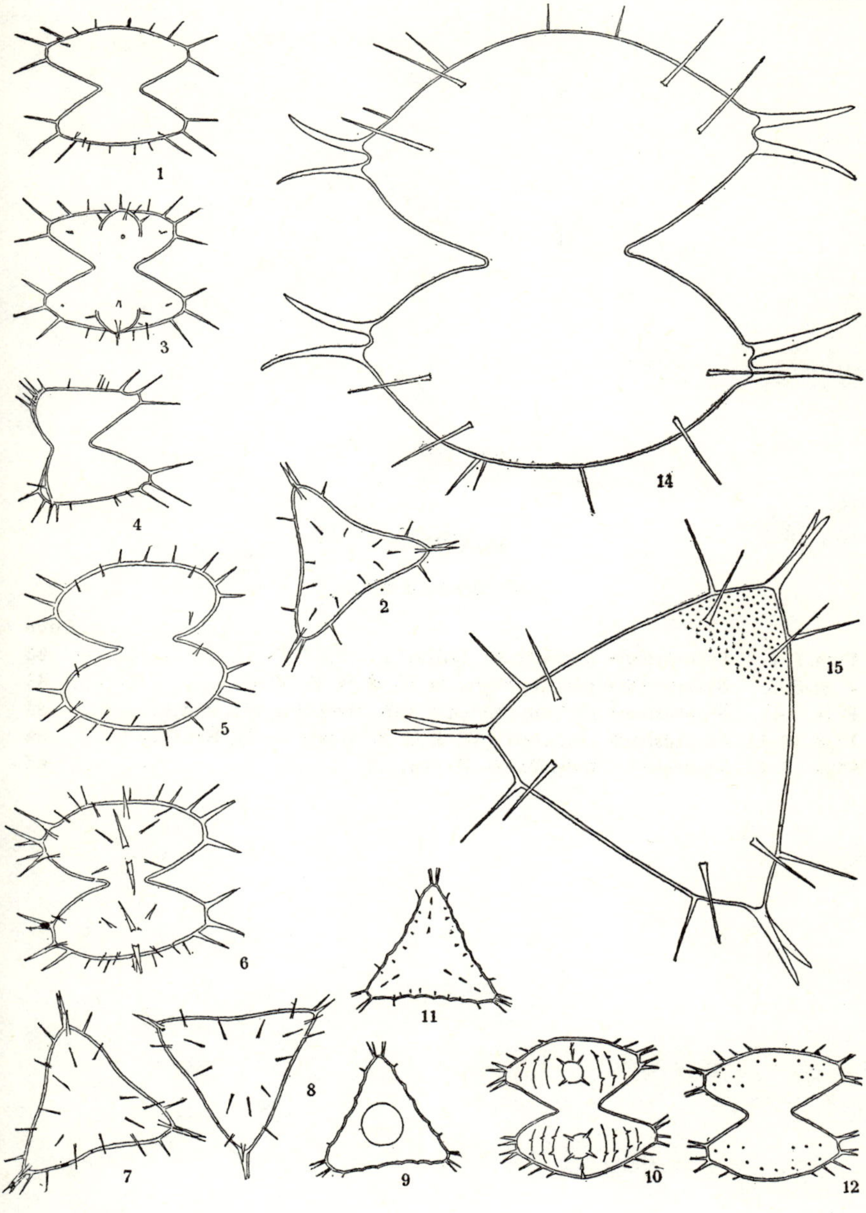

PLATE 72.

All figures x 500.

PLATE 72

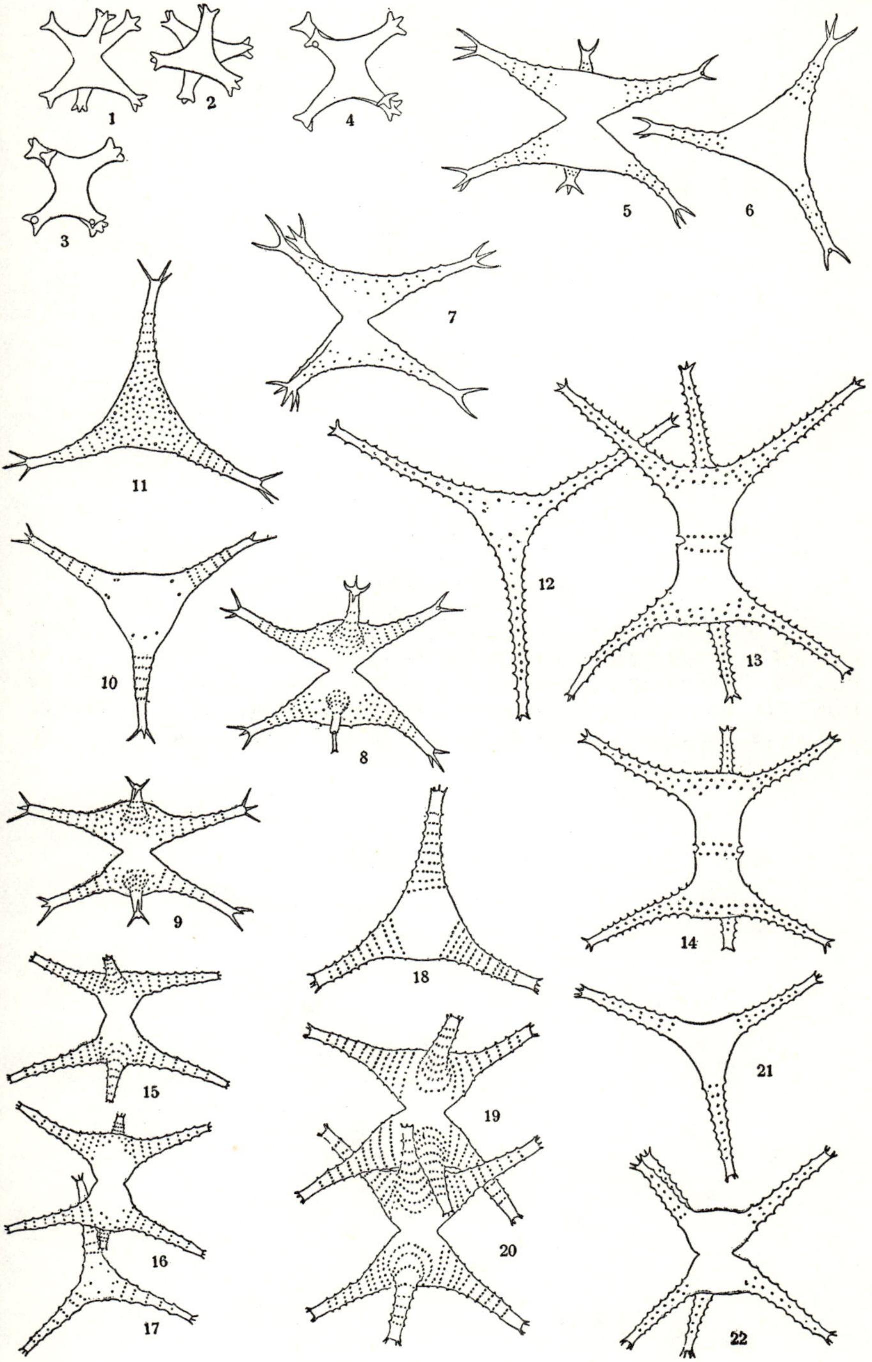

PLATE 73.

All figures x 500.

		Page
Figs. 1–2.	*Staurastrum paradoxum* Meyen	85
Figs. 3–6.	*Staurastrum paradoxum* var. *longipes* Nordst	86
Figs. 7–15.	*Staurastrum paradoxum* var. *parvum* W. West	87
Figs. 16–18.	*Staurastrum gracile* Ralfs	88

PLATE 73

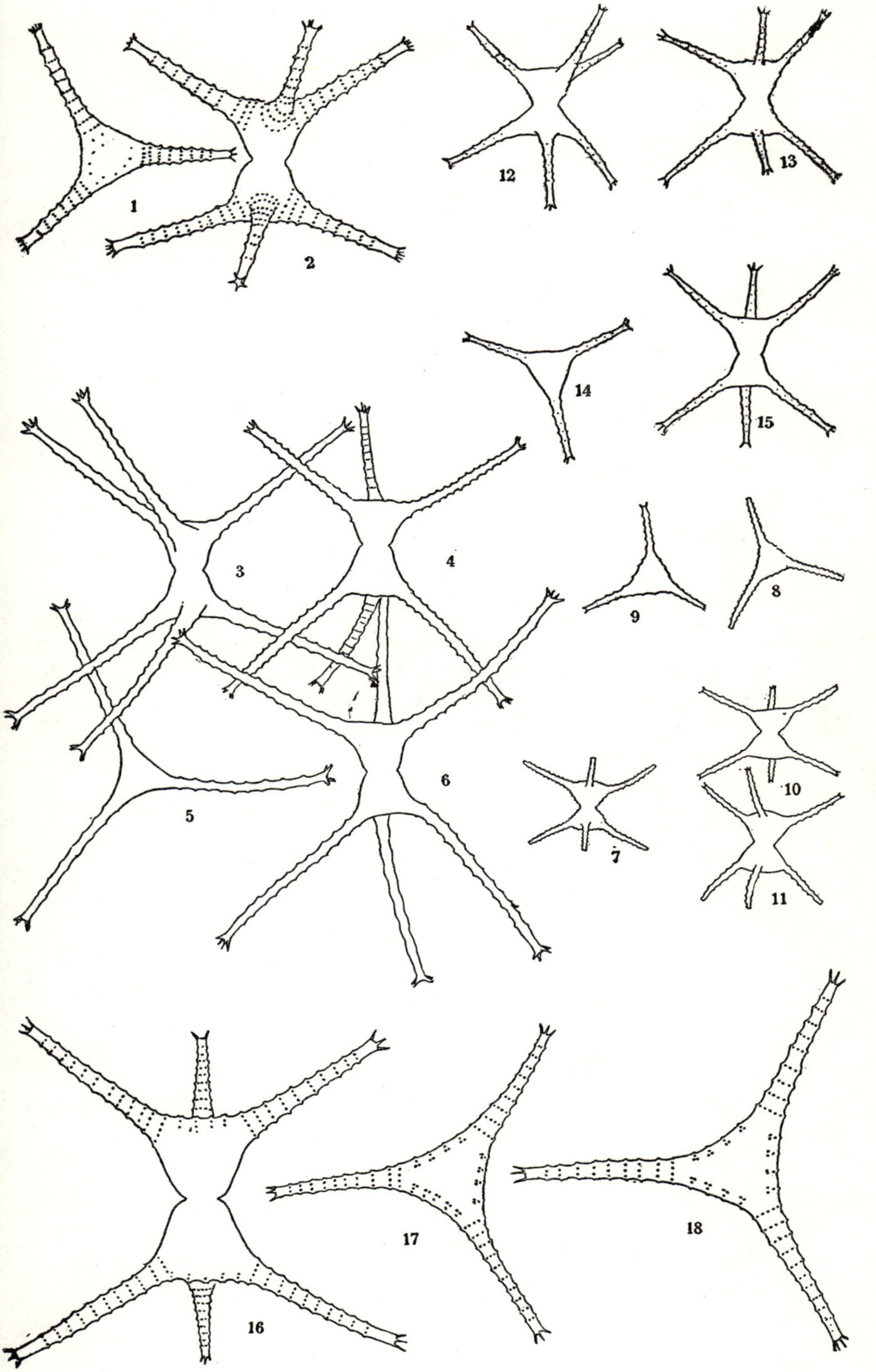

PLATE 74.

All figures x 500.

PLATE 74

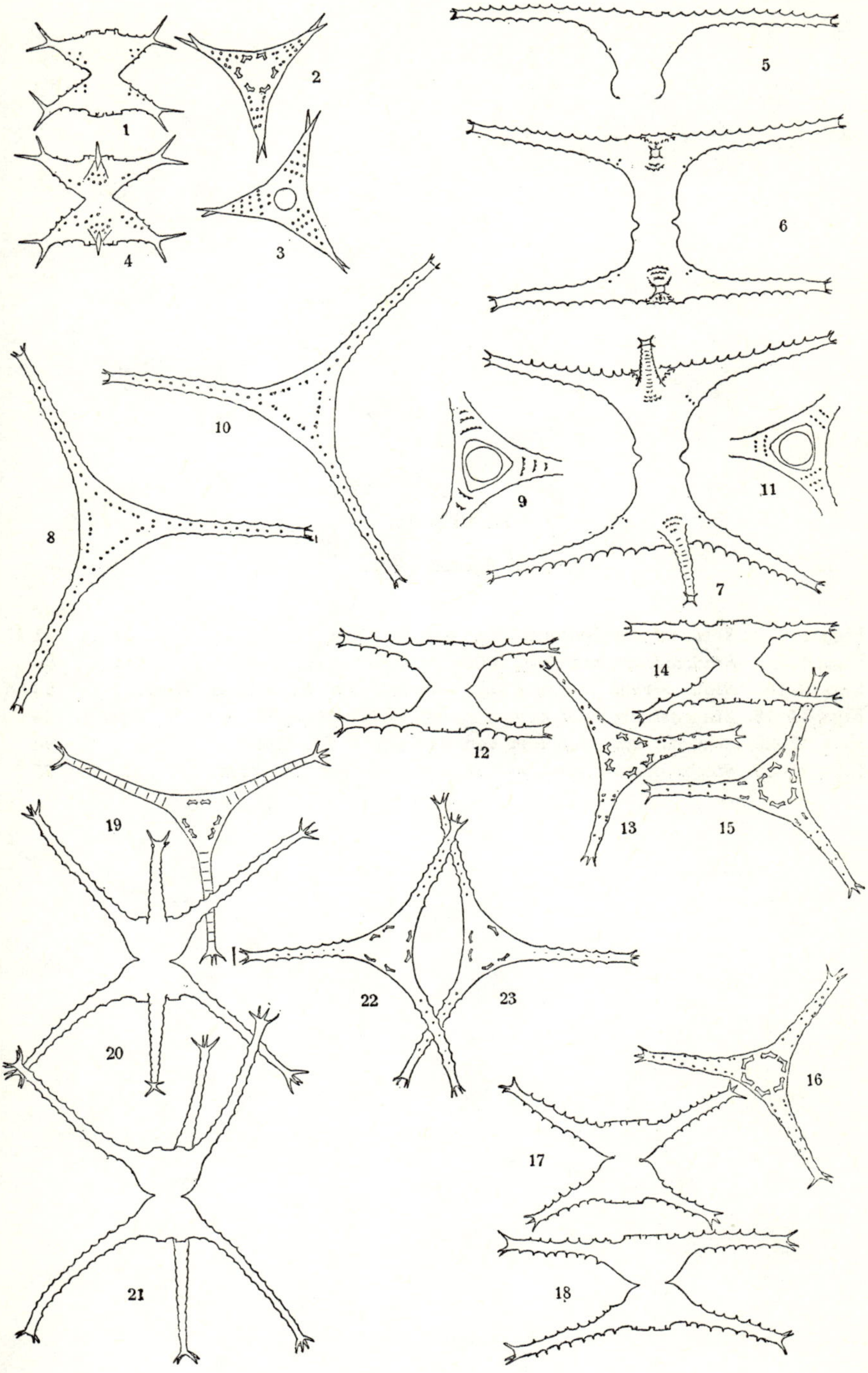

PLATE 75.

All figures x 500.

PLATE 75

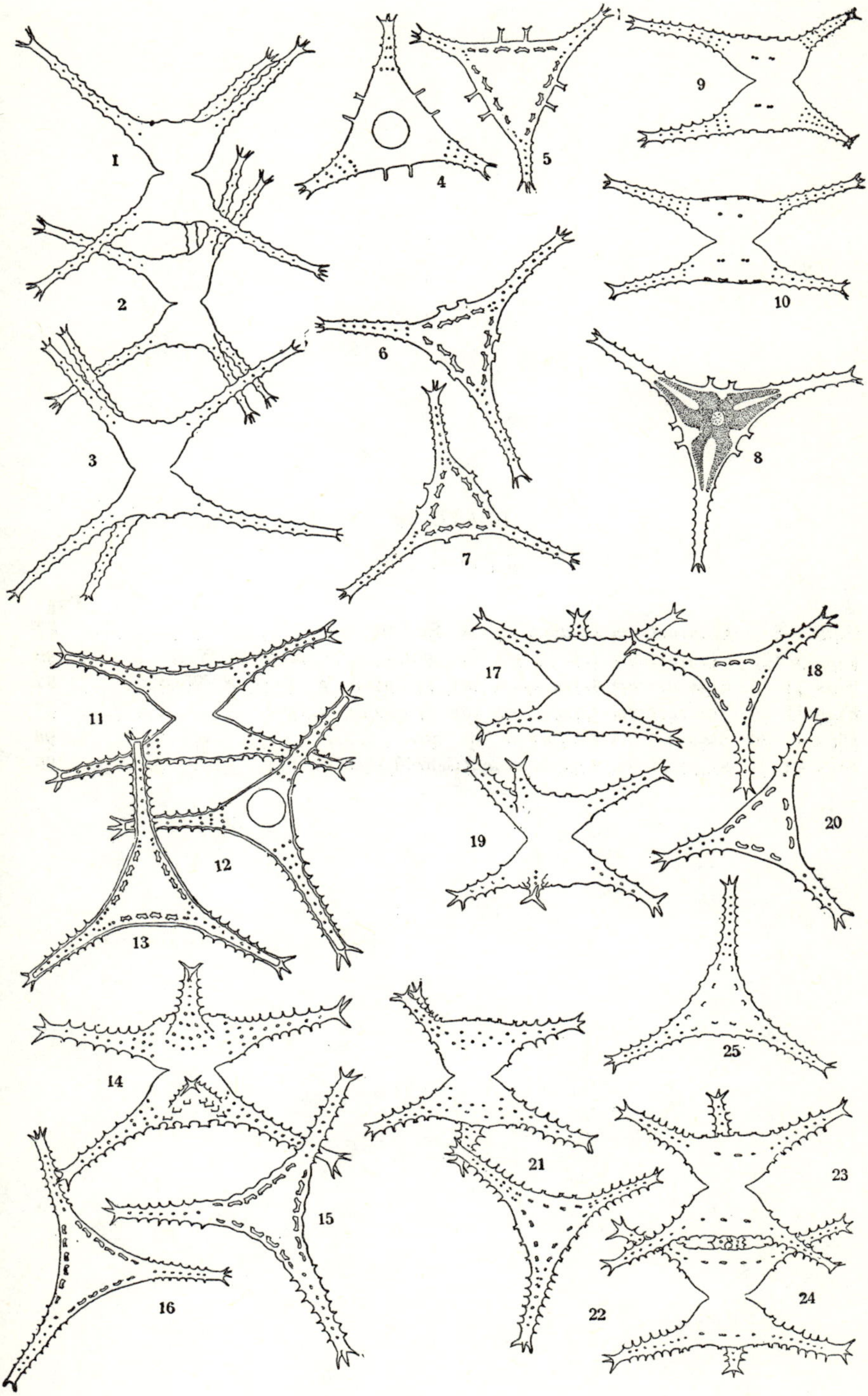

PLATE 76.

All figures x 500.

PLATE 76

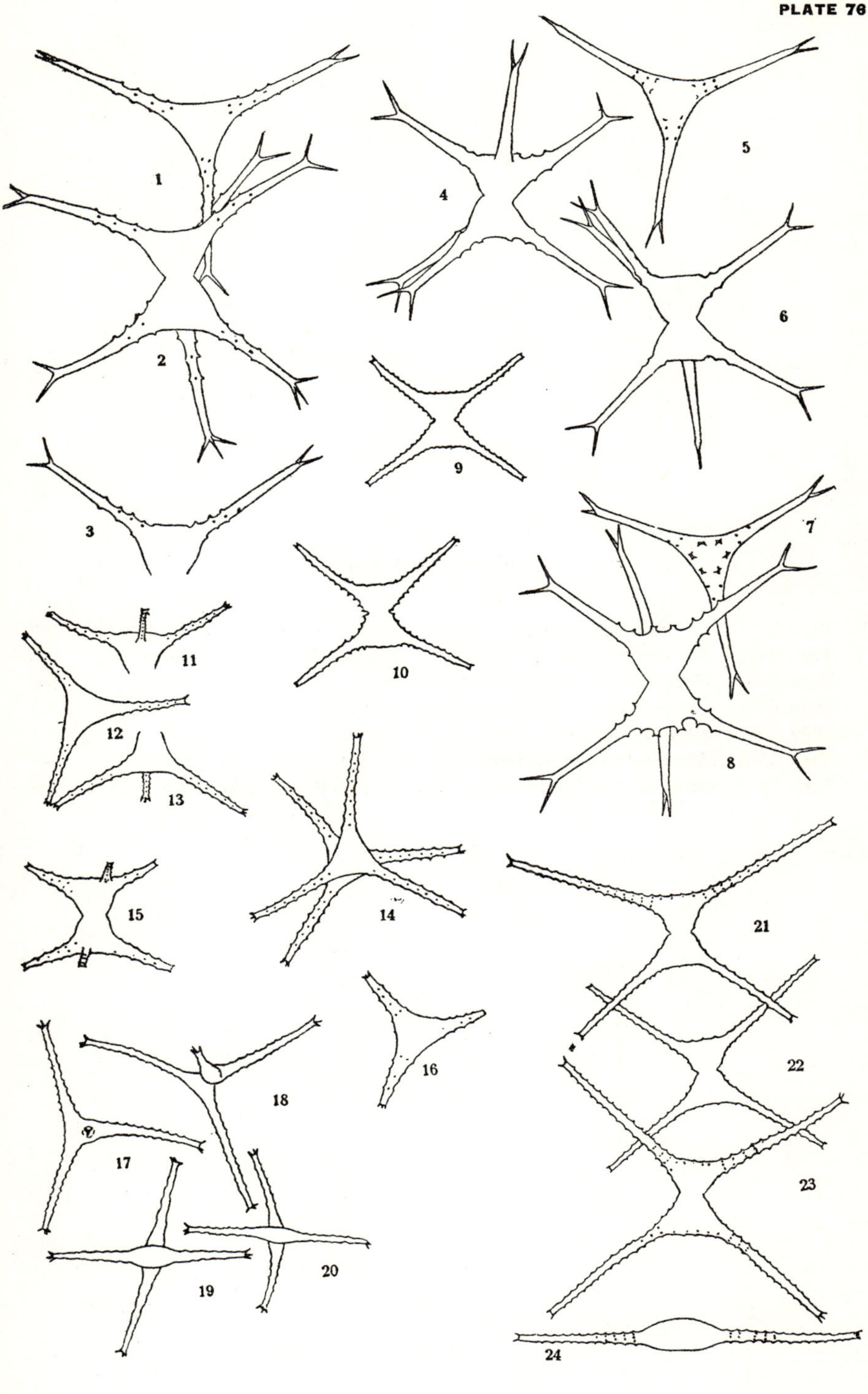

PLATE 77.

All figures x 500.

		Page
Fig. 1.	*Staurastrum Chaetoceras* (Schröd.) comb. nov.	99
Fig. 2.	*Staurastrum americanum* (W. & G. S. West) G. M. Smith	100
Figs. 3–5.	*Staurastrum grallatorium* Nordst.	100
Figs. 6–7.	*Staurastrum grallatorium* var. *forcipigerum* Lag.	101
Figs. 8–9.	*Staurastrum anchora* W. & G. S. West	101
Figs. 10–14.	*Staurastrum leptocladum* var. *denticulatum* G. M. Smith	103
Fig. 15.	*Staurastrum leptocladum* var. *insigne* W. & G. S. West	103

PLATE 77

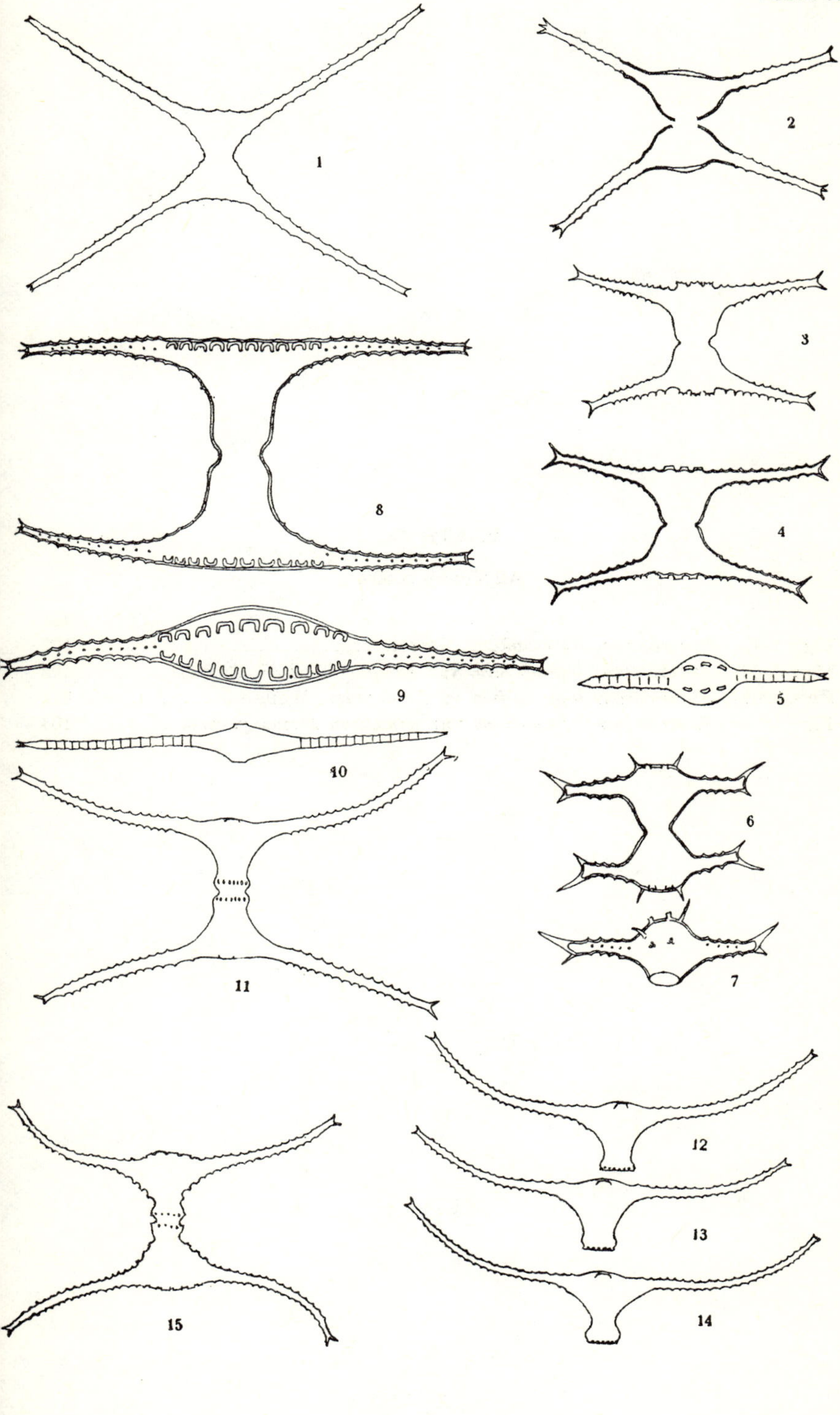

PLATE 78.

All figures x 500.

		Page
Figs. 1–7.	*Staurastrum leptocladum* Nordst.	102
Figs. 8–11.	*Staurastrum leptocladum* var. *insigne* W. & G. S. West	103
Figs. 12–14.	*Staurastrum leptocladum* var. *sinuatum* Wolle	104
Fig. 15.	*Staurastrum leptocladum* var. *sinuatum* forma *planum*	104

PLATE 78

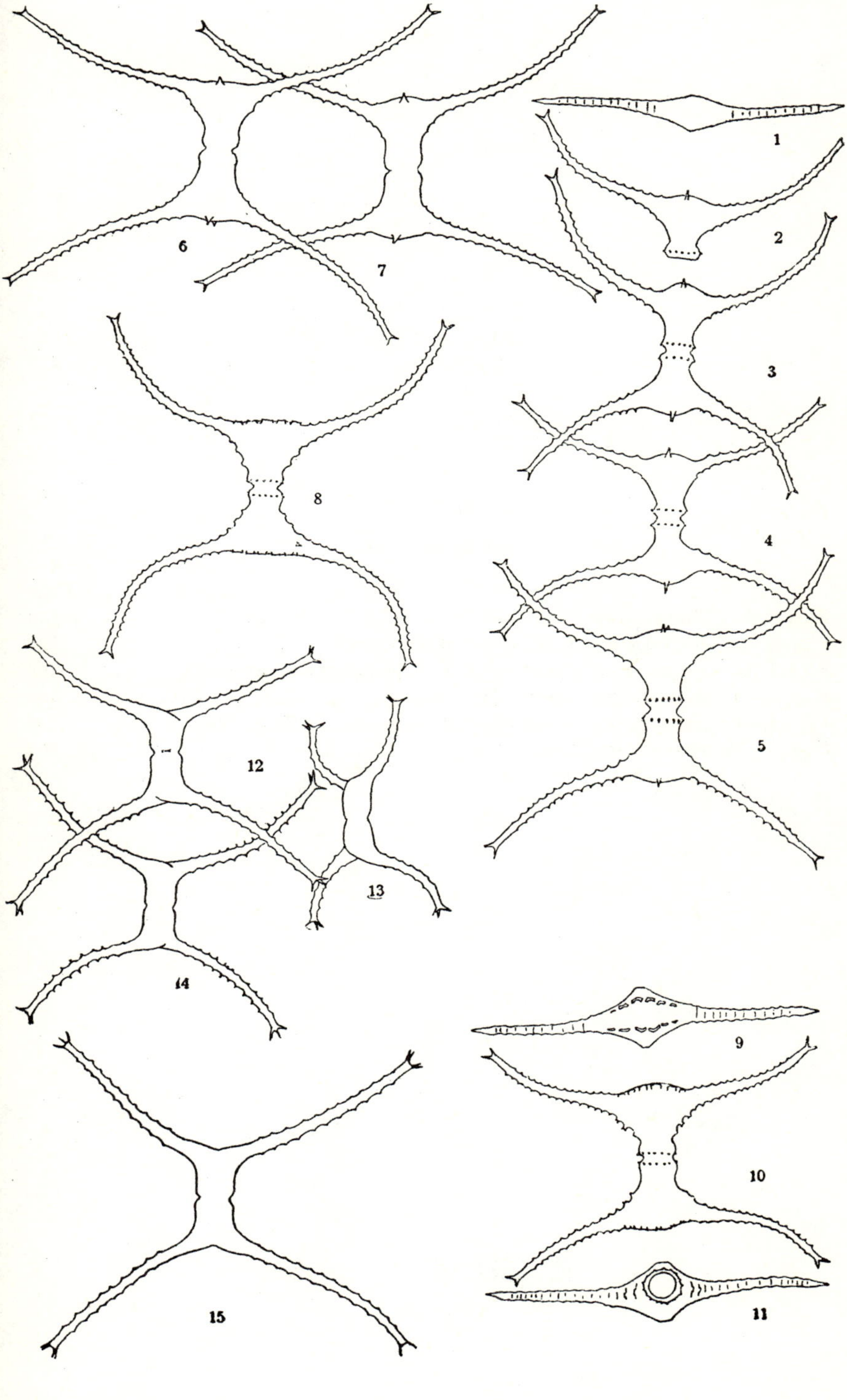

PLATE 79.

All figures x 500.

PLATE 79

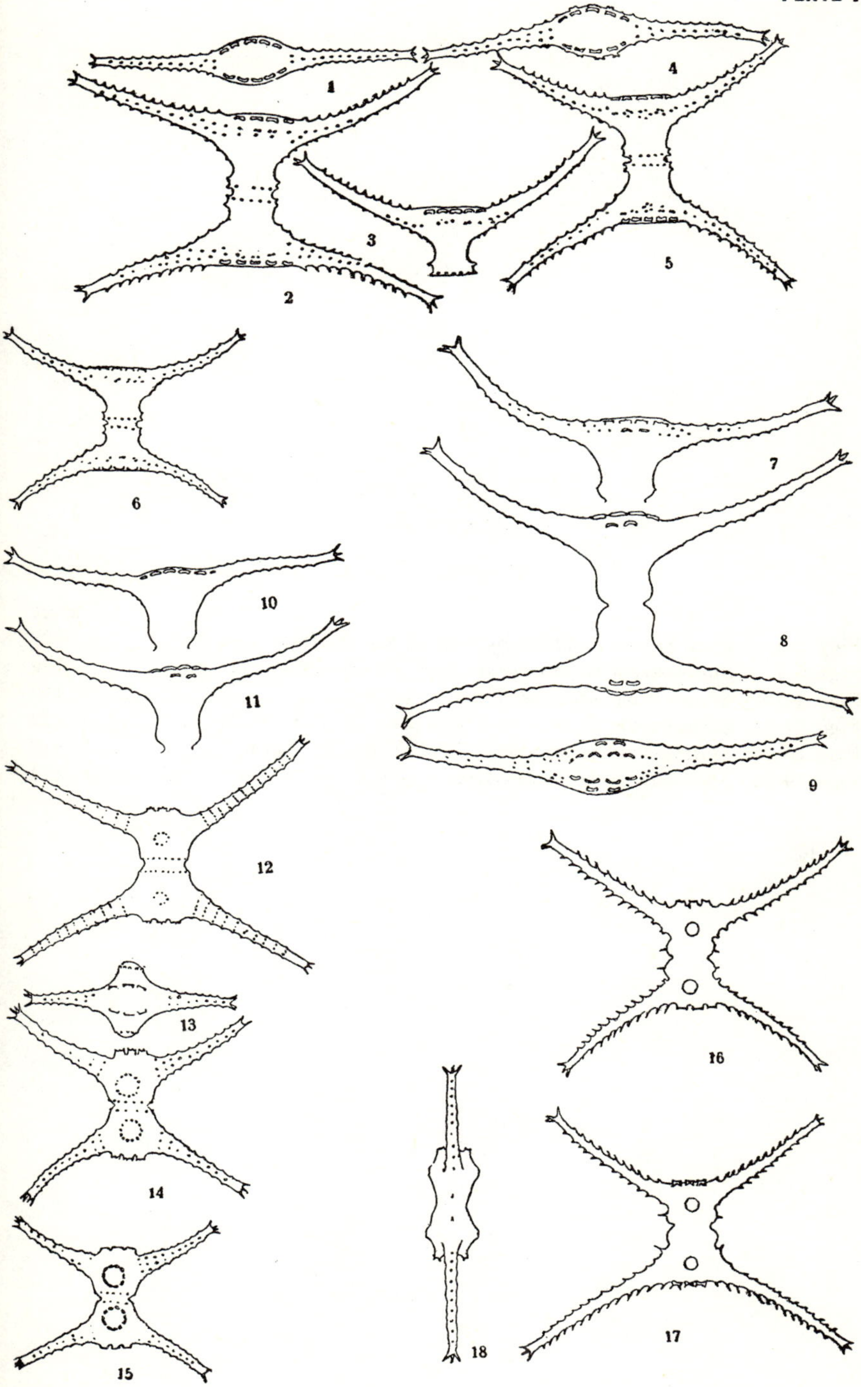

PLATE 80.

All figures x 500.

PLATE 80

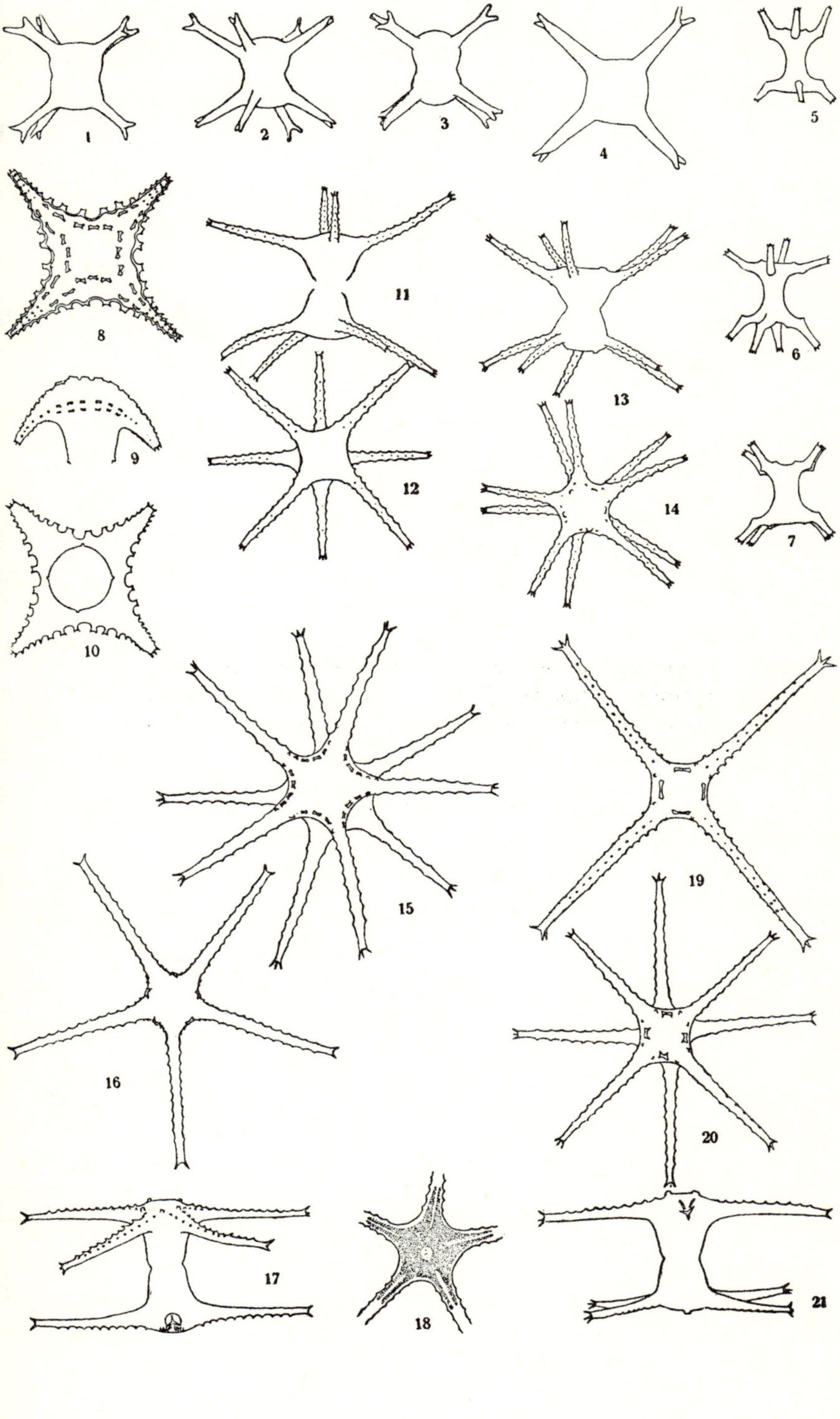

PLATE 81.

All figures x 500.

PLATE 81

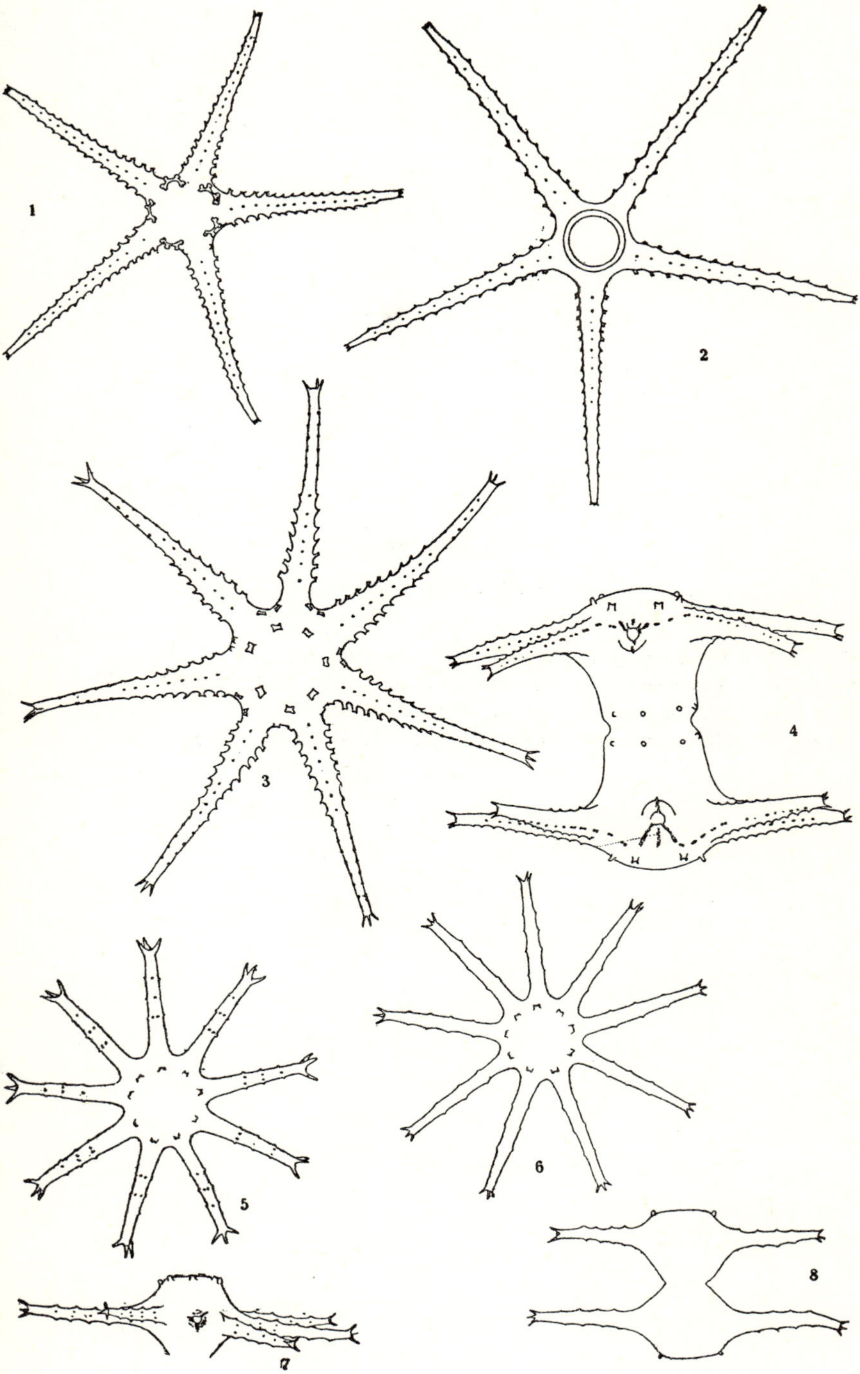

PLATE 82.

All figures x 500.

PLATE 82

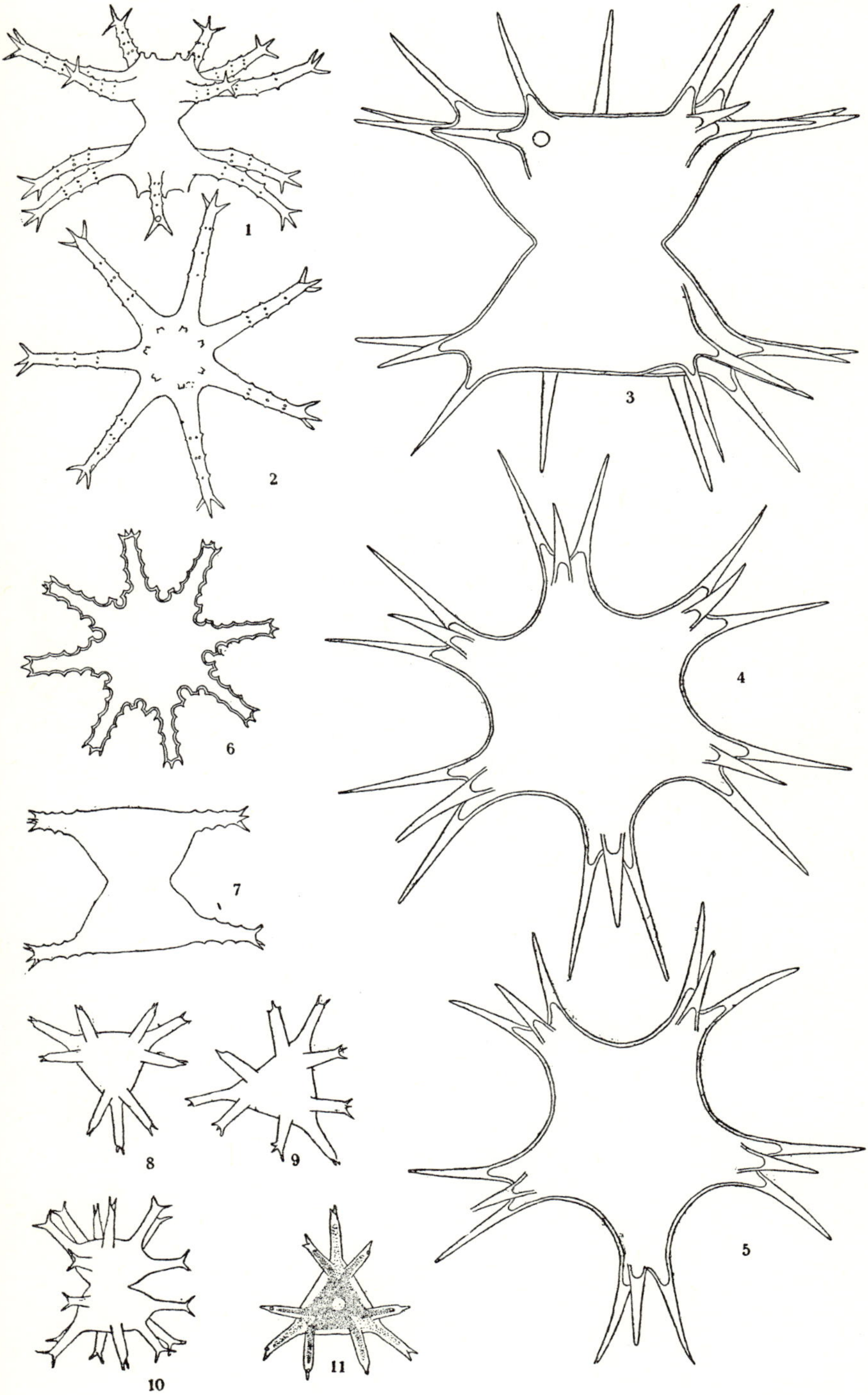

PLATE 83.

All figures x 500.

PLATE 83

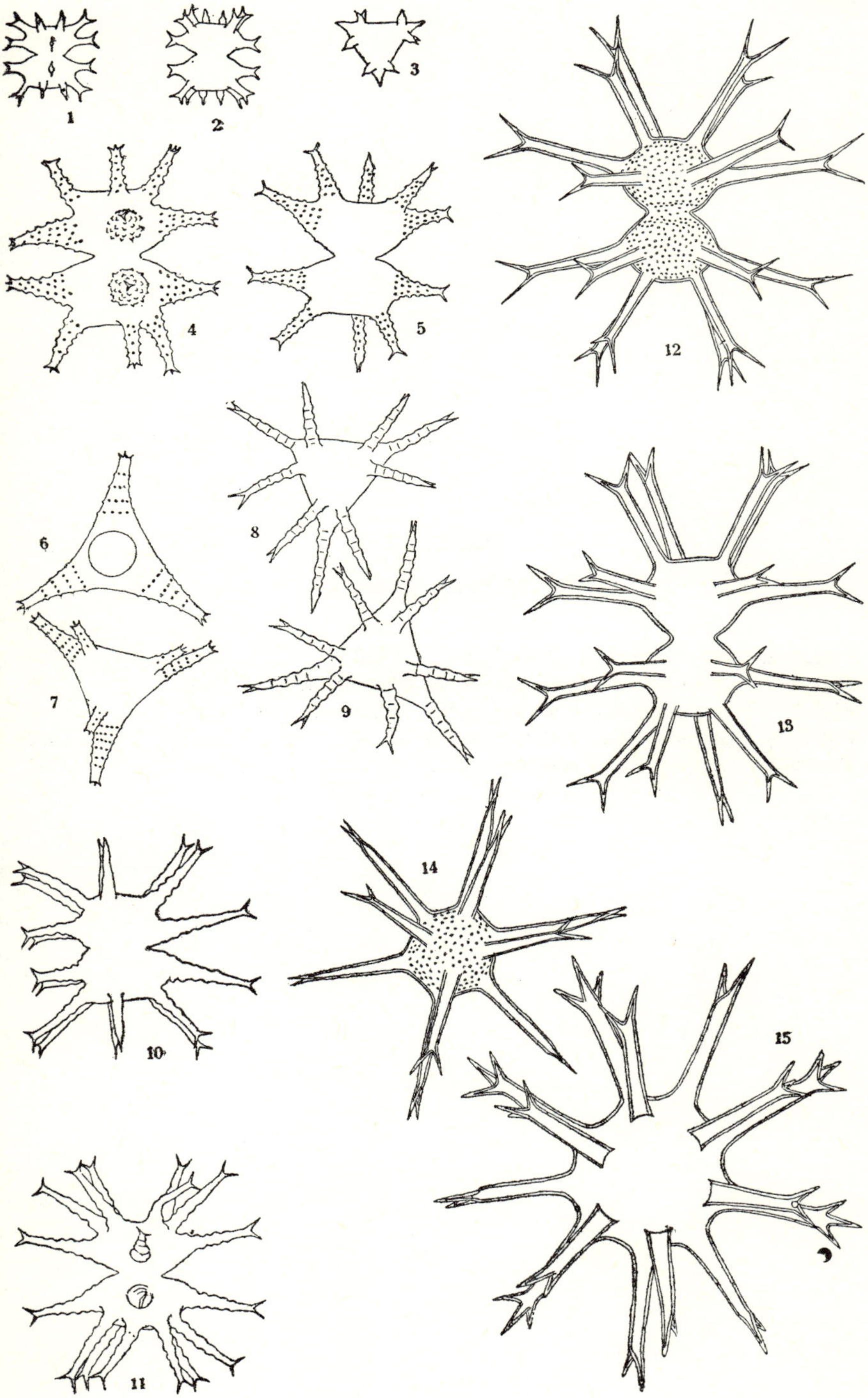

PLATE 84.

All figures x 500.

PLATE 84

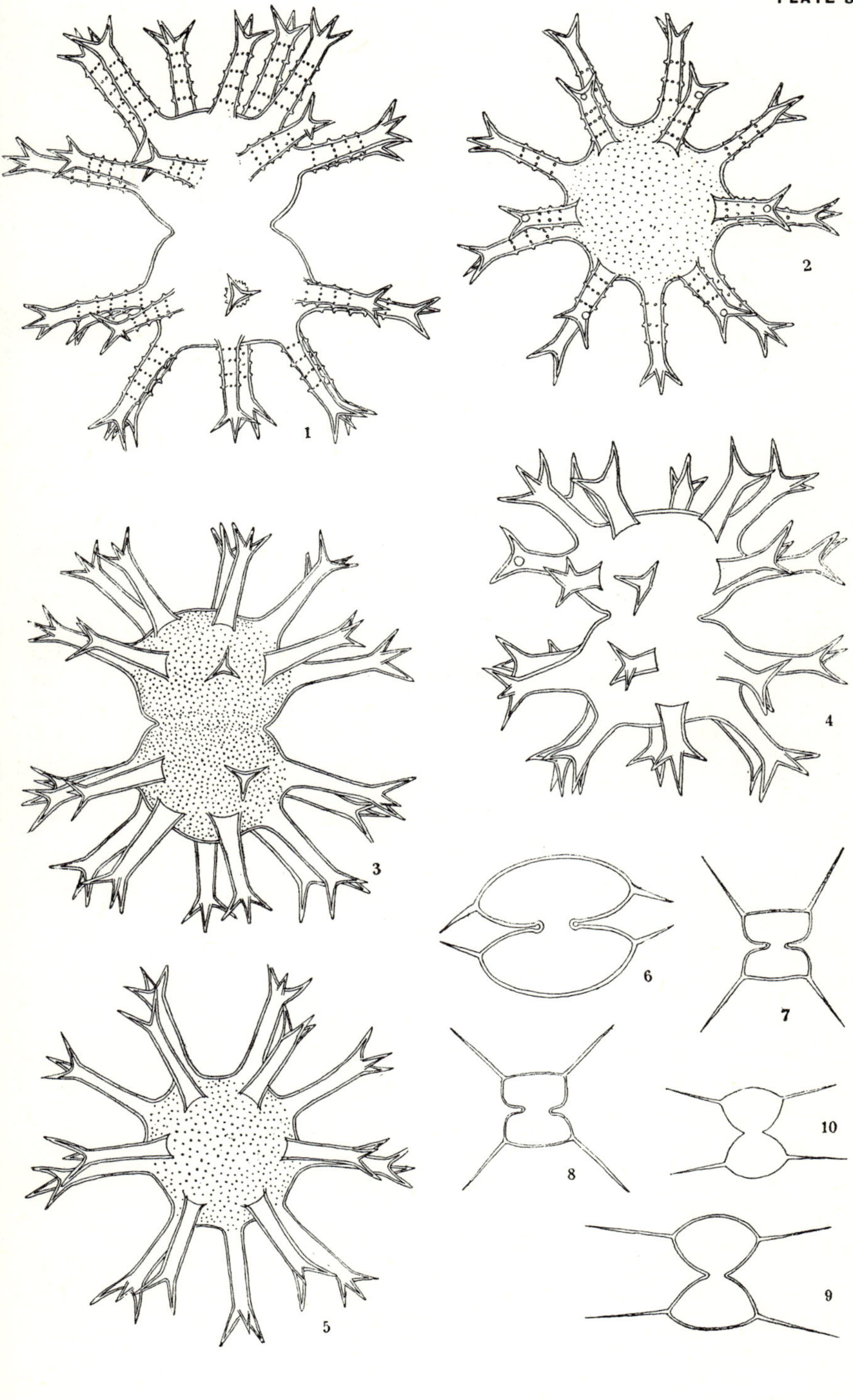

PLATE 85.

All figures x 600.

		Page
Figs. 1–3.	*Arthrodesmus subulatus* var. *Nordstedtii* var. nov.	127
Fig. 4.	*Arthrodesmus michiganensis* Johns.	128
Fig. 5.	*Arthrodesmus phimus* Turn.	128
Figs. 6–8.	*Arthrodesmus constrictus* G. M. Smith.	129
Figs. 9–12.	*Arthrodesmus Ralfsii* W. West.	129
Figs. 13–17.	*Arthrodesmus Ralfsii* var. *Brebissonii* (Racib.) comb. nov.	130
Fig. 18.	*Arthrodesmus quiriferus* W. & G. S. West.	131
Figs. 19–22.	*Arthrodesmus incus* (De Bréb.) Hass.	131
Figs. 23–26.	*Arthrodesmus incus* var. *extensus* Anders.	132
Figs. 27–29.	*Arthrodesmus incus* var. *praelongus* var. nov.	132

PLATE 85

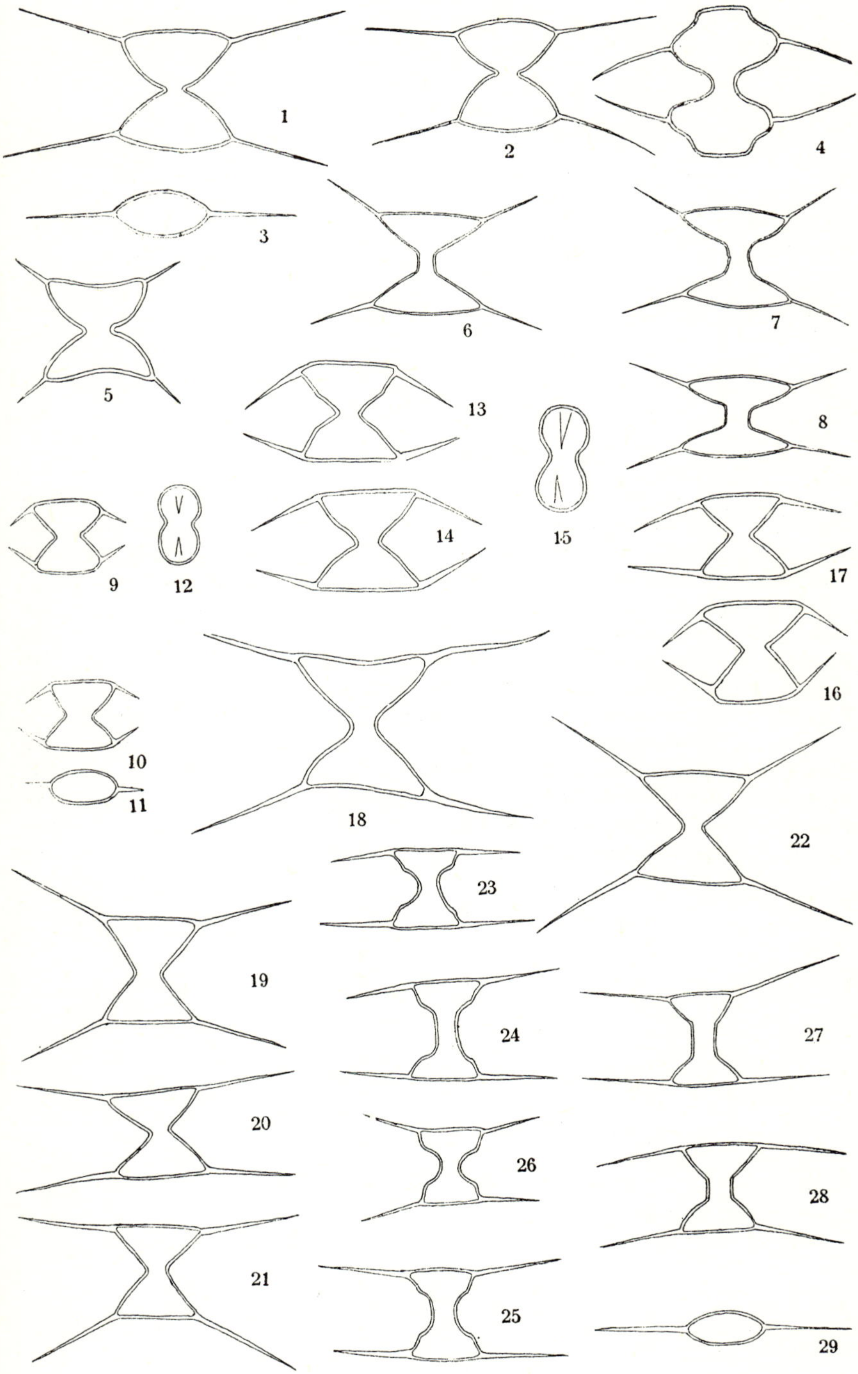

PLATE 86.

All figures x 600.

PLATE 86

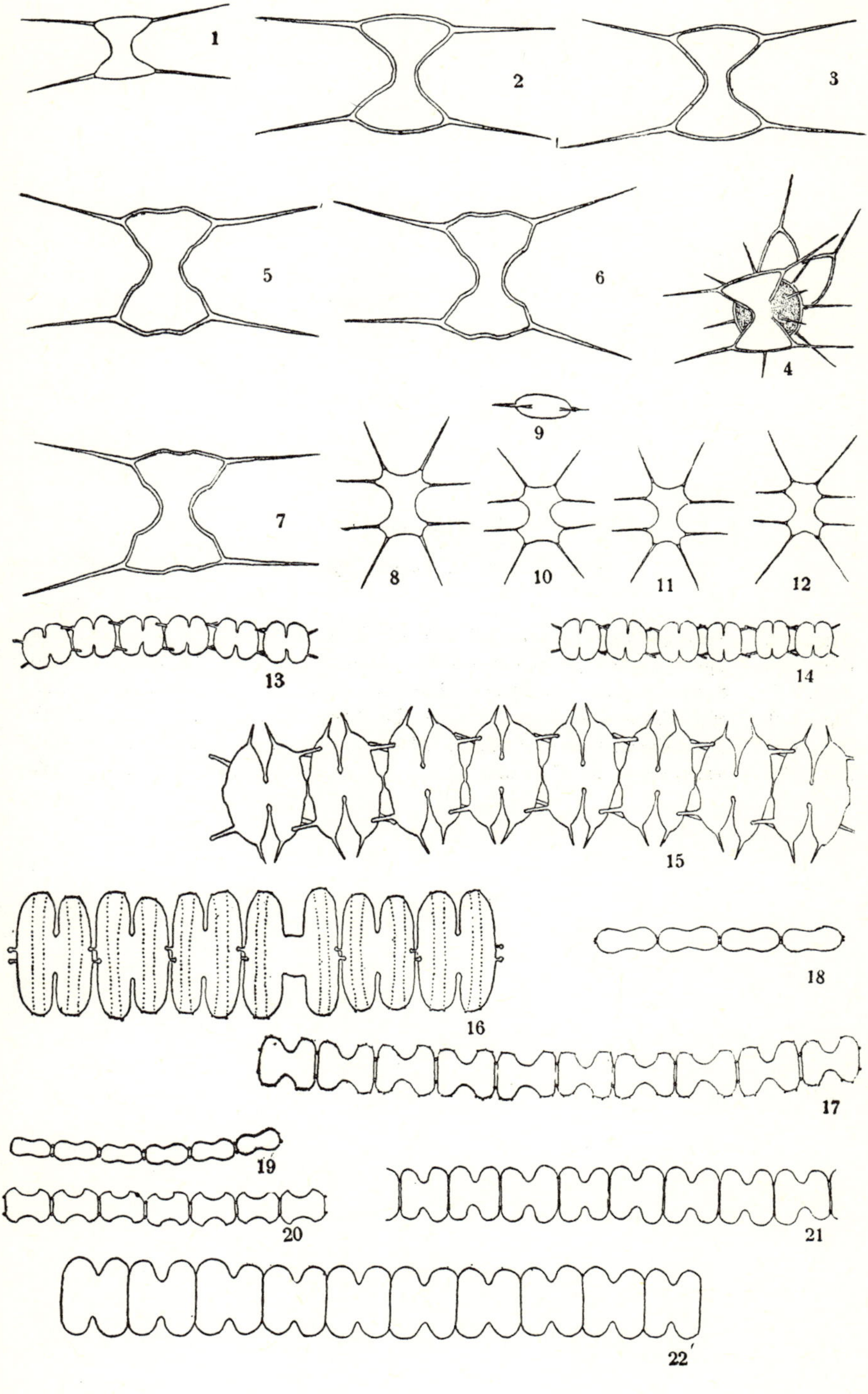

PLATE 87.

Figs. 1–2, 4–11, x 600; Fig. 3, x 300.

PLATE 87

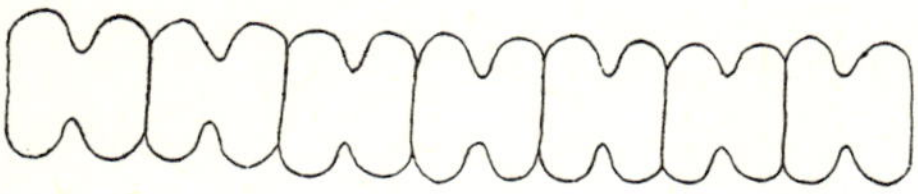
1

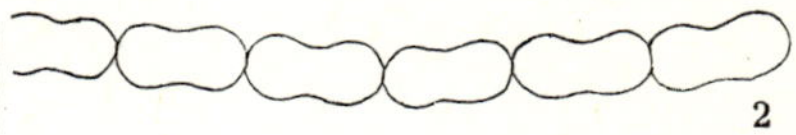
2

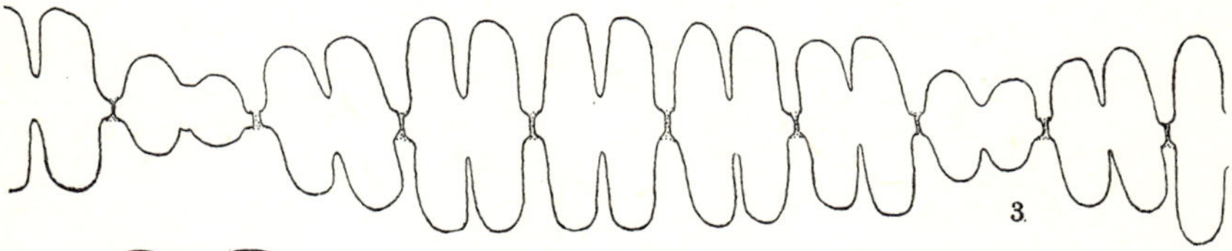
3

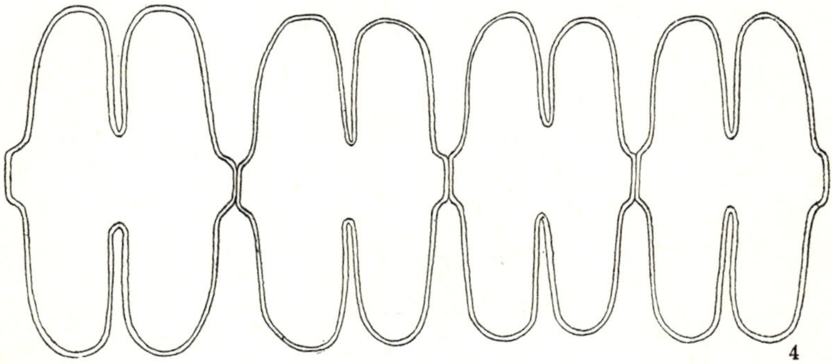
4

5

6

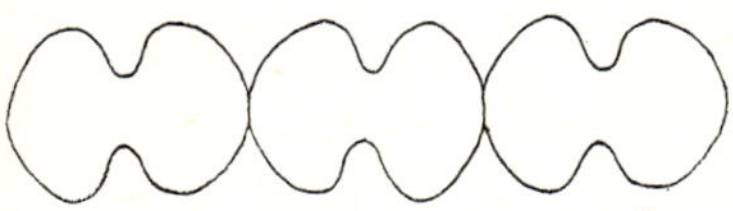
7

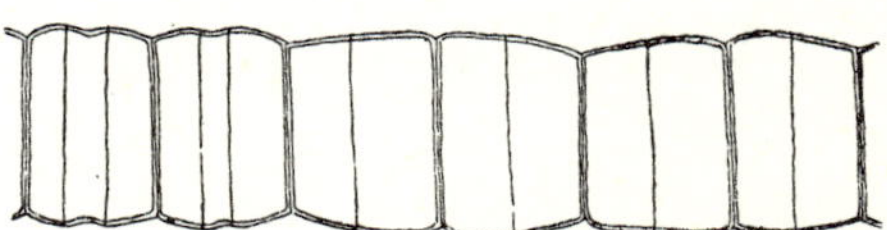
9

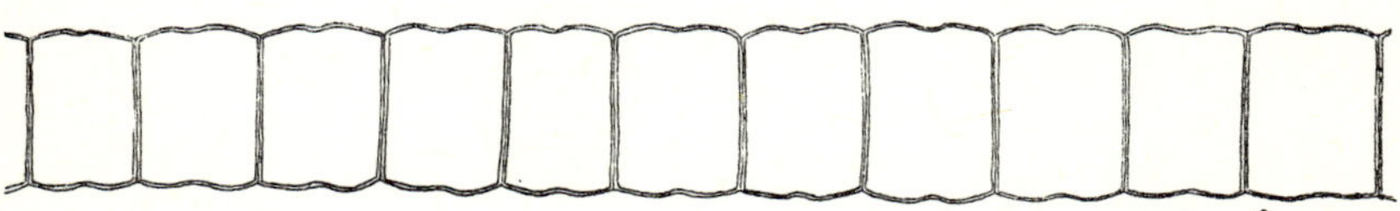
8

10

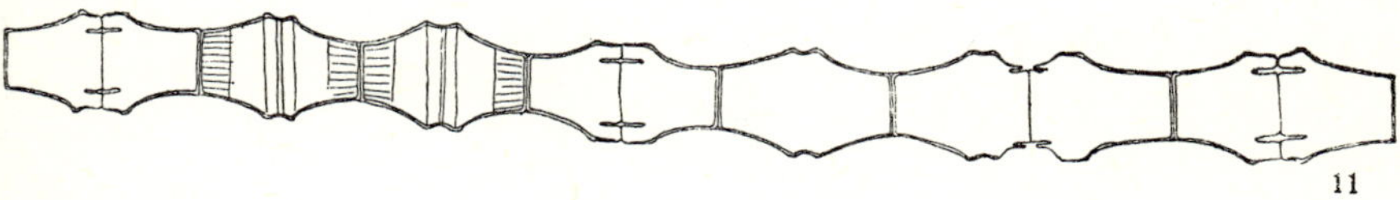
11

PLATE 88.

All figures x 600.

		Page
Figs. 1–2.	*Desmidium Swartzii* Ag.	144
Figs. 3–4.	*Desmidium Aptogonium* De Bréb.	144
Figs. 5–7.	*Desmidium Baileyi* (Ralfs) Nordst.	145
Fig. 8.	*Desmidium Grevillii* (Ktz.) De Bary	145

PLATE 88

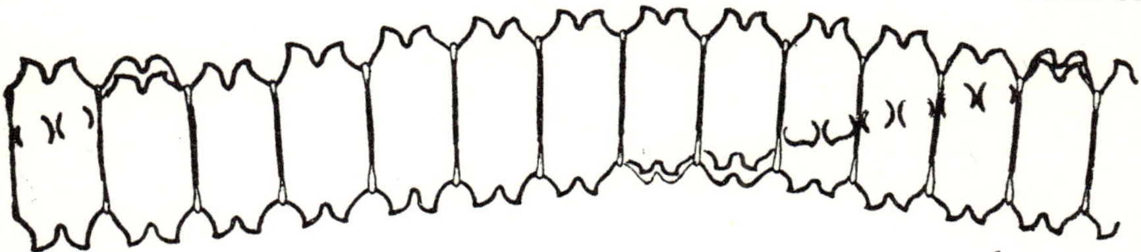

1

3

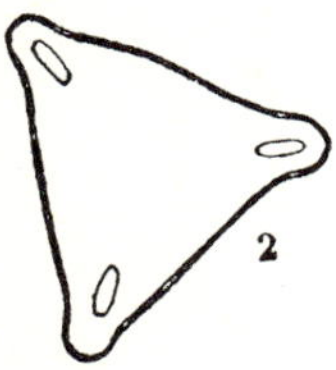

2

5

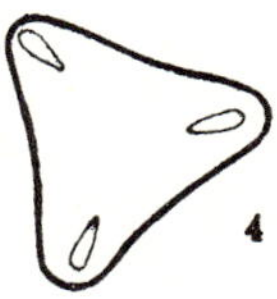

4

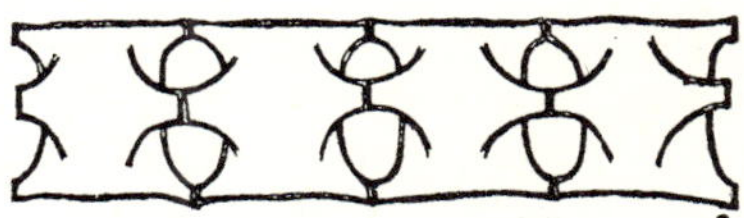

6

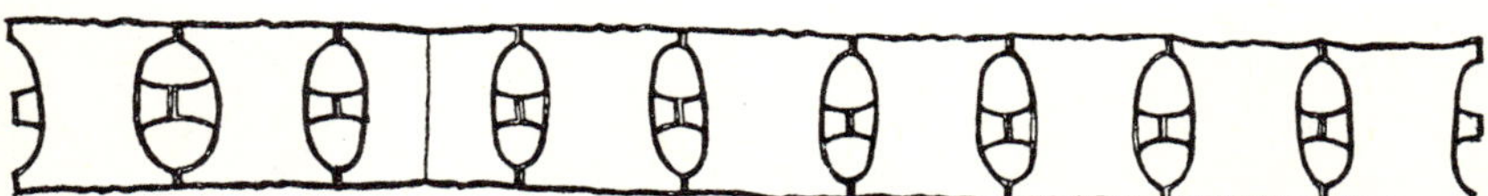

7

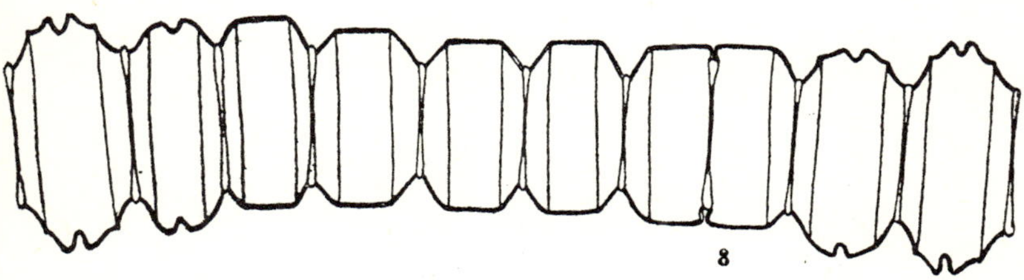

8

INDEX